PRIME

2008 PhD Research in Microelectronics and Electronics

PROCEEDINGS

Istanbul, Turkey
June 22 – 25, 2008

Organization

Conference Sponsors

BOĞAZİÇİ ÜNİVERSİTESİ VAKFI

 Springer

TÜBİTAK

 IEEE

IEEE
Circuit and Systems
Society

intersil

BROADCOM.

National Semiconductor
The Sight & Sound of Information

MKR-IC, Mikroelektronik
R&D Design Center
Mikroelektronik ARGE ve Tasarım Ltd.

Copyright and Reprint Permission: Abstracting is permitted with credit to the source. Libraries are permitted to photocopy beyond the limit of U.S. copyright law for private use of patrons those articles in this volume that carry a code at the bottom of the first page, provided the per-copy fee indicated in the code is paid through Copyright Clearance Center, 222 Rosewood Drive, Danvers, MA 01923. For other copying, reprint or republication permission, write to IEEE Copyrights Manager, IEEE Operations Center, 445 Hoes Lane, P.O. Box 1331, Piscataway, NJ 08855-1331. All rights reserved. Copyright ©2008 by the Institute of Electrical and Electronics Engineers

IEEE Catalog Number: CFP08622-PRT
ISBN: 978-1-4244-1983-8
Library of Congress: 2007943122

Foreword

Welcome to Istanbul, Turkey and the 4th Ph.D. Research in Microelectronics and Electronics (PRIME) Conference 2008. We are pleased to have you here on the South Campus of Boğaziçi University, which is located on one of the hills of Istanbul on the European side, next to the Rumeli Hisar Fortress. The campus overlooks the section of the beautiful Istanbul Strait (Bosphorus) in between the two suspension bridges, Boğaziçi and Fatih Sultan Mehmet bridges, which connect the two continents, Asia and Europe together. Like the bridges connecting the two continents, we hope that this conference would connect the Ph.D. research in different countries together as well as the students to industry.

Previously, PRIME conferences were held in Lausanne, Switzerland in 2005, in Otranto, Italy in 2006 and in Bordeaux, France in 2007. This year, the electronics design laboratory, BETA, of the Department of Electrical and Electronics Engineering of Boğaziçi University has organized this conference. Like the previous years, participants of the conference came from different regions of Europe, North America and Asia. This year's participation is mostly from Turkey (23 papers), Italy (19 papers) and France (8 papers). The list is followed by Switzerland (3), USA (3), Finland (2), Ireland (2), Iran (2), England (1), Poland (1), Spain (1), Austria (1), Israel (1), Tunisia (1) and Romania (1). We believe these numbers will continue to grow in the coming PRIME conferences.

Company fair is a very important part of this conference. It starts immediately after the opening ceremony. Six companies have joined the company fair this year. We hoped to create a warm environment in the company fair so that constructive connections can be built between Ph.D. students and companies.

The main topics of PRIME presentations are on very large scale integration (VLSI), system on chip (SoC), analog and digital integrated circuits (ICs), micro electro mechanical systems (MEMS), microelectronic and micro electro mechanic devices, analog and digital signal processing, visual signal processing, computer aided design (CAD), integrated power ICs and sensor systems. This year 9 papers are presented in MEMS, 10 in RF circuit design, 9 in analog design, 10 in CAD, 5 in devices, 5 in sensor interfaces, 5 in signal processing, 5 in digital design, 4 in wireless systems, 3 in modeling and 3 in ADC.

The Technical Program Committee is formed by 65 internationally recognized experts in their fields working in academic institutes and industry. This committee selected 67 oral presentations out of 97 applications, corresponding to an overall acceptance rate of 69%. We would like to thank all the members of this committee for being prompt and diligent in the reviewing process of the submitted abstracts. Based on these reviewing scores, accepted papers are rated. Top 10% received the gold leaf certificates, top 10-20% silver leaf and top 20-30% bronze leaf certificates.

We would like to thank Prof. Franco Maloberti for his guidance and advices on organizing this conference. We want to thank TÜBİTAK (The Scientific and Technological Research Council of Turkey) for providing funds for printing costs and travel funds for presenters from Turkey. We also want to thank to the Region-8 IEEE Circuits and Systems Society (CAS) for providing travel funds for presenters coming from Eastern European and Middle Eastern countries. Finally, we want to acknowledge the technical sponsor IEEE Circuits and Systems (CAS) Society for support and the advertisement of this conference.

We wish you a pleasant stay in Istanbul and hope that you would leave this city with new friendships, collaborations, stimulated ideas and rejoiced look on life and research.

Best Regards,

Prof. Dr. Günhan Dündar	Asst. Prof. Dr. Şenol Mutlu	Asst. Prof. Dr. Arda Deniz Yalçınkaya
General Chair	Technical Program Co-Chair	Technical Program Co-Chair

General Chair

Günhan Dündar, *Boğaziçi University, Turkey*

Technical Program Chairs

Arda D. Yalçınkaya, *Boğaziçi University, Turkey*
Şenol Mutlu, *Boğaziçi University, Turkey*

Local Arrangement Chair

Erdinç Tan, *K2 Event, Turkey*

Publicity Chair

Selçuk Talay, *University of Pavia, Italy*

PRIME Steering Committee

Franco Maloberti, *University of Pavia, Italy (Chairman)*
Andrea Baschirotto, *University of Lecce, Italy*
Catherine Dehollain, *EPFL, Switzerland*
Alberto Gola, *Infineon, Italy*
Frank Henkel, *IMST GmbH, Germany*
Peter Kennedy, *University College Cork, Ireland*

PRIME Industrial Advisory Committee

Klaas Bult, *Broadcom*
Peter Mole, *Intersil*
Dennis Monticelli, *National Semiconductor*
Wolfgang Pribyl, *AMS*
Francesco Rezzi, *Marvell*
Susan Sanchez, *International Rectifier*

Table of Contents

Session Name:	**CAD 1**	
Session Time / Location:	**SUNDAY, JUNE 22 / VYKM 2**	
Session Chair:	**Sule OZEV, Duke University, USA**	

15:30 – 15:50   *N. KAHRAMAN, T.YILDIRIM*   P1
Yildiz Technical University, Istanbul, Turkey
"Technology Independent Circuit Sizing for Fundamental Analog Circuits"

15:50 – 16:10   *E. DENIZ[1], G. DUNDAR[2]*   P5
[1]Dogus University, Istanbul, Turkey
[2]Bogazici University, Istanbul, Turkey
"Hybrid Approach for Performance Estimation; Embedded Tool for Analog Design Automation Systems"

16:10 – 16:30   *E. YAHYA[1,2], M. RENAUDIN[3]*   P9
[1]CIS Group, TIMA Laboratory, Grenoble, France
[2]Banha High Institute of Technology, Banha, Egypt
[3]TIEMPO SAS, Grenoble, France
"Optimal Asynchronous Linear-Pipelines"

16:30 – 16:50   *R. Vural ACAR, T. YILDIRIM*   P13
Yildiz Technical University, Istanbul, Turkey
"Optimization of Integrated Circuits using an Artificial Intelligence Algorithm"

Session Name:	**DEVICES**	
Session Time / Location:	**SUNDAY, JUNE 22 / VYKM 3**	
Session Chair:	**Peter KENNEDY, Univ. College Cork, Ireland**	

15:30 – 15:50   *A.O. SEVIM, S. MUTLU*   P17
Bogazici University, Istanbul, Turkey
"Post Fabrication Electric Field Treatment of Polymer Light Emitting and Photovoltaic Devices"

15:50 – 16:10   *H. TRANG[1], R. PATRICE[1], V. MARIE-HELENE[1], B. PHILIPPE[2]*   P21
[1]CEA/LETI Grenoble, France
[2]INPG/UJF/CNRS Grenoble, France
"Effect of thin Polyimide film on Performance of AlN/SiO2/Si SAW Device"

16:10 – 16:30   *O. GUL, S.S. KALLEMPUDI, H. BASAGA, U. SEZERMAN, Y. GURBUZ*   P25
Sabanci University, Istanbul, Turkey
"Biosensors For the Detection of Cardiovascular Risk Markers in Human Serum "

16:30 – 16:50   *A.T. CIFTLIK, H. KULAH*   P29
Middle East Technical University, Ankara, Turkey
"A Direct Injection Method for Blood Cells into Microchannels from Pure Blood Droplets with Switchable"

Session Name:	ANALOG DESIGN 1
Session Time / Location:	MONDAY, JUNE 23 / VYKM 2
Session Chair:	Franco MALOBERTI, University of Pavia, Pavia, Italy

09:00 – 09:20 *M. DEI[1], P. BRUSCHI[1], M. PIOTTO[2]*   P33
[1]University of Pisa, Pisa, Italy
[2]IEIIT Pisa CNR, Pisa, Italy
"A Four Quadrant CMOS Analog Multiplier Based on the Non Ideal MOSFET I-V Characteristics"

09:20 – 09:40 *C. CAKIR, O. CICEKOGLU*   P37
Bogazici University, Istanbul, Turkey
"Low-Voltage High-Performance CMOS Current Differencing Buffered Amplifier (CDBA)"

09:40 – 10:00 *R. KEPENEK, I.E. OCAK, H. KULAH, T. AKIN*   P41
Middle East Technical University, Ankara, Turkey
"A µg Resolution Microaccelerometer System with A Second-Order Σ-Δ Readout Circuitry"

10:00 – 10:20 *A. CITO[1,4], M. De MATTEIS[1], S. D'AMICO[1], A. BASCHIROTTO[1], P. DELIZIA[1],* P45
R. REIS[2], W. ZHAO[2], C. AZEREDO-LEME[2], A. TAVARES[2]
[1]University of Salento,Salento, Italy
[2]Chipidea Microelectronica, Portugal
[3]University of Milano Bicocca, Milano, Italy
[4]University of Pavia, Pavia, Italy
"A CMOS-90nm 15.6mW 20dBm-In-Band-IIP3 Analog Baseband Chain for UWB Receivers"

10:20 – 10:40 *U. GUVENC[1], A. ZEKI[2]*   P49
[1]UEKAE, Tubitak, Gebze, Turkey
[2]Istanbul Technical University, Istanbul, Turkey
"A New Tunable Linear Current Mirror"

Session Name:	SIGNAL PROCESSING
Session Time / Location:	MONDAY, JUNE 23 / VYKM 3
Session Chair:	Bulent SANKUR, Bogazici University, Turkey

09:00 – 09:20 *G. FISCELLI, C. CUCCHIARA, A. DI STEFANO, C. G. GIACONIA* P53
University of Palermo, Palermo, Italy
"Wide Bandwidth Impedance Meter using Low Rate Random Sampling"

09:20 – 09:40 *C. VURAL, S. KAZAN*   P57
Sakarya University, Sakarya, Turkey
"Robust Digital Image Watermarking Based on Normalization and Complex Wavelet Transform"

09:40 – 10:00 *A. PARUZEL, E. HERMANOWICZ*   P61
Gdansk University of Technology, Gdansk, Poland
"Efficient Fractional Delay Hilbert Transform Filter in the Farrow Structure"

10:00 – 10:20 *V. RATNER, Y.Y. ZEEVI*   P65
Technion, Israel Institute of Technology, Haifa, Israel
"Image Representation and Enhancement on Elastic Manifolds"

10:20 – 10:40 *K. AVCI, A. NACAROGLU*   P69
Gaziantep University, Gaziantep, Turkey
"A New Window Based on Exponential Function"

Session Name:	MODELING
Session Time / Location:	MONDAY, JUNE 23 / VYKM 2
Session Chair:	A. RODRIGUEZ-VAZQUEZ, Universidad de Sevilla, Spain

11:00 – 11:20   *S. LENCI, A. NANNINI, F. PIERI*   P73
University of Pisa, Pisa, Italy
"A Model for Piezoresistance in Torsional MEMS Springs"

11:20 – 11:40   *T. MAEHNE, A. VACHOUX, Y. LEBLEBICI*   P77
EPFL, Lausanne, Switzerland
"Development of a Bond Graph Based Model of Computation for SystemC-AMS"

11:40 – 12:00   *M.VASILEVSKI, N. BEILLEAU, H. ABOUSHADY, F. PECHEUX*   P81
UPMC, Paris, France
"Efficient and Refined Modeling of Wireless Sensor Network Nodes Using SystemC-AMS"

12:00 – 12:20   *E. SALMAN, E.G. FRIEDMAN*   P85
University of Rochester, Rochester, NY, USA
"Methodology for Placing Localized Guard Rings to Reduce Substrate Noise in Mixed-Signal Circuits"

Session Name:	MEMS 1
Session Time / Location:	MONDAY, JUNE 23 / VYKM 3
Session Chair:	Tayfun AKIN, Middle East Technical University, Turkey

11:00 – 11:20   *J.L. LOPEZ, F. TORRES, G. MURILLO, J.GINER, J.TEVA, J.VERD,*   P89
A.URANGA, G. ABADAL, N.BARNIOL
UAB, Barcelona, Spain
"Double-Ended Tuning Fork Resonator in 0.35um CMOS Technology for RF Applications"

11:20 – 11:40   *Y. D. GOKDEL, B. SARIOGLU, A.D. YALCINKAYA*   P93
Bogazici University, Istanbul, Turkey
"LED Integrated Miniaturized Polymer MEMS Display"

11:40 – 12:00   *S. A. JAWED[1], D. CATTIN[2], N. MASSARI[1], M. GOTTARDI[1],*   P97
A.BASCHIROTTO[3]
1ITC-IRST, Trento, Italy
2University of Trento, Trento, Italy
3University of Milano – Bicocca, Milano, Italy
"A MEMS Microphone Interface with Force-Balancing and Charge-Control"

12:00 – 12:20   *F. TOY, O. FERHANOGLU, H. TORUN, F. L. DEGERTEKIN*, H. UREY*   P101
Koc University, Istanbul, Turkey
*Georgia Institute of Technology, Atlanta-USA
"MOEMS Thermal Imaging Camera"

Session Name: DIGITAL DESIGN
Session Time / Location: TUESDAY, JUNE 24 / VYKM 2
Session Chair: Yusuf LEBLEBICI, EPFL, Switzerland

11:00 – 11:20   *D.J. WILLINGHAM, I. KALE*   P105
University of Westminster, London, Great Britain
"An Asynchrobatic, radix-four, carry look-ahead adder"

11:20 – 11:40   *H. K. ZRIDA[1], M. ABID[1], A. C. AMMRI[2], A. JEMAI[2]*   P109
[1]ENIS Institute Sfax University, Tunisia
[2]INSAT, Carthage University, Tunisia
"A YAPI-KPN Parallel Model of a H264/AVC Video Encoder"

11:40 – 12:00   *S. LAABIDI, B. ROBISSON, M. AGOYAN*   P113
EMSE, Gardanne, France
"An evaluation methodology for the security of cryptosystems"

12:00 – 12:20   *M.M. OZBILEN, M. GOK*   P117
Mersin University, Mersin, Turkey
"A Multi-Precision Floating-Point Adder"

Session Name: MEMS 2
Session Time / Location: TUESDAY, JUNE 24 / VYKM 3
Session Chair: Hakan UREY, Koc University, Turkey

11:00 – 11:20   *A. CABONI[1], M. BARBARO[1], A. HOMSY[2], P. VAN DER WAL[2], V.*   P121
LINDER[2], N. DE ROOIJ[2]
[1]University of Cagliari, Cagliari, Italy
[2]Institute of Microtechnology, University of Neuchatel, Switzerland
"Integration of a microfluidic flow cell on a CMOS biosensor for DNA detection"

11:20 – 11:40   *M. YILMAZ[1], E. ALACA[2], A.D. YALCINKAYA[3], Y. LEBLEBICI[4]*   P125
[1]Columbia University, NY, USA
[2]Koc University, Istanbul, Turkey
[3]Bogazici University, Istanbul, Turkey
[4]EPFL, Lausanne, Switzerland
"Design and Integration of a Bimorph Thermal Microactuator with Electrostatically Actuated MicroTweezers"

11:40 – 12:00   *A. SUMMANWAR[1,2,3], F. NEUILLY[2], T. BOUROUINA[3]*   P129
[1]Université Paris-Est, Marne la Vallée, France
[2]NXP Semiconductors, Caen, France
[3]ESIEE, Noisy-le-Grand, France
"Elimination of notching phenomenon which occurs while performing deep silicon etching and stopping on an insulating layer"

12:00 – 12:20   *I. SARI, T. BALKAN, H. KULAH*   P133
Middle East Technical University, Ankara, Turkey
"A Micro Power Generator with Planar Coils on Parylene Cantilevers"

Session Name:	CAD 2
Session Time / Location:	TUESDAY, JUNE 24 / VYKM 2
Session Chair:	Berna ORS YALCIN, Yildiz Technical University, Istanbul, Turkey

15:30 – 15:50   *S. BAYAR, A. YURDAKUL*   P137
Bogazici University, Istanbul, Turkey
"Self-Reconfiguration on Spartan-III FPGAs with Compressed Partial Bitstreams via a Parallel Configuration Access Port (cPCAP) Core"

15:50 – 16:10   *H. PARVEZ, H. MRABET, H. MEHREZ*   P141
Université Pierre et Marie Curie, *Paris, France*
"Generic Techniques and CAD tools for automated generation of FPGA Layout"

16:10 – 16:30   *Y. YALCIN[1], G. DUNDAR[1], B. M. WILAMOWSKI[2]*   P145
[1]*Bogazici University, Istanbul, Turkey*
[2]Auburn University, Alabama, USA
"Design and Optimization of PWL Circuits Used in Fuzzy Logic Hardware"

16:30 – 16:50   *G. Di GUGLIELMO,*   P149
University of Verona, Verona, Italy
"On the validation of embedded systems through functional ATPG"

16:50 – 17:10   *K.ATASU, T.TODMAN, O. MENCER, W. LUK*   P153
Imperial College, London, England
"Optimal Implementation of Combinational Logic on Look-up Tables"

Session Name:	RF 1
Session Time / Location:	TUESDAY, JUNE 24 / VYKM 3
Session Chair:	Catherine DEHOLLAIN, EPFL, Lausanne, Switzerland

15:30 – 15:50   *V. VALENTA, G. BAUDOIN, M.VILLEGAS*   P157
ESIEE, Paris, France
"Phase Noise Behaviour of Fractional-N Synthesizers with $\Delta\Sigma$ Dithering for Multi-Radio Mobile Terminals"

15:50 – 16:10   *M. VOLTTI, T. KOIVISTO, E. TIILIHARJU*   P161
University of Turku, Turku, Finland
"Statistical performance of IIP2 in Active and Passive Mixers"

16:10 – 16:30   *F. CANNONE, G. AVITABILE, D. CASCELLA*   P165
Politecnico di Bari, Bari, Italy
"Fully integrated coarse-fine wideband distributed Voltage Controlled Oscillator"

16:30 – 16:50   *M. KAYNAK, I. TEKIN, Y. GURBUZ*   P169
Sabanci University, Istanbul, Turkey
"A Matching Circuit Tuned, Multi-Band (WLAN and WiMAX), Class – A Power Amplifier Using 0.25µm-SiGe HBT Technology"

16:50 – 17:10   *M. De MATTEIS[1], S. D'AMICO[1], A. BASCHIROTTO[1], P. DELIZIA[1], C.AZEREDO-LEME[2], A. TAVARES[2]*   P173
[1]*University of Salento,Salento, Italy*
[2]*Chipidea Microelectronica, Portugal*
"A 90nm-CMOS 1.8mW 87dB-SNR 3rd Order Analog Filter for GSM Receiver"

Session Name:	RF 2	
Session Time / Location:	WEDNESDAY, JUNE 25 / VYKM 2	
Session Chair:	Barbaros SEKERKIRAN, ETA-IC, Turkey	

09:20 – 09:40   *S. DANESHGAR, M. P. KENNEDY*   P177
Univ. of College-Cork, Cork City, Ireland
"Analysis and Design of an LC Oscillator-based Injection-Locked Frequency Divider"

09:40 – 10:00   *C. KLAPF[1], A. MISSONI[1], W.PRIBYL[1], G. HOLWEG[2], G. HOFER[2]*   P181
[1]Graz University of Technology, Graz, Austria
[2]Infineon AG, Design Center Graz Graz, Austria
"Analyses and Design of Low Power Clock Generators for RFID TAGs"

10:00 – 10:20   *N. DEMIREL[1], E. KERHERVE[1], D.PACHE[2], R. PLANA[3]*   P185
[1]IMS, Bordeaux, France
[2]STMicroelectronics R&D Crolles, France
[3]LAAS-CNRS, University of Toulouse, France
"A 24 GHz, 18 dBm Fully Integrated Power Amplifier in a 0.13um SiGe HBT Technology"

10:20 – 10:40   *P.E. THOPPAY, C. DEHOLLAIN, M. J. DECLERCQ*   P189
EPFL, Lausanne, Switzerland
"Noise analysis in Super-regenerative receiver systems"

Session Name:	SENSOR INTERFACES	
Session Time / Location:	WEDNESDAY, JUNE 25 / VYKM 3	
Session Chair:	Denis MONTICELLI, NSC	

09:20 – 09:40   *P. DELIZIA[1], S. D'AMICO[1], A. BASCHIROTTO[12]*   P193
[1]University of Salento, Italy
[2]University of Milano Bicocca, Italy
"A Low-Power Readout Circuit for Lab-on-a-chip Applications"

09:40 – 10:00   *A. LOMBARDI[1], M. GRASSI[1], L. BRUNO[1], P. MALCOVATI[1], A. BASCHIROTTO[2]*   P197
[1]University of Pavia, Pavia, Italy
[2]University of Milano Bicocca, Italy
"An Interface Circuit for Temperature Control and Read-Out of Metal Oxide Gas Sensors"

10:00 – 10:20   *G. FERRARI, F. GOZZINI, M. SAMPIETRO*   P201
Politecnico di Milano, Milan, Italy
"Transimpedance amplifier for very high sensitivity current detection over 5MHz bandwidth"

10:20 – 10:40   *M. DEI[1], P. BRUSCHI[1], M. PIOTTO[2]*   P205
[1]University of Pisa, Pisa, Italy
[2]IEIIT Pisa CNR, Pisa, Italy
"Design of CMOS Chopper Amplifiers for Thermal Sensor Interfacing"

Session Name:	**ANALOG DESIGN 2**
Session Time / Location:	**WEDNESDAY, JUNE 25 / VYKM 2**
Session Chair:	**Ali ZEKI, Istanbul Technical University, Istanbul, Turkey**

11:00 – 11:20   *F. A. AMOROSO, A. PUGLIESE, G.CAPPUCCINO, G. COCORULLO*   P209
Universita Della Calabria, Italy
"Slewing Investigation and Improved Design Rules for SC Circuits Employing Two-Stage Amplifiers with Current-Buffer Miller Compensation"

11:20 – 11:40   *A. CITO[14], M. De MATTEIS[1], S. D'AMICO[1], A. BASCHIROTTO[1], P. DELIZIA[1], R. REIS[2], W. ZHAO[2], C. AZEREDO-LEME[2], A. TAVARES[2]*   P213
[1]University of Salento,Salento, Italy
[2]Chipidea Microelectronica, Portugal
[3]University of Milano Bicocca, Milano, Italy
[4]University of Pavia, Pavia, Italy
"A CMOS 90nm 4mW 15dBm-IIP3 Base-band Programmable Gain Amplifier for UWB Receivers"

11:40 – 12:00   *E. ARSLAN, A. MORGUL*   P217
Bogazici University, Istanbul, Turkey
"Wideband Self-Biased CMOS CCII"

12:00 – 12:20   *T. KOIVISTO, J. MAUNU, E. TIILIHARJU*   P221
University of Turku, Turku, Finland
"An Analog Baseband Chain for DS-UWB Radio Receiver"

Session Name:	**WIRELESS**
Session Time / Location:	**WEDNESDAY, JUNE 25 / VYKM 3**
Session Chair:	**Kivanc MIHCAK, Bogazici University, Turkey**

11:00 – 11:20   *E. S. ERDOGAN, S. OZEV, L. M. COLLINS*   P225
Duke University, Durham, NC, USA
"Online SNR Detection for Dynamic Power Management in Wireless Ad-Hoc Networks"

11:20 – 11:40   *K. M. SILAY, C. DEHOLLAIN Dehollain , M. J. DECLERCQ*   P229
EPFL, Lausanne, Switzerland
"Improvement of Power Efficiency of Inductive Links for Implantable Devices"

11:40 – 12:00   *O. Z. BATUR, G. DUNDAR, M. KOCA*   P233
Bogazici University, Istanbul, Turkey
"Measurements of Impulsive Noise in Broad-band Wireless Communication Channels"

12:00 – 12:20   *R.G. YUDANTO, D. BURDESE, M. MULASSANO, L. REYNERI*   P237
Politecnico di Torino, Torino, Italy
"Architecture of Automatic Monitoring System for Fresh Food Quality using Wireless Sensor Network"

Session Name:	ADC
Session Time / Location:	WEDNESDAY, JUNE 25 / VYKM 2
Session Chair:	Selcuk TALAY, University of Pavia, Pavia, Italy

14:00 – 14:20   Z. YE, M. P. KENNEDY   P241
Univ. of College-Cork, Cork City, Ireland
"Hardware Reduction in Digital MASH Delta-Sigma Modulators via Error Masking"

14:20 – 14:40   F. ERARIO, A. AGNES, E. BONIZZONI, F. MALOBERTI   P245
University of Pavia, Pavia, Italy
"Design of an Ultra-Low Power Time Interleaved SAR Converter"

14:40 – 15:00   B. KAYAALTI, O. CERID, G. DUNDAR   P249
Bogazici University, Istanbul, Turkey
"A Design Methodology for Asynchronous Sigma-Delta converters"

Session Name:	OTHERS
Session Time / Location:	WEDNESDAY, JUNE 25 / VYKM 3
Session Chair:	Yasemin KAHYA, Bogazici University, Turkey

14:00 – 14:20   A. CONVERTINO, A. GIORGIO, S. GIUSTO, R. MARANI, A. G. PERRI   P253
Politecnico di Bari, Bari, Italy
"Innovative Devices in Biomedical Electonics"

14:20 – 14:40   F. Y. YAMANER[1], L. CENKERAMADDI[2], A. BOZKURT[1]   P257
[1]Sabanci University, Istanbul, Turkey
[2]Norwegian University of Science and Technology, Trondheim, Norway
"Front-end IC Design for Intravascular Ultrasound"

14:40 – 15:00   B. METIN, O. CICEKOGLU   P261
Bogazici University, Istanbul, Turkey
"Tunable All-pass Filter with a Single Inverting Voltage Buffer"

Technology Independent Circuit Sizing for Fundamental Analog Circuits Using Artificial Neural Networks

Nihan Kahraman, Tulay Yildirim
Electronics and Communication Department
Yildiz Technical University
Istanbul, TURKEY
{nicoskun, tulay}@yildiz.edu.tr

Abstract— This study introduces technology independent neural network modeling for fundamental blocks of analog integrated circuits. The circuits modeled here are basic current mirror structures and a differential amplifier which serves as the input stage to most op-amps. Here if a designer defines the output specifications of the circuit, the neural network gives the channel widths (W) of all transistors in the circuit. It must be noted that the neural network in this novel approach is trained with the database including simulations using 1.5μm, 0.5μm, 0.35μm and 0.25μm technology SPICE parameters and the test data is constituted with simulations using only 0.18μm technology SPICE parameters which are not applied to the neural network for training beforehand. This shows that neural network is able to give the transistor sizes of circuit for a new unknown technology, independent on the SPICE parameters. As artificial neural network (ANN) structures, General Regression Neural Network (GRNN) and Multilayer Perceptron (MLP) having back propagation algorithm are used. Using new channel widths and lengths obtained from neural network's output, SPICE simulations of current mirrors and differential amplifier give the desired circuit output specifications for new technology.

I. INTRODUCTION

The analog circuit sizing is a technique which determines device sizes or parameters, e.g. transistor size, resistance and capacitance in given circuit topology, and in ordinary it can be expressed as an optimization problem [1]. In optimization process the routine of searching the solution and evaluating the circuit performance is repeated. In general numerous number of this iteration continues until it terminated. One difficulty of this technique is that it costs much computational effort due to their circuit analysis during optimization process. Another difficulty is that as integrated circuits' dimensions are reduced or a new device is developed, the old device or circuit models are no longer suitable. Consequently, new models have to be developed in order to use and predict the performances of the new device or circuit.

There is a growing interest in applying the potential of neural networks to many new fields especially in CAD of VLSI circuits. In this paper, we propose an optimization method which costs less computational cost than conventional methods for optimization.

Neural network algorithms are used for so many modeling themes in either microwave circuits or microelectronic device characterizations so far [2-4]. A few researchers are also used ANN to predict the outputs of analog and digital integrated circuits while the transistors' sizes are changing. However few researchers have been tried to predict the transistors' sizes for basic integrated circuits where the circuit DC&AC outputs are used as inputs for ANN [5].

II. CIRCUIT SIZING

In contrast to other modeling researches, the output specifications of IC's are predicted here for new technology designs using ANN. Current mirrors and differential amplifier are chosen as fundamental circuits of analog systems. For all circuits, the design constraints are determined and several simulations are done for different channel length and channel width of all transistors that meet the constraints using Cadence Spectre Analog Environment. Furthermore the simulations are performed using five different BSIM3 technology SPICE parameters for MOS transistors. These are 1.5μm AMIS, 0.5μm AMIS, 0.35μm TSMC, 0.25 μm TSMC and 0.18μm TSMC SPICE parameters [6]. This means that the fundamental circuits are designed for different transistor sizes (W, L) and for five different technologies mentioned above those decreasing channel lengths, respectively. Eventually, a large database is developed for neural network. The novel thing is that the neural network is trained with the database including simulations using 1.5μm, 0.5μm, 0.35μm and 0.25μm technology SPICE parameters and the test data is constituted with simulations using only 0.18μm technology SPICE parameters which were not applied to the neural network for training beforehand.

This research has been supported by Yildiz Technical University Scientific Projects Coordination Department. Project Number : 26-04-03-01.

The fundamental analog circuits are diversified using different circuit topologies. For current mirrors, four different topologies, basic current mirror, cascade current mirror, Wilson and Regulated Wilson current mirrors are chosen and standard topology is used for the differential amplifier.

Using new channel widths and lengths obtained from neural network's outputs, circuits are simulated with Cadence again and simulation results are compared with design constraints for new technology.

A. Current Mirrors

A current mirror (CM) is a circuit designed to duplicate a current through one active device by controlling the current in another active device of a circuit while keeping the output current constant in spite of loading. The current mirror is used to provide bias currents and active loads to circuits. Theoretically, an ideal current mirror is simply an ideal current amplifier with unity current gain.

In this paper four current mirror circuits are used for optimization via neural networks. These are simple CM, cascade CM, Wilson CM and Regulated Wilson CM which are shown in Figure 1 a, b, c, d respectively.

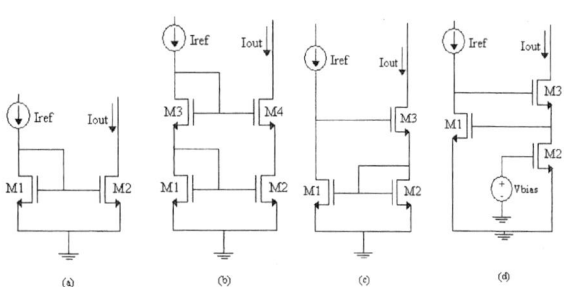

Figure 1. Current mirror circuits
a) Simple CM b) Cascade CM
c) Wilson CM d) Regulated Wilson CM

The ANN has 6 inputs and 4 outputs. Two of inputs are binary coded current mirror structure ('00' for simple CM, '01' for cascode CM, '10' for Wilson CM and '11' for Regulated Wilson CM) where the others are minimum channel length for the technology going to be used, reference current, desired output current and compliance voltage. The lowest output voltage that results in correct mirror behavior, the compliance voltage, is $V_{OUT} = V_{GS}$ for the output transistor at the output current level with VDG = 0 V [7]. The outputs of the neural network are the channel widths of all transistors. Note that the outputs W3 and W4 should be zero if the designer chooses the simple current mirror or W4 should be zero if Wilson current mirror is chosen and so on. Figure 2 shows the input-output mapping of neural network.

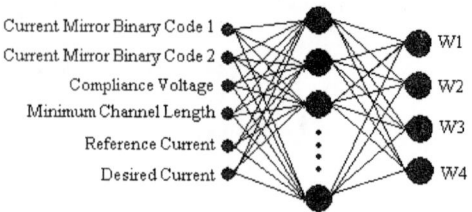

Figure 2. Input-output mapping of NN for transistor size estimation of current mirrors

B. Differential Amplifier

Among the basic analog circuits differential amplifiers play a very important role because of their excellent performances as input amplifiers and the straightforward application possibility of feedback to the input. It is also very compatible with integrated circuit technology and serves as the input stage to most op-amps. The differential amplifier with active load and single ended output shown in Figure 3 is the most common version of the differential amplifier in CMOS analog circuits.

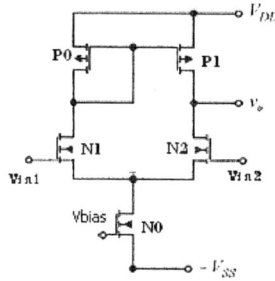

Figure 3. CMOS single ended differential amplifier

Differential amplifier design parameters are the W/L values of N0, N1, N2, P0, P1 and the current in N0 which is defined by external voltage Vbias. In large signal analysis, under quiescent conditions (V_D=0), the two currents in N1 and N2 are equal and sum to the current in N0. The current of N1 will determine the current of P0. Ideally, this current is mirrored in P1. If V_{GSN1}=V_{GSN2} and N1 and N2 are matched, the currents through N1 and N2 are equal. Thus the current that P1 sources to N2 should be equal to the current those N2 requires, causing output current to be zero.

The design constraints determined for dif-amp are open-loop gain (Av), gain-bandwidth product (GBW), input common mode range (ICMR), slew rate (SR) and power dissipation (Pdiss). These are also inputs for neural network structure.

CMOS differential amplifier is also simulated using five different SPICE parameters mentioned beforehand. Simulation results were used to form the inputs of NN, where the outputs give the sizes of each transistor. Channel lengths are accepted as minimum values for each technology. Figure 4 shows the inputs and the outputs of the NN model used.

978-1-4244-1983-8/08/$25.00 ©2008 IEEE

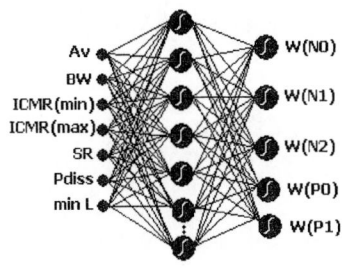

Figure 4. Input-output mapping of NN for transistor size estimation of differential amplifier

Cadence Spectre Analog Environment has a tool, 'Cadence Optimization Tool', that the designer can also optimize the circuit to obtain the desired outputs if he has the schematic of the circuit. It first assigns initial W/L values to each transistor and iterates the simulations for one of the given output specifications not all of them. After a lot of iterations it gives the optimum W/L values for each transistor. Figure 5 shows an example of these iterations.

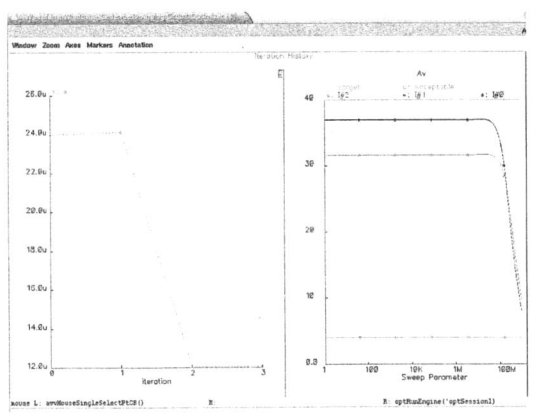

Figure 5. CADENCE optimization tool iterations for desired gain of 40 dB

III. NEURAL NETWORK MODELING

In ANN modeling of VLSI circuits, complex semiconductor equations are not required and the parameter extraction step, which represents a difficult and time consuming method, can also be omitted.

An artificial NN consists of some basic elements called neurons. Each neuron includes a set of inputs, weight coefficients (called synaptic weights), and an activation function. Neurons form the layers coming together called input layer, an output layer, and some optional intermediate layer(s), hidden layers. The input layer is made up of the sensory units that collect the information from its environment. The hidden layers apply a transformation from the input space to the output space. To obtain the desired output for any given input, the coefficients should be determined by training the network where sets of inputs with the corresponding outputs are given to the network through a training algorithm [8].

In this work, General Regression Neural Network (GRNN) and Multilayer Perceptron (MLP) having back propagation algorithm are used for transistor size optimization in fundamental circuits of analog integrated systems.

A. General Regression Neural Network (GRNN)

As Specht [9] outlines in his training procedure, all training variables are normalized so that the mean is zero and the variance is one. The probability density functions (PDF's) are Gaussian functions with centers that are defined by the training data. The PDF's are functions of the Euclidean distance in the input space between the current input and each of the training points. The network prediction is a pdf weighted average of the training targets. If the widths of the Gaussian nodes are equal and scalar, the output of the network would be defined by

$$Y(x) = \frac{\sum_{i=1}^{n} Y_i \exp(-\frac{D_i^2}{2\sigma^2})}{\sum_{i=1}^{n} \exp(-\frac{D_i^2}{2\sigma^2})} \quad (4)$$

$$D_i^2 = (x - x_i)^T (x - x_i) \quad (5)$$

where n is the number of nodes, X the input vector. And Xi and Yi are the ith input and target training vectors respectively. The spread 'σ' is determined empirically by minimizing the mean square error on the training data [10].

B. Multilayer Perceptron (MLP)

Multilayer networks are sometimes called layered networks. They can implement arbitrary complex input/output mappings or decision surfaces separating pattern classes. The most common learning algorithm for MLP is error back propagation, in which synaptic strengths are systematically modified so that the response of the network increasingly approximates the desired response, can be interpreted as a modeling problem.

The MLP model consists of a network of processing elements or nodes arranged in layers. Typically it requires three or more layers of processing nodes: an input layer which accepts the input variables used in the classification procedure, one or more hidden layers, and an output layer with one node per class. The principle of the network is that when data from an input pattern is presented at the input layer the network nodes perform calculations in the successive layers until an output value is computed at each of the output nodes. The most common learning algorithm for MLP is backpropagation, in which synaptic weights are modified due to an algorithm so that the response of the network approximates the desired response increasingly.

978-1-4244-1983-8/08/$25.00 ©2008 IEEE

The network processes the inputs and compares its resulting outputs against the desired outputs. Errors are then propagated back through the system, causing the system to adjust the weights which control the network.

IV. METHOD AND RESULTS FOR TECHNOLOGY INDEPENDENT DESIGN

Firstly, DC simulations of current mirrors were performed for various channel widths of all transistors that meet different current gains using Cadence Spectre Analog Environment using five different BSIM3 technology SPICE parameters for MOS transistors. These were 1.5μm AMIS, 0.5μm AMIS, 0.35μm TSMC, 0.25 μm TSMC and 0.18μm TSMC SPICE parameters. A large database, including 4620 simulations for simple CM, 10632 simulations for cascode CM, 8480 simulations for Wilson CM and 9086 simulations for Regulated Wilson CM, was constituted in order to apply to neural network. Eventually 32548 samples were applied to train the NN for CMs and 1245 samples were applied to test the NN including only simulation results for 0.18μm TSMC SPICE parameters. GRNN was used for current mirror sizing and spread parameter was selected as 0.8. 87 of the test results gave the wrong transistor sizing which means GRNN can estimate the current mirror circuits transistor sizes for new technology with 94% accuracy.

Secondly, CMOS differential amplifier was modeled using MLP. First, 100 simulations were performed for differential amplifier using Cadence Spectre Analog Environment including the output specifications, DC Gain (Av), Bandwidth (BW), Input Common Mode Range (ICMR), Slew Rate and power dissipation (Pdiss), for various W/L values for each transistor. These simulation results were used to form a database as inputs of MLP, where the outputs of neural network give the sizes of each transistor. Channel lengths were accepted as minimum values for each technology. 80 samples from the simulations were applied to NN for training where 20 samples were applied for testing. 7 input neurons, 10 hidden layer neurons and 5 output neurons were used in MLP structure. Tangent sigmoid functions and linear function were used as activation functions for hidden layer and output layer, respectively. Learning rate and momentum coefficient were chosen as 0.6 and 0.8, respectively, for 1500 iterations. Accepting 10% tolerance at the output specifications of the circuit (Av, GBW, SR, etc.), test database estimates the transistors' sizes with 90% accuracy. This shows that the neural network gives the channel widths (W) of all transistors in differential amplifier for 0.18μm technology when a designer gives the circuit output specifications and minimum channel length value although it only learns from other technologies.

Furthermore, the accuracy of the Cadence optimization tool is compared with NN. As an output specification Av should be 40 for the differential amplifier, the optimization tool gives W/L values as 24μm/0.18μm. To monitor the result of the optimization tool, W/L values were applied to Cadence Spectre Schematic as 24μm/0.18μm and the circuit was simulated again. Cadence optimization tool yielded Av=36. However, this tool can only optimize the circuit if the designer gives schematic including the technology SPICE parameters. Here, it is important that, the neural network gives transistor sizes in the circuit for an unknown technology. It is trained with the database including simulations using only 1.5μm, 0.5μm, 0.35μm and 0.25μm technology parameters and the test data are constituted with simulations using only 0.18μm technology parameters which are not applied to the neural network for training previously.

V. CONCLUSION

The aim of this study is to predict the transistor sizes for fundamental circuits in new technologies, which corresponds the design constraints, without knowing the SPICE parameters, using neural networks. As technology builds up day by day, and transistor sizes decreases, the usage of neural network to predict the dimensions of circuits in new technologies can be an innovation on the design of VLSI circuits.

REFERENCES

[1] Tomohiro, F.; Osamu, I, "Analog circuit sizing with dynamic search window", Circuits and Systems, Proceedings of ISCAS 2006 pp:4

[2] Creech G.L., Paul B., Lesniak C., Jenkins T., Lee R., Brown K., "Artificial Neural Networks For Accurate High Frequency CAD Applications", Proceedings of ISCAS '96, Vol. 3 pp:317 - 320

[3] Dos Santos A.L., Romariz A.R.S.,. De Carvalho P.H.P, "Neural Model Of Electrical Devices For Circuit Simulation", International Microwave and Optoelectronics Conference 'Linking to the Next Century' Proceedings, pp:253 - 258 vol.1

[4] Peik S.E., Coutts G. , Mansour R.R., "Applications Of Neural Networks In Microwave Circuit Modeling", IEEE Canadian Conference on Electrical and Comp. Eng., 1998, Vol. 2, pp:928 - 931

[5] Zaabab. A.H., Zhang Q.J., Nakhla M., "Application Of Neural Networks In Circuit Analysis", Neural Networks, 1995. Proceedings., IEEE International Conference on Volume 1, pp:423 - 426 vol.1

[6] http://www.mosis.com/test/

[7] Baker R.J., CMOS Circuit Design, Layout and Simulation, Second Edition, New York: Wiley-IEEE, 2005 ISBN 0-471-70055-X.

[8] Hatami S., Azizi M. Y., Bahrami H. R., Motavalizadeh D., and. Afzali-Kusha A, "Accurate and Efficient Modeling of SOI MOSFET With Technology Independent Neural Networks", IEEE Transactions On Computer-Aided Design Of Integrated Circuits And Systems, Vol. 23, No. 11, 2004

[9] Specht, D.F. (1991). A general regression neural network. IEEE Transactions on Neural Networks,2(16)568-576]

[10] Sinclair, M.J.; Musavi, M.T.; Qiao, M., "Radial basis function neural network as predictive process control model", Circuits and Systems, ISCAS '95., Page(s):1948-1951

978-1-4244-1983-8/08/$25.00 ©2008 IEEE

Hybrid Approach for Performance Estimation; Embedded Tool for Analog Design Automation Systems

Engin Deniz
Department of Electronics and Communication Engineering
Dogus University
Istanbul, Turkey
edeniz@dogus.edu.tr

Gunhan Dundar
Department of Electrical and Electronics Engineering
Bogazici University
Istanbul, Turkey
dundar@boun.edu.tr

Abstract—**A novel approach for performance estimation is described which gives the designer access to the design space boundaries of a circuit topology so that the tradeoff analysis of competing performances can be evaluated by optimum design points and Pareto curves with an acceptable execution time and transistor-level accuracy can be obtained. Recently, there is a strong emphasis on numerical techniques for performance estimation models however neither of them satisfies the following three significant issues; accuracy, time-consumption and topology-independent modeling. This new approach allows using different performance estimation methods together in order to speed up the overall design automation system with an acceptable time and accuracy for any given topology.**

I. INTRODUCTION

Time to market is an important factor in integrated circuit (IC) manufacturing and a big pressure on all designers. Nowadays any IC has mixed-signal applications with analog components which require the consideration of a variety of technological and physical effects and thus dominating the total design time. On the other hand, analog design lacks Computer Aided Design (CAD) Tools which reduce the effort for the design, and which are necessary to overcome the complexity problem of the system with respect to size and performance demands. An Analog Design Automation (ADA) System such as proposed in [1] could be a solution to this problem.

Utilization of a Performance Estimator (PE) is important in order to accelerate the operation of the overall ADA system. Generally, analog circuits have several transistors and each of them has many independent inputs that can be controlled. Related to this, there are numerous possibilities most of which do not meet the performance specifications because most of these parameters are competing with each other such as gain and bandwidth, slew rate and stability. Unfortunately, too much time will be wasted on undesired results. The operation can be made faster using a PE, which will give hints for optimization thus avoiding unwanted

sections of the design space and obtaining a tradeoff between competing performance parameters.

EKV transistor model which is the abbreviation of the surnames of Christian C. Enz, Francois Krummenacher and Eric A. Vittoz, has already been used in extracting a PE model which has been presented in [2-3] and applied as an embedded system in a sigma-delta analog-to-digital converter design automation tool in [4]. This approach satisfies the accuracy and execution time expectations; however, it is extracted manually by analytical equations for one specific circuit; as the schematic of the circuit is changed, the model must be re-extracted.

In this work, the main goal is to develop an effective and accurate PE model with reasonable model extraction time for any given topology. To obtain such behavior, different methodologies will be investigated with the purpose of limiting the boundary conditions for the performance specifications. The paper is organized as follows; background of the problem and literature survey is addressed in section II. In section III, hybrid approach for PE is described. The hybrid model is validated in section IV. Conclusions are drawn in section V.

II. BACKGROUND OF THE PROBLEM

A. Definition of the Problem

The PE must be able to handle any circuit topology, accepting some performance parameters as variables, and estimating the remaining parameters. One can denote gain, bandwidth, slew rate, output resistance, etc... as performance parameters (p_p) of any analog circuit. Independent circuit parameters (i_p) influencing the performance can be indicated as width, length of the transistor, current, bias voltages, etc...The problem is to find an expression or at least estimation for a p_p in terms of each other without having to calculate an i_p. In general, functions which links the p_p to the i_p, are non-linear and most of the

978-1-4244-1983-8/08/$25.00 ©2008 IEEE

time it is very complicated to express them analytically. From this problem statement therefore, it is obvious that to develop a PE tool is a very difficult task.

B. Literature Survey

PE modeling approaches can be grouped into two main areas as shown in Fig. 1.

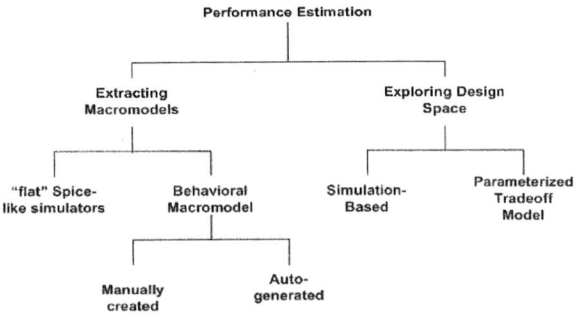

Figure 1. Performance estimation methods found in recent literature

Generally speaking, macromodels are extracted for one specific circuit that is fully designed at cell-level; as the design variables are changed, the model must be re-extracted. Unfortunately, extracting a macromodel is a necessary but not sufficient solution to the problem. Therefore, exploring the design space plays an important role for analog synthesis.

One simple solution would be to form a lookup table for all combinations of the parameters i_p once and to search from the table required p_p combinations every time. This method can be called the Brute Force (BF) approach. Assume that the selected technology provides a lower bound of $0.5\mu m$ for the transistor width and an upper bound of 1mm and each time sizes are swept by $1\mu m$ steps; it can be easily calculated that 1000 data points are needed for a single transistor. It follows then a simple 10 transistor opamp requires 10000 data points to be calculated. Therefore, intelligent experiment design approaches are applied in order to avoid huge-sized lookup table. In the open literature, most of the PE tools use a given set of design variables and calculates a set of circuit parameters based on SPICE simulation. After gathering the data points, regression methods can be applied in order to construct the Pareto curves such as in [5-6], mathematical templates such as posynomial, wavelet, radial basis, support vector machines, etc... are used for fitting the data [7-9] or symbolic analysis is applied such as proposed in [10]. On the other hand, there are some approaches which have less simulation time but they are topology-specific. Using analytical equations which are specific to the topology can be a good solution to extract the circuit model in reasonable simulation time and accuracy such as proposed in [2-3] and [11-12]. However, those equations are derived manually for a given analog block and for each new circuit; analog designers have to derive new equations so that this approach becomes impractical for complex analog blocks. In

this paper, a hybrid method which benefits from different PE methods is described.

III. HYBRID APPROACH FOR PE

Reasonable simulation time and high accuracy property of topology-specific analytical equations and topology-free design space exploration property can be used together. The flowchart of PE tool is illustrated in Fig.2.

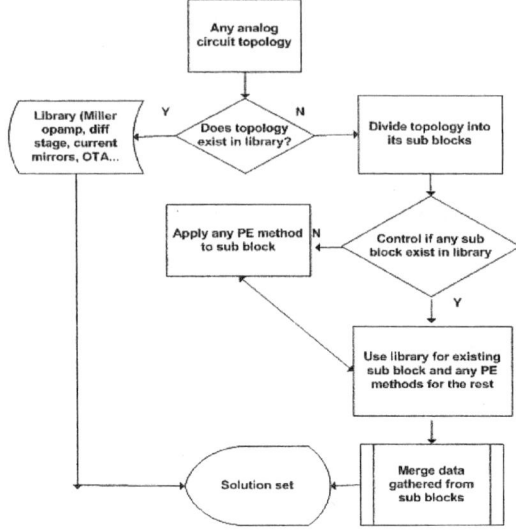

Figure 2. Flowchart of proposed PE

Any given topology will be checked if it exists in the library or not. If the topology exists in the library, PE model is already obtained so that there is no need construction time of modeling. Otherwise, any approach can be applied as summarized in the literature survey given above, but simulation time is still a problem when compared to our approach in [1]. Therefore, the topology will be divided into its subblocks such as differential pair (diff), current mirrors, etc...Then each subblock is checked if it exists in the library. If yes, a predefined PE model of a subblock can be used, otherwise any PE model will be applied to a subblock. As a result, execution time will be reduced because any topology-free PE approach is applied only to a subblock instead of main analog block. Remember that the simulation time of PE model increases more than linearly as the elements of analog block increases.

A. EKV-Based Modeling (Topology-Specific)

This method [2] is based on using the analytical equations of EKV mosfet model to estimate the behavior of some basic blocks and later on utilizing the block estimations to arrive at the final circuit performance estimations. The estimator is coded in C programming language and model equations are implemented to the code which takes p_p as input and gives us the estimation results as output. This

978-1-4244-1983-8/08/$25.00 ©2008 IEEE

method is improved by increasing the accuracy of analytical equations resulting gmID and gds parameters.

B. Brute-Force Modeling (Topology-Free)

Considering the input space and gathering output and performance information from SPICE, a lookup table is formed. This table has a large memory which inevitably increases the simulation time more than linearly as the number of elements in analog block increases. However, its simple algorithm may be useful in some analog blocks which have two or three elements. In this paper, BF approach is applied to a simple output stage. First of all, a wide range of transistor width for each transistor is used in SPICE simulation; as expected it takes too much time (2 days) to finalize the output file. Therefore, a very simple rule which reduces the input range is applied so that simulation time reduces to five minutes. Finally, solution set is obtained and Pareto curves are constructed.

IV. AN APPLICATION OF HYBRID PE

A lead compensated basic two stage (BTS) Miller opamp (Fig.3) is taken as an example for the proposed hybrid approach. Assume that this opamp does not exist in the library. Then, it is divided into its subblocks; diff stage and output stage. Assume diff stage exists in library; however, the output stage does not. Therefore, predefined EKV-based diff stage model is used and BF approach is applied to the output block. Eventually, data from each subblock are merged together in order to satisfy the performance criteria of the BTS opamp. Verification of this BTS opamp application can be provided by comparing the results from hybrid PE approach and EKV-based PE approach described in [2]. In other words, combining the results of diff stage modeled by EKV-based approach and the results of output stage modeled by BF approach must satisfy the results of BTS opamp modeled by EKV-based approach.

A. Analysis of BTS opamp

Assume that BTS opamp has the following performance criteria; minimum gain is 2000, 3dB frequency is 10 kHz, output capacitance is 1pF and minimum slew rate is 2 V/μs. EKV-based macromodel finds only 10 solutions out of 444360 candidates. All performance criteria are satisfied and ac and dc behavior of the opamp are taken into account while estimating those results. While evaluating the solution set, optimum solution sets could be more useful for the end-users. For example, layout issues could be more critical than the other parameters, and then minimum area values are taken into account for the same current and voltage values. Bias voltage values and gain of input stage (A_{V1}) are given in Table 1 which represents the EKV-based optimum solution set related to area. Currents flowing from M_5 and M_1 are 7μA and 14μA respectively. Power consumption of input and output stage is calculated as 46.2 μW. It should be reminded that those results are obtained only in few minutes.

TABLE I. LOOK UP TABLE FOR BTS OPAMP

V_{G3} (V)	V_{bias} (V)	A_{V1}	Power (μV)	Area (μm)^2
0.672	2.50	66.8	92.4	287.4
0.683	2.49	66.8	92.4	287.4
0.672	2.50	66.8	92.4	287.4
0.683	2.49	66.8	92.4	287.4
0.672	2.50	66.8	92.4	287.4
0.683	2.49	66.8	92.4	287.4
0.672	2.50	66.8	92.4	287.4
0.683	2.49	66.8	92.4	287.4
0.672	2.50	66.8	92.4	287.4
0.683	2.49	66.8	92.4	287.4

B. Analysis of Single Differential Stage

The boundary of input space of the diff stage is limited due to the performance criteria of BTS opamp. A lookup table is formed and then only a specific region which is given in Table 2 is selected for the application. It is observed that gain and power values are equal to the results that are found in analysis of BTS opamp for the same bias voltages.

C. Analysis of Output Stage

All possible circuit parameters are taken into account and in addition to this, some constraints such as limiting the bias voltage values due to the diff stage are also evaluated for the example. Referring to the Fig.3, output voltage of diff stage (V_{DSM3}) is equal to the input voltage of output stage (V_{GSM1}). Furthermore, after gathering the data, some of them have to be eliminated because the input space results some infeasible solutions such as undesired output voltage (V_{DSM1}). Finally, 2881 candidates reduce to 410 and a lookup table is obtained to generate optimum solution sets for the end-users. For this example, a specific region which satisfies the results collected from the BTS opamp is selected in order to verify the application. Table 3 represents the verification of the results. First column shows the estimated circuit parameters which are critical for this example. Second column indicates the results from BTS opamp which is assumed to be modeled by EKV-based approach. Third and last column illustrates the diff and output stage respectively. It is easily seen that the addition of area values of different subblock is equal to the area value of the overall system.

TABLE II. LOOK UP TABLE FOR DIFF STAGE

V_{out} (V)	V_{bias} (V)	Gain	Power (μV)	Area (μm)^2
0.67	2.50	66.8	45.5	269
0.68	2.49	66.8	46.2	273
0.67	2.50	66.8	46.2	273
0.68	2.49	66.8	46.2	273
0.67	2.50	66.8	46.8	277
0.68	2.49	66.8	46.8	277
0.67	2.50	66.8	46.8	277

TABLE III. VERIFICATION TABLE OF OVERALL SYSTEM

Estimated Values	BTS opamp	Diff stage	Output stage
Output voltage of M3 Input voltage of M1 (V)	between 0.67 and 0.68	0.67 0.68	0.67 0.68
Gate voltage of M7 and M2 (V)	between 2.49 and 2.50	2.49 2.5	2.49 2.5
Current (μA)	7 for diff stage 14 for output stage	7	14
Area (μm)^2	281-284	269 272	12.2

V. CONCLUSION AND FUTURE WORK

A PE tool which uses both EKV-based approach and BF approach is proposed. Performance response graph, which gives an idea of circuit tradeoffs to the designers are obtained. Fig. 4 which is obtained from lookup table of BTS opamp illustrates an example of an optimal solution due to area issues with respect to gain and power values. As a result, two approaches are applied successfully to an analog block. This encourages the future work such as using different approaches together which are more intelligent than BF in order to increase the accuracy and simulation time. The future work will be proceeding in several directions. One direction is the incorporation of more subblocks and more analog circuits which are modeled by EKV-based approach in order to find more predefined subblocks from the library during the estimation process. The second direction is investigation on different approaches about extracting Pareto curves. The third and last direction is the addition of those different approaches into the estimator.

REFERENCES

[1] S. Balkır, G. Dundar, and A. S. Ogrenci, Analog VLSI Design Automation, CRC Press, 20032

[2] E. Deniz, G. Dundar, " Performance estimator for an analog design automation system using EKV-modeled analog circuits", Proc. of ECCTD, 2005.

[3] E. Deniz, G. Dundar, " Mosfet modeling with EKV 2.6 and analog circuit design stragety for performance estimator tool", Proc. of ELECO, 2005.

[4] S. Talay, E. Deniz, G. Dundar, "A Sigma-Delta ADC design automation tool with embedded performance estimator", Proc of MIXDES, 2006.

[5] B. De Smedt, G. Gielen, "WATSON: Design space boundary exploration and model generation for analog and RF IC design" Proc of IEEE CAD of integrated circuit and systems, vol. 22, no. 2, February 20003.

[6] R. Harjani, J. Shao, "Feasibility and performance region modeling analog and digital circuits", Proc of analog integrated circuits and signal processing, vol. 10, pp. 23-43, 1996.

[7] W. Daems, G. Gielen, W.Sansen, "Simulation based generation of posynomial performance models for sizing of analog integrated circuits" Proc of IEEE CAD of integrated circuit and systems, vol. 22, no. 5, May 20003.

[8] J. Tao, X. Zeng, D. Zhou, "Analog behavioral modeling by multicompanding and wavelet collocation method, Proc. of analog integrated circuits and signal processing, 2006

[9] T. Kiely, G. Gielen, "Performance modeling of analog integrated circuits using least-squares support vector machines, Proc. of DATE, 2004

[10] T. McConaghy, T. Eeckelaert, G.Gielen, "CAFFEIN: Template-free symbolic model generation of analog circuits via canonical form functions and genetic programming", Proc. of ESSCIRC, 2005

[11] H. Taher, D. Schreurs, B. Nauwelaers, "Black box modeling at the circuit level: Opamp as a case study", Proc. of IEEE MELECON 2006

[12] R. L. Oliveira Pinto, F. Maloberti, "X Ray and Blue Print: Tools for mosfet analog circuit design addresing short-chanel effects", Proc. of IEEE ISCAS, 2004

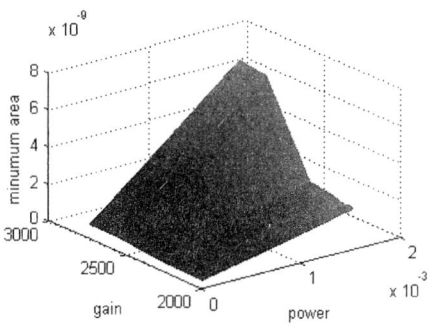

Figure 4. BTS opamp-minimum area due to gain and power

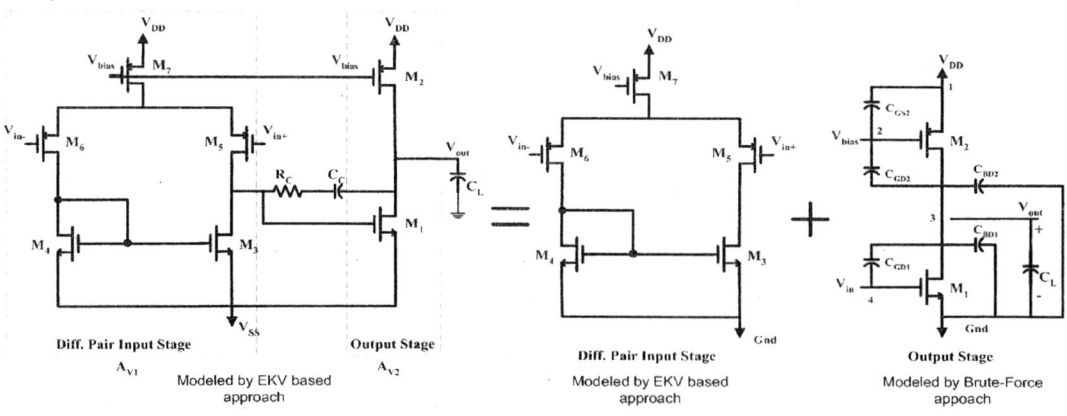

Optimal Asynchronous Linear-Pipelines

Eslam Yahya[1,2] and Marc Renaudin[3]

[1]CIS Group, TIMA Laboratory, Grenoble, France
[2]Banha High Institute of Technology, Banha, Egypt
Eslam.Yahya@imag.fr

[3]TIEMPO SAS
Grenoble, France
Marc.Renaudin@tiempo-ic.com

Abstract— **This paper introduces a new methodology for determining the minimum number of registers needed to pipeline a linear asynchronous pipeline so that the final cycle time meets some constraints. Moreover, the methodology defines the optimum placing for the registers; ends with an optimally pipelined circuit.**

I. INTRODUCTION

As an emerging technology, asynchronous circuits provide design solutions which are low power, low EMI and robust against process and environment variations [5]. However, the development of analysis and optimization methods of asynchronous circuits is a hard task due to their complex behavior.

Building an efficient design needs accurate timing analysis. In [1] and [6], some performance analysis methods are introduced. These methods are slow and not flexible especially when time variability is considered. As a result, the use of them to guide optimization algorithms is impractical. In [4], a very fast performance analysis is proposed. Compared to the similar work as [1] and [6], the work in [4] is three orders of magnitudes faster. Moreover, it shows high flexibility when time variable delays are considered for the pipeline stages.

Generally speaking, asynchronous circuit optimization is being done by means of manual or semi-automated techniques. In [7], a method based on pipeline optimization is introduced. The optimization technique introduced here in this paper can be classified in the same category. The idea is to find the minimum pipelining degree to satisfy the performance constraints. Consequently, a minimum number of registers is used which reduces the pipeline area and power consumption for a given performance. The problem addressed in this paper is the problem of a designer having asynchronous linear pipeline [2,3] where time delay variability, due to data dependency, structure and PVT (Process-Voltage-Temperature) can be taken into account. The designer needs to answer the following question: "Targeting certain performance, what is the minimum number of asynchronous registers that can be used? And, what is the optimum placing of these registers that not only satisfies the target performance, but also results in the maximum possible performance using this number of registers?" The answer to this question is the contribution of this paper.

Section II gives a background on the underlying performance analysis method. The proposed optimization algorithm is specified in Section III. Some experimental results are shown in Section IV. Section V shows the application of the optimization algorithm to an asynchronous link between two synchronous processors. Finally, Section VI states the paper conclusion and future works.

II. PERFORMANCE ANALYSIS METHOD

Optimization techniques presented in this paper is based on the performance analysis method introduced in [4]. As a result, a brief description of this analysis method is shown in this section.

The analysis method is based on four main components:

1) Circuit Model: it is constructed using a Dependency Graph (DG), which is a graph-based method where the nodes of the graph correspond to specific rising or falling transitions of the circuit signals, and the arcs depict the dependencies between signal transitions [8].

2) Delay Model: the Delay Token Vector is introduced. It is a token vector consisting in a list of delay pairs [T_{Eval}, T_{Reset}]. Each pair expresses the evaluation and reset delays (4 phase protocol). This model can simply represent time fixed delays as well as time variable delays. If the component has nondeterministic time-variable delays then the TV is filled-in with randomly-ordered delay pairs satisfying the probability distribution of the delays.

3) Analytical Model: Analyzing the performance of the pipeline needs to derive equations from the circuit model. These equations are derived in terms of delay TVs values which are determined by the delay model. These equations are called the Analytical Model.

4) Performance Analysis: many performance metrics can be efficiently extracted by solving the analytical model. In this paper only the Cycle Time, equivalently the throughput is considered.

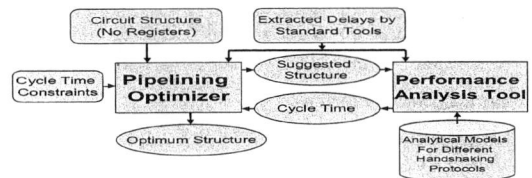

Figure 1: The Tool Flow

Fig.1 shows the connections between the optimizer and the performance analysis tool. As shown in the figure, delay information for all circuit components are extracted by standard tools. All delay time variability can be included (Data dependency, Structure and/or PVT). Circuit structure (containing no registers) and performance constraints are passed to the optimizer. It analyzes the original structure and passes a suggested structure to the performance analysis tool, which responds by the Cycle Time (CT) of that suggested structure. The optimizer compares the resultant CT with the target CT and decides for the next step, etc... Finally, the optimizer finds the optimal structure and passes it as an output.

It is obvious that the optimizer is calling the performance analysis tool for many times until the optimal structure is found. That means that computation time needed to find the optimal structure is not only determined by the optimizer but also by the time response of the underlying performance analysis tool.

III. OPTIMIZATION ALGORITHM

Gauging the circuit performance gives important information. However, the ultimate goal is to use this information for optimizing the circuit. Most of the performance analysis methods applied to asynchronous circuits are based on average delays or static timing delays. Optimization of asynchronous circuits using these methods does not exploit the ability of asynchronous circuits to reduce pessimism. Applying timing analysis methods which are based on probabilistic time variable delays is the right choice to efficiently optimize asynchronous circuits. Previously introduced methods which are able to support timing variability are a bit slow. As a result, applying optimization algorithms which are guided by these methods is not practical. The fast and efficient performance analysis method introduced through the previous section constitutes a basement for practical optimization method.

As shown in Fig.2, asynchronous linear pipeline consists of a number of Function Blocks (F), a transmitter (TX) and a receiver (RX). In this pipeline there are some possible places, denoted (P), to place the registers. These places are predetermined and fixed inside the structure. That means that our algorithm does not perform retiming (this will be considered in a future work). The tool is searching the minimum number of registers which are needed to satisfy a given performance and find where to insert these registers between the function blocks. In Fig.2, there are seven places (P=7), TX and RX are supposed to implicitly contain a register to handle the interaction with the environment. Register1 (R1) and Rigester2 (R2) are placed in places P2 and P6 respectively. The Stage consists of one register with all its preceding Function Blocks. As an example, Stage2 (S2) consists of R2 plus (F5, F4, F3, F2).

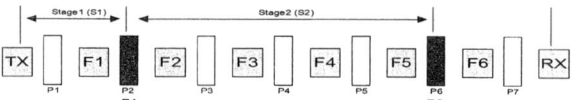

Figure 2: Linear Pipeline

Each component in the circuit, (F, R, TX and RX), has its own Delay Token Vector (TV). In case of time variable delays this TV consists in several delay pairs. The number of pairs is defined as TV-Length and denoted by (M). Each component will have Average, Min and Max delays which are representing the average value of the whole TV, the minimum value of the whole TV and the maximum value of the whole TV, respectively. Taking function block F3 as an example these delays will be respectively denoted as $(D^{F3}_{Avg}, D^{F3}_{Min}, D^{F3}_{Max})$.

The problem the optimizer is solving is as follows. Given a linear pipeline structure that has possible places "P" and a cycle time in case of no registers are placed "CT_{NR}", the optimizer finds a minimum number of registers "η" for which the pipeline's CT satisfies a target cycle time constraints "CT_T". Moreover, the optimizer finds the optimum placing for the "η" registers among the

possible places "P". It ends with "η_{Opt}" such that the pipeline's CT is not only satisfying CT_T, but is also the minimum cycle time, Max throughput, that can be achieved using the "η" registers.

The only way to be sure that the optimization algorithm ends with the optimum solution is to test all the possible architectures. A straightforward way to do that is to use an exhaustive search for all the possible η (from 1 to P), and for each η test all the possible placing combinations "C^{η}_P". With such a Brute Force (BF) algorithm, it is guaranteed that the optimizer will find "η_{Opt}" for any given CT_T. The complexity (number of iterations) of the BF algorithm is shown in Eq.1.

$$BF_C = 2^P \qquad (1)$$

Hereafter is the pseudo code of the BF algorithm.

```
{
    For (η = 1 → η = P)
    {
        Test all Cᵖη and pick the one giving the least CT "η_Opt";
    }
    Out the min η_Opt where CT ≤ CT_T ;
};
```

The problem of finding the minimum number of registers "η" and finding the optimum placing "η_{Opt}" can be seen as a two-dimensional problem. The first dimension is to find η where: $\eta \in [1,P]$. This dimension appears in the above pseudo code in the main outer loop. As a result we call it Outer Loop "OL". For each suggested η, there is an inner loop enumerating all possibilities for placing η registers among P places. This loop is the second dimension of the problem and it is called the Inner Loop (IL). Optimizing both OL and IL can of course significantly enhance the algorithm performance.

If we can prove that adding registers to the pipeline is affecting the CT monotonically, then we can stop the analysis once the pipeline CT is less than or equal to CT_T. And still formally sure that the resultant η is the optimum one with no need to cover all possible $\eta \in [1,P]$.

Therom1: For any linear pipeline, if the max internal delay of the registers "R" is lower than the min delay of the Function Blocks "F", then adding registers to a linear pipeline is increasing the throughput monotonically. In other words, increasing the number of registers is decreasing the cycle time of the pipeline.

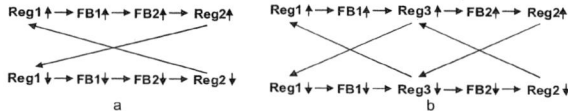

Figure 3: Dependency Graph for Stage Pipelining

Proof: Fig.3.a shows the dependency graph of a stage which contains 2 cascaded function blocks (F1,F2). The CT of the stage is:

$$CT_1 = D^{R1} + D^{F1} + D^{F2} + D^{R2}$$

If register R_3 is added in between F1 & F2, as in Fig.3.b, the CT becomes:

$$CT_2 = Max[(D^{R1} + D^{F1} + D^{R3}); (D^{R3} + D^{F2} + D^{R2})]$$

978-1-4244-1983-8/08/$25.00 ©2008 IEEE

If: $D^{R3} \leq D^{F2} + D^{R2}$ & $D^{R3} \leq D^{R1} + D^{F1}$

$\therefore CT_2 \leq CT_1$

If $\underset{i=1}{\overset{i=\eta}{MaxD}}{}_{Max}^{Ri} < \underset{j=1}{\overset{j=P}{MinD}}{}_{Min}^{Fj}$ (2)

Then $\eta\uparrow \rightarrow CT\downarrow$ and $\eta\downarrow \rightarrow CT\uparrow$

i.e, adding a register to the pipeline is a monotonic problem.

If the condition in Eq.2 is satisfied, pipelining is guaranteed to affect the cycle time monotonically. This is an important property which helps the designers to simplify their circuit optimization and it is one of the paper contributions. From Therome1, one optimization can be applied to the "OL" of the BF algorithm. That the algorithm starts to scan from $\eta = 1$ and fully tests the "IL"; if the minimum achievable CT is less than or equal to CT_T the algorithm breaks; if not it tests $\eta = 2$ ...etc. in this way, we avoid unnecessary testing for greater values of η. Meanwhile, we are formally sure that the resultant η is the "η_{Opt}". Hereafter, the pseudo code of the optimized algorithm, it will be denoted by Efficient Algorithm "EA".

For ($\eta = 1 \rightarrow \eta = P$)

{

     Test all C_P^{η} and pick the one giving the least CT (η_{Opt}) ;

     If ($CT \leq CT_T$) Break ;

};

The Efficient Algorithm "EA" calls the performance analysis tool with each tested structure. Consequently, its complexity determines the final execution time needed for optimizing the circuit. Eq.3 gives the complexity of the EA.

$$EA_C = \sum_{i=1}^{i=\eta} C_P^i = \sum_{i=1}^{i=\eta} \frac{P!}{i! \times (P-i)!} \quad (3)$$

This EA optimizer is implemented in our tool. Many test cases were conducted and we verified that the algorithm always finds the optimum solution. Some test cases are shown in Section IV, to highlight the algorithm performance.

IV. TEST CASE RESULTS FOR THE OPTMIZATION ALGORITHM

Many test cases were conducted to test the algorithm and its implementation. In one test case, a pipeline as the one depicted in Fig.2, is designed. This pipeline contains 11-function blocks plus TX and RX, that means that P=12. Delay TVs are time variable probabilistic delays, different probabilistic distributions, (Random, Uniform, Gaussian, Exponential, ..) are used. The design is analyzed and optimized targeting many different CTs. Fig.4 shows a comparison between the BF algorithm and the EA in terms of number of iterations needed to determine η_{Opt}. The X-axis represents 36 different CT values in which the CT_T is decreasing. The Y-axis represents the number of iterations

Compared to BF, EA has much better performance for high CT, equivalently lower η. It is explained by the fact that when η is low, EA gets the solution early and prunes many unnecessary iterations. Same contribution can be derived from Eq.1 and Eq.2.

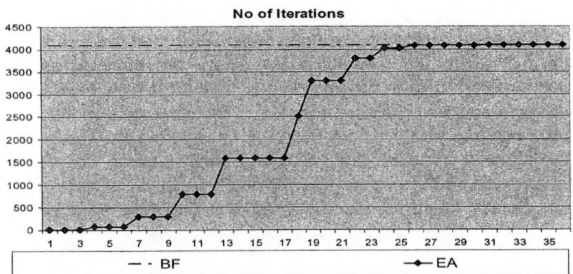

Figure 4: Number of Iterations for BF and EA

Table1: Total Iterations (BF vs EA)

Technique	Iterations	Gain% WRT BF
Brute Force (BF)	4095	
Efficient Alg (EA)	2346	42.7%

Table 1, shows a comparison between the BF and the EA in terms of the total number of iterations in the test case of Fig.4. The EA introduces a gain of 42.7% in terms of number of iterations compared to the BF algorithm.

Regarding the algorithm computation time, the EA needs 703.2 Sec to find η_{opt} for 12 possible places (P=12) where each function block has a probabilistic time variable delays with a TV length of M=10^5 tokens (execution times are measured on A SPARC III machine with 2GB of Ram). To the best of our knowledge, the nearest work to the presented work in this paper is the work introduced in [7]. The main difference between the two works is that in [7], only average delays are supported. It means that each component in the circuit is assigned a single delay value, and no time variability is supported. In contrast, our work supports all types of delay. Regarding the performance, it is stated that the algorithm needs 650 Millisec to pipeline a circuit having 15 places with 2 registers initially placed. In comparison, our tool needs 354 Microsec to solve the test case example if average delays are considered with no initially placed registers. The comparison remains difficult because the computers used are not identical; however, based on these data, our method seems to be much faster.

V. OPTIMIZING A COMMUNICATION LINK BETWEEN TWO SYNCHRONOUS MICROPROCESSORS

In this test case, two synchronous microprocessors are communicating asynchronously. As shown in Fig5, both processors have Sync/Async interfaces to handle the conversion process. The goal is to establish a communication channel to transfer data from Processor-A (MP-A), which is acting as TX, to Processor-B (MP-B), which is acting as RX. The clock frequency of MP-A is 500 MHz where the clock frequency of MP-B is 400 MHz. Both processors emit data bursts. Output/Input characteristics of MP-A/MP-B are modeled using deterministic delays. The Output/Input TVs are filled so that the average data rate on both sides is 100 MHz. However, bursts on both sides are reshuffled so that their occurrences are different in time. This makes the task of the FIFO more complex and relevant.

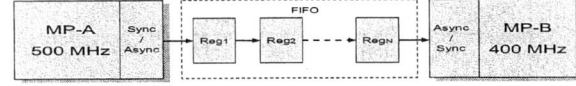

Figure5: Two Microprocessors Communicating Asynchronously

The physical route between the two MPs is modeled as a linear pipeline where the interconnection delays are modeled as the function block delays. This pipeline is predefined with 16 possible places to add asynchronous registers. That gives us a problem of linear pipeline with P=16 and identical function blocks. The goal is to optimally pipeline this communication channel to satisfy different given CTs. Note that pipelining such an asynchronous link, locally amplifies the signals and at the same time inserts registers in a FIFO like manner which speeds up the communication throughput.

First of all, the communication rate between the two processors, when they directly communicate without any register, is determined using the performance analysis tool. The average CT obtained is 71.185ns, i.e. a communication rate of 14 MHz even though both processors can afford a 100 MHz communication rate. The limitation in the communication rate comes from the unmatched bursts and the interconnection delays. This result shows that the communication channel needs registers to speed up the rate. Using a 90nm technology, we are able to design a register operating at a communication rate of 500 MHz. With this technology, the process variability effect on delays is estimated to ± 25%. Probabilistic time variable delays are assigned to the pipeline registers and the interconnection delays which are (modeled as function blocks). The Out/In timing characteristics of the two processors are modeled as deterministic delays in TX and RX.

The designer question now is: targeting a given cycle time "CT_T", what is the minimum number of registers he should use to pipeline the channel between the processors? And, what is the optimum placing of these registers which not only satisfies the CT_T, but also results in the minimum possible CT using this number of registers?

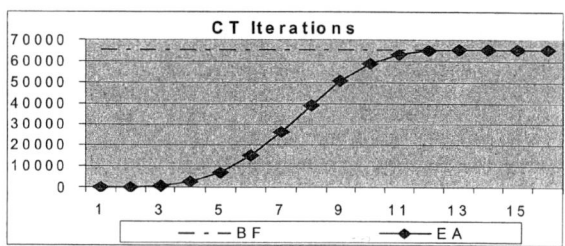

Figure6: BF and EA Iterations

Table2: Communication channel optimization

η	CT_T (ns)	RA	η	CT_T (ns)	RA
1	40.9	16	9	14.2	50642
2	29.8	136	10	13.7	58650
3	25	696	11	13.1	63018
4	21.7	2516	12	12.7	64838
5	19.1	6884	13	12.5	65398
6	17.1	14892	14	12.2	65518
7	16.4	26332	15	11.6	65534
8	15	39202	16	11	65535

Different CT_T are targeted, CT_T is gradually decreased from the maximum (no register) to the minimum achievable with 16 registers. Table2 shows the values of CT_T with the corresponding η. A comparison between the BF algorithm and the EA is depicted in Fig.6. While considering process effect, the optimizer gives us the ability to optimize for many different target CTs for this link. Without the optimizer it was very hard, may be impossible, to have these optimum results.

VI. CONCLUSIONS AND FUTURE WORKS

This paper addressed the problem of optimizing linear asynchronous pipelines by controlling the number of registers. The ultimate goal was to design/implement a fast tool enables the designers to optimize their pipelines while considering time variable delays. The target is to find the minimum number of registers to be used to satisfy given performance constraints. Moreover, the method finds the optimum placing of these registers which not only satisfies the target performance, but also results in the maximum achievable performance using this number of registers.

To accomplish these targets, an optimal algorithm "BF" is analyzed and implemented. The condition guaranteeing that adding registers to the pipeline is monotonically decreasing the cycle time is stated and proved. Based on that, an Efficient Algorithm "EA" is proposed the algorithm shows around 40% enhancement compared to the BF algorithm. Many test cases are conducted. They show the correctness and efficiency of the implemented optimizer. Compared to recent similar works, our optimizer shows better flexibility and accuracy in delay modeling and faster execution time. In addition to asynchronous linear pipelines and links, our method shows high efficiency in optimizing self timed rings and individual branches in non-linear pipelines.

Many extensions of this work are in progress. For example, investigating some heuristics helps to prune more unpromising iterations are in progress. An important extension of this work which consists in optimizing nonlinear pipelines is in progress too.

REFERENCES

[1] A. Xie, S. Kim, and P. A. Beerel. Bounding average time separations of events in stochastic timed petri nets with choice. In Proceedings of ASYNC, Spain (1999).

[2] Andrew Lines: Pipelined Asynchronous Circuits. Master Thesis, Caltech (1995)

[3] Eslam Yahya, Marc Renaudin: QDI Latches Characteristics and Asynchronous Linear-Pipeline Performance Analysis, PATMOS France (2006). Lecture Notes in Computer Science, ISBN 978-3-540-39094-7.

[4] Eslam Yahya and Marc Renaudin: Performance Modeling and Analysis of Asynchronous Linear-Pipeline with Time Variable Delays. In Proceedings of ICECS, Marrakech 2007.

[5] Jens SparsØ, Steve Furber: Principles of Asynchronous Circuit Design. A System Perspective. Kluwer Academic Publishers (2001). ISBN 0-7923-7613-7.

[6] Peggy McGee, Steven Nowick, E.G. Coffman: Efficient performance analysis of asynchronous systems based on periodicity. 3rd IEEE / ACM / IFIP CODE (2005). ISBN:1-59593-161-9.

[7] Sangyun Kim and Peter Beerel: Pipeline Optimization for Asynchronous Circuits: Complexity Analysis and an Efficient Optimal Algorithm. IEEE Transactions on international conference on Computer-aided design (2006). ISBN:0-7803-6448-1.

[8] Ted Williams: Performance of Itrative Computation in Self-Timed Rings. Journal of VLSI Signal Processing, 7, 17-31(1994)

Optimization of Integrated Circuits using an Artificial Intelligence Algorithm

Revna Acar Vural, Tulay Yildirim

Yildiz Technical University Electronics and Communication Engineering Department
Besiktas, Istanbul, TURKEY
{ racar, tulay}@yildiz.edu.tr

Abstract—Generating the optimized solution for obtaining a particular combination of design criteria is very affordable and time consuming. Optimization algorithms should carry out optimization process with high accuracy in a short time. An artificial intelligence algorithm that does not require complex equations and long simulation time for optimization of circuit topologies would be developed. Due to simple structure and fast response time, particle swarm intelligence algorithm is an ideal candidate as an optimizer. Algorithm would be trained with a data set consists of transistor dimensions and performance measurements, in other words algorithm would learn the circuit topology with all nonidealities and nonlinear occasions. Once trained, algorithm would not calculate complex circuit equations again and again; so faster response of optimization result would be obtained.

I. INTRODUCTION

Field of integrated circuit design optimization has been one of the most important and challenging topics in VLSI design process. Generating the optimized solution for obtaining a particular combination of design criteria is very affordable and time consuming. Selection of independent design parameters is quite important in optimization process. Ideally all design parameters are accepted as variables and the optimum solution is searched. However, tremendous growth of search space makes the search process inefficient In addition; there are several relations that should hold between certain L, *W*, and *W/L* ratios to make the search space smooth and the optimization process reliable. These constraints are part of the design knowledge that human experts usually rely on, and that can be derived from the operating principles of the circuit [1].

The purpose of this work is to develop an artificial intelligence algorithm that does not require complex equations and long simulation time for optimization of circuit topologies. This algorithm should also carry out optimization process in a shorter time and/or with more success than random methods. Optimization of circuit topologies that has been designed for particular process parameters will also be investigated in case that these topologies are designed with different process parameters.

There have been several researches for optimization of circuit design in the literature. An OP-AMP compiler was developed for layout optimization in [2]. Topology invention and parameter optimization in OP-AMPs were performed by genetic programming and Current-Flow analysis methods [3]. For CMOS OP-AMP synthesis, a tool called DARWIN which is on the basis of a genetic algorithm is used for transistor dimension optimization [4]. Geometric programming was also used in [5] for determining the transistor sizes and component values for CMOS OP-AMPs. Reference [6] introduced methodologies for optimization of supply and threshold voltages and channel lengths of CMOS structures for power minimization. Convex optimization algorithm was used for optimizing transistor dimensions in order to minimize the delay in the critical path [7]. Optimization of OTA for improving the DC circuit performance was performed by genetic algorithm [8]. Human immune algorithm was developed for determination of analogue active filter component values [9].

The methods explained above uses some circuit descriptions such as complex circuit equations and MOS model equations, set of current-flow lists, Ebers-Mol model, Elmore delay model etc... In this work, different from those methods, no circuit equation or MOS model would be used for circuit representation. Artificial intelligence algorithm would be trained with a set of circuit parameters and performance criteria. All nonidealities and nonlinear situations that were met during design process would be represented in the data set. As trained with the data set, algorithm would not calculate complicate equations again and again and faster results would be obtained.

Simpler structure and better result providing in case of parameter growth makes particle swarm optimization [10] an ideal candidate for optimization of circuit topologies. Since there has been no research on this topic, regardless of complex circuit equations, obtaining faster and more accurate results with this algorithm would be an innovation in circuit optimization.

II. METHOD AND PROBLEM DESCRIPTION

First, analogue circuit topologies would be investigated and designed in a simulation environment. Channel widths and/or lengths of MOSFETs in the design would be changed to obtain the performance measurements as shown in the figure below.

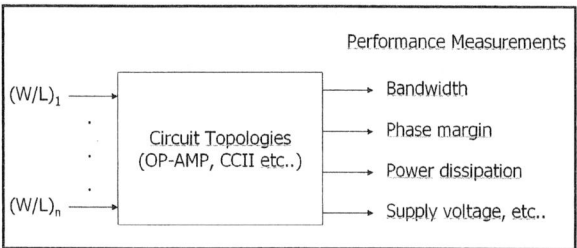

Figure 1.   Obtaining data set in a simulation environment

Artificial intelligence algorithm would be trained with data set. Thus, all nonidealities and nonlinear occasions would be learned by algorithm. As shown in the figure below, design criteria of analogue integrated circuit would be applied to the input of algorithm, while the outputs would be the optimum solution of circuit topology. This optimum solution is the minimum sized transistor dimensions of circuit topology for the design criteria.

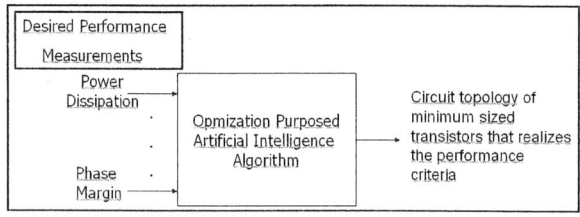

Figure 2.   Schematic of problem description

III. PARTICLE SWARM OPTIMIZATION

Particle Swarm Optimization (PSO) is an evolutionary computation technique developed by Kennedy and Eberhart in 1995 [10]. It is inspired by the swarming behavior of birds and insects. A population of particles exists in the n-dimensional search space that the optimization problem lives in. Each particle has a certain amount of knowledge, and will move about the search space based on this knowledge. The particle has some inertia attributed to it and so will continue to have a component of motion in the direction it is moving. It also knows where in the search space the best solution it has encountered is and finally it knows where the best solution is. The particle will then modify its direction such that it has additional components towards its own best position. For ith particle, in dth dimension, the algorithm is

$$v_{i,d}(n+1) = wv_{i,d}(n)+c_1 r_{1,d}(p_{i,d}-x_{i,d}(n))+c_2 r_{2,d}(p_{g,d}-x_{i,d}(n)),$$

$$x_{i,d}(n+1)=x_{i,d}(n)+v_{i,d}(n+1), \qquad (1)$$

where, $v_{i,d}$ represents the velocity of the particle at time step (n+1) and $x_{i,d}$ represents the position of the particle. The initial position and velocity of the particles are chosen randomly. The values of w, c_1 and c_2 are fixed for a given run of the PSO and are set by the outer genetic algorithm. $r_{1,d}$ and $r_{2,d}$ are uniformly distributed random numbers. $p_{i,d}$ and $p_{g,d}$ are the best position the ith particle has currently found and the best position found by any article respectively [11].

PSO, like the other evolutionary computation algorithms, can be applied to solve most optimization problems and problems that can be converted to optimization problems. The most important difference from the classic optimization techniques is that PSO does not require any derivation which simplifies PSO structure. Because of the parameter number to be adjusted is quite small; application of PSO is very easy. Since neighborhood structure is realized by uncomplicated processes such as adding and subtracting, good results has been obtained without using complicated neighborhood structures [12,13]

IV. ANALOGUE CIRCUIT EXAMPLE

In order to obtain a data set by adjusting the transistor dimensions, an analogue circuit structure, exponential function generator circuit, was designed using 1.5 micron YITAL parameters and it is a novel circuit structure. It is based on Taylor series expansion as given in (2).

$$e^{-x} = 1 - \frac{1}{1!}x + \frac{1}{2!}x^2 - \cdots + \frac{1}{n!}x^n - \cdots \qquad (2)$$

If $|x| \ll 1$ than,

$$e^{-x} \cong 1 - x + \frac{1}{2}x^2 \qquad (3)$$

Taylor series expansion simplifies to equation (3). Exponential function circuit generator realizes this equation using CMOS Current-mode structure as given below. Vdd and Vss were set to 3.3V and -3.3V respectively.

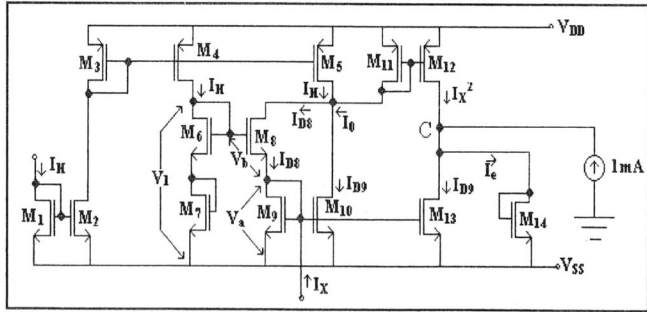

Figure 3.   Exponential Function Generator Circuit

978-1-4244-1983-8/08/$25.00 ©2008 IEEE

Exponential function generator circuit has two inputs and one output as notated in (3).

$$I_e = 1 - I_X + \frac{I_X^2}{2} \qquad (3)$$

In order to obtain (4), a circuit structure, which is based on squarer/divided circuit [14], was designed. At point C in Fig.3, I_x which is mirrored by M9, M_{10} and M_{13} is subtracted from summation of current source I_A and I_X^2, and input-output relationship as given in (3) is obtained. For realizing the constraint of

$$I_X^2 = \frac{I_X I_X}{2} \qquad (4)$$

I_H was chosen as 1mA and transistor sizes were chosen as shown in Table I.

TABLE I.   TRANSISTOR DIMENSIONS

MOSFET	Dimension-W/L (μm/μm)
M_1, M_2	2.5/1.5
M_3, M_4, M_5	7/1.5
M_6, M_7	25/1.5
M_8, M_9, M_{10}, M_{13}	4.5/4.5
M_{11}, M_{12}	8/1.5
M_{14}	6/4.5

Simulations were performed using HSPICE and DC performance measurements were obtained. Using transistor dimensions as given in Table 1, power dissipation was 17.4 mW and output range was 8.5 dB with %5.8 error.

V.   CONCLUSION

Simulation results which are performance measurements are obtained for the transistor dimensions given above. As noticed before by adjusting transistor dimensions, numerous different values of performance measurements (for this example, power and output range) will be obtained which form the data set. This data set would reflect all nonidealities and nonlinear occasions of the circuit topology and it would be used in the training of artificial intelligence algorithm. This algorithm would learn from patterns in data set and optimizes for the best solution. PSO seems to be an ideal candidate as an optimizer due to its simple structure and fast processing. In our further work, results would be compared to classical methods and if algorithm would be more successful in case of speed and/or accuracy, optimization of topologies with different process parameters than the one used for design process would be investigated.

REFERENCES

[1] R. K. Brayton, G. Hachtel, and A. Sangiovanni-Vincentelli, A Survey of Optimization Techniques for Integrated Circuit Design, Proceedings of the IEEE, Vol. 69, No. 10, pp. 1336-1361, Oct. 1981.

[2] H. Onodera, H. Kanbara, and K. Tamaru, "Operational amplifier compilation with performance optimization," *IEEE J. Solid-State Circuits*, vol. 25, pp. 466–473, Apr. 1990.

[3] T. Sripramong, C. Toumazou, "The Invention of CMOS Amplifiers using Genetic Programming and Current-Flow Analysis," IEEE Trans. on Computer-Aided Design of Integrated Circuits and Systems, vol. 21, no. 11, Nov. 2002.   (4.20)

[4] W. Kruiskamp and D. Leenaerts, "DARWIN: CMOS op amp synthesis by means of a genetic algorithm", in *Proc. 32nd Annu. Design Automation Conf.*, pp. 433–438, 1995.

[5] Maria del Mar Hershenson, Stephen P. Boyd, and Thomas H. Lee, "Optimal Design of a CMOS Op-amp via Geometric Programming," *Applied and Computational Control, Signals, and Circuits*, vol. 2, Birkhauser, 2000.

[6] A J Bhavnagarwala, Vivek K. De, Blanca Austin and James D. Meindl, 'Optimal Circuit Design for Low Power CMOS GSI', Proc of the 9th IEEE Int'l ASIC Conference & Exhibit, PP 313-317, Sept 1996

[7] S. Sapatnekar, V. B. Rao, P. Vaidya, and S.-M. Kang, "An exact solution to the transistor sizing problem for CMOS circuits using convex optimization," *IEEE Trans. Computer-Aided Design*, vol. 12, pp. 1621–1634, 1993.

[8] M. Wójcikowski, J. Glinianowicz, and M. Bialko, "System for optimization of electronic circuits using genetic algorithm," in *Proc. IEEE Int. Conf. Electronics, Circuits Syst.*, 1996, pp. 247–250.

[9] A. Kalinli, "Optimal Circuit Design Using Immune Algorithm", 3rd International Conference on Artificial Immune Systems, ICARIS 2004, *Lecture Notes in Computer Science*, Springer-Verlag Berlin Heidelberg, LNCS 3239, 42-52, 2004.

[10] J. Kennedy, R.C. Eberhart, "Particle Swarm Optimization", Proc. IEEE Int. Conf. Neural Networks, pp.1942-1948, Perth, Australia Nov. 1995

[11] C. R. Mouser and S. A. Dunn, "Comparing genetic algorithms and particle swarm optimization for an inverse problem exercise," *The Australian & New Zealand Industrial and Applied Mathematics Journal*, vol. 46, part C, pp. C89–C101, 2005.

[12] M. Clerc, "The Particle Swarm - Explosion, Stability, and Convergence in a Multidimensional Complex Space," in IEEE Transactions on Evolutionary Vol. 6 No. I, 2002, pp. 58-73.

[13] T. Kiink, J.S. Vesterstroem and J. Riget, "Particle Swam Optimization with Spatial Particle Extension," in Proceedings of the IEEE Congress on Evolutiorian Coniputation (CEC2002), , pp. 1474-1479, 2002.

[14] B. Liu, C. Chen ve J. Tsao, J "A Modular Current-Mode Classifier Circuit for Template Matching Application", IEEE Transactions on Circuits and Systems-II: Analog and Digital Signal Processing, 47(2):145-151, 2000.

BLANK PAGE

Post Fabrication Electric Field Treatment of Polymer Light Emitting and Photovoltaic Devices

Ali Osman Sevim

System and Control Engineering
Institute for Graduate Studies in Sciences and Engineering
Bogazici University
34342 Istanbul, Turkey
ali.sevim@boun.edu.tr

Senol Mutlu

Department of Electrical and Electronics Engineering
College of Engineering
Bogazici University
34342 Istanbul, Turkey
senol.mutlu@boun.edu.tr

Abstract—Conjugated polymers are semiconductors that offer flexibility, simplicity and lower cost in the fabrication of polymer light emitting diodes (PLEDs), polymer transistors and circuits, polymer photodetectors and solar cells. They have a potential to lead to a technology that can monolithically integrate all the semiconductor devices mentioned above as well as polymer sensors and actuators. This paper presents a novel method as a post-fabrication treatment and its improvements over the performance of the light emitting and photovoltaic properties of PLEDs. Investigated PLED is made of indium tin oxide (ITO), poly(3,4-ethylenedioxythiophene) poly(styrenesulfonate) (PEDOT:PSS), poly[2-methoxy-5-(2'-ethyl-hexyloxy)-1,4-phenylene vinylene] (MEH-PPV) and aluminum (Al). Following fabrication at room conditions, heat treatment is performed at 130C° for one hour and electric field treatment is realized with voltage levels from 0 to -8V under 0.2 atm vacuum. Heat treatment after fabrication restores the light emitting function of otherwise not functioning PLEDs, which degrade due to exposure to oxygen and water vapors at normal room conditions. Electric field treatment reduces the turn-on voltage of PLEDs when they are used as light emitting devices. Electric field treatment of -1 volt reduces the turn-on voltage to 3 volts from 10 volts, which is the case for the devices with heat treatment only. It also improves open circuit voltages and short circuit currents of PLEDs by an order of magnitude when they are used as photo-detectors or photocells. Devices treated with heat only show a short circuit current of around 0.5 nA and open circuit voltage of 5 mV under 500 mW/m² light intensity. These values improve to 5 nA and 55 mV respectively after the devices are electrically treated with -1 volt. Electric field treatment after the thermal treatment also improves the stability and uniformity of the devices. *(Abstract)*

I. INTRODUCTION

Conjugated polymers are semiconductors that have attracted increasing attention in the past ten years, which enabled fabrication of light emitting diodes [1], transistors [2], electronic circuits [3] and photovoltaic cells [4]. They have a potential to realize a system that can integrate all the devices mentioned above monolithically. They can

The Scientific & Technological Research Council of Turkey funds this research, 106E013, INtegrated Polymer MIcro Systems (INPOMIS).

even be integrated to polymer micro-electromechanical systems (MEMS) to acquire sensing and actuation capabilities [5]. These kinds of polymer based systems are flexible, lightweight, easy to fabricate and have low cost. They can be produced in large areas [6]. However, when compared to their inorganic counterparts, polymer semiconducting devices currently have lower performances, less reliability, shorter lifetimes [7]. The main reason for this is the fast degradation nature of these polymers under oxygen and water vapor environment. Therefore, polymer semiconducting devices still need more research on the refinement of their fabrication, characterization and testing of their performances and post fabrication treatments for performance improvement.

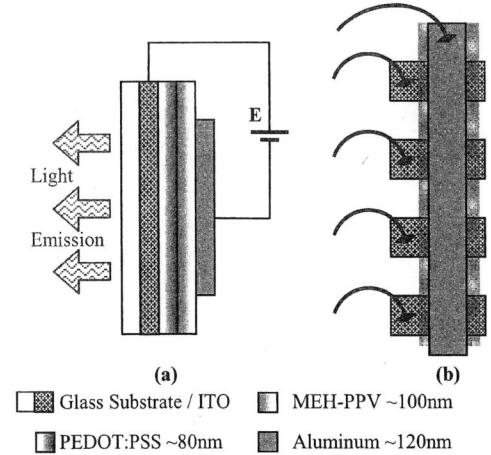

Figure 1. a) Cross sectional view of PLED b) Top view of the wire bonded samples

Several groups have studied performance improvement and stability of organic materials used in polymer light emitting devices [8-11] and solar cells [12, 13]. In these studies, in order to increase the performance of polymer or to prevent degradation of semiconducting devices all fabrication steps are done in an inert environment like nitrogen. Similarly they have applied different treatment methodologies at some point of the fabrication process like UV ozone or chemical treatments applied to ITO or

978-1-4244-1983-8/08/$25.00 ©2008 IEEE

conducting polymers. [8-13] However, all these extra precautions make the production harder and more costly. This paper presents a novel method, a post-fabrication heat treatment followed by an electric field treatment, and its influences on the performance of ITO/PEDOT:PSS/MEH-PPV/Al PLEDs. Previously, a similar post fabrication treatment has been shown to work on polymer solar cells [14] or electroluminescent devices based on dendrimers, small molecule devices and polymers [15-17]. However, this is the first time a post fabrication thermal treatment followed by an electric field treatment in reverse bias is applied to PLEDs made with PEDOT:PSS and MEH-PPV and it is shown that it modifies their turn-on voltages and improves their photovoltaic properties.

II. EXPERIMENT

A. Fabrication

All fabrication steps are realized in ordinary room with the relative humidity of %40-50 and the temperature of 21-26°C. To understand the effects of treatments, the samples are selected from same wafer in close proximity. ITO coated PET sheets (Aldrich) with a sheet resistivity of 35ohm/□ are cut into 4" diameter wafer sized circles and attached to 1mm thick glass wafers with silicone gel. The samples are cleaned using ultrasonic cleaner in acetone, isopropyl alcohol and deionized water for 3 minutes. To eliminate the hydrophobic nature of aqueous dispersion of PEDOT:PSS (Sigma) on PET surface and to clean the surface of the ITO from organic residuals, oxygen plasma with a power of 1.3W is applied for 15 minutes under 300mTorr vacuum. After the filtration of PEDOT:PSS with a 0.25 μm syringe filter, a film of approximately 80nm thickness is obtained by spin-coating the solution at 1200rpm speed for 30 seconds. To increase the conductivity of the PEDOT:PSS film, the sample is baked at 110°C under nitrogen, N_2, environment for one hour and under 0.2 barr vacuum for one hour in sequence.

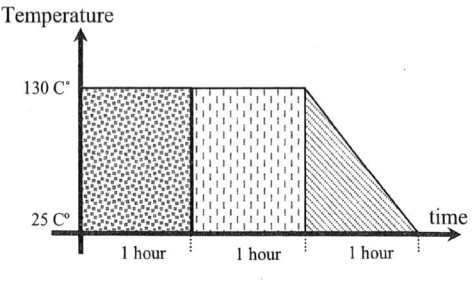

Figure 2.   Temperatures of the treatment stages

MEH-PPV is dissolved in toluene, 4mg / mL and stirred at 50°C on hot plate for at least 4 hours until it is fully dissolved. After filtering the mixture with a 0.25 μm Teflon syringe filter, MEH-PPV solution is spin coated at 1000rpm to obtain approximately 100nm of film thickness. At the end, film is baked at 65ºC for one hour.

To deposit aluminum electrode, a shadow mask is prepared from a 50μm thick copper foil. Vacuum chamber with a base pressure of 10^{-6} Torr is used to evaporate 120nm of aluminum layer. A shadow mask with 3 mm wide holes is aligned perpendicular to the patterned ITO. As a result $3x3mm^2$ square test cells are obtained as shown in Figure 1.

B. Treatment

Post fabrication treatments developed in this work consist of heat treatment and the electric field treatment immediately following that. Heat treatment is done for a specific duration on a hotplate at a specific temperature and in a low vacuum nitrogen environment. This treatment removes most of the absorbed oxygen and water vapors inside the devices. This is immediately followed by electric field treatment. Electric field treatment is done at the same vacuum and temperature values as the heat treatment. In addition, a negative voltage is applied to the devices for a specific duration making them reverse biased. Finally, this treatment is finished by a cooling stage as shown in Figure 2.

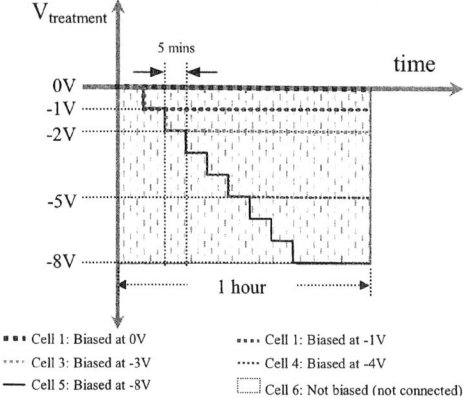

Figure 3.   Biasing details of electric field treatment

The temperature of annealing must be higher than the glass transition temperature of the polymers. Optimum temperature for heat treatment is found to be 150°C from the I-V and luminance measurements. However, 130°C is used due to physical stress and operation difficulties of PET substrates. The time required for the annealing is offered as one hour, since any further treatment did not cause any further improvement [17]. Electric field treatment is realized with voltage levels from 0 to -8V. After one hour of heat treatment under 0.2 bar vacuum at 130°C in dark environment, samples are biased with different voltage levels. The durations of the treatment phases and the temperatures are given in Figure 2. The bias voltage of each sample is decreased in steps of 1V/5min as shown in Figure 3. In the electric field treatment stage, in addition to reference cell which is not connected to a certain bias supply, -1, -2, -3, -5, -8 potential levels are tested to determine its effect. The

978-1-4244-1983-8/08/$25.00 ©2008 IEEE

reference cell represents the case in which only heat treatment is applied to the device.

At the end of treatment processes, samples are packaged immediately and cooled down to room temperature without breaking the vacuum. A couple millimeter thick layer of hot melt silicone (Henkel) is used to encapsulate the devices. The thickness of the encapsulation is made quite thick to eliminate water or oxygen diffusion to semiconductor polymers.

III. RESULTS AND DISCUSSION

In this work, heat treatment under vacuum is used for the purpose of reviving the functionalities of PLEDs contaminated with oxygen and water vapor in the environment. Different from a typical fabrication process in literature; fabrication is done at ordinary room conditions without using glove-boxes. Therefore, during each steps of the fabrication, polymer material is polluted by humidity and oxygen in the environment and the resulting device did not emit light when forward biased or did not behave as a photo detector or solar cell when tested in a solar simulator. Heat treatment under vacuum followed by packaging under the same conditions restored the light emitting characteristics of PLEDs. However, the devices did not show any photovoltaic properties.

The effects of electric field treatments are tested on both light emitting and photovoltaic properties of PLEDs. Its effect on light emission can be seen in Figure 4. For the devices that have been heat treated but not electric field treated known as unbiased cell, shown on the column named NC (not connected) in that figure, the turn voltage is relatively high and the uniformity is very poor. By applying negative electric field to the junctions after the fabrication and heat treatment, the turn-on voltage of the polymer light emitting devices falls down. These turn-on voltages increase slowly with the magnitude of the negative treatment voltage. For example, for a device that experienced only heat treatment (NC column in Figure 4) has a turn-on voltage of around 11 volts. However, a device that experienced both heat treatment and an electric field treatment with -1 volts (-1 column in Figure 4), has a turn-on voltage of around 4 volts. A device that is treated with -5 volts (-5 column in Figure 4), has a turn-on voltage of around 5 volts. Another advantage of electric field treatment is that the uniformity of the light emitting surface is improved.

Photovoltaic response of the PLEDs is tested by measuring open circuit voltages and short circuit currents under both dark environment and the light source of a solar simulator with an intensity of 500W/m². Figure 5 shows the open circuit voltage response of the heat treated but not electrically treated PET/ITO/MEH-PPV/PEDOT:PSS/Al PLED. The measured voltage is noisy and its average value when exposed to light is very small. On the other hand, after electric field treatment is applied the output voltage of the polymer device as shown

in the same figure, is stabilized at a certain level and its average value (55 mV for -1 volt treatment) is an order of magnitude higher than the untreated (heat treated but not electric field treated) ones (5 mV).

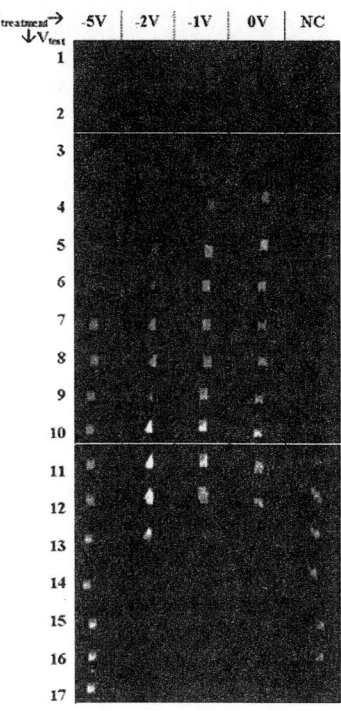

Figure 4. Electric field treated rows of PLED samples

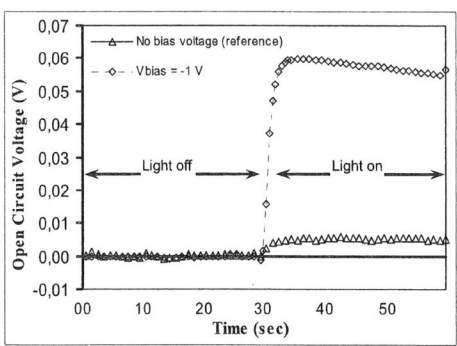

Figure 5. Effect of electric field treatment to the open circuit voltages

Similar to the open circuit voltage characteristics of the treated and untreated samples, short circuit current performances are enhanced significantly after the electric field treatment. The results are shown in Figure 6. Before the electric field treatment, the short circuit response of the device when exposed to light can not be distinguished from the dark response. However, after the electric field treatment is applied, the intensity of the light shined on the device can be easily detected from its short circuit response (6 nA for -1 volt treatment). The average open circuit voltage and average short circuit current obtained

978-1-4244-1983-8/08/$25.00 ©2008 IEEE   19

from electric field treated samples are strongly depended on the treatment voltages. These improvements in the photovoltaic operations of PLEDs are significant since they show that when right electric field treatment is applied to PLEDs after fabrication, they can also be used as photodetectors as well as light emitting devices.

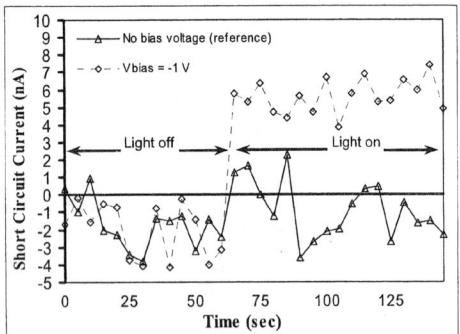

Figure 6. Effect of electric field treatment to the short circuit currents.

The light emitting property of a PLED can be restored with a heat treatment only step since it removes most of oxygen and water vapors absorbed inside PLEDs during fabrication steps. Further improvement of both the light emitting and photovoltaic functions of PLED by the electric field treatment after heat treatment is still under investigation. However, we predict that this is due to the alignment of the polymer chains under electric field. Since the polymer chains become mobile above the glass transition temperatures, they are reoriented by the applied electric field resulting in a more uniform structure. This new arrangement improves the chemical bonding between interfacial junctions, increases the mobility of the carriers and lowers the junction barriers.

IV. CONCLUSION

In this work, the effects of heat treatment and a following electric field treatment on PLEDs have been investigated. Treatment experiments are tested for three different fabrications and concurrent results have been observed. After the fabrication of square test cells heat treatment is performed at 130C° for one hour. Electric field treatment is realized afterwards with voltage levels from 0 to -8V. In the fabrication of PLEDs, no preventive measures are taken against exposure to oxygen and water vapors, which make the fabrication easier. Heat treatment after fabrication followed by packaging in the same conditions restores the light emitting properties of otherwise not functioning PLEDs by removing the absorbed oxygen and water vapors inside them. Devices that did not receive any post fabrication treatment did not show any light emitting or photovoltaic properties.

Electric field treatment after heat treatment is applied to PLEDs made of PEDOT:PSS and MEH-PPV in reverse bias for the first time in this work. Light emitting and photovoltaic properties of PLEDs are improved compared

to devices that are treated with heat only. Electric field treatment after heat treatment lowers the turn on voltage levels when used as a light emitting device. Turn-on voltage of a PLED drops to 3 volts when treated with -1 volt in comparison to 10 volts when treated with heat only. When PLED is used as a photovoltaic device, its average value of the short circuit current and open circuit voltage increases by one order of magnitude compared to the values of the devices treated with heat only. Under 500mW/m^2 light intensity, 5 mV open circuit voltage and 0.5 nA short circuit current values with heat treatment only improve to 55 mV and 5 nA respectively after electrically treated with -1 volt. Furthermore, the uniformity of the light emitting surface increases significantly after electric field treatment.

ACKNOWLEDGEMENT

This work has been fully supported by Scientific & Technological Research Council of Turkey under the project name INtegrated Polymer MIcro Systems (INPOMIS) with project number 106E013.

REFERENCES

[1] N. K. Patel, S. Cinà, J. H. Burroughes, IEEE J. Selected Topics in QE, vol. 8, no. 2, April 2002.

[2] J. Krumm, E. Eckert, W. H. Glauert, A. Ullmann, W. Fix, W. Clemens, IEEE Electron Device Letters, vol. 25, no. 6, pp.346-361, June 2004.

[3] C. J. Drury, C. M. J. Mutsaers, C. M. Hart, M. Matters, & D.M. de Leeuw, Appl. Phys. Lett., vol. 73, pp.108-110, 1998.

[4] Y. Kim, K. Lee, N. E. Coates, D. Moses, T. Nguyen, M. Dante, A. J. Heeger1, Science, vol. 317. no. 5835, pp. 222 – 225, July 2000.

[5] C. Liu, Advanced Materials, vol. 19, no. 22, pp. 3783-3790, 2007.

[6] F. Padinger, C. J. Brabec, T. Fromherz1, J. C. Hummelen, N. S. Sariciftci, Opto-Electronics Review 8(4), pp.280-283, 2000.

[7] J. Huang, X. Wang, A. J. deMello, J. C. deMello, D. D. C. Bradley, J. Mater. Chem., vol 17, pp. 3551-3554, 2007.

[8] M. Atreyaa, S. Lia, E. T. Kanga, K. G. Neoha, Z. H. Mab, K. L. Tanb, W. Huangc, Polymer Degradation and Stability, vol. 65, is. 2, pp.287-296, 1999.

[9] A. Petr, F. Zhang, H. Peisert, M. Knupfer, L. Dunsch, Chemical Physics Letters vol. 385, pp. 140–143, January 2004.

[10] Z. Y. Zhong, Y. D. Jiang, Journal of Colloid and Interface Science, vol. 302, pp. 613–619, August 2006.

[11] C. Tengstedt, A. Kanciurzewska, M. P. de Jong, S. Braun, W. R. Salaneck, M. Fahlman, Thin Solid Films, vol. 515, is. 4, pp.2085–2090, August 2006.

[12] M. Y. Songa, D. K. Kimb, S. M. Jo, D. Y. Kim, Synthetic Metals , vol. 155, pp. 635–638, 2005.

[13] H. Jin, Y. Hou, X. Meng, Y. Li, Q. Shi, F. Teng, Solid State Communications, vol. 142, pp. 181–184, 2007.

[14] F. Padinger, R. S. Rittberger, N. S. Sariciftci, Adv. Func. Matter, vol. 13, no.1, 2003.

[15] D. Ma1, J. M. Lupton, R. Beavington, P.L. Burn, I. D. W. Samuel1, J. Phys. D: Appl. Phys. vol. 35, pp.520-523, 2002.

[16] M. Yahiro, D. Zou, T. Tsutsui, Synthetic Metals, vol.s 111-112, pp. 245-247, 2000.

[17] T. W. Lee, O. O. Park, Appl. Phys. Lett., vol. 77, is. 21, pp. 3334-3336, 2000.

Effect of thin Polyimide film on Performance of AlN/SiO2/Si SAW Device

HOANG Trang, REY Patrice, VAUDAINE Marie-Helene

Heterogeneous Silicon Integration Department
CEA/LETI
Grenoble, France
trang.hoang@cea.fr

BENECH Philippe

Institute of Microelectronics, Electromagnetism and Photonics
INPG/UJF/CNRS
Grenoble, France

Abstract— **The design, modeling and charaterization of SAW device are presented. The attenuation rate of polyimide absorber is proportional to the length and the thickness of the absorber film. With thin polyimide films (95nm), the wave attenuation effects of polyimide films are almost removed. Besides, the thin polyimide film seems to affect the center frequency a little, in which the center frequency is proportional to the length of thin polyimide film. So, because of the limitation of the device size, to increase the attenuation rate of the reflected wave, the thickness of the absorber film must been increased**

I. INTRODUCTION

Surface Acoustic Wave (SAW) devices play a key role in today's telecommunication systems and are widely used as electronic filters, resonators, delay lines, convolvers or wireless identification systems (ID tags). A SAW device includes a piezoelectric substrate, transmitting and receiving transducers provided on the surface of the substrate for propagating waves between the transducers, and absorbers provided between the transducers and an end of the substrate for absorbing unwanted acoustic waves. When an alternating electrical potential is applied to the electrodes of the transmitting transducer through terminal plates, an alternating electric field is generated that causes vibrations in the substrate material. These vibrations give rise to acoustic waves, which propagate along the surface of the substrate. And after that, the received acoustic waves can be converted to an electrical signal at the receiving transducer because of inverse piezoelectric effect.

In these SAW devices, the transmitting transducer launches the surface acoustic waves in opposite directions simultaneously while the receiving transducer receives the waves traveling in either direction. This is a critical issue in most SAW devices because in addition to responding to surface acoustic waves traveling directly from the transmitting to receiving transducer, the transducers respond to surface acoustic wave reflected from the ends of the substrate. These waves produce unwanted signals, for example, as spurious signal in the time domain, ripples of the frequency response domain that distort the main, desired

signal, so adversely affecting the performance of the SAW device. Therefore, the edges of substrates are often provided with wax, silicones or other viscous organic materials which absorb the surface acoustic wave.

A very useful material is polyimide, a viscous organic material which maintains its absorptive acoustic properties after curing [2][3]. Polyimide films are patternable on wafers with high accuracy and they do not degrade in the high temperature sealing of packages. With SAW device using LiNbO3, the attenuation of 3.0-3.5 μm thick polyimide films was measured to be in the range of 10-20dB/100μm at frequencies in the range of 300 MHz-500 MHz [2]. Generally, the attenuation rate of absorber is proportional to the length and the thickness of the absorber film. So, because of the limitation of the device size, to increase the attenuation rate of the reflected wave, the thickness of the absorber film must been increased. Ref [3] showed the attenuation rate of 14 μm, 20 μm thick polyimide films, while ref [2] presented the rate of 3.0-3.5 μm thick polyimide films.

In this paper, the effects of thin polyimide films (95nm) with different length on the performance of AlN/SiO2/Si SAW device are studied. Experimental results of SAW devices in the frequency range 565 MHz -585MHz are measured with Vector Network Analyzer HP 8753E.

This paper is organized as follows. Section II presents briefly the analytic method for SAW device, transfer function. Experimental results are introduced in section III in different lengths of polyimide films, in three wafers. Conclusion is given in section IV.

II. ANALYTIC METHOD FOR SAW DEVICE

The most important parameter for SAW device design is the center frequency, which is determined by the period of the IDT fingers and the acoustic velocity. The governing equation that determines the operation frequency is

$$f_0 = v_{SAW}/\lambda \qquad (1)$$

Where

λ is the wavelength at f_0, determined by the periodicity of the IDT. For the technology being used in this paper

$$\lambda = p = \text{finger width} \times 4 \qquad (2)$$

where the finger width is determined by the design rule of the technology which sets the minimum metal to metal distance.

f_0 center frequency of the device;

v_{SAW} surface acoustic wave velocity.

Matrix method that is proposed by Cambel and Jones (1968) and Ingerbrigtsen (1969) is used to determine velocity and frequency. So, electromechanical coupling factor K is also given by

$$K^2 = 2(v_0 - v_s)/v_0 \qquad (3)$$

where v_0, v_s are respectively velocities when the plane where IDTs are located is electrically open or short-circuit.

There are four main analytic methods used for SAW device: 1. The equivalent-circuit model; 2. The Coupling-Of-Mode (COM) model; 3. The S-matrix model; 4. The P-matrix model.

The equivalent-circuit model is chosen because it can determine the frequency response, impedance parameters and transfer characteristics of SAW device. This allows the designer to determine quickly the major dimensions and parameters in number of fingers, finger width, aperature, delay line distance.

There are two different circuit models for analysing IDT [6]. Both of them are based on the bulk-wave three-port models originally published by Mason [1]. The crossed field model is selected for the modeling of the devices because it was shown in the literature that the crossed field model yielded better agreement than the experiment when compared to the in-line model when K is small. In our device, AlN is used as piezoelectric layer and K in any our different configurations are less than 7% (Fig. 1).

Each IDT is represented by a three-port network shown in Fig. 2 [6], in which an approximation of no reflected wave is used, this means an ideal absorber is used. So, the equivalent circuit for SAW device is showned in Fig. 3.

G_0 the electrical characteristic admittance of a one-period IDT, $G_0 = 1/Z_0 = K^2 C_s f_0$ (mho) $\qquad (4)$

Z_0 electrical characteristic impedance;

K electromechanical coupling factor;

C_s static capacitance of one periodic section;

N the number of sections in IDT;

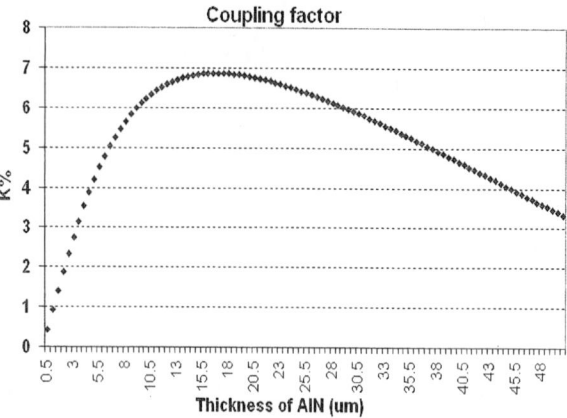

Figure 1. Electromechanical coupling factor in structure AlN/SiO2/Si with different thicknesses of AlN

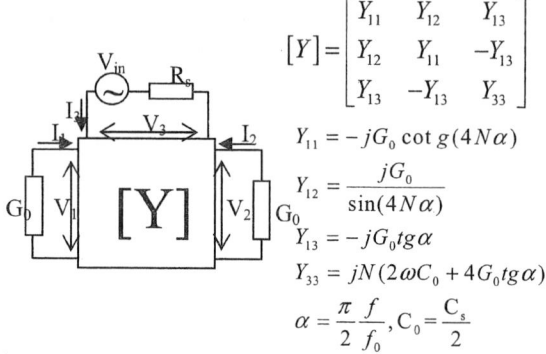

Figure 2. Three-port equivalent admittance network representation for an IDT in the crossed field model

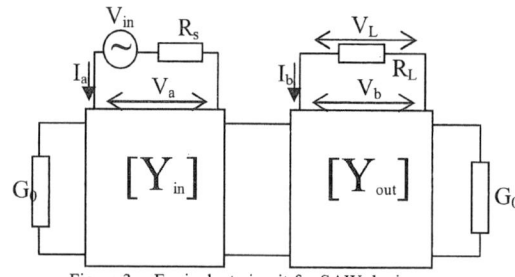

Figure 3. Equivalent circuit for SAW device

Applying the circuit theory for the current voltage relations on input and output, transfer function is calculated for SAW device (Fig.3) and is modeled in Matlab

$$H(f) = \frac{V_L}{V_{in}} = \frac{y_{ab} R_L}{(1 + y_{aa} R_s)(1 + y_{bb} R_L) - y_{ab}^2 R_s R_L} \quad (5)$$

Where

$$y_{aa} = j2\pi f C_T + G_a(f)$$

$$y_{bb} = j2\pi f C_T^{out} + G_a^{out}(f)$$

$$y_{ab} = 8NMG_0 \frac{\sin x}{x} \frac{\sin y}{y} e^{j\pi(1-(N+M)\frac{f-f_0}{f_0}}$$

$$G_a(f) = 8N^2 G_0 \left(\frac{\sin x}{x}\right)^2 ; x = N\pi \frac{f-f_o}{f_0}$$

$$G_a^{out}(f) = 8M^2 G_0 \left(\frac{\sin y}{y}\right)^2 ; y = M\pi \frac{f-f_o}{f_0}$$

$$C_T = NC_s ; C_T^{out} = MC_s$$

N, M the number of sections in input and output IDTs respectively.

III. EXPERIMENTAL RESULTS

The experimental devices are SAW devices: AlN (1.97 μm)/SiO2 (0.5 μm)/Si substrate (P, 1-50 ohm.cm), IDT metal is AlCu, finger width = 2 μm, thickness of AlCu is 730 nm, thickness of polyimide film is 95 nm, N=M=50.

The length of polyimide is the number of wavelength (l=a.λ), where a=0 (no polyimide absorber), a=10, 20, 30, 40 (see Fig. 5)

Figure 4.   Image SEM of polyimide on AlN

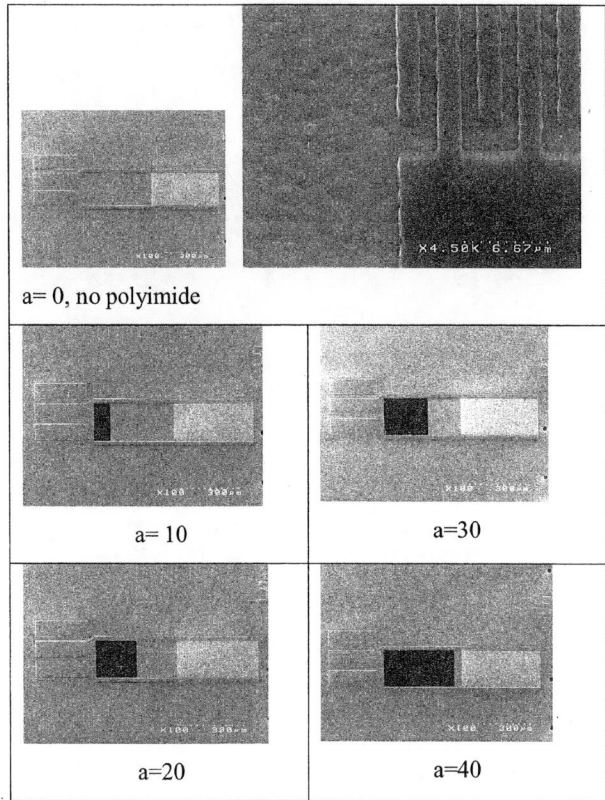

Figure 5.   Image SEM of SAW devices with different length of polyimide

The S matrixes of the devices have been measured by an HP 8753E network analyzer with frequency range 565-585 MHz, number of points is 1600. The devices in 3 wafers have been measured as followed

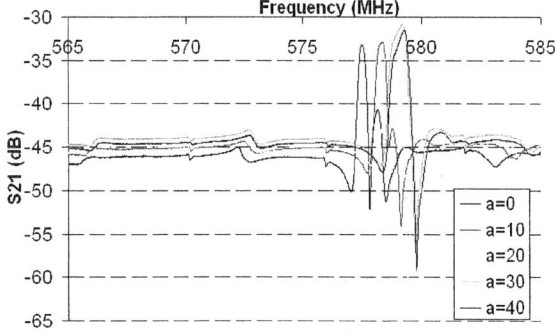

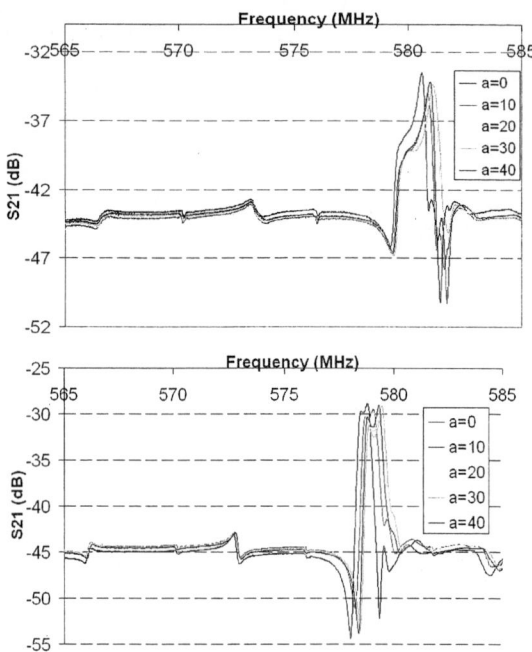

Figure 6. S21 (dB) in devices with different lengths of polyimide in (a) wafer 1, (b) wafer 2 and (c) wafer 3

The center frequency is simulated: f0= 584 MHz

While, the center frequency in experiments is 577.5 MHz-579.3 MHz. The comparison between center frequencies of devices with different lengths of polyimide in three wafers is shown in Table I.

TABLE I. COMPARISON BETWEEN CENTER FREQUENCIES OF DEVICES WITH DIFFERENT LENGTHS OF POLYIMIDE IN THREE WAFERS

f0 (MHz)	a=0	a=10	a=20	a=30	a=40
wafer 1	577.5375	578.2625	578.8625	579.1875	579.3
wafer 2	580.625	580.875	581.1375	581.1125	581.0125
Wafer 3	578.7625	579.0125	579.3625	579.4375	579.4875

We see that with thin polyimide films, the wave attenuation effects of polyimide films are almost removed. Besides, the thin polyimide film seems to affect the center frequency a little, in which the center frequency is proportional to the length of thin polyimide film. This is confirmed by using circuit model in Fig. 3. In this model, an ideal absorber is used, there is no reflected wave. By applying the circuit theory for matching condition, this means two terminal admittances have value of G0. In this study, thin polyimide film doesn't absorb absolutely the wave, that is equivalent to using other values of two terminal

admittances in the circuit model. So, the effect of thin polyimide film on the center frequency of SAW devices are modeled as in Fig. 7.

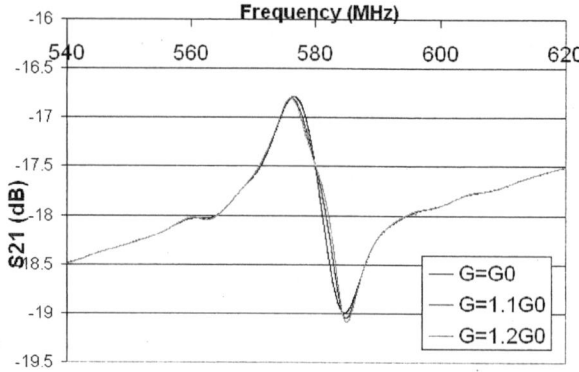

Figure 7. Simulation of S21(dB), effects of polyimide films on the f0

IV. CONCLUSION

The design, modeling and charaterization of SAW device are presented. The attenuation rate of absorber is proportional to the length and the thickness of the absorber film. With thin polyimide films (95nm), the wave attenuation effects of polyimide films are almost removed. Besides, the thin polyimide film seems to affect the center frequency a little, in which the center frequency is proportional to the length of thin polyimide film. These effects are confirmed by using the circuit model. So, because of the limitation of the device size, to increase the attenuation rate of the reflected wave, the thickness of the absorber film must been increased

REFERENCES

[1] Warren P.Mason, "Electromechanical Transducer and Wave Filters", second edition, D.Van Nostrand Company Inc, 1948.

[2] C.A.Johnsen, T.L.Bagwell, J.L.Henderson, P.C.Bray, "Polyimide as An Acoustic Absorber For High Frequency SAW Application", IEEE Ultrasonics Symposium, 1988, pp. 279-284.

[3] W.C.Qian, A.Venema, "An Acoustic Absorption Film for SAW Devices", Sensors and Actuators A, 25-27 (1991), pp.535-539.

[4] N.Tirole, A.Choujaa, D.Hauden, G.Martin, "Lamb waves pressure sensor using an AlN/Si structure", IEEE Ultrasonics Symposium, 1993, pp. 371-374.

[5] G.Schimetta, F.Dollinger, G.Scholl, R.Weigel, "Optimized design and fabrication of a wireless pressure and temperature sensor unit based on SAW transponder technology", IEEE Microwave Symposium Digest, 2001, pp. 335-358.

[6] W.R.Smith, H.M.Gerard, J.H.Collins, T.M.Reeder, H.J.Shaw, "Analysis of Interdigital Surface Wave Transducers by Use of an Equivalent Circuit Model", IEEE Transactions On MicroWave Theory and Techniques, Vol. MTT-17, No.11, Nov 1969, pp. 856-864.

[7] A.N.Nordin, M.E.Zaghloul, "Modeling and Fabrication of CMOS Surface Acoustic Wave Resonators", IEEE Transactions on MicroWave Theory and Techniques, Vol.55, No.5, May 2007, pp. 992-1001.

BIOSENSORS FOR THE DETECTION OF CARDIOVASCULAR RISK MARKERS IN HUMAN SERUM

Ozgur Gul, Sreenivasa S. Kallempudi, Huveyda Basaga, Ugur Sezerman and Yasar Gurbuz

Sabanci University

Faculty of Engineering and Natural Sciences, Istanbul 34956, Turkey

Tel: ++90(216) 483 9533, Fax: ++90(216) 483 9550, e-mail: yasar@sabanciuniv.edu

Abstract— This paper presents two affinity-based biosensor implementations for the detection and quantification of cardiovascular risk proteins/markers, such as C-Reactive Protein, Myoglobin, TNF-α, Serum Amyloid, using specific antibodies. The first method is based on interdigitated microelectrodes while the other is based microarrays. The signal transduction in micro-IDEs from biological-signal to the electronic/electrical signal occurs via fringe fields and dielectric change, induced by the capture of antigens by already immobilized antibody between the microelectrodes. Being process/integration-compatible with microelectronic/fabrication technologies, this method provides many advantages for the utilization in portable and/or remote sensing applications, such as smaller size, lower power consumption, small sample consumption, disposability, etc. The other method also analyzes the capability and reliability of sandwich-type antibody microarrays for the detection quantification of cardiovascular risk markers, using optical scanner system for the detection of fluorescent-labeled, sandwiched-structured antibodies on the chemically modified glass substrates. Our findings in microarray study provides the advantages for lab-based detection and quantification applications such as being able to detect and quantify multiple markers at once and also better dynamic ranges over some of the commercial ELISA test-kits.

INTRODUCTION

Biosensors for fast, direct, and label-free measurements are attractive for different applications. Affinity sensors allow the detection and quantification of target molecules in complex mixtures by affinity-based interactions. Immobilized antibody molecules are the probes that bind to specific protein molecules (targets) in biological fluids. Currently, we are working on two different approaches to quantify marker proteins in human serum. First method is capacitance based and the second is optical based. In both approach, we are using *label-free* detection mechanisms.

Originally, protein assays were developed in the enzyme linked immunosorbent assay (ELISA) format in microtiter plate format. Although ELISA has been long standing standard for quantitative analysis, this technique suffers from relatively low throughput, due to lack of multiplexing ability and high reagent and sample consumption [1,2].

In this paper, we have analyzed sandwich type antibody microarrays for the detection and quantification of CVD marker proteins in human serum. Proteins in complex mixtures such as human serum can be quantified without labeling samples. Sandwich model allow us to detect and quantify proteins without labeling them first. Without

labeling proteins, we eliminate the problems that arose from labeling of antigen binding sites on proteins. Besides low sample and reagent consumption of microarray approach, multiplexing ability will generate data that are more reliable by looking more than one marker at a time. Also in this paper, we have analyzed label-free, micro-interdigitated electrodes in the form of capacitance, used to detect and quantify Cardiovascular risk markers (C-Reactive Protein as an example) via changes in the dielectric (or capacitance) properties between the electrodes upon antigen binding. Besides the inexpensive production and low sample consumption, main advantages of interdigitated, capacitive immunosensors are the label free detection mechanism and the fast measurements. Unlike other detection mechanisms used to quantify protein concentrations, capacitive method does not need labeled samples, similar to the microarrays, also presented in this study.

MICRO-INTERDIGITATED ELECTRODES

Multi-finger types of capacitors are widely used in microstrip technology. The capacitor itself is defined between the two ports as shown in Fig. 1 (a). To synthesize IDC structures equations formulated earlier were used [3]. The simulation of this type of structures at high frequencies is not easy and electromagnetic simulators like MOMENTUM/HFSS should be used for high accuracy. In this work, modeling and simulation of interdigital capacitors are performed using ADS (Advance Design System) MOMENTUM® and HFSS® tools. For the ease of fabrication, image reversal techniques (for lift-off) were employed instead of etching.

Interdigitated electrodes in the form of capacitance on glass slide can be used to measure the changes in the dielectric properties of the interdigated capacitances upon antigen binding (Figure 1)[4]. Besides the inexpensive production and low sample consumption, main advantages of interdigitated, capacitive immunosensors are the label free detection mechanism and the fast measurements. Unlike other detection mechanisms used to quantify protein concentrations, capacitive method does not need labeled samples. Upon binding of protein molecules to the immobilized antibody molecules on the surface of electrodes, dielectric properties between the interdigitated electrodes changes and measurement can be taken immediately after hybridization (Figure 2).

Surface activation and immobilization is performed by using the following, 2% of (3-glycidoxypropyl)-trimethoxy silane (GPTS) solution was used to coat SiO_2 surface (which is sputtered on finger structures). After one hour silanization reaction, sensors was washed several times with ethanol and dried using centrifuge. 2μl of C-Reactive Protein antigen at 0.5 mg/ml concentration was added onto surface and incubated for two hours at room temperature. After immobilization step, biosensor surface was blocked using 2% BSA for nonspecific binding of antigen. After several times washing with PBS-T, sensors were stored at 4°C until use.

The binding of an analyte to an immobilized molecule can be detected without any labeling or secondary reactions using interdigitated, capacitive electrodes. The detection mechanism of the sensor is based on the change in the dielectric constant of the interdigitated capacitance. This change is correlated to the bound protein molecules to the antibodies on the surface (Figure 3). The dielectric constant of the antibodies changes the dielectric constant between two conductor fingers and this result in alternating capacitance. Interdigitated electrode arrays are used in order to increase the surface area of the conductors for higher sensitivity to binding events. AFM and electrochemical methods are used to measure the changes, but impedance analysis gives faster results.

PROTEIN MICROARRAYS

In the second approach, we are using antibody microarrays to quantify marker protein inhuman serum (Figure 4). We have analyzed the detection and quantification capability and reliability of sandwich-type antibody microarrays for cardiovascular risk markers. We used optical scanner system for the detection of fluorescent-labeled, sandwiched-structured antibodies on the chemically modified glass substrates (Figure 5).

C-Reactive Protein, TNF-α, Serum Amyloid A, Myoglobin, and Fatty Acid Binding Protein (FABP) antigens were mixed and serial dilutions were prepared using diluent buffer. Serial dilutions of antigens were used to determine dynamic range of each antibody. Standard curves (Figure 6), for C-reactive Protein (a), Myoglobin (b), Serum Amyloid A (c), and TNF-α (d), were calculated using spot intensity values after background subtraction. 12-μm squares were selected from corners between spots on the glass-slide for background calculations. Final spot intensities were calculated by subtracting reference spot intensity values from each antibody's spot intensity. We have used four-parameter "logistic fit" for standard curve analysis. R^2 values were also calculated as 0.9795, 0.9584, 0.9955, and 0.9966 for C-Reactive Protein, Myoglobin, Serum Amyloid A, and TNF-α, respectively. With the process and structure presented in this study, baseline and elevated levels of C-Reactive Protein, Myoglobin, TNF-α, Serum Amyloid A protein, known cardiovascular risk markers in human serum, can be detected. Four-parameter logistic fit algorithm gave high R^2 values and dynamic working ranges are similar to the commercially available ELISA kits for these antibodies. This finding strongly suggests that these antibody pairs can be used to detect and quantify marker proteins in human serum. As microarray systems are versatile, other marker specific antibodies can also be added to the panel and hence multiplex assays for quantification could be possible. We have observed that detection limits of sandwich-type antibody microarray is enough to detect serum levels, both baseline and elevated, of marker proteins used in this study. In our method, we have obtained standard curves similar to reported ELISA ranges, as also summarized in Table I, for the following antibodies; C-Reactive Protein, TNF-α, Serum Amyloid A and Myoglobin. These antibody pairs can be used in multiplex assays.

CONCLUSIONS

In this paper, we presented two affinity-based biosensor implementations for the detection and quantification of cardiovascular risk proteins/markers, such as C-Reactive Protein, Myoglobin, TNF-α, Serum Amyloid, using specific antibodies. The first method is based on micro-interdigitated microelectrodes (micro-IDEs) and is suitable for portable, remote sensing and disposable application of biosensors while the other is based microarrays and sutable for lab-based detection and quantification of biomolecules. The signal transduction in micro-IDEs from biological-signal (C-Reactive Protein in this example) to the electronic/electrical signal occurs via dielectric / capacitance change, induced by the capture of antigens by already immobilized antibody between the microelectrodes. Highly repeatable, reproducible and sensitive with a wide dynamic range of operation obtained for the detection and quantification of C-Reactive Proteins using micro-interdigitated electrodes. Being process/integration-compatible with microelectronic/fabrication technologies, this method provides many advantages for the utilization in medical, environmental monitoring, defense, and process control (portable and/or remote sensing) applications, such as smaller size, lower power consumption, small sample consumption, disposability, etc. The other method also analyzes the capability and reliability of sandwich-type antibody microarrays for the detection quantification of cardiovascular risk markers, using optical scanner system for the detection of fluorescent-labeled, sandwiched-structured antibodies on the chemically modified glass substrates. With the process and structure presented in this study, baseline and elevated levels of C-Reactive Protein, Myoglobin, TNF-α, Serum Amyloid A protein, known cardiovascular risk markers in human serum, can be detected. Four-parameter logistic fit algorithm gave high R^2 values and dynamic working ranges are similar to the commercially available ELISA kits for these antibodies. This finding strongly suggests that these antibody pairs can be used to detect and quantify marker proteins in human serum. As microarray systems are versatile, other marker specific antibodies can also be added to the panel and hence multiplex assays for quantification could be possible. We have observed that detection limits of sandwich-type antibody microarray is enough to detect serum levels, both baseline and elevated, of marker proteins used in this study. Our findings in microarray study provides the advantages for lab-based, where higher sensitivity/selectivity/responsivity is desired, such as medical diagnostics, detection and quantification applications such as being able to detect and quantify multiple markers at once and also better dynamic ranges over some of the commercial ELISA test-kits.

FIGURES AND TABLES

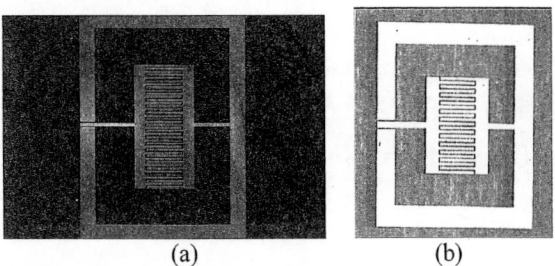

(a)   (b)

Figure 1: *Simulated current distribution across the interdigitated electrodes (a) and a representative/fabricated interdigitated capacitors (b).*

Figure 2: *Antibody immobilization confirmed by hybridization with Alexa-488 labeled anti-mouse antibodies.*

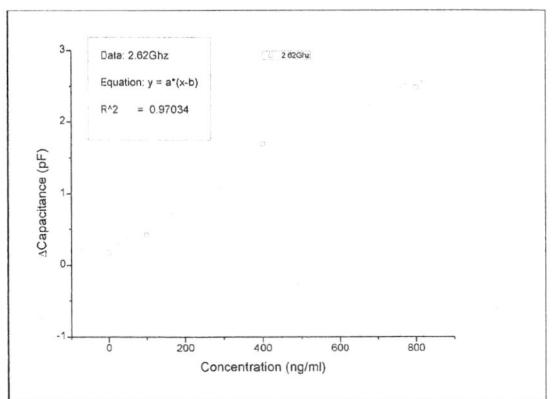

Figure 3: *Capacitance change vs. antigen concentration at 2.62GHz.*

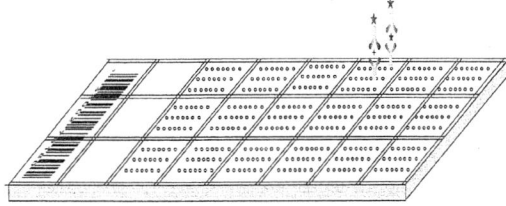

Figure 4: *Microarray microplate apparatus, used in this study, allow us to print twenty-four subarray onto glass slide in 3x8 format.*

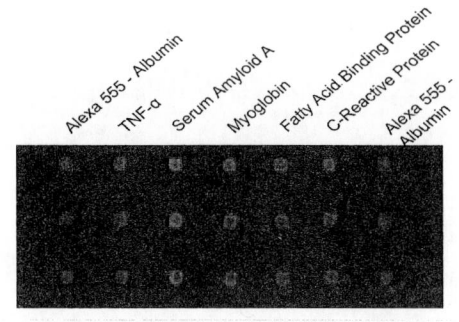

Figure 5: *Lane 1 and 7 are Alexa 555 labeled albumin. Lane 2: TNF-α, Lane 3: Serum Amyloid A, Lane 4: Myoglobin, Lane 5: Fatty Acid Binding Protein, Lane 6: C-Reactive Protein. This subarray formation in each well was used for all experiments. Alexa-488 labeled rabbit anti-mouse antibody was used to check quality of printed slides. This fluorescent antibody bind to immobilized antigen specific antibodies.*

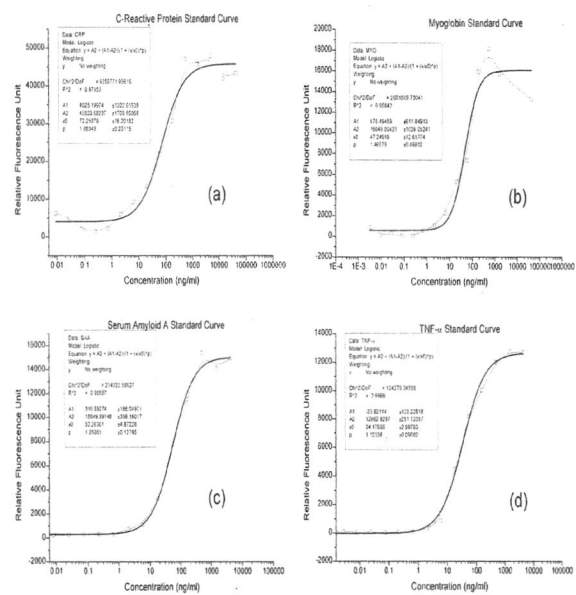

Figure 6: *C-Reactive Protein, Myoglobin, Serum Amyloid A, and TNF-α antigens were used to generate standard curves. Serial dilution started from 40 µg/ml concentration to 0.9 pg/ml by three fold dilution with dilution buffer. After background subtraction, relative fluorescence unit versus concentration curves were generated. Only detectable ranges were taken into consideration.*

978-1-4244-1983-8/08/$25.00 ©2008 IEEE

Table I: *Comparison of detection ranges of commercially available ELISA kits with sandwich-type antibody-microarray.*

Protein Markers	Normal Serum Levels	Elevated Serum Levels	Detection Range of commercial ELISA	Vendor	Sandwich-type antibody microarray detection ranges (this study)
C-Reactive Protein	? 1 µg/ml	? 10 µg/ml	0 – 100 µg/ml 1.9 – 150 ng/ml 10 – 500 ng/ml	(Life Diagnostics Inc.) (ALPCO Diagnostics) (Diagnostic Systems Lab.)	*2.2 – 531 ng/ml*
Myoglobin	? 110 ng/ml	? 110 ng/ml	5 – 1000 ng/ml 5 – 1000 ng/ml 25 – 250 ng/ml	(Life Diagnostics Inc.) (Oxis International Inc.) (ALPCO Diagnostics)	*6.5 – 531 ng/ml*
Serum Amyloid A	~ 3 µg/ml	? 1 µg/ml	1.1 – 80 ng/ml 9.4 – 600 ng/ml	(US. Biological) (Biosource Int.)	*5.9 – 478 ng/ml*
TNF-α	NA	NA	10 – 2000 pg/ml 15.6 – 1000 pg/ml 15.63 - 1000 pg/ml	(CytoLab Ltd.) (Biosource Int.) (Assay Design)	*5.9 – 1434 ng/ml*
Fatty Acid Binding Protein	? 16 ng/ml	? 16 ng/ml	1 – 100 ng/ml 0.5 – 25 ng/ml 1 – 100 ng/ml	(Life Diagnostics Inc.) (Biovendor Lab. M. Inc) (Oxis International Inc.)	*ND*

REFERENCE

[1] M. F. Elshal and J. P. McCoy, "Multiplex bead array assays: performance evaluation and comparison of sensitivity to ELISA," *Methods*, vol. 38, pp. 317-23, 2006.

[2] E. Eteshola and D. Leckband, "Development and characterization of an ELISA assay in PDMS microfluidic channels," *Sensors and Actuators B: Chemical*, vol. 72, pp. 129-133, 2001.

[3] Gevorgian *et al.*: "CAD Models for multilayered substrate interdigital capacitors, " *IEEE Transactions on Microwave Theory and Techniques,* vol. 44, no. 6, June 1996.

[4] Christine Bergren, Capacitive Biosensors, *Electroanalysis* 13: 173-180, 2001.

A Direct Injection Method for Blood Cells into Microchannels from Pure Blood Droplets with switchable in-situ Distillation of Erythrocytes

Ata Tuna Ciftlik and Haluk Kulah

Electrical and Electronics Engineering Department
Middle East Technical University
Ankara, TURKEY

Abstract — **This paper represents a direct injection method for leukocytes into microfluidic channels using negative dielectrophoresis. The blood sample is a heparinated pure blood droplet and injection is realized on-chip without any external microfluidic connections to the microchip. During injection erythrocytes are immobilized in-situ at injection reservoir electrodes. The method is easily applicable to most BioMEMS applications requiring direct injection of cells from pure blood and to eliminate pressure driven control.**

I. INTRODUCTION

Increasing number of microsystems analyzing blood cells are being introduced into literature in recent years [1-6]. Other than high throughput, major goal of these blood analyser systems is to eliminate need for stationary laboratory equipment and realize a standalone system for use in remote places. Unfortunately, sample injecting microfluidic ports connecting these total analyzers to external world has been a major obstacle to realize stand-alone systems. Pressure driven analyte input ports require external pumps as well as increased sample amount significantly. Moreover, fluidic ports without leakage are needed to be implemented on chip by post-processing. An injection method which draws cells into microfluidic channels from wild blood sample can easily eliminate mentioned problems.

On the other hand, most of these microsystems target cells other than erythrocytes and demonstrate a need to separate information carrying targets from erythrocytes first. These microsystems include but not limited to: fully integrated genetic analyzers [1], cancer detectors targeting metastatic cells [2,3], HIV detectors using CD4+ count [4]. Erythrocyte separation carried out in synchrony with injection would increase throughput and eliminate the need for extra separation procedures needed with similar devices.

This paper represents a blood cell injection method from heparinated pure blood droplets using dielectrophoresis. The erythrocytes are immobilized on the injection reservoir electrodes while other cell content of the blood is injected into micro channels. While allowing portability, the method

This work is supported by The Scientific and Technological Research Council of Turkey under grant number 104S605

improves throughput in erythrocyte separating processes with in-situ immobilization on the injection reservoir. Moreover, dielectrophoresis is a common method in cell-handling microsystems and can easily be integrated to blood-analyzing devices.

II. THEORY AND DESIGN

A. Dielectrophoresis

Dielectrophoretic force is defined as the force exerted on an uncharged dielectric particle in presence of non-uniform electric field, where the medium in which the particle is suspended has a dielectric constant different than the particle or particles. Dielectrophoresis (DEP), as expected, is the manipulation or motion of particles representing aforementioned properties using non-uniform electric field [7]. Dielectrophoretic force on exerted on a particle placed in a spatially non-uniform electric field is given by:

$$< F_{DEP} >= \pi \varepsilon_m \mathbf{R}^3 \text{Re}(\mathbf{f_{CM}}) \nabla |\mathbf{E}|^2 \qquad (1)$$

$$f_{CM} = \frac{\varepsilon_p^* - \varepsilon_m^*}{\varepsilon_p^* + 2\varepsilon_m^*} \qquad (2)$$

where ε_p^* and ε_m^* are relative complex permittivity of particle and medium respectively, $\mathbf{f_{CM}}$ is called Clausius-Mossotti factor, $\mathbf{R}$ is the particle radius and $\mathbf{E}$ is the electric field.

Rather very useful property of dielectrophoresis is the direct dependence of DEP force on Clausius-Mossotti factor, f_{CM}. Consider equation 3, which is a general expression of complex permittivity:

$$\varepsilon^* = \varepsilon - j\frac{\sigma}{\omega} \qquad (3)$$

where σ is the conductivity, ε is the permittivity and ω is the angular frequency. From this effect and the subtractive property of $\mathbf{f_{CM}}$, there exists a frequency such that the Clausius-Mossotti factor becomes zero, and hence the DEP forces. It is also evident that, at one side of this zero-force frequency the force is negative and at the other side, force is positive. This zero force frequency is called DEP crossover frequency and can be expressed as in eqn. 4 [7].

$$f_{cr} = \frac{\sqrt{2}}{8\pi R C_m} \sqrt{(4\sigma_{med} - RG_m)^2 - (3RG_m)^2} \quad (7)$$

where C_m and G_m are capacitance and conductance of membrane respectively and σ_{med} is medium conductivity.

In this work, rather than using frequency selectivity, which may be troublesome to implement on-chip, the Radius selectivity of crossover frequency modified with medium conductivity is exploited. The non-spherical, disk-shape geometry of erythrocytes results in different crossover frequencies as compared to other cells representing uniform spherical structure and higher radiuses. This can be used to differentiate cells under DC electric fields. The medium conductivity of cell suspension, namely the ringer solution, can be set to a value such that the erythrocytes experience a negative dielectrophoretic force whereas other cells experience positive forces with respect to electric field gradient. Now, let there exist two cells with the same membrane conductivity G_m but differ in effective radius with values R_1 and R_2 conditioned as $R_1 < R_2$. Then, equation (7) suggests that if the medium conductivity σ_{med} is modified to a value such that

$$R_1 < \frac{\sigma_{med}}{G_m} < R_2 \quad (7)$$

then the particles with higher radius R_2 have imaginary crossover frequency while the particles with lower radius R_1 have a real crossover frequency. As a result, for a given radius, there is a medium conductivity such that cells of radius R_2 experience always positive (or negative) dielectrophoresis forces over all frequency spectrum, while cells with radius R_1 demonstrate a real crossover frequency.

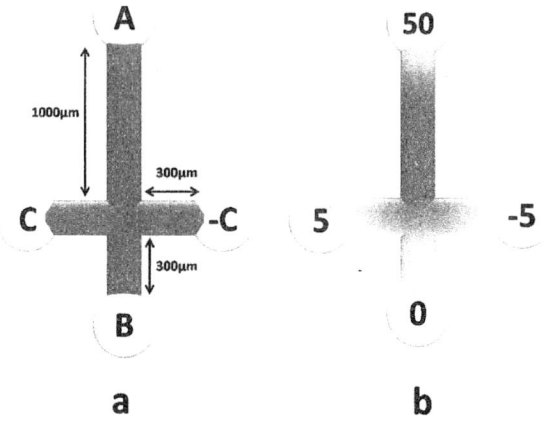

Figure 1 – a) The schematic of the cross shaped microchannel used for cell injection from droplet. The reservoir B is used to place blood droplets, while the reservoirs C, -C are used for generating electric field gradients. b) The voltages applied for injection with in-situ distillation of erythrocytes and the magnitude of electric field inside microchannels

For in-situ distillation of erythrocytes during injection from other blood cells, the medium conductivity can be adjusted to leave crossover frequency of injected target cells on the imaginary side. By this configuration, the erythrocytes can be left at the injection side with negative dielectrophoretic forces while the attractive force injects targets into the channel. From now on, it is assumed that equation (7) holds such that erythrocytes experience negative dielectrophoresis at DC potentials, whereas target cell positive.

B. Design

The microchip schematic containing cross shaped microchannels are shown in Figure 1a, and the electrodes are labeled for simplicity. The reservoir B is the blood injection reservoir and electrodes at reservoirs A, B, C and –C are used to generate electric field gradients for injection. Figure 1b illustrates the electrical potentials applied to reservoirs and the resulting electric field intensity inside microchannels. In-plane motion of cells having positive $\mathbf{f_{CM}}$ will be in the positive electric field gradient towards the electrode A, the collection reservoir electrode. On the other hand, there is a negative electric field towards the electrodes at reservoirs C and –C attracting particles with negative $\mathbf{f_{CM}}$.

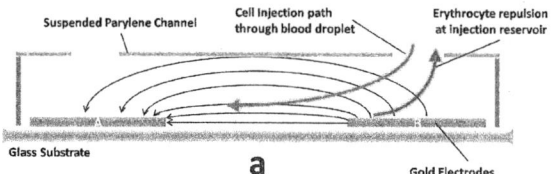

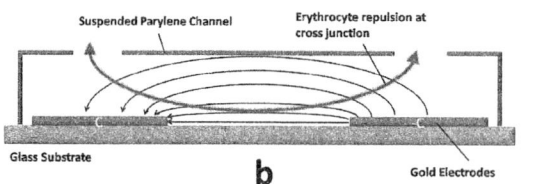

Figure 2 - a) Cross section of device through A-B line and forces in the z-direction. Positive gradients in z direction forces cells to flow into reservoir and microchannels while repulsing erythrocytes b) Cross section of device through –C-C line and erythrocyte pump-out at junction point.

Simultaneously, out of plane gradients are also generated due to the planar nature of the electrodes (see part III). There exists a positive electric field gradient at the reservoir electrodes in the z direction, as shown in figure 2-a, which facilitates motion of target cells in the z direction, injecting them through the reservoir hole present in the device. Conversely, this positive gradient forces erythrocytes to remain in the reservoir B. Hence, the reservoir B accumulates erythrocytes as injection continues, up to a point that overall drag of the target cells due to collision results in a red blood cell leakage since the out of plane gradient generated by fringing fields are not enough to stand against. As the number of accumulated cells grows in reservoir B,

978-1-4244-1983-8/08/$25.00 ©2008 IEEE   30

this leakage also grow into a large number of erythrocytes, so an additional separation mechanism is needed.

Consider figure 1b, in which the –C-C line is illustrated to have a negative electric field gradient. The erythrocytes tending to flow down in the negative voltage gradient perpendicular to the main stream flow line, so that they can be separated from other cells in the junction. This has also shown in the figure 2b. Moreover, as depicted in figure 2b, the voltage gradients are positive in z direction due to planar electrodes places on the substrate. This arrangement further improves the separation efficiency and decrease the drag forces on erythrocytes as they tend to levitate from the substrate while targets tend to drop down to substrate.

Consequently, when overall three dimensional forces considered, the blood cells are injected from droplet left on reservoir B to collection channel ending with reservoir A, realizing in-situ distillation of erythrocytes. Overall device operation is illustrated in figure 3.

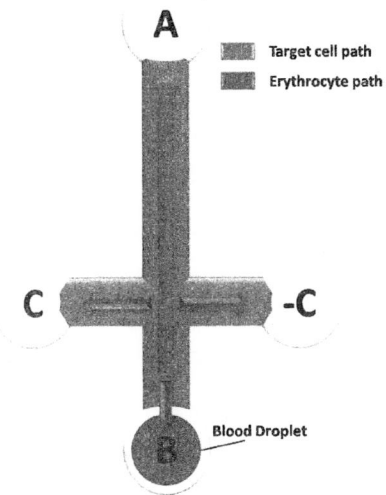

Figure 3 - Overall device operation with defined path of erythrocytes and target cells

III. FABRICATION PROCESS

The chip fabrication is realized by parylene suspended channel process and is illustrated in figure 4. The fabrication process is realized as follows: 1) 5μm parylene is deposited onto a glass substrate, 2) 0.5nm titanium and 5nm gold metal is sputtered onto parylene. Here, titanium is sputtered under gold due to poor adhesion of gold onto parylene. Then metal is patterned in wet etchant.   3) A 25μm thick photoresist (AZ9260) has spun on the wafer and developed as channel sacrificial layer. 4) 20 μm parylene is deposited. 5) Again, a 50μm thick resist (AZ9260) has spun and developed to define reservoir openings. This resist is chosen thick since parylene RIE is a 40 min long process. 6) The parylene is etched in RIE. 7) The sacrificial photoresist is dissolved with acetone to release the channel. A methanol critical point dry may be needed to overcome channel collapse.

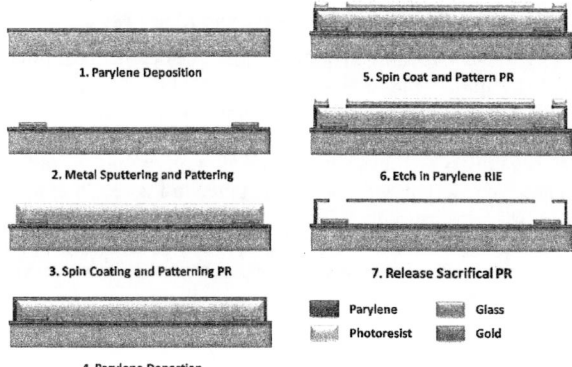

Figure 4 - The fabrication flow of device with a total of three masks: metal mask, channel mask and opening mask [11].

It is possible to fabricate chip with a number of different processes, like PDMS mould [8] or glass-glass anodic bonding [9] processes. However, the parylene is used as a structural material due to its proven biocompatibility and long term stability of surface characteristics [10]. Moreover, substrate adhesion of blood cells is found to be minimum for parylene structures when compared to PDMS and Glass [10]. Besides, among given fabrication processes, parylene suspended channel is the only one capable of lifting of from the substrate and to be used as flexible devices. The fabricated device and a detailed SEM image of the reservoir opening are shown in figure 5.

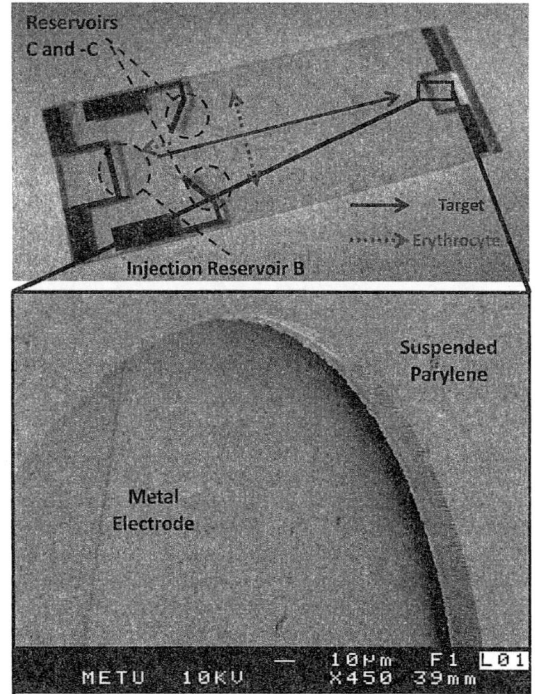

Figure 5 - The photograph of fabricated device and the reservoir opening.

978-1-4244-1983-8/08/$25.00 ©2008 IEEE

IV. TEST METHODS AND RESULTS

The device is tested using a normal heparinated human blood sample without any dilution. First, the device is filled with Ringer Solution whose conductivity is adjusted to satisfy equation (7). The fluid uptake is carried out passively with capillary action. Then, a 5μL sample blood droplet is placed to reservoir with potentials given in Figure 1b applied to reservoir electrodes. The motion of the cells is recorded with a camera mounted to Olympus SZX-12 microscope. Figure 6a shows a photograph of A channel in a successful injection which is rich in white blood cells.

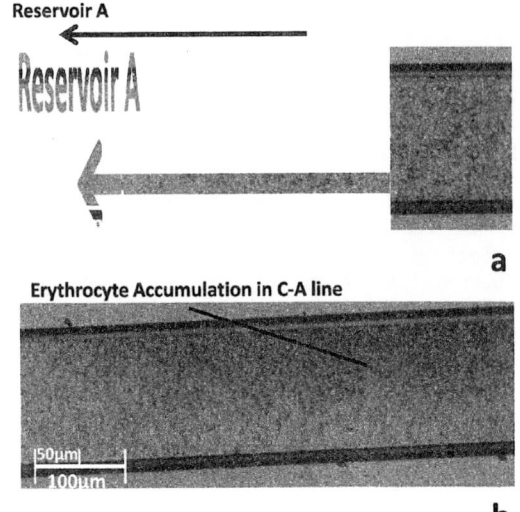

Figure 6 – a) Cells rich in leukocytes in the A channel. b) Erythrocyte leakage due to low –C-C voltage and their accumulation due to electric field asymmetry.

In figure 6b, -2 and 2 volts to the –C and C nodes are applied, which is lower than the normal value. As expected, there is an excessive erythrocyte leakage to the A channel but with an asymmetry. The erythrocytes tend to align themselves to C-A line, due to asymmetry of two electric fields from C-A and –C-A. Conversely, the leukocytes are aligned to opposite side. (Note that the C and –C polarity is reversed during tests.)

V. CONCLUSIONS

A device realizing direct injection of blood cells into microchannels from blood droplets placed into input reservoirs is represented. DC field dielectrophoresis is utilized to inject cells into microchannels from sample blood droplets. During injection, the microchip realizes in-situ erythrocyte separation by carrying crossover frequency of target cells to imaginary side with adjusting medium conductivity. By applying voltages to reservoir electrodes in cross shaped device, cells are attracted into channels with both in-plane and out-of-plane electric field gradients generated in the device. Erythrocyte distillation is carried on

both input reservoir and at cross junction with negative dielectrophoresis. Device is fabricated with a 3 mask process with parylene suspended channels. Parylene is used as a structural material due to its superiorities to other available material in terms of cell adhesion and surface stability. Device is tested with heparinated normal human blood samples without any dilution. When appropriate potential is applied, cells rich in leukocytes are injected into collection reservoir successfully. When erythrocyte separation potentials are decreased, an increase in erythrocyte leakage to collection channel is observed. This property can be exploited to switch erythrocyte separation property of the device.

The device can be used to eliminate external microfluidic connections and hence to realize real stand-alone systems. Moreover, port connections requiring post-processing on microdevices can also be eliminated. The erythrocyte distillation of the device can be employed or deployed depending on application.

ACKNOWLEDGMENT

Authors like to thank Sertan Sukas for device fabrication, Assist. Prof. Dr. Ayse Elif ERSON from Biology Department for supplement of materials and Mr. Gurkan Yilmaz for his help in the tests.

REFERENCES

[1] C. J. Easley et al, "A fully integrated microfluidic genetic analysis system with sample-in–answer-out capability", PNAS, vol. 103, pp. 19272-1927, December 2006.

[2] F. F. Becker et al, "Separation of human breast cancer cells from blood by differential dielectric affinity", PNAS, vol. 92, pp. 860-864, January 1995.

[3] S. Nagrath et al, "Isolation of rare circulating tumour cells in cancer patients by microchip technology", Nature, Vol. 450, 1235-1239, December 2007

[4] X. Cheng et al, "A microfluidic device for practical label-free CD4+ T cell counting of HIV-infected subjects", Lab Chip, 2007, vol. 7, pp. 170–178.

[5] D. S. Kim et al, "Disposable integrated microfluidic biochip for blood typing by plastic microinjection moulding", Lab Chip, 2006, vol. 6, 794–802.

[6] C. J. Huang et al. "Integrated microfluidic systems for automaticglucose sensing and insulin injection", Sensors and Actuators B, vol. 122, 2007, pp. 461–468

[7] H. Morgan et al. "Single cell dielectric spectroscopy", J. Phys. D: Appl. Phys., vol. 40, 2007, pp. 61–70

[8] H.-M. Wu et al, "An Integrated Microfluidic Blood Sampler for Determination of Blood Input Function in Quantitative Mouse microPET Studies", 2005 IEEE Nuclear Science Symposium Conference Record, M03-184.

[9] F. A. Shaikh and V. M. Ugaz, "Collection, focusing, and metering of DNAin microchannels using addressable electrode arrays for portable low-power bioanalysis", PNAS, vol. 103,no. 13, pp. 4825–4830, March 2006

[10] T. Y. Chang et al, "Cell and Protein Compatibility of Parylene-C Surfaces", Langmuir, 2007, vol. 23, pp. 11718-11725

[11] S. Sukas, A.E. Erson, C. Sert and H. Kulah, "A parylene based double-channel micro-electrophoresis system for rapid mutation detection", Proceedings of MicroTas 2007 Conference, vol. 1, pp. 634-636, October 2007.

A Four Quadrant CMOS Analog Multiplier Based on the Non Ideal MOSFET I-V Characteristics

Michele Dei, Nicolò Nizza, Paolo Bruschi
Dipartimento di Ingegneria dell'Informazione
Università di Pisa
via G. Caruso 16, I-56122 Pisa, Italy
e-mail: michele.dei@iet.unipi.it

Massimo Piotto
IEIIT Pisa
CNR
via G. Caruso 16, I-56122 Pisa, Italy

Abstract—**This paper concerns an analog CMOS multiplier based on a novel approach to compensate for non idealities of the MOSFET square law approximation. A numerical algorithm has been implemented to find the optimum sizing of the active devices, starting from the process characteristics. The effectiveness of the proposed configuration has been demonstrated by means of electrical simulations performed on a prototype cell, designed using 0.32 μm – 3.3 V CMOS devices from the STMicroelectronic process BCD6s.**

I. INTRODUCTION

The analog multiplier is a versatile cell with applications in the field of signal processing, continuously variable gain amplifiers, neural networks and all-analog compact control loops. In last two decades, a large number of different architectures for the design of four quadrant analog CMOS multipliers have been proposed. Unfortunately, none of them seems to match the linearity, speed and wide range characteristics of the popular bipolar Gilbert multiplier. At present, the choice of the best CMOS multiplier topology for a given application is still a difficult task for the designer. CMOS multipliers can be roughly divided into (i) strong inversion [1] and (ii) weak inversion multipliers [2]. The latter exploit the bipolar-like exponential characteristic of MOSFET in weak inversion and subthreshold region. They offer simple topologies and low power consumption but suffer from reduced speed and small dynamic range. Strong inversion multipliers, either in saturation or triode region, strongly rely on the square law approximation of the drain current *vs.* gate voltage characteristic. In modern processes, field induced mobility degeneration [3] and, for short channel devices, velocity saturation produce important deviations from the ideal square law. The net effect on the multiplier characteristics is increased non linearity on the X and/or Y port.

In this paper we present an alternative approach based on an architecture similar to the Gilbert multiplier, where the bipolar differential pairs present in the original topology are replaced by the cascade of two CMOS differential pairs. Differently from the above mentioned CMOS multipliers,

the proposed configuration can be easily sized to compensate for deviations from the square law approximation in order to obtain, similarly to bipolar pairs, a linear transconductance *vs.* tail current relationship over a wide current range. Following the proposed approach, a temperature compensated, four quadrant analog multiplier, has been designed using a commercial process. Estimated performances obtained by means of electrical simulations are presented.

II. TRANSCONDUCTOR BASED MULTIPLIER

A. MOS implementation of Gilbert Multiplier

Figure 1 shows a block representation of a transconductor based multiplier. Block A and block B are current controlled fully differential transconductors having the following response with reference to the dashed box in Fig. 1:

$$I_2 - I_1 = k(I_{IN} + I_0) \cdot V_x, \qquad (1)$$

where I_1, I_2 are output currents, V_x is the input differential voltage, I_{IN} is the input control current, while I_0 and k are constant terms. Block C is a fully differential transconductor, with constant transconductance G_m. Its output currents I_A and I_B in Fig. 1 are related to a common mode current I_C and a differential component $G_m V_y$ as given by the following relationships:

$$I_A = I_C + \frac{1}{2}G_m V_y; \qquad I_B = I_C - \frac{1}{2}G_m V_y, \qquad (2)$$

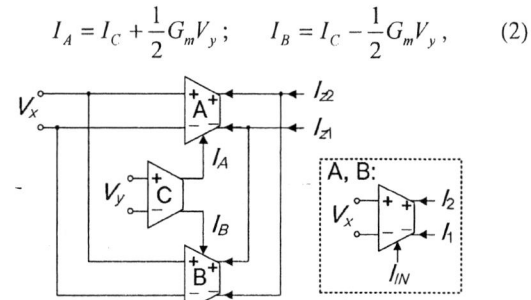

Figure 1. Block representation of a transconductor based multiplier.

The output signal is represented by the current difference $I_{z2} - I_{z1}$ that, according to Eqs. (1, 2), results in the four quadrant multiplier law:

$$I_{z2} - I_{z1} = (I_2 - I_1)_A - (I_2 - I_1)_B = kG_m \cdot V_y V_x. \quad (3)$$

The well known Gilbert multiplier is actually a particular case of Fig. 1 where A and B are bipolar differential pairs [4]. For this case we have:

$$I_2 - I_1 = I_{IN} \tanh\left(\frac{V_x}{2V_T}\right), \quad (4)$$

where $V_T = k_B T / q$ is the thermal voltage. In this case the linearity with respect to I_{IN} is guaranteed by the known g_m vs I_C relationship of bipolar devices that holds with good approximation over an I_{IN} range of several decades. Linearity on the V_x port is obtained by proper predistortion of the input voltage.

We will now consider what happens if we simply replace the bipolar differential pair with a simple MOSFET differential pair. Using the square law approximations for the drain currents and considering small input differential voltage, we would have:

$$I_2 - I_1 = I_{D2} - I_{D1} = V_x \sqrt{\beta I_{IN}} \quad (5)$$

where, as usual, $\beta = \mu C_{ox} W / L$, μ is the carriers mobility, C_{ox} the gate capacitance per unit area and W/L the transistor aspect ratio. Although this configuration has been used in programmable current mirrors [5], the non linear dependence of the output currents on I_{IN} limits the use of simple MOSFET differential pairs to the case of small I_{IN} variations around the d.c. bias value. In terms of signal dynamic range, such a limitation is particularly detrimental, since the thermal noise floor (current spectral density) is proportional to the bias current, which, for the discussion above, should be much larger than the maximum signal amplitude, to preserve linearity.

In principle it is possible to exploit the subthreshold [6] exponential behavior of MOSFETs to mimic bipolar devices. In practice, the usable current range is much smaller and the frequency response is sacrificed. Partial mitigation of these restrictions can be obtained by extending the operation to moderate inversion by means of a more complicated topology [7].

B. Linear Current Controlled Transconductors

The schematic of the proposed linear current controlled transconductor is shown in Fig. 2. The circuit is based on the cascade of two differential pairs, formed by M1a-M2a and M1b-M2b, respectively. The drain terminals of MOSFETs M1a-M2a are connected to a differential resistive load R_L. In order to control the common mode voltage between nodes n1-n2 a simple common mode feedback (CMFB) circuit, formed by M1c-M2c and M5-M6, is added. The current I_{IN} is

fed to the CMFB circuit by the mirror M1-M4, enhancing the rejection of common mode variations even at high frequency.

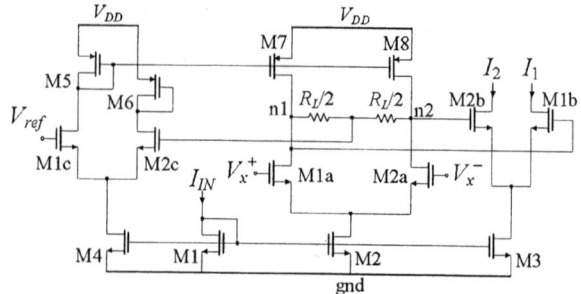

Figure 2. Schematic view of the proposed linear current controlled transconductor.

For a small differential voltage applied to the gate terminals of M1a and M2a, we easily get:

$$I_2 - I_1 = (g_{ma} g_{mb}) R_L \cdot V_x, \quad (6)$$

where g_{ma} and g_{mb} are the transconductances of the M1a-M2a and the M1b-M2b pairs, respectively.

First, let us consider that the two mentioned differential pairs are identical, thus $g_{mb} = g_{ma} = g_n$. Applying the square law approximation of the drain current we would get:

$$I_2 - I_1 = g_m^2(I_{IN}) \cdot R_L V_x = \beta_n I_{IN} R_L V_x, \quad (7)$$

indicating a linear dependence on I_{IN}, as required.

Electrical simulations have been performed using the electrical simulator ELDO™ (Mentor Graphics) with CMOS device models from the STMicroelectronics process BCD6s (0.32 μm, Bipolar-CMOS-DMOS). For this test, M1a-M2a and M1b-M2b have been sized with $W = L = 16$ μm. A significant residual non-linearity is visible in Fig. 3(a), showing the g_m^2 curve as a function of I_{IN} and its linear fit.

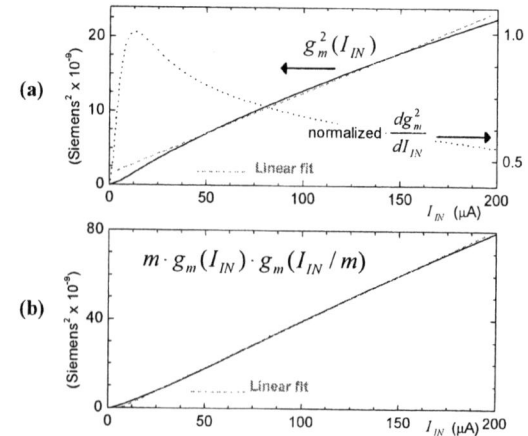

Figure 3. Plot of: (a) g_m^2 vs. I_{IN} for a differential pair with $W/L = 16/16$; (b) $m \cdot g_m(I_{IN}) \cdot g_m(I_{IN} / m)$ vs I_{IN} in the optimum condition $m = 10$.

In terms of derivative, shown also in Fig. 3(a), two different regions can be distinguished: a low current region, where the derivative increases, and a high current region where the derivative decreases. The former is well explained by the transition to weak inversion while the behavior at high currents originates from field induced mobility degeneration [3]. Both phenomena are precisely taken into account in the PHILIPS-9 model provided with the process design-kit.

To obtain a satisfactory linear response it is necessary to compensate for this non linearity. The idea is to make the first differential pair (M1a, M2a) operate in the high current region and the second pair (M1b, M2b) in the low current region, trying to balance the two opposite tendencies. To this aim, M1b and M2b are now a parallel of m transistors identical to M1a and M2a, so that each element of the parallel receives a current m times smaller than the transistors of the first pair. Eq. (7) becomes:

$$I_2 - I_1 = m \cdot g_m(I_{IN}) \cdot g_m(I_{IN}/m) \cdot R_L V_x , \qquad (8)$$

An automatic procedure has been implemented in the MATLAB environment to find the optimum m which minimizes the non linearity of function (8). The procedure uses a spline interpolation of the simulated g_m vs I_{IN} curve. The rest value of I_{IN} was chosen equal to 100 μA. For each integer value of the parameter m, the procedure calculates: (i) the linear approximation of Eq. (8) around the rest point and (ii) the I_{IN} interval where the deviation from linearity is less than 1% (linearity interval). The optimum m is that for which the linearity interval is maximum.

Applying the procedure to the case mentioned above, (process, MOSFET size and resting point) an optimum value $m = 10$ turned out. The corresponding plot of $m \cdot g_m(I_{IN}) \cdot g_m(I_{IN}/m)$ is shown in Fig. 3(b) along with the linear fit. Thanks to this operation, the linearity error in the interval $20\,\mu A \div 180\,\mu A$ is reduced from 5 % ($m = 1$, Fig. 3(a)) to 1% ($m = 10$, Fig. 3(b)). A linear approximation of Eq. (8) in the mentioned interval of currents gives:

$$I_2 - I_1 = \beta_{eq} I_{IN} R_L V_x , \qquad (9)$$

where β_{eq} can be approximated by $(\beta_1\beta_2)^{-1/2}$ where β_1 and β_2 refer to M1a-M2a and M1b-M2b pairs, respectively.

Resistors $R_L / 2$ have been implemented using n-MOSFETs with their gates connected to V_{DD}. These devices operates in triode region providing an equivalent resistance $R_L = 2/\beta_T(V_{DD} - V_{ref} - V_t)$, independent of V_x and V_y signals. From Eq. (9):

$$I_2 - I_1 = \frac{2\beta_{eq}}{\beta_T} \cdot \frac{V_x}{V_{DD} - V_{ref} - V_t} \cdot I_{IN} . \qquad (10)$$

Equation (10) implies a first order cancellation of process variations and temperature effects on β_{eq}.

C. Linear Transconductor Topology

The transconductor C of Fig. 1 has been implemented with the simple source degenerated topology shown in Fig. 4.

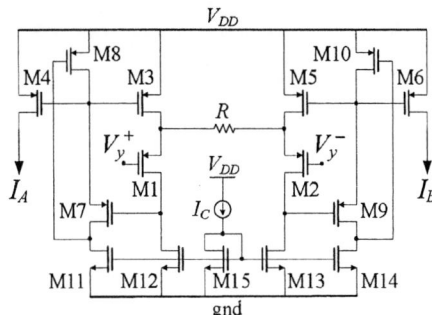

Figure 4.   Schematic view of the linear transconductor.

M1 and M2 are precision voltage shifter based on the feedback loop involving M7, M8, M3 and M9, M10, M5, respectively. The two loops impose M1 and M2 to operate at constant V_{GS}, fixed by the bias current I_C, fed by M12 and M13, respectively. As a consequence, the differential voltage V_y is precisely replicated across the resistor R. It can be easily shown that Eq. (2) holds for the output currents I_A and I_B with $G_m = 2/R$.

III.   SIMULATION RESULTS

The multiplier prototype has been designed using the 0.32 μm / 3.3 V CMOS devices of the Bipolar-CMOS-DMOS "BCD6s" process of STMicroelectronics. In order to obtain a single/ended voltage output the currents I_{z2} and I_{z1} are subtracted by means of simple current mirrors and fed to a resistive load as shown in Fig. 5.

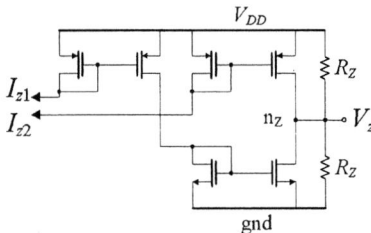

Figure 5.   Current subtraction network performing the current to voltage conversion.

The following response is obtained:

$$V_z = \frac{V_{DD}}{2} + K_M V_x V_y , \qquad (11)$$

where, considering Eqs. (2, 3, 10), the multiplier gain K_M is given by:

$$K_M = \frac{R_Z}{R} \frac{2\beta_{eq}}{\beta_T} \frac{1}{V_{DD} - V_{ref} - V_t} . \qquad (12)$$

A power supply of 3.3 V has been used for all the simulations and V_{ref} has been set 1.65 V. Resistors R and R_Z has been set 10 kΩ and 25 kΩ, respectively, while MOSFETs replacing R_L have $W = L = 1$ μm. With these values the multiplier gain K_M results 2.92 V⁻¹.

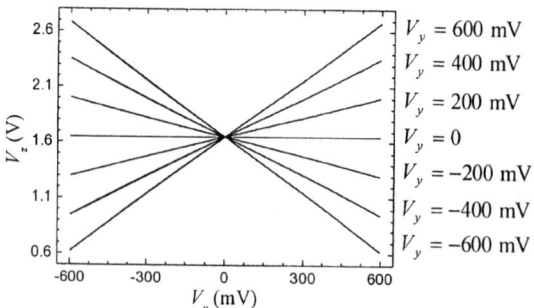

Figure 6. Simulated d.c. characteristics of the multiplier.

Figure 6 shows the simulated d.c. characteristics of the multiplier. A 3 % total linearity error over V_x and V_y ranges of 1.2 V peak-to-peak has been obtained. Temperature sensitivity has been extracted by means of d.c. simulations for various values of the inputs. The resulting maximum multiplier gain sensitivity is $2 \cdot 10^{-3}$ °C⁻¹ in the interval 0÷70 °C. The effects of process errors were studied by means of Monte Carlo simulations (20 runs). We found that multiplier characteristics were well described by the following expression:

$$V_z = \frac{V_{DD}}{2} + V_0 + \left(K_M + \Delta K_M\right)\left(V_x + V_{X0}\right)\left(V_y + V_{Y0}\right), \quad (13)$$

indicating that linearity with respect to both X and Y ports is maintained. Clearly, offsets (V_0, V_{X0}, V_{Y0}) and gain error (ΔK_M) appear. The estimated standard deviations of these parameters are reported in Table I.

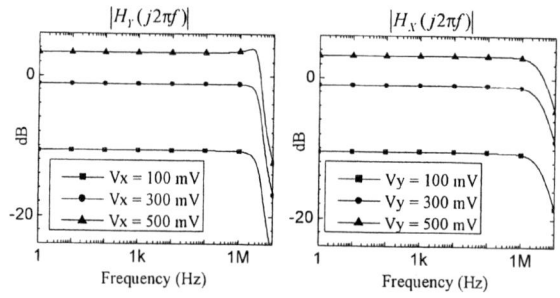

Figure 7. AC bandwith for the Y and X port.

Figure 7 shows the AC amplitude response of the circuit for a signal fed to the Y input with constant V_x, and vice versa; the resulting bandwidth values are reported in Table I.

IV. CONCLUSIONS

A four quadrant multiplier designed in CMOS technology has been presented. The multiplier architecture is based on linear current controlled transconductors obtained by a cascade of two MOSFET differential pairs, properly sized by means of an empirical procedure. The result of electrical simulations clearly proves that input ranges obtainable with the proposed approach are considerably wide. In the presented implementation, all MOSFETs operate in strong inversion or, at least, at the lower boundary of the strong inversion region. Monte Carlo simulations, preformed on the prototype cell, demonstrated the robustness of the linearization procedure with respect to process variations. The optimization procedure can be easily automated to be applied to different design choices. Reasonably, the degree of linearity that can be actually achieved will depend on the process and transistor size considered. In particular, extension of the technique to shorter channel length devices is being investigated in order to improve the multiplier bandwidth.

TABLE I. SUMMARY OF PERFORMACE

Power consumption @ 3.3 V	4.56 mW
Gain temperature sensitivity 0÷70 °C (total relative variation)	15 %
Offsets[a] (see Eq. (12))	$\sigma(V_0) = 19.1$ mV $\sigma(V_{X0}) = 5.5$ mV $\sigma(V_{Y0}) = 28.8$ mV
Relative variation of the multiplier gain[a]	$\sigma(\Delta K_M / K_M) = 3.75$ %
AC bandwith	$B_x = 4.2$ MHz $B_y = 3.7$ MHz

a. Estimated by means of 20 runs of Monte Carlo analysis.

ACKNOWLEDGMENT

The authors would like to thanks the R & D group of the STMicroelectronics of Cornaredo (MI, Italy) for providing the design kit of the BCD6s process.

REFERENCES

[1] Han G., Sánchez-Sinencio E., `CMOS Transconductance Multipliers: A Tutorial', *IEEE Trans. Circuits Syst. II – Analog And Digital Signal Processing*, 1998, **45**, (12), pp. 1550-1563.

[2] Cheng-Chieh Chang and Shen-Iuan Liu `Weak inversion four-quadrant multiplier and two-quadrant divider', *Electronics Letters*, 1998 **34** (22), pp. 2079-2080

[3] B. Razavi, `Design of Analog CMOS Integrated Circuits', McGrow-Hill, Singapore 2001, pp. 584-586.

[4] Gilbert B., `A Precise Four-Quadrant Multiplier with Subnanosecond Response', *IEEE J. Solid-State Circuits*, 1968, **SC-3**, (4) pp. 365-373.

[5] Ramírez-Angulo J., Garimella S.R.S., López-Martín A.J. and Carvajal R.G., `Gain programmable current mirrors based on current steering', *Electronics Letters*, 2006, **42**, (10), pp. 559-560

[6] Walke R.L., Quigley S.F., Webb P.W., `Design of an analogue subthreshold multiplier suitable for implementing an artificial neural network' *IEE Proceedings*, 1992, **139**, (2) pp. 261-264.

[7] López-Martín A.J., Ramírez-Angulo J., Durbha C., and Carvajal R.G., `Highly linear programmable balanced current scaling technique in moderate inversion', *IEEE Trans. Circuits Syst. II - Express Briefs*, 2006, **53**, (4), pp. 283-285.

Low-Voltage High-Performance CMOS Current Differencing Buffered Amplifier (CDBA)

Cem CAKIR
Department of Electrical and Electronics Engineering
Bogazici University, 34342, Bebek
Istanbul, Turkey
cem.cakir@boun.edu.tr

Oguzhan CICEKOGLU
Department of Electrical and Electronics Engineering
Bogazici University, 34342, Bebek
Istanbul, Turkey
cicekogl@boun.edu.tr

Abstract—**This paper presents a new realization of a low-voltage, high-performance CMOS current-differencing buffered amplifier (CDBA). The proposed circuit can be operated with the power supplies down to ±0.75V and it also consumes less power than its counterparts that have been reported so far. Moreover this CDBA has good voltage and current accuracies. UMC 0.18 μm CMOS technology is used for the simulations. The performance of the CDBA is verified with the HSPICE. To show the performance of the circuit, a second-order current-mode notch filter is selected from the literature. The results are in good agreement with the theoretical ones.**

I. INTRODUCTION

The current differencing buffered amplifier (CDBA) was introduced by Acar and Özoğuz to provide further possibilities in the circuit synthesis [1]. The fact that the device can operate in both current and voltage mode, provides flexibility and enables a variety of circuit designs. In addition, it is free from many parasitic capacitances and appropriate for high frequency operation. So far there are several implementation schemes for CMOS technology that have been reported in the literature [2-6]. However, the terminal resistances of the CMOS-based CDBAs are quite high, in the order of several hundred ohms, and their voltage and current transfer ratios are much less than one. Moreover, most of the existing CDBAs are operated at high supply voltages. Advances in integrated circuit technology make the devices in an IC form very small and the power supply voltage of the circuits must be restricted to a low value. Furthermore, with the increasing demands for battery-operated portable equipment, it is clear that low-voltage operation becomes more and more important. Thus a CDBA with very low input terminal resistances that can be operated with low supply voltage is preferable.

II. CIRCUIT DESCRIPTION

The block diagram and equivalent circuit of the CDBA is shown in Fig. 1. The CDBA basically consists of two fundamental building blocks, current subtractor and voltage follower. The current and voltage characteristics of the CDBA can be described by the following matrix equation;

$$\begin{bmatrix} I_z \\ V_w \\ V_p \\ V_n \end{bmatrix} = \begin{bmatrix} 0 & 0 & 1 & -1 \\ 1 & 0 & 0 & 0 \\ 0 & 0 & 0 & 0 \\ 0 & 0 & 0 & 0 \end{bmatrix} \begin{bmatrix} V_z \\ I_w \\ I_p \\ I_n \end{bmatrix} \qquad (1)$$

It is clear that p and n are current-mode input ports which have ideally zero impedance. Current of the port-z is equal to the difference of the input currents, I_p and I_n. Therefore it is defined as the current output which has ideally infinite impedance. Moreover the voltage of port-w follows that of port-z. Hence port-w is the voltage output that should have zero impedance.

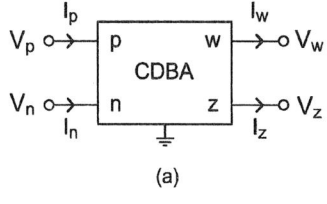

(a)

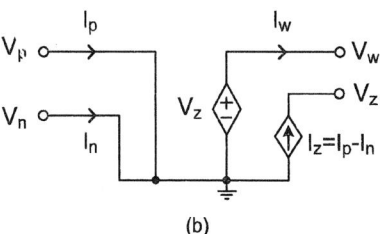

(b)

Figure 1. CDBA (a) Circuit symbol (b) Equivalent circuit

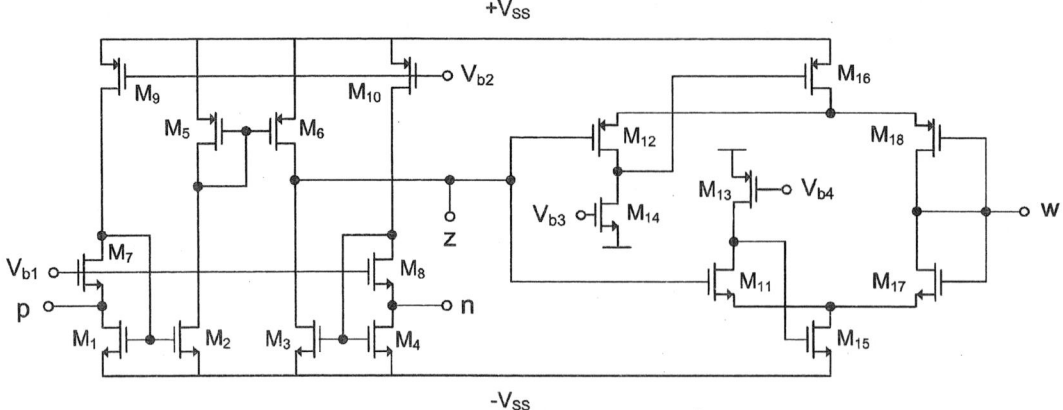

Figure 2. Proposed current differencing buffered amplifier (CDBA)

III. PROPOSED LOW-VOLTAGE CDBA CIRCUIT

Fig. 2 shows the complete schematic of the proposed low-voltage CDBA circuit, which is based on the use of the current differencing circuit (M_1–M_{10}) and the voltage buffer (M_{11}–M_{18}). Proposed circuit is supplied by the voltage of ±0.75 V. UMC 0.18 µm CMOS technology is used for the simulations. The aspect ratios of the transistors are shown in TABLE I.

The current subtractor circuit consists of the transistors M_1 to M_{10}. This circuit is based on the flipped voltage follower current sources (FVFCS) which give rise to very low input resistances at the input ports [7,8]. Fig. 3 shows that the impedances of ports p and n are equal to 50 Ω for a wide frequency range.

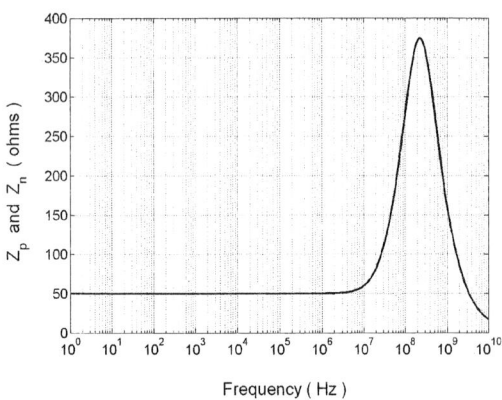

Figure 3. Frequency variation of the input impedance magnitudes

Output stage of the proposed CDBA which offers low output impedance and a moderate output swing can be seen in Fig. 2. This circuit is a class AB voltage buffer which is based on the differential FVF (DFVF) topology [9]. It uses two complementary DFVF cells, M_{12}-M_{16} and M_{11}-M_{15}, with current sources, M_{13} and M_{14}. Fig. 4 shows the frequency characteristics of the port z and w impedances which are equal to 102 kΩ and 158 Ω respectively.

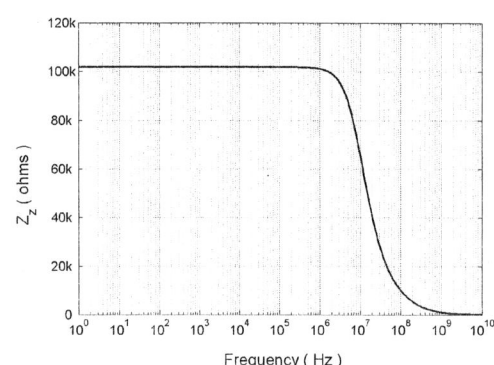

(a) Frequency variation of the port-z impedance magnitude

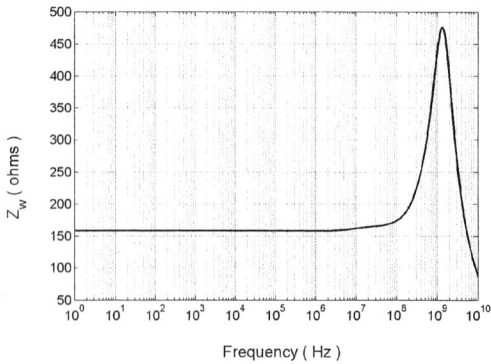

(b) Frequency variation of the port-w impedance magnitude

Figure 4. Frequency variations of the port z and w impedance magnitudes

TABLE I. MOS TRANSISTOR DIMENSIONS

Transistors	W/L(μm/μm)
M_1, M_2, M_3, M_4	5.4/1.80
M_5, M_6	90/1.80
M_7, M_8	180/0.90
M_9, M_{10}	99/1.80
M_{11}	6/0.24
M_{12}	18/0.24
M_{13}	36/1.80
M_{14}	5.4/1.80
M_{15}	30/0.24
M_{16}	99/0.24
M_{17}	30/0.24
M_{18}	99/0.24

In TABLE II, the results of the simulation for the proposed CDBA are summarized.

TABLE II. PERFORMANCE OF THE PROPOSED CDBA

Summary of CDBA Performance	
Supply voltage (V_{ss})	±0.75V
Power dissipation	1.2 mW
Current transfer ratio, $\alpha = I_z / (I_p - I_n)$	0.978
Voltage transfer ratio, $\beta = V_w / V_z$	0.970
Port p and n resistances	50 Ω
Port-z resistance	102 kΩ
Port-w resistance	158 Ω
Offset current on port-z	0.14 μA

Fig. 5 illustrates the AC transfer characteristics of the proposed CDBA. The current and voltage transfer ratios, α and β are found to be 0.978 and 0.970 respectively.

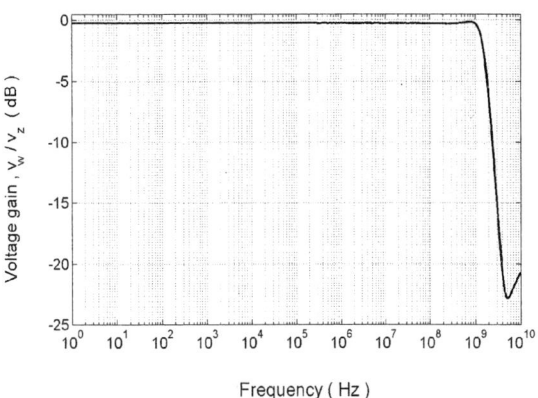

(a) The frequency response of the voltage transfer ratio

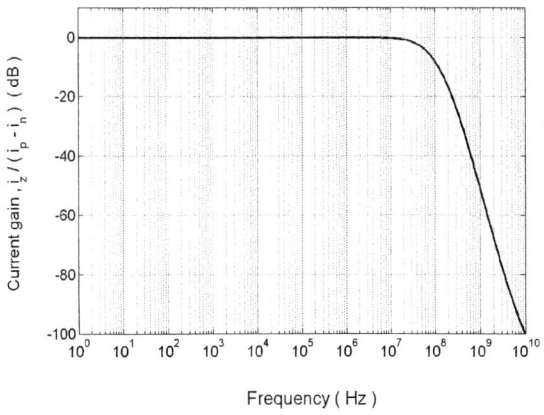

(b) The frequency response of the current transfer ratio

Figure 5. The AC transfer characteristics of the proposed CDBA

IV. DESIGN EXAMPLE

As an application example, a current-mode second-order notch filter circuit [10] is chosen. The current transfer function of the circuit is given as follows:

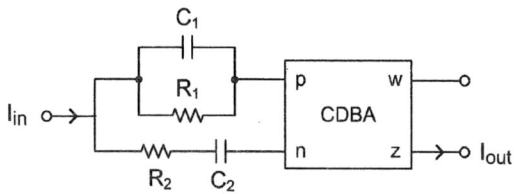

Figure 6. Current-mode second-order notch filter

$$\frac{I_{out}}{I_{in}} = \frac{s^2 + s\left(\dfrac{1}{C_1 R_1} + \dfrac{1}{C_2 R_2} - \dfrac{1}{C_1 R_2}\right) + \dfrac{1}{C_1 C_2 R_1 R_2}}{s^2 + s\left(\dfrac{1}{C_1 R_1} + \dfrac{1}{C_2 R_2} + \dfrac{1}{C_1 R_2}\right) + \dfrac{1}{C_1 C_2 R_1 R_2}} \quad (2)$$

Therefore matching condition for the realization of the second-order notch filter will be $C_1 R_1 + C_2 R_2 = C_2 R_1$. The natural frequency, ω_0 and quality factor, Q for the filter can be expressed as;

$$\omega_0 = \sqrt{\frac{1}{C_1 C_2 R_1 R_2}} \quad (3)$$

$$Q = \frac{\sqrt{C_1 C_2 R_1 R_2}}{C_1 R_1 + C_2 R_2 + C_2 R_1} \quad (4)$$

Sensitivity analysis of the filter with respect to passive elements yields;

978-1-4244-1983-8/08/$25.00 ©2008 IEEE

$$S_{R_1}^{\omega_o} = S_{R_2}^{\omega_o} = S_{C_1}^{\omega_o} = S_{C_2}^{\omega_o} = -\frac{1}{2}$$

$$S_{R_1}^Q = -S_{R_2}^Q = \frac{1}{2} \frac{C_1 R_1 + C_2 R_1 - C_2 R_2}{C_1 R_1 + C_2 (R_1 + R_2)} \qquad (5)$$

$$S_{C_1}^Q = -S_{C_2}^Q = -\frac{1}{2} \frac{C_1 R_1 - C_2 (R_1 + R_2)}{C_1 R_1 + C_2 (R_1 + R_2)}$$

To verify the theoretical analysis, this filter is simulated by using UMC 0.18 μm CMOS process parameters. HSPICE simulations are performed by using the circuit proposed in Fig. 6.

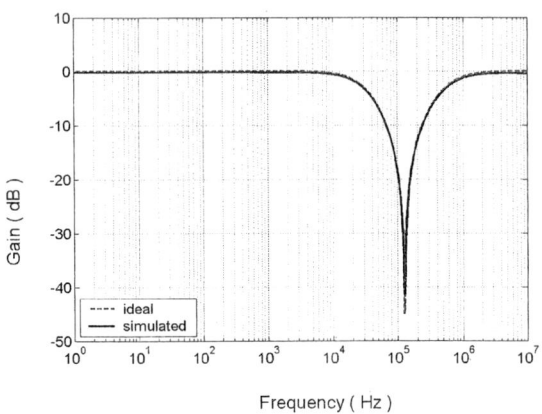

(a) Gain response of the second-order notch filter

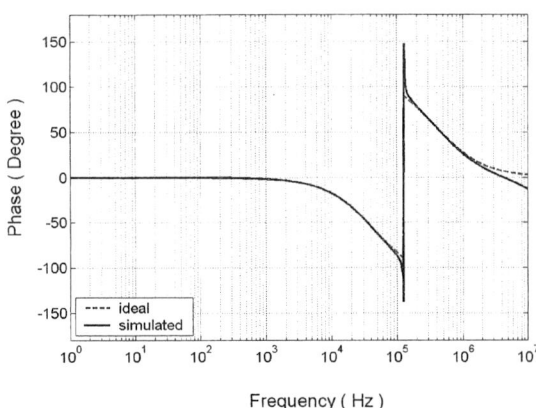

(b) Phase response of the second-order notch filter

Figure 7. Gain and phase responses of the second-order notch filter

Fig. 7 shows simulated gain and phase responses of the notch filter compared to the ideal case. By taking the

matching condition into consideration, external component values are chosen as R_1=50 kΩ, R_2=25 kΩ, C_1=25 pF and C_2=50 pF. Then the center frequency of the circuit is found as f_c=127 kHz which is in close agreement with the theoretical one.

V. CONCLUSION

In this paper, a novel CMOS current-differencing buffered amplifier (CDBA) is presented which is suitable for low-voltage operation. The proposed circuit operates with the power supplies of ±0.75 V. It also has low power consumption which is about 1.2 mW. UMC 0.18 μm spice parameters are used for the simulations. HSPICE simulation results show that the proposed CDBA has terminal resistances of R_p=R_n=50 Ω, R_z=102 kΩ and R_w=158 Ω respectively. Moreover it provides higher current and voltage transfer ratios which are α=0.978 and β=0.970. The feasibility of the proposed CDBA is shown on a second-order current-mode notch filter which has a center frequency of f_c=127 kHz.

REFERENCES

[1] C. Acar and S. Ozoguz, "A new versatile building block: current differencing buffered amplifier suitable for analog signal processing filters", Microelectronics Journal, 30, pp. 157-160, 1999.

[2] S. Ozoguz, A. Toker and C. Acar, "Current-mode continuous-time fully-integrated universal filter using CDBAs", Electronics Letters, 35, pp. 97–98, 1999.

[3] H. Sedef and C. Acar, "On the realization of voltage-mode filters using CDBA", Frequenz, 54, pp. 198–202, 2000.

[4] N. Tarim and H. Kuntman, "A high performance current differencing buffered amplifier", Proc. of the 13th International Conference on Microelectronics, Rabat, Morocco, pp. 153–156, 2001.

[5] C. Acar and H. Sedef, "Realization of nth-order current transfer function using current differencing buffered amplifiers", International Journal of Electronics, 90, pp. 277–283, 2003.

[6] W. Tangsrirat, K. Klahan, T. Dumawipata and W. Surakampontorn, "Low-voltage NMOS-based current differencing buffered amplifier and its application to current-mode ladder filter design", International Journal of Electronics, Volume 93, Number 11, pp. 777-791(15), 2006.

[7] R. G. Carvajal, et al, "The flipped voltage follower: a useful cell for low-voltage low-power circuit design", IEEE Transactions on Circuits and Systems I: Fundamental Theory and Applications, Regular Papers Vol. 52, Issue 7, pp. 1276–1291, 2005.

[8] A. Uygur, H. Kuntman, "Low-voltage current differencing transconductance amplifier in a novel allpass configuration", MELECON'06, Proc. of the 13th IEEE Mediterranean Electrotechnical Conference, Spain, pp. 23-26, 2006.

[9] R. G. Carvajal, A. Torralba, J. Ramírez-Angulo, J. Tombs and F. Muñoz, "Compact low-power high slew-rate CMOS buffer for large capacitive loads," Electronics Letters, vol.38, no. 32, pp. 1348-1349, 2002.

[10] U. Cam, "A novel current-mode second-order notch filter configuration employing single CDBA and reduced number of passive components", Int. Journal Computers & Electrical Engineering, 30, pp. 147-151, 2004.

A µg Resolution Microaccelerometer System with A Second-Order Σ-Δ Readout Circuitry

Reha Kepenek, Ilker Ender Ocak, Haluk Kulah, and Tayfun Akin

Department of Electrical and Electronics Engineering
Middle East Technical University
Ankara, TURKEY

Abstract—**This paper reports a 2nd order electromechanical sigma-delta accelerometer system. Accelerometer is fabricated using Dissolved Wafer Process, and has a structural thickness of 15µm. A large proof mass is used to decrease the mechanical noise of the accelerometer and 306 fingers per side are used to increase the sensitivity and operation range of the accelerometer. In order to obtain a high resolution, low noise accelerometer system, a fully differential, closed loop, oversampled sigma-delta capacitive readout circuit is designed and implemented. The chip includes a switched-capacitor charge integrator and a comparator, and can be used in either open-loop or closed-loop mode. The readout circuit has more than 115dB dynamic range and can resolve less than 3aF/√Hz. A digital filtration and decimation circuitry is also implemented to signal process the output bit stream of the readout circuit. The Σ-Δ second order closed loop readout circuit consumes 16mW power from a 5 V supply and the complete accelerometer system has a 0.3% non-linearity in ±1 g range, 86µg bias drift, 74µg/√Hz of noise level and maximum operation range of ±18.5g.**

INTRODUCTION

High precision accelerometers with micro-g (µg, g=9.8m/s^2) resolution have many applications, including inertial navigation and guidance, microgravity measurements in space, tilt control and platform stabilization, seismometry, and GPS-aided navigators for the consumer market. Most of these applications require digital output with force-feedback operation, large dynamic range, and high sensitivity with inertial grade resolution. The attractive features of MEMS as applied to inertial systems are its potentially low cost, drastically reduced size and weight, and low power dissipation, all of which are prerequisites for the development of next generation navigation systems. Recently, capacitive accelerometers have become very attractive for high precision µg applications due to their high sensitivity, low temperature sensitivity, low power, wide dynamic range, and simple structure [1, 2]. To date, only a few inertial-grade silicon accelerometers have been reported [3-9]. Majority of these accelerometers lack of achieving high measurement ranges. The accelerometer system presented in this paper achieves micro-g resolution with an operation range of more than 15g.

This paper reports a capacitive accelerometer system, composed of a lateral accelerometer and the interface electronics, and can be operated both in open-loop and closed-loop. In the closed-loop mode of operation, the interface chip forms a second-order electromechanical oversampled sigma-delta modulator with the sensor, and provides direct digital output and force feedback of the proof mass simultaneously. The overall system achieves low noise, high sensitivity, and high dynamic range.

ACCELEROMETER STRUCTURE

It has been especially challenging to design precision accelerometers sensitive in lateral axes because of the difficulty in achieving small capacitive gaps over large area. The accelerometer implemented in Dissolved Wafer Process (DWP) is sensitive in lateral axes, and designed for minimum mechanical noise, maximum operation range, and high sensitivity based on the allowable features of the process and device dimensions. Figure 1 shows the structure of the lateral accelerometer. In order to decrease the mechanical noise of the accelerometer, a large proof mass is designed (3300µmx1802µm), and suspended with 6 double folded springs. Sensor has 182 capacitive fingers on each side and there is another set of 124 fingers at the center of the proof mass to increase the sensitivity and maximum operation range of the accelerometer.

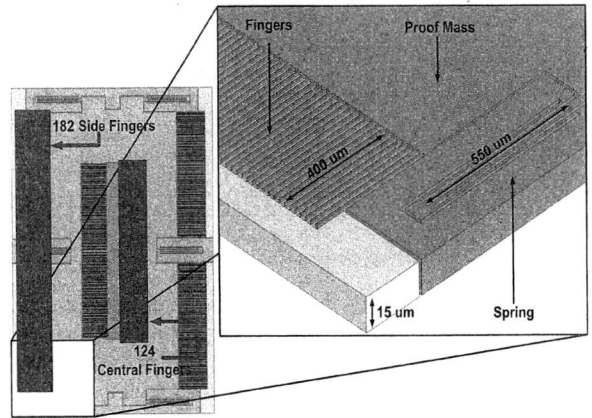

Figure 1. The structure of the SOG accelerometer

978-1-4244-1983-8/08/$25.00 ©2008 IEEE

The fabrication process of the accelerometer requires 3 masks. First, a glass substrate is etched to form anchor regions of the accelerometer (figure 2a). After the formation of the anchor regions chromium and gold is sputtered on the glass wafer and patterned to form the electrical connections (figure 2b). Then, a 100μm thick silicon wafer is doped with boron (figure 2c) and etched reactively to form the structural layer (figure 2d). Next, silicon and glass wafers are anodically bonded (figure 2e) and the undoped silicon is completely etched (figure 2f).

Due to the limitations on the boron diffusion depth, the structural layer has low thickness (15 μm). However, this provides smaller finger gaps (~1.1μm) to be fabricated with DRIE. The large proof mass and long fingers (400μm) results in a high sensitivity (1.12x10^{-5} F/m), low Brownian noise (5.26 μg/√Hz at atmospheric pressure) and high maximum operation range (~20 g) in closed loop operation. Table 1 shows the fabricated accelerometer parameters.

Figure 2. Fabrication process of the SOG accelerometer.

TABLE I
FABRICATED SENSOR PARAMETERS

Parameter	Value
Mass of proof mass	0.18 milli-g
Resonant frequency	1.72 kHz
Proof mass thickness	15 μm
Sensing gap	1.1 μm
Sense capacitance	16.2 pF
Sense fingers length	400 μm (side) / 350 μm (center)
Number of fingers	182 @ each side/ 124 @ the center
Sensitivity	1.12x10^{-5} F/m
Brownian noise	5.26 μg/√Hz

INTERFACE ELECTRONICS

In order to measure the acceleration signals from the accelerometer, a capacitive readout circuit is designed and implemented. The readout circuit forms a fully differential switched capacitor Σ-Δ modulator together with the sensor. The circuit can operate either in open loop or closed loop mode. In open loop, the capacitance difference of the accelerometer is measured, and the output is generated as an

analog voltage with a high sampling rate of 500 kHz. In this case, the output is proportional to the capacitance difference for a limited range of inputs. In the closed loop case; the output is a one-bit Pulse Density Modulated (PDM) digital stream and it is applied as an electrostatic feedback to the proof mass to hold it stationary independent from the incoming acceleration. In other words, the proof mass is not moving at all during the operation and hence the linear range is extended significantly.

Figure 3 shows the detailed schematic view of the readout electronics which has four main blocks: the switched capacitor front-end, start-up, bias generator, and clock generator. The front-end circuit includes a charge integrator structure with the switched capacitor network, the operational trans-conductance amplifier (OTA), and a comparator. OTA used in the charge integrator is a critical component for the sampling rate and noise considerations. For this purpose a folded cascode fully-differential OTA is designed to ensure high power supply rejection ratio, high gain, and low noise. The readout circuit both reads capacitive difference and applies feedback in one sampling period. In readout phase, a differential voltage, due to the sense capacitor difference, is generated at the output of the charge integrator stage. The integrator output is then given to the latching comparator, which generates a one-bit PDM for electrostatic force-feedback at the rising edge of the applied clock signal, and holds it during the feedback phase. In this phase, a 5V potential difference is applied between one of the electrodes and the proof mass to provide pulling. Besides pulling the proof mass to its rest position, this PDM bit stream is also used as the closed-loop system output.

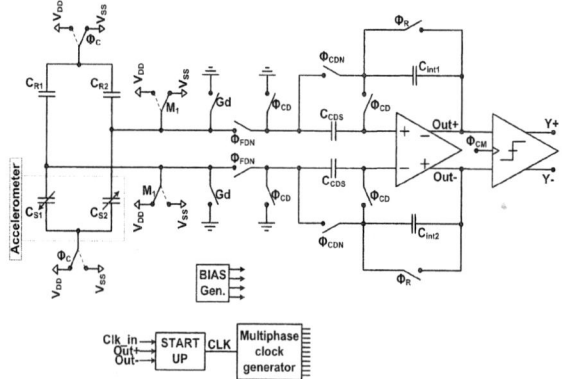

Figure 3. Schematic of the sigma delta readout circuit.

The bias generator block provides temperature and supply independent biasing for the folded cascode OTA and for other circuit components. The start-up circuit is needed to avoid high accelerations and deflections of proof mass at the start up of the system. Another major block is the multi phase clock generator, which supplies various clock signals with various duty cycles and phases to provide proper switched capacitor operation.

A summary of the circuit parameters are given in Table II, which are obtained by testing the interface circuit alone in open loop mode.

TABLE II
TEST RESULTS OF THE CMOS INTERFACE READOUT CIRCUIT
(WITH C_{INT} = 4pF)

Parameter	Value
Circuit Sensitivity	1.3 V/pF
Circuit Noise (open-loop)	0.81 µV/√Hz
Minimum Resolvable Capacitance	2.2 aF
Open-Loop Dynamic Range	~115 dB

The model of the overall electromechanical system is constructed in Matlab Simulink environment, as shown in figure 4. Here, the accelerometer is modeled as a low pass filter with second order transfer function. Interface electronics is also modeled by Simulink components and stability is ensured with extensive simulations.

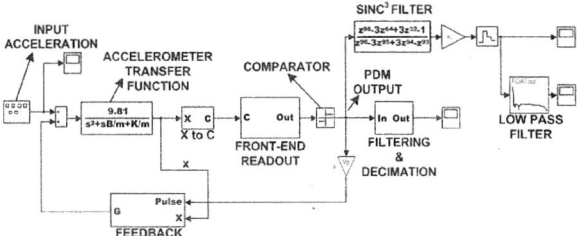

Figure 4. Matlab simulink model of the accelerometer system.

In order to extract the meaningful acceleration data from the oversampled PDM bit stream, low pass filtering and decimation is required. For these purposes, a Sinc demodulator filter is implemented in this study. Sinc filters have the advantage of easy implementation, as they employ only addition and subtraction blocks. The order of the Sinc filter is chosen as 3 (one plus the order of the modulator) to keep the noise in the operation range below 0.5dB [10]. Figure 5 shows the structure of the $Sinc^3$ filter. The $Sinc^3$ filter is composed of 3 stages of integration and 3 stages of differentiation. In this system the bit stream frequency is 500 kHz and a $Sinc^3$ filter with a decimation factor of N=40 is used to have an output signal at 16 kHz. Besides implementing the $Sinc^3$ filter in a PIC, another PIC is used to decimate the 16 kHz data rate to 815 Hz. The second decimation with N=16 is performed by using a $Sinc^2$ filter structure. This second microcontroller has the capability of scaling the data and adjusting the offset level.

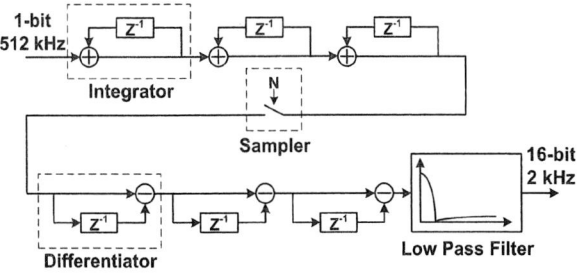

Figure 5. Block diagram of the $Sinc^3$ filter.

Figure 6 gives the Simulink simulation results showing the PDM bit stream and the $Sinc^3$ filter output in response to a 2g sinusoidal input acceleration.

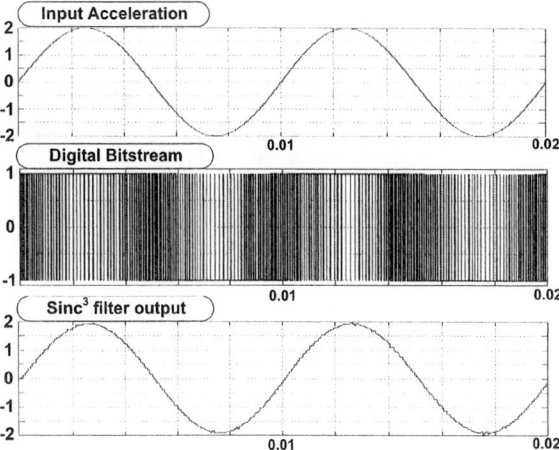

Figure 6: Simulation results of the Simulink model with the $Sinc^3$ filter.

IMPLEMENTATION AND TEST RESULTS

Figure 7 shows the test PCB employing all the system components including the accelerometer, readout circuit, and the external sinc demodulator.

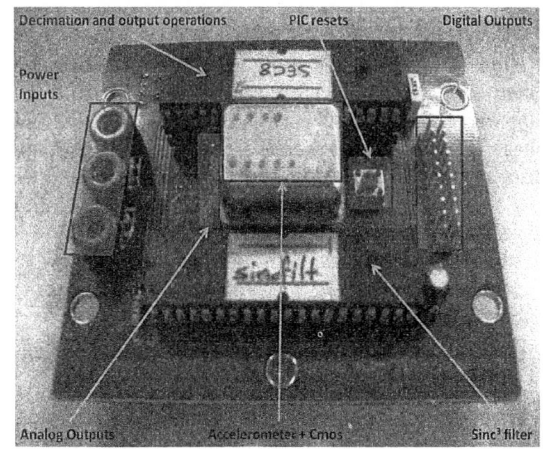

Figure 7. Accelerometer system with accelerometer, Σ-Δ modulator and Sinc filters on a PCB.

The output of the system is a 16 bit digital signal at 815 Hz which is collected with a data acquisition card during the tests. Circuit operates with three power supplies which are +5V, +2.5V, and 0V and consumes 160mW power during its normal operation.

For the linearity tests, accelerometer is mounted on a rotating table which has 1 degree rotation accuracy. The table is rotated for 30 degrees at each step and 1 minute of data is collected 12 times until the full cycle completes to 360 degrees. Figure 8 shows the data collected for linearity tests. By finding the average values of the accelerometer output at each level and plotting these values with respect to the applied input accelerations, scale factor is measured as 1660.15 and non-linearity is found as %0.3.

To extract the bias drift and noise of the accelerometer, Allan-variance method is used. In this method accelerometer is placed on a rigid plane and the output voltage is collected for more than 2 hours. After the data acquisition is complete, this data is processed and Allan-variance graph is plotted. From this graph bias drift and noise level are found as 74 μg and 86μg/√Hz, respectively.

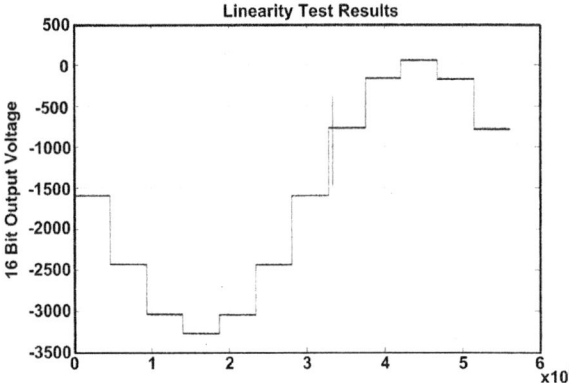

Figure 8: Linearity test result in ±1g range.

High operation range tests are done by mounting the system on a centrifuge table and rotating the table up to ±10g with 2g steps. At each acceleration step the output voltage is recorded and averaged to find the corresponding sensor output. Figure 9 shows the output voltage between ±10g's with 2g step size, and the linearity of the accelerometer in ±10g range is found as 0.45%. Tests also showed that the maximum operation range of the accelerometer is ±18.5g's. Table III summarizes the performance parameters of the system.

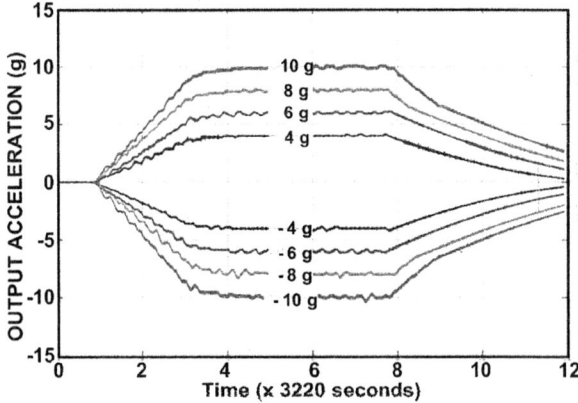

Figure 9: ±10g acceleration test results measured with 2g steps.

TABLE III
ACCELEROMETER SYSTEM PERFORMANCE PARAMETERS

Performance Parameter	Test Result
Non – Linearity in ±1g	<0.30 %
Non – Linearity in ±10g	<0.45 %
Bias Drift	74 μg
Noise	86 μg
Maximum Operation Range	±18.5 g

CONCLUSION

This paper reports a high-sensitivity, low-noise, closed-loop capacitive accelerometer system with micro-g resolution. Accelerometer, fabricated with DWP, has 306 fingers per side to achieve high sensitivity, and large proof mass to decrease the Brownian noise. The system operates as a 2nd-order electromechanical sigma delta modulator together with the interface electronics. The interface electronics can be operated either in open loop mode or in closed loop mode. The fully differential, switched capacitor circuit uses the Correlated Double Sampling technique to reduce the flicker noise and the offset. In addition to these, the circuit components are designed to have minimum level of noise. Hence, the readout circuit accomplishes more than 115dB dynamic range and can resolve less than 3aF/√Hz. The hybrid system has been tested in closed loop mode with digital Sinc demodulators and decimation circuitry with a 16 bit digital output at 815 Hz. The closed loop sigma-delta modulator consumes 16mW from a 0-5V supply. The overall system has a 0.3% non-linearity in ±1 g range, 86μg bias drift, 74μg/√Hz of noise level and maximum operation range of ±18.5g.

REFERENCES

[1] N. Yazdi, F. Ayazi and K. Najafi, "Micromachined Inertial Sensors," *IEEE Proc.* 86, 1998, pp. 1640-1659.

[2] B. V. Amini, R. Abdolvand, and F. Ayazi, "Sub-Micro-Gravity Capacitive SOI Microaccelerometers," *Tran. 2005*, pp. 515-518.

[3] J. Chae, H. Kulah and K. Najafi, "Hybrid Silicon-On-Glass (SOG) Lateral Micro-Accelerometer with CMOS Readout Circuitry," *MEMS 2002*, pp. 623-626.

[4] B. Amini, S. Pourkamali and F. Ayazi, "A High Resolution, Stictionless CMOS Compatible SOI Accelerometer with a Low Noise, Low Power 0.25μm CMOS Interface" *MEMS 2004* pp572-575.

[5] H. Kulah, J. Chae and K. Najafi, "Noise Analysis and Characterization of A Sigma-Delta Capacitive Silicon Microaccelerometer," *Transducers 2003*, pp. 95-98.

[6] X. Jiang, F. Wang, M. Kraft and B. E. Boser "An Integrated Surface Micromachined Capacitive Lateral Accelerometer with 2 μg/√Hz Resolution," *Solid State Sensor and Actuator Workshop, 2002*, pp. 202-205.

[7] M. A. Lemkin, M. A. Ortiz, N. Wongkomet, B. E. Boser and J. H. Smith "A 3-Axis Surface Micromachined ΣΔ Accelerometer," *ISSCC 1997*, pp. 202-203.

[8] R. Gallorini and N. Abouchi, "A Capacitance Meter based on an Oversampling Sigma-Delta Modulator and Its Application to Capacitive Sensor Interface," *ICECS 2001*, pp. 1537-1540.

[9] T. Kajita, U. Moon and G. C. Temes, "A Noise Shaping Accelerometer Interface Circuit for Two Chip Implementation," *ISCAS 2000*, pp. 337-340.

[10] S. R. Norsworty, R. Schreier, G. C. Temes "Delta-Sigma Data Converters, Theory, Design and Simulation," IEEE Press, 1996.

978-1-4244-1983-8/08/$25.00 ©2008 IEEE

A CMOS-90nm 15.6mW 20dBm-In-Band-IIP3 Analog Baseband Chain for UWB Receivers

A. Cito[1,4], M. De Matteis[1], S. D'Amico[1], A. Baschirotto[1], P. Delizia[1]

[1]Dep. of Innovation Engineering, University of Salento– Italy
[3]Dep. of Physics - University of Milano Bicocca, Italy
[4]Dep. Of Electric Engineering, University of Pavia, Italy

Ricardo Reis[2], Wen-Hu Zhao[2], Carlos Azeredo-Leme[2], A. Tavares[2],

[2]Chipidea Microelectronica – Portugal

Abstract – **In this paper the design of the analog baseband chain for UWB Receivers in a standard 90nm CMOS technology is presented. The baseband chain is composed by the 5th order Elliptic analog low-pass filter and a Programmable Gain Amplifier (with up to 60dB gain). Due to the large UWB signal bandwidth (250MHz), the design has been carried-out using a MATLAB model taking into account the capacitance parasitics effect and the opamp finite bandwidth impact on the filter transfer function. From a single 1.2V supply voltage, the total power consumption is 13mA per channel, the in-band and out-of-band IIP3 is 20dBm and the maximum IRN for G>15dB is 9nV/√Hz.**

I. Introduction

UWB systems have recently received great interest from the market and from the researchers. The UWB standard is employed in short-range applications in the so-called Wireless Personal Area Network (WPAN) systems. Typically two different approaches can be used. The first one is based on sub-nanosecond impulses and features relatively low data rate. The second one features an higher data rate by an efficient multiband division of the UWB available spectrum. Fig. I shows the architecture of an UWB Receiver under development to be used with the multiband OFDM alliance (MBOA) standard. It is based on subdivision of the large available bandwidth in sub-bands of 528MHz [1]. In this paper, the design of the analog baseband blocks for the UWB Receiver in a standard 90nm technology with a single 1.2V supply is proposed. In particular the complete design of the PGAs and of the Filter will be presented.

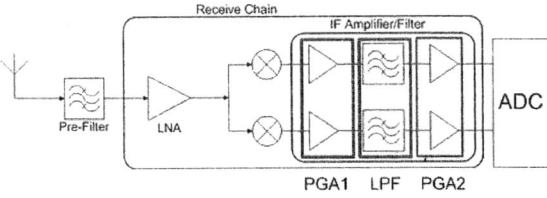

Fig. I – UWB TRANSCEIVER

The request of larger bandwidth for UWB application disagrees with the poorer performance of analog devices (larger parasitic capacitance and lower gain transistor) in scaled technology. In fact in the active RC cells, no-controlled complex poles can be generated when the parasitic capacitances of the opamps input stage and output stage are comparable with the Miller compensation capacitance. High quality factor poles in the PGA – that is the most critical block for the linearity – could increase the in-band distortion, when strong out-band interferers are present, worsening the linearity performance. Furthermore, due to the large bandwidth the opamp finite gain A_o and UGBW f_u, can not be neglected. In fact the maximum f_u is limited by the output stage parasitic capacitances, so that it is comparable with the PGA and filter f@-3dB. In this conditions the opamp finite frequency response has to be considered in the design, in order to comply with the IRN, transfer function accuracy and linearity requirements.

For these reasons the design of each block has been developed using a proper Matlab model. The Matlab model includes the parasitic cap C_p, but also the effects related with the opamp finite bandwidth and gain. The Matlab procedure gives as output results the resistance and capacitance values, the minimum requirements for the opamp bandwidth, dc gain and phase margin for each cell in order to comply with the IRN, transfer function accuracy requirements and to avoid linearity performance degradation.

II. Baseband chain UWB Specifications

The general specifications of the UWB Receiver baseband chain are reported in Table I.

TABLE I - UWB RECEIVERS BASEBAND CHAIN SPECIFICATIONS

Dc-gain (G)	0÷60dB
-3dB frequency (f_{-3dB})	264MHz
Attenuation @ 660MHz	-40dB
In-band IIP3 - (280MHz&300MHz)	20dBm
Out-bandIIP3 - (480MHz&810MHz)	20dBm
IRN (for G>15dB)	10nV/√Hz

The f_{3dB} frequency is 264MHz considering the 528MHz for each UWB sub-band. Second the interferers scenario represented in the

Fig. II, the linearity performance is dominated by the WLAN interferers (out-of-band) and unwanted UWB interferers (in-band). The Filter and PGA must achieve an attenuation at least of 40dB to reject the strong out-band WLAN interferer.

978-1-4244-1983-8/08/$25.00 ©2008 IEEE

The maximum required IRN is 10nV/√Hz. This spec is referred to the chain large gain condition. In fact when the gain is low the input signal power is higher, and an IRN increasing does not degrade the SNR. Considering the large linearity (IIP3=20dBm for both linearity test) the closed-loop topology appears the most feasible solution in order to comply with the linearity requirement. Furthermore the low IRN is stringent only for the first cell of the chain considering the required high gain (60dB).

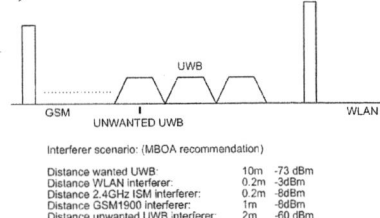

Fig. II –UWB Interferers Scenario

The functional scheme of the baseband chain is shown in Fig. III. It is composed by the Low Pass Filter (LPF) and by the Programmable Gain Amplifier (PGA). The PGA1 is embedded in the LPF that features a total 45dB gain. The PGA2 stage performs 15dB gain. Details about gain level, and transfer function parameters for each cell are given in Table II.

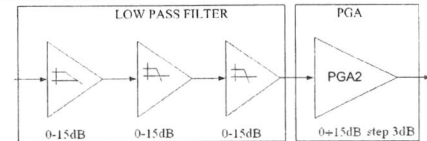

Fig. III – Baseband UWB Receiver Functional Scheme

TABLE II - FILTER & PGA TRANSFER FUNCTION PARAMETERS

	G	f_o	Q	Order
Cell1	0dB - 15dB	135MHz	-	1st
Cell2	0dB - 15dB	250MHz	1.3	2nd
Cell3	0dB - 15dB	250MHz	1.3	2nd
Filter	0dB - 45dB	255MHz	-	5th
	G	Max Attenuation @ 264MHz	-	-
PGA	0dB - 15dB	≤ 0.5dB	-	-

To relax the opamp f_u requirements the overall cut-off frequency is fixed at 255MHz. Furthermore the PGA does not feature signal drop at 255MHz, so that the in-band signal at the PGA input is not significant attenuated.

III. LPF and PGA Design

The filter used in the UWB receiver baseband chain is a 5th order low pass filter composed by the cascade of three active cells. The schematic of each cell is shown in the Fig. IV. The *Cell1* is a 1st-order cell. The *Cell2* and *Cell3* are 2nd-order Active-Gm-RC cells [2]. The PGA schematic is presented in the Fig. V.

The design of the above closed-loop configurations operating in the UWB Receiver frequency range is critical for the presence of two basic no-idealities:

- the opamp finite unity-gain bandwidth (f_u)

- the input stage opamp parasitic capacitance (C_p).

In *Cell1* the opamp bandwidth f_u is comparable with the maximum input signal frequency, 264MHz, so the in-band loop gain decreases, the linearity performance and the transfer function accuracy can be affected. The opamp input stage MOS parasitic cap is not negligible for the large in-band frequency (264MHz) and the 90nm CMOS scaled technology where the oxide thickness is reduced. Designing a large bandwidth opamp requires large input differential pair transistors, and this results in a large input parasitic cap. Thus the two non-idealities are in trade-off. In the following part of this section the Matlab design procedure for each cell and the results are presented.

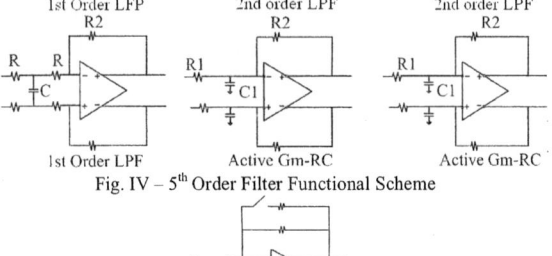

Fig. IV – 5th Order Filter Functional Scheme

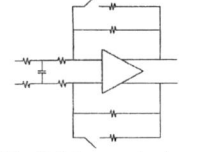

Fig. V – Cell 1 general schematic

A. Cell1

The Cell1 gain (0dB and 15dB) is selected using the feedback resistance, Fig. V. The pole is synthesized using the input RC net and its frequency does not change with the gain. In the design of Cell1, the passive components values and the opamp characteristics, in terms of f_u, phase margin and minimum dc gain, are defined by means of a MATLAB model. This model considers the parasitic capacitance due to the opamp input stage transistors MOS and the finite bandwidth of the opamp.

The effect of the parasitic capacitance is not so dominant in the Cell1, because the pole is at 135MHz i.e. at lower frequency respect to the pole frequency due to the C_p capacitance. The most dominant no-ideality in this cell are the opamp finite bandwidth and dc gain. A low opamp dc gain can critically affect the Cell1 gain. To maintain the maximum gain drop below the 3% of the nominal value an opamp dc gain higher than 45dB is necessary. Furthermore the cell transfer function is sensitive to the opamp finite bandwidth because at high frequency the loop gain decreases. The basic steps of the Matlab procedure are the following.

- The system level specifications of Cell1 are fixed: IRN, G, f_o.
- The opamp minimum dc gain and phase margin are defined. In this design Ao≥45dB, PM≥50°.
- The maximum parasitic cap is fixed around 150fF. That means to have an aspect ratio for the opamp input pair MOS around 50μ/300n.

978-1-4244-1983-8/08/$25.00 ©2008 IEEE   46

- The opamp f_u changes between 1GHz and 3GHz and the cut-off frequency deviation with respect to the nominal value is calculated for each opamp f_u.
- The minimum f_u is selected in order to maintain the maximum cut-off frequency deviation below the 10%.

In Table III the results of the MATLAB procedure are reported.

TABLE III - MATLAB MODEL CELL1 PARAMETERS

Opamp f_u	Opamp Cc	C1	R1	R2
2.9GHz	200fF	3.3pF	1.4kΩ	7.9kΩ

B. Cell2 & Cell3 Design

The basic formulas of Active Gm-RC biquadratic cell are the following.

$$(1) \quad G = \frac{R2}{R1}, \quad \omega_o = \sqrt{\frac{\omega_u}{C1 \cdot R2}}, \quad Q = \frac{1}{1+G} \sqrt{\omega_u C1 \cdot R2}$$

The Cell2 gain (0dB and 15dB) is selected using the feedback resistance, but in order to align the biquad cell cut-off frequency a readjusting of the f_u and C1 is needed. The MATLAB model for this cell considers the effect of the parasitic capacitance on the filter transfer function. In fact at high frequency the C1 can be comparable with C_p. The C1 error generates a deviation on the biquadratic cell Q and f_o. The algorithm used for the MATLAB procedure is the following:

- The basic specs of Cell2 are fixed: IRN, G, Q, f_o.
- The opamp minimum dc gain and phase margin are defined. In this design Ao≥45dB, PM≥50°. The opamp f_u is obtained by the biquadratic cell specs.
- The C_p variation affects the frequency response as shown in Fig. VI. The C_p value is selected in order to maintain the maximum f_o and Q deviation below a target value (10% in this design). In the Fig. VII the f_o percent error vs. C_p is plotted.

The procedure output results are the R C values and the opamp characteristics, as indicated in the Table IV.

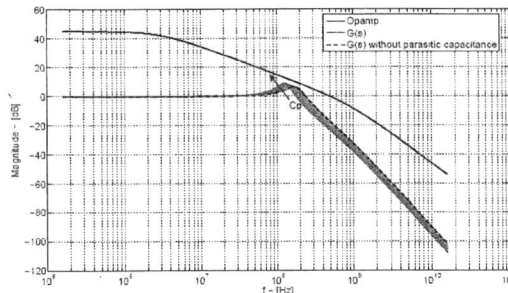

Fig. VI – Cell2 frequency response vs. C_p

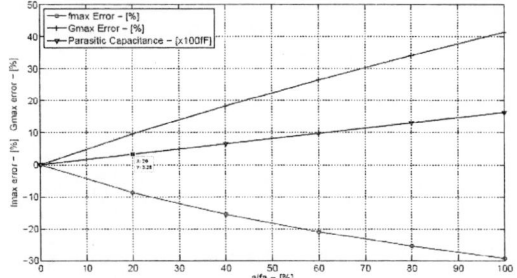

Fig. VII – Cell2 Gmax and fmax deviation vs. C1 error

TABLE IV - MATLAB MODEL CELL2 CELL3 PARAMETERS

	Cell1	Cell2 & Cell3	
Gain		0dB	15dB
Opamp f_u	2.9GHz	650MHz	2.15GHz
Opamp Cc	200fF	1.5pF	200fF
C1	3.3pF	0.6pF	0.4pF
R1	1.4kΩ	2.09kΩ	2.09kΩ
R2	7.9kΩ	2.09kΩ	11.7kΩ

C. PGA design

The PGA schematic is represented in the Fig. VIII. In this structure, the signal current does not flow through the non-linear resistance of the switches, thus avoiding linearity degradation. The equivalent circuit of the PGA used in the MATLAB model is shown in Fig. XI..

The opamp input stage parasitic capacitance introduces one pole whose frequency is comparable with the opamp f_u. This can generate complex poles with high quality factor. In order to obtain a flat frequency response in the PGA the most common solution is to place a feedback capacitance in parallel with the R2 resistance. But considering the strategy used for the programmable gain, the R2 changes with the PGA gain and the feedback cap should be readjusted in order to maintain constant the PGA pole. That increases the switches in the schematic and in general the complexity of the PGA.

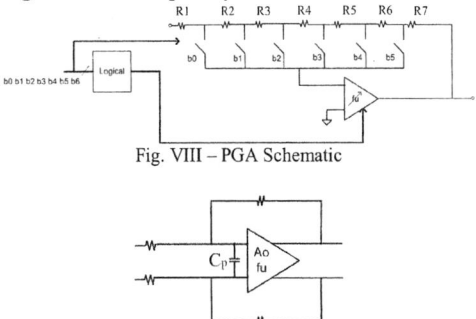

Fig. VIII – PGA Schematic

Fig. IX – PGA equivalent circuit

The solution adopted in this design optimizes the trade-off between the opamp unity gain frequency and the maximum C_p parasitic cap in order to synthesize the flat frequency response. So no additional switches are added. The Matlab algorithm used for the PGA is run for each gain – 0÷15dB with 3dB per step – and produces the opamp f_u for each gain in order to maintain the flat frequency response. In the MATLAB procedure the maximum C_p is fixed around 150fF. The output results of the algorithm are reported in Table V.

TABLE V - MATLAB MODEL PGA PARAMETERS

PGA Gain	Opamp fu	Cc	R1	R2
0dB	500MHz	1.7pF	3.1552 kΩ	3.1552 kΩ
3dB			3.6947 kΩ	2.6157 kΩ
6dB	800MHz	1.3pF	4.2036 kΩ	2.1068 kΩ
9dB			4.6578 kΩ	1.6526 kΩ
12dB	1GHz	500fF	5.0435 kΩ	1.2669 kΩ
15dB			5.3576 kΩ	0.9527 kΩ

IV . Simulation Results

Starting from the results produced by the Matlab models the design of the baseband chain has been developed using a 90nm CMOS technology. The opamp used for each cell of the Filter and for the PGA is shown in Fig. X. A large opamp gain can be achieved by adopting two stage structures.

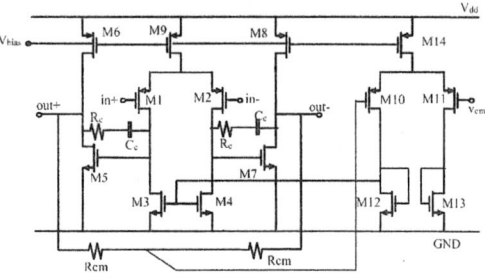

Fig. X – Class A Opamp Schematic

The simulation results for different gains are reported in Fig. XI and Fig. XII, in terms of frequency response and IRN respectively. The linearity performance of the filter and PGA chain are evaluated in terms of IIP3. In Fig. XIII the IIP3 is calculated for the unwanted UWB signal, where the input signal is a dual tone signal at 280MHz and 300MHz frequency. The Filter Gain is 0dB. In the Fig. XIV the IIP3 is calculated for the WLAN interferes case, when the input signal is a dual tone signal at 480MHz and 810MHz. In this case the gain is 60dB.

V. Conclusion

A complete design of the analog baseband part of the UWB receiver has been described in this paper. The design has been developed using a MATLAB model for each cell of the filter and PGA. The simulation results are reported in the Table VI. The total current consumption is 13mA per channel, the maximum IRN for chain gain of 60dB is lower then 10nV/√Hz and the IIP3 is higher than 20dBm for both interferers, WLAN and unwanted UWB. The technology used for the design is the CMOS 90nm.

TABLE VI – UWB BASEBAND CHAIN PERFORMANCE SUMMARY

Parameter	Simulation Results
Technology	90nmCMOS
VDD	1.2 V
Gmax	58.2dB
f_0	255MHz
Attenuation@660MHz	-60dB
Group Delay	2.5ns

IRN$_{max}$ for G>20dB	9.6÷30 nV√Hz
Out-of-band IIP3 - (480MHz&810MHz)	29dBm
In-band IIP3 - (280MHz&300MHz)	22dBm
I$_{tot}$	13mA
Output Load	5kΩ

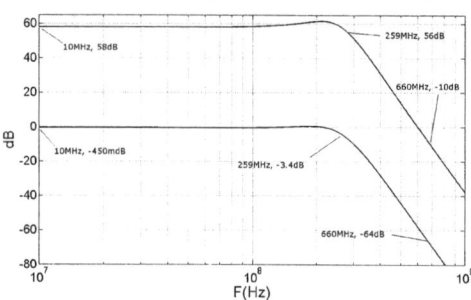

Fig. XI – Simulation Results – Frequency Rsponse

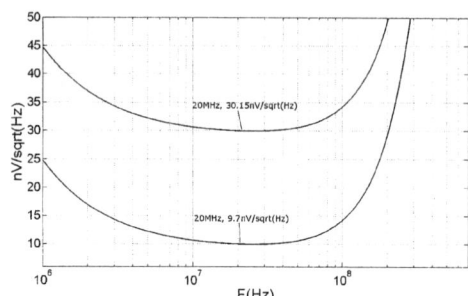

Fig. XII - Simulation Results – IRN

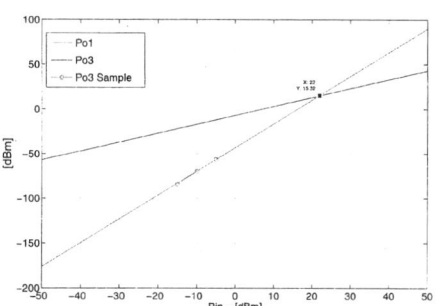

Fig. XIII – IIP3 for the Unwanted UWB signal

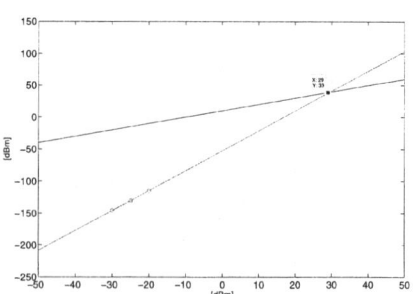

Fig. XIV – IIP3 for the WLAN interferers

Reference

[1] "Multi-band OFDM Physical Layer Proposal for IEEE 802.15" Task Group 3a", IEEE P802.15 Working Group for WPAN, March 2004.

[2] S. D'Amicoet all, "A 4th-order active-Gm-RC reconfigurable (UMTS/WLAN) filter", IEEE JSSC July 2006.

Tunable Linear Current Mirror

Umut Guvenc

UEKAE – TUBITAK
Kocaeli, Turkey
Institute of Science and Technology
Istanbul Technical University
Istanbul, Turkey
umutguvenc@uekae.tubitak.gov.tr

Ali Zeki

Faculty of Electrical and Electronic Engineering
Istanbul Technical University
Istanbul, Turkey
zekia@itu.edu.tr

Abstract—**A new tunable linear current mirror topology using a BJT differential pair as a current divider is proposed which is to be an alternative to tunable devices used in analog electronics. The structure has the advantage of simplicity, wide tuning range of five decades and a linear operating region of two decades.**

I. INTRODUCTION

A new tunable linear current mirror topology is proposed as an alternative to the tunable devices which are required in analog electronic systems, especially in filtering applications for which high performances cannot be obtained by using conventional techniques. The new structure is promising wider linear operation and wider tuning ranges compared to its counterparts [1-3]. It is thought that the proposed structure would find a very wide application areas since current mirrors are one of the main building blocks of integrated active components. The proposed topology has a simple structure consisting of a BJT differential pair and a single MOSFET satisfying the biasing current. Due to electrical characteristic of the BJT differential pair, the circuit can satisfy a wide linear operation range of two decades with a THD of 1 percent and the gain of the current mirror can be tuned within a wide range of five decades.

II. STRUCTURE

The circuit schematic of the proposed current mirror is shown in Fig. 1. It contains only three transistors, two BJTs of differential pair and a MOSFET as the biasing transistor. The circuit has two voltage inputs, one is the control voltage (VC) input and the other one is the reference bias voltage (VB) input. By applying a voltage difference between the V_{bias} and V_{tune} inputs, the gain of the circuit can be tuned. The aim of using BJTs in differential pair instead of MOSFETs is to use the voltage-current relation of BJTs given in (1) to satisfy constant ratio between input and output currents of the circuit when a constant input voltage is applied. When the voltage-current characteristic equation of the BJTs is evaluated, the gain of the current mirror is obtained as in (2) [4-6]. Since the constancy of the current

gain upon current level change is satisfied by the exponential voltage-current relation of the BJTs, these transistors cannot be replaced by MOSFETs.

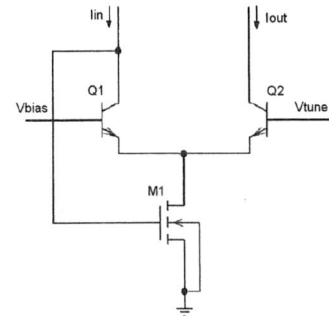

Figure 1.   Schematic of new current mirror structure.

$$I_C = I_S e^{\frac{V_{BE}}{V_T}} \tag{1}$$

$$\frac{I_{out}}{I_{in}} = \frac{I_S e^{\frac{V_{tune}-V_E}{V_T}}}{I_S e^{\frac{V_{bias}-V_E}{V_T}}} = e^{\frac{V_{tune}-V_{bias}}{V_T}} \tag{2}$$

$$I_{MOS} = I_{in} + I_{out} = I_{in}\left(1 + e^{\frac{V_{tune}-V_{bias}}{V_T}}\right) \tag{3}$$

The MOSFET supplying the tail current of the differential pair is used in a negative feedback configuration that is satisfying the biasing voltage of this transistor to supply the desired total input and output currents. Any increase (decrease) in the input current would result in the increase (decrease) in the input voltage level which is the V_{GS} voltage of the NMOS transistor. Thus, the tail current of the differential pair increases (decreases) which is the sum of input and output currents. Since the ratio between the input and output currents is adjusted by the difference of V_{tune} and

V_{bias} voltages as given in (2), the current flowing through the MOSFET is determined by input current as in (3).

V_{bias} and the size of the MOSFET should be chosen such that, the BJT Q_1 remains in forward active region and the MOSFET remains in pinch-off region, while the total current ($I_{in} + I_{out}$) takes its minimum and maximum possible values. Therefore, V_{bias} should be selected as shown in (4), (5) and (6). Although it seems that there is a very narrow range in which the V_{bias} can be selected, the limit at which the MOSFET transistor enters the triode region can be relaxed since the circuit still works when the MOSFET is in triode region. However, the loop gain becomes smaller in this situation. Selecting of the V_{bias} voltage can also be relaxed by using a MOSFET in source follower configuration as a lever shifter in the V_{GS} feedback of the MOSFET supplying the tail current.

$$V_{THn} > V_{GS} - V_{bias} - V_{BE} > V_{CEsat} \qquad (4)$$

$$V_{GS\min} - V_{CEsat} + V_{BE} > V_{bias} \qquad (5)$$

$$V_{bias} > V_{GS\max} - V_{THn} + V_{BE} \qquad (6)$$

In the ideal case, the gain of the circuit as given in (2) depends only on V_{bias}, V_{tune} and V_T voltages and does not change with the level of the current. However, it is seen that when the early effects of the BJTs are added to the gain function as in (7) and (8) [4], nonlinearity occurs since the voltage drop between collector and base terminals of BJTs change with the level of the current steering. Therefore, the higher the early voltage V_A is, the lower the dependency of the gain to the current level thus the nonlinearity. In (7) and (8), V_{CBout} and V_{CBin} are collector-base voltages of the BJTs on the output and input branches respectively.

$$\frac{I_{out}}{I_{in}} = \frac{I_S e^{\frac{V_{tune}-V_E}{V_T}} \left(1 + \frac{V_{CBout}}{V_A}\right)}{I_S e^{\frac{V_{bias}-V_E}{V_T}} \left(1 + \frac{V_{CBin}}{V_A}\right)} \qquad (7)$$

$$\frac{I_{out}}{I_{in}} = e^{\frac{V_{tune}-V_{bias}}{V_T}} \left(\frac{V_A + V_{CBout}}{V_A + V_{CBin}}\right) \qquad (8)$$

III. SIMULATIONS

In this section, simulation results concerning the performance properties of the proposed current mirror are given. Although the circuit has BJTs and should be used with BiCMOS technologies for better accuracy, simulations were run with a standard CMOS technology with parasitic BJTs to show the applicability of the circuit with CMOS technologies. To run the simulations, PNP version of the circuit is constructed by using the lateral parasitic PNP transistors of the 0.35μm CMOS process of AMS.

According to the DC simulation result shown in Fig. 2, the gain of the circuit can be controlled over a wide range of five decades, from 0.003 to 300 within a linear operation region of two decades. The linear operation region of the circuit varies according to the input current when the gain of the circuit is higher than unity.

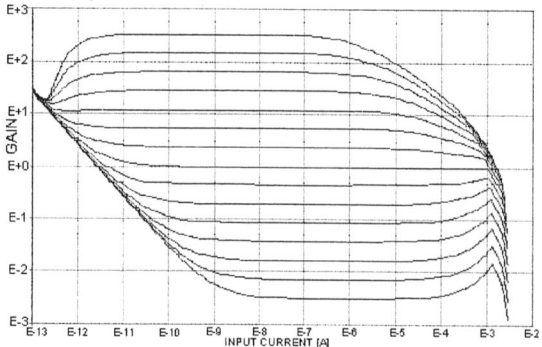

Figure 2. Gain of the tunable current mirror according to the input current for different V_{tune} values.

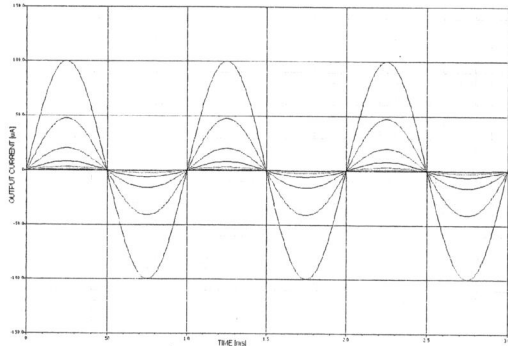

(a)

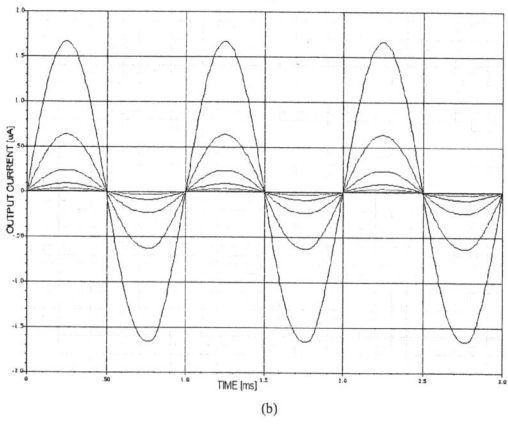

(b)

Figure 3. Transient simulation for gains (a) smaller than unity and (b) greater than unity.

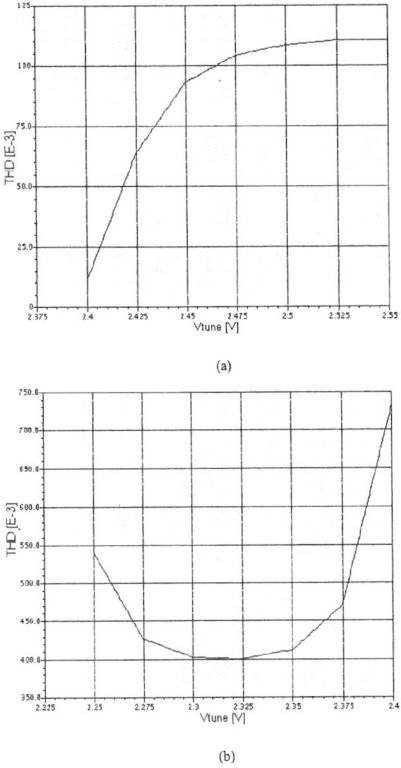

(a)

(b)

Figure 4. THD for gains (a) smaller than unity and (b) greater than unity.

To show the linearity of the circuit, transient simulations were run with a sinusoidal input source swinging between 200pA and 20nA for the gains smaller than unity and with another sinusoidal source swinging between 20nA and 200μA for the gains greater than unity. The resultant graphics were given in Fig. 3a and Fig. 3b for gain smaller than unity and greater than unity respectively. THD values of the circuit has also been extracted from the transient simulations and given in Fig. 4a and Fig. 4b. These results show that the circuit can operate with a very low THD within a region greater than two decades whose boundaries vary due to high gain of the circuit. For the gains closer to unity the circuit can operate linearly with very low THD within even more than three decades of input current. The THD values given in Fig.4 are obtained from different simulations in which the applied input signal swing varies according to the gain. The frequency of the sinusoidal input current source is 1 kHz and the biasing voltage V_{bias} used in these simulations is 2.4 V. The V_{tune} voltage is swept from 2.25 V to 2.55 V.

IV. CONCLUSION

A new tunable linear current mirror topology has been introduced as an alternative to commonly used tunable devices and its simulation results were given. The tunable current mirror structure provides a wide tuning range and a wide linear operation region by using the linear DC transition characteristic of BJT differential pair. According to the simulation results, the circuit can operate linearly for two decades of input current with a maximum THD of 0.73 and the gain of the circuit can be tuned from 0.003 to 300.

REFERENCES

[1] Tsividis, Y.P., "Integrated continuous-time filter design-an overview", IEEE Journal of Solid-State Circuits, 29, No. 3, 166-175, 1994.

[2] Palmisano, G. and Pennisi, S., "New tunable transconductor for filtering applications", ISCAS, 1, 196-199, 2001.

[3] Voo, T. and Toumazou, C., "Efficient tunable continuous-time integrated current-mode filter designs", ISCAS, 1, 93-96, 1996.

[4] Leblebici, D., Analog Elektronik Devreleri, I.T.U Press, Istanbul, 2001.

[5] Kuntman, H.H., Analog Tümdevre Tasarımı, Birsen Press, Istanbul, 1998.

[6] Kuntman, H.H., Analog MOS Tümdevre Tekniği, I.T.U Press, Istanbul, 1997.

978-1-4244-1983-8/08/$25.00 ©2008 IEEE

BLANK PAGE

Wide Bandwidth Impedance Meter using Low Rate Random Sampling

Giuseppe Fiscelli, Carlo Cucchiara, Antonio Di Stefano, Costantino G. Giaconia

Dipartimento di Ingegneria Elettrica, Elettronica e delle Telecomunicazioni, Università degli Studi di Palermo

Viale delle Scienze, Block 9, 90128

Palermo, Italy

giuseppe.fiscelli@dieet.unipa.it, antonio.distefano@dieet.unipa.it, costantino.giaconia@unipa.it

Abstract— **A novel impedance measurement method based on random sampling of voltage and current signals is proposed. This technique dramatically reduces the sampling frequency requirements, thus circumventing the limitations imposed by maximum speed of the analog to digital converter and the signal processing unit. The lowering of the sampling frequencies allows the design and the implementation of an almost all digital architecture by using a simple microprocessor based embedded system and a digital frequency synthesizer. The basic principles are presented, and the implemented algorithms are described. Experimental results show the instrument performances compared to others commercial alternatives.**

I. INTRODUCTION

Wide frequency range impedance measurement is useful in several application including non-invasive blood impedance measurements [1], electro-impedance spectroscopy [2] and electronic device characterization as well as wide band speaker or transducers impedance measurements for audio or ultrasound applications [3].

These kind of measures require systems capable of a wide frequency range span, typically from a few Hz up to 10MHz or more, maintaining a certified degree of accuracy.

A number of methods have been described in literature, such as: bridge methods, resonant methods, network analysis methods and auto balancing bridge methods [4]. This last is one of the most efficient and precise when applied to wide band measurements. One of the key advantage of this method is - above all - its suitability for full automation by means of a digital system. In fact it only needs the measure of few electrical parameters, while the others may require more complex operations such as a change of the circuit topology and/or components in order to perform the measures. Even though, if high frequency operation are desired, a particular care must be posed on the choice of some critical device, mainly the Analog to Digital Converters (ADCs) and processors. Both in fact must be capable of handling

(generating or processing) data with the required throughput, this in turn may rise the implementation complexity and costs. This paper presents a novel method for wide band impedance measurements employing the auto-balancing method together with a random sampling technique. This allows dramatic simplification of the circuit implementation and measure procedure. In fact only a low cost processor, low speed ADCs (far below the maximum employed frequency) and no filters are required. By using this method an instrument has been realized, capable of measuring impedance ranging from 10Ω up to 100 $K\Omega$ in a frequency range spanning from 10Hz to 10MHz, and maintaining an accuracy better than 1%.

The paper is organized as follows: in section II the auto-balancing bridge method is described, in section III the random sampling technique is analysed, while in section IV the implemented instrument is illustrated and some experimental results are also showed. Then some conclusion follows in section V.

II. AUTO-BALANCING BRIDGE METHOD

This method is based on the voltage and current measurement made across the impedance under test. As shown in Figure 1a a sinusoidal generator working at the desired frequency is used to supply the device under test (DUT), while the current to voltage converter, implemented by the operational amplifier (OA) and the resistor (R), allows to determine the current passing through it (that is proportional to the voltage in point O). Since the OA negative input terminal behaves like a virtual ground, the voltage in point H is equal to the voltage across the DUT. In order to evaluate the DUT impedance (Z) it is then sufficient to measure the voltage in points H (V_H) and O (V_O) and using the following expression:

$$V_H / Z(\omega) = - (V_O/R) \qquad (1)$$

Another circuit is showed in figure 1b where the DUT is connected to ground.

978-1-4244-1983-8/08/$25.00 ©2008 IEEE

In this case the relationship between the voltage and the current is:

$$V_H / Z(\omega) = (V_H - V_O)/R \qquad (2)$$

As mentioned above V_H and V_O are sinusoidal voltage with frequency equal to the oscillator one. By estimating the amplitude and relative phase of these signals it is possible to evaluate the impedance magnitude and phase components (i.e. the real and imaginary part).

Even if this method is quite straightforward it requires considerable care in measuring the V_H and V_O voltage. As already said these signals may cover a wide range of frequencies, so the system have to handle them properly. In particular if these signals have to be processed by a digital system they have to be sampled with an adequate sampling frequency according to the Nyquist theorem [5]. This may result in a very high sampling frequency that in turn requires an high speed processing capability and memory resources. Moreover scaling the sampling frequency according to the generator makes difficult to design proper dynamic anti-alias filters (either analog or digital).

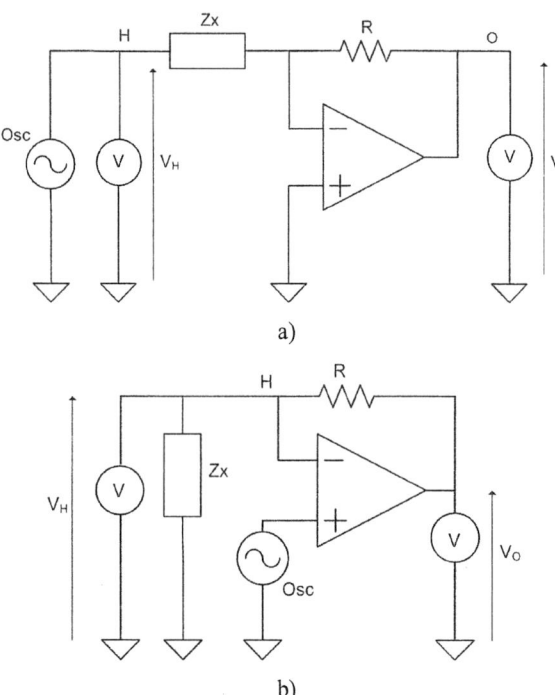

Figure 1. Auto-Balancing Bridge Schematics
a) with virtual ground and floating impedance
b) with grounded impedance

All these factors make less convenient this method for low cost, portable and low power application. These situation can be easily solved with the proposed method employing a random sampling scheme as discussed in the following paragraph.

III. RANDOM SAMPLING

The sinusoidal nature of V_H and V_O allows their full characterization starting from the knowledge of their amplitudes and phases. If a traditional uniform sampling technique is used, these features can be determined either by simultaneously acquiring a number of period of both signals and performing an FFT; either by sampling them in just one period, if the sampling rate is much higher than the frequency of the signals. A third option is to evaluate only the amplitude of the signals and indirectly obtain the phase. Supposing to employ two ADCs to sample both sinusoidal signals synchronously ($V_H(t)$ and $V_O(t)$ in the follows), we can obtain their amplitude by evaluating the maximum value (V_{HMAX} and V_{OMAX}) in each samples set. The phase can be derived by using the following expression:

$$V_O^* = V_{OMAX} \cdot \cos(\theta) \qquad (3)$$

$$\theta = arcos(V_O^* / V_{OMAX}) \qquad (4)$$

where V_O^* is the value of $V_O(t)$ sampled together with V_{HMAX}, and θ is the relative phase between V_O^* and V_{HMAX}, thus the one introduced by DUT.

It can be noted that in order to evaluate these quantities it is not required a uniform sampling, but it is sufficient only to obtain the maximum value (i.e. V_{OMAX}) and the corresponding other value (V_O^*). This can be done by sampling both signals at random intervals and evaluating the maximum value in the sample set. The acquisition of instantaneous values, by means of ADCs with adequate bandwidth, allows to search for the maximum value without fulfilling the Nyquist criterion as explained hereafter (since the signals are not to be reconstructed), thus the mean sampling periods can be much greater than the signal period.

This sampling scheme presents a number of advantages over the uniform sampling, the most important being the chance to use a very low average sampling rate (in the order of $1/10000^{th}$ of measure frequency). Moreover no anti-alias filters are required and there is no need to store samples (maximum points in fact can be evaluated in real time).

In order to establish the minimum number of sample to acquire for detecting the maximum values with a good probability, a study has been carried out by using a statistical model and the consequent numerical simulations. A number of signals at different frequencies have been considered, and for each of them some samples have been picked up at random intervals T(t), with:

$$T(t) = T_S + \delta(t) \qquad (5)$$

where T_S is a constant period and $\delta(t)$ is a random offset evenly distributed in the interval $[0, T_S]$. Simulations allowed to evaluate the minimum required number of samples to be taken into account in order to surely acquire a maximum with an error bound of 0.01%.

This can be obtained by evaluating the number of random trials needed to extract the first maximum. This study ended up to the so called probability density function (PDF) whose behaviour is show in Figure 2.

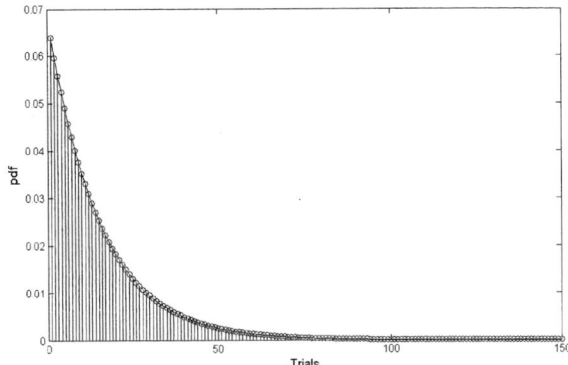

Figure 2. Probability Density Function

Surprisingly, this PDF has a negative exponential behaviour. By applying the least minimum square method this result has been fitted with the following exponential function:

$$p(n) = K \cdot e^{-N \cdot \lambda} \qquad (6)$$

where K and λ can be retrieved from the data and are, respectively 0.0683 e 0.0659. By integrating this expression the following probability function, $P(n)$, can be obtained:

$$P(n) = 1 - e^{-N \cdot \lambda} \qquad (7)$$

that allows to evaluate the probability of reading a maximum after N trials, as showed by the Figure 3.

The probability of reading a maximum is beyond 99% when the number of trials reaches 80 samples. In order to achieve a better accuracy, the implemented system acquires 1000 samples so peeking up at least 10 maximum values, then the actual maximum is computed as the average of the read values.

IV. IMPLEMENTED INSTRUMENT

Starting from the above consideration, an embedded system has been designed and implemented for testing the effectiveness of the proposed method. It is based on a low cost 8bit Zilog Z8!Encore microcontroller [6], two Texas Instruments ADCs (ADS7886) [7] and an integrated direct digital synthesizer (DDS) from Analog Devices (AD9832) [8] capable of generating signals up to 20MHz with 0.1Hz resolution and a low total harmonic distortion.

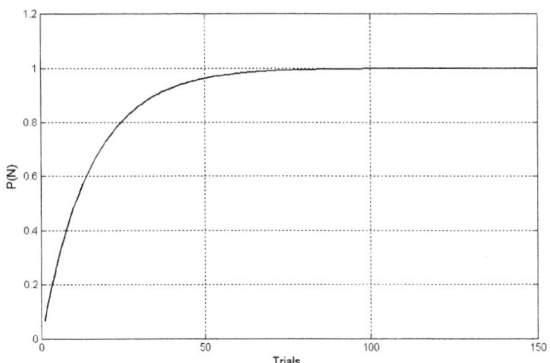

Figure 3. Probability of reading a maximum after N trials.

The chosen ADCs are 12 bit successive approximation converters sampling up to 1MSPS, with a full power bandwidth of 15MHz. These components are connected to the system according to the functional layout shown in Figure 4.

The DDS is controlled by the Z8 so to generate a sinusoidal signal at the desired frequency which feeds the impedance through a variable gain amplifier used to fully exploit the ADCs dynamic range. The two ADCs are controlled in order to perform the simultaneous sampling of the voltage and current across the impedance. The average sampling frequency was about 32KHz. Acquired samples are directly processed in order to find the maxima without the need of storing them, as already explained.

The sampled values are processed by using a 40bit fixed point arithmetic in order to calculate the impedance value. The 40bit data depth are a good trade-off between the precision needed for the mathematical manipulations of the sampled data and the performance of the overall digital system in terms of speed. A number of routine in the system firmware automates the most common measurement procedures. The results are finally visualized in an alphanumeric display and the total time used to perform a measurement is below 0.1 sec.

V. EXPERIMENTAL RESULTS

A number of tests have been performed in order to characterize the system performances. In particular the system has been tested for inductance measures. By measuring many known inductances, the perceptual error has been determined at various frequencies. These results have been compared with the one of an high end system available in the market (Agilent AD4294A) [9]. These data have been plotted in an impedance versus frequencies graph, as showed in Figure 5.

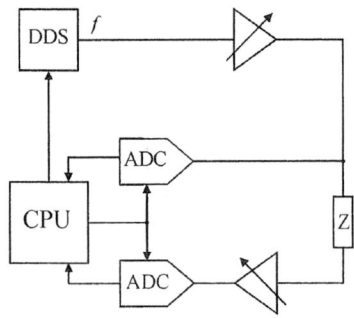

Figure 4. System schematic.

In this graph inductance lies along diagonal lines. Region delimited by dotted lines includes points where the Agilent instrument achieves an error equal or lower than the indicated value. The dashed line instead shows the theoretical area where the precision of our system stays within an error of 0.6% while the bold line bounds the region in which the real measurement achieves the above mentioned error.

The authors consider these results are quite remarkable, especially if compared with the low cost of the proposed system. In fact the principal on witch the measurement method lies are independent from its implementation; hence the use of more sophisticated devices such as a 32 bit microprocessor, ADCs with a better precision and DDS with a larger bandwidth can be improve the overall performance, while causing a very low rise of the total cost of the system.

VI. CONCLUSIONS

A simple impedance measurement system has been designed and implemented. Its performance are encouraging especially because the used method, based on a random and low rate sampling, seems to be not affected by the scaling of the measure frequency, that could be very high, making this method even suitable for impedance measures in RF field. The remarkable sampling frequencies lowering allows the implementation of simple, microprocessor based, portable system characterized by low power consumptions.

ACKNOWLEDGMENT

The Authors wish to acknowledge Stefano Mangione, from the TLC group of the same Department, for the helpful discussions regarding the statistical behaviour of the sampled data.

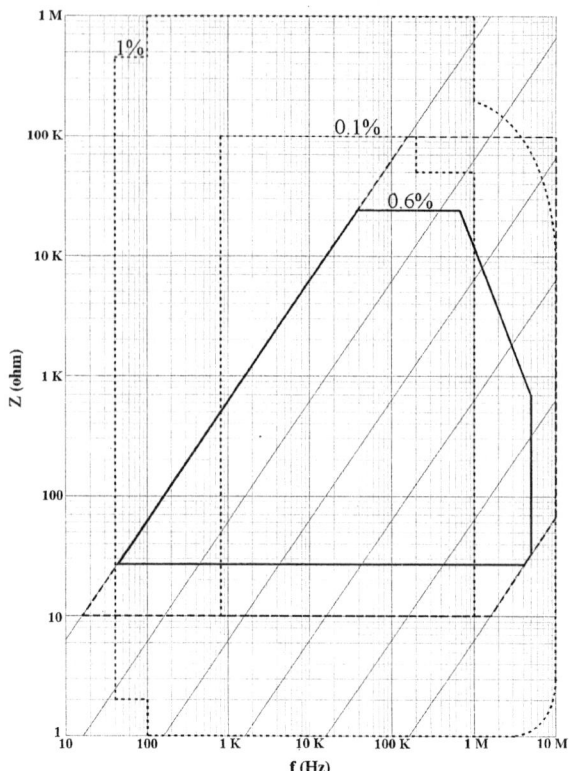

Figure 5. A Comparison Experimental Results.

REFERENCES

[1] T. Dai and A Adler, "Blood Impedance Characterization from Pulsatile Measurements", Canadian Conference on Electrical and Computer Engineering, May 2006.

[2] C. E. F. Amaral and B. Wolf, "Effects of Glucose in Blood and Skin Impedance Spectroscopy", AFRICON 2007, October 2007.

[3] S. Brennan, "Measuring a Loudspeaker Impedance Profile Using the AD5933", Analog Devices, AN-843.

[4] C. C. Huang, F. Wong, D. Kim, "Different electrical measuring techniques of package and interconnect parasitics for high speed VLSI devices", Northcon94 Conference Record, October 1994.

[5] A. V. Oppenheim and R. W. Schafer, "Discrete Time Signal Processing", Prentice-Hall, pp. 447-448, 1989.

[6] Z8 Encore Microcontroller with Flash Memory and 10 bit A/D Converter, Product Specification PS0176, www.zilog.com.

[7] ADS7889, 12 Bit, 1 Mbps, MicroPower, Miniature SAR Analog to Digital Converter, Data Sheet Texas Instruments, www.ti.com.

[8] AD9832, CMOS Complete DDS, Data Sheet Analog Devices, www.analog.com.

[9] Agilent 4284A Precision LCR Meter, Data Sheet Agilent, www.agilent.com

Robust Digital Image Watermarking Based on Normalization and Complex Wavelet Transform

Cabir Vural[1]
[1]Electrical-Electronics Engineering, Sakarya University, 54187 Esentepe, Sakarya, Turkey.
cvural@sakarya.edu.tr,

Serap Kazan[2]
[2]Computer Engineering, Sakarya University, 54187 Esentepe, Sakarya, Turkey.
scakar@sakarya.edu.tr

Abstract-In this study, a new digital image watermarking algorithm based on moment-based image normalization and the two dimensional dual tree complex wavelet transform (2D DT-CWT) was developed. Normalization provides robustness against geometrical distortions, whereas 2D DT-CWT increases robustness for attacks such as additive noise, linear and nonlinear filtering, JPEG compression. It was accomplished that added watermark achieves both transparency and robustness requirements by taking the properties of the human visual system account.

I. INTRODUCTION

The need for digital multimedia copyright protection and copy protection emerged in the last decade as a result of extensive use and dissemination of digital multimedia data. Encryption and digital watermarking algorithms have been developed in order to fulfill this need. Digital data is protected during transmission from the sender to the receiver in case of encryption. Once the data is decrypted, protection is no longer available. In case of watermarking, on the other hand, a hidden signal called *watermark* that is always present and provides protection is added to the digital multimedia data. Digital watermarking has numerous applications such as copyright protection, fingerprinting, copy protection, broadcast monitoring, data authentication, indexing, and data hiding [1].

Many digital image watermarking algorithms are sensitive to image manipulations such as geometric distortions, compression, linear and non-linear filtering. In other words, if a watermarked image undergoes a manipulation, watermark can not be usually detected reliably. Various digital image watermarking algorithms have been proposed in the literature to solve this robustness problem. Ruanaidh and Pun [2] proposed to use Fourier transform and Log Polar Mapping (LPM) together. Fourier transform provides robustness against translation whereas LPM provides invariance for rotation and scaling. The method does not consider compression and filtering artifacts. Furthermore, LPM and inverse LPM are lossy operations that cause degradations in the watermark image.

The method based on Zernike moments [3] provides good performance in terms of noise robustness and data fidelity. However, reconstructing a digital image from its Zernike moments, which is a lossy procedure, is computationally prohibitive. Two Dimensional (2-D) Discrete Wavelet Transform (DWT) based method [4] takes the properties of Human Visual System (HVS) into account and achieves good performance against image compression (including JPEG and JPEG 2000), median filtering, cropping. However, it is not resistant to geometrical distortions such as rotation, scale and translation.

The fact that moment based image normalization is robust to geometrical distortions was demonstrated in [5-6]. In this study, a new digital image watermarking algorithm is developed by using moment based image normalization and 2D CWT. It is possible to add a watermark to a digital image in spatial domain or in a suitable frequency domain. That frequency domain methods are superior to spatial domain methods was shown in numerous studies. Among the frequency domain methods, 2D Discrete Wavelet Transform (2D-DWT) based methods usually give the best results. However 2D-DWT is not shift invariant and has poor selectivity for diagonal features. 2D-CWT was developed in order to overcome these shortcomings of 2D-DWT. Using 2D CWT instead of 2D-DWT, therefore, may increase robustness of the watermark against intentional or unintentional attacks. Implementation of 2D-CWT is computationally costly. Recently, a dual tree implementation of 2D CWT, which reduced computational complexity, called 2D DT-CWT was developed. The interested reader can find details about 2D CWT and 2D DT-CWT in [7]. In this study 2D DT-CWT implementation of 2D CWT was used.

The paper is organized as follows. A summary of the moment based image normalization and 2D DT-CWT is given in Section 2. The proposed image watermarking algorithm is explained in Section 3. Robustness of the proposed method against several attacks is investigated in Section 4 via computer simulations. Finally, Section 5 concludes the paper.

978-1-4244-1983-8/08/$25.00 ©2008 IEEE

II. MOMENT BASED IMAGE NORMALIZATION AND 2D DT-CWT

A. Moment Based Image Normalization

Normalization is used for the purpose of increasing watermark robustness for geometrical distortions such as rotation, translation, and scaling. Normalized image is obtained by using geometric moments m_{pq} and central moments $\mu_{p,q}$, which are defined by,

$$m_{p,q} = \sum_{x=0}^{M-1} \sum_{y=0}^{N-1} x^p y^q f(x,y) \tag{1}$$

$$\mu_{p,q} = \sum_{x=0}^{M-1} \sum_{y=0}^{N-1} (x - \bar{x})^p (y - \bar{y})^q f(x,y) \tag{2}$$

where $p,q=0,1,..$, $\bar{x} = m_{1,0} / m_{0,0}$ and $\bar{y} = m_{0,1} / m_{0,0}$, and MxN is the support of the host image $f(x,y)$ to be normalized. The normalization is achieved by applying the following tree transformations to the host image. In order to obtain translation invariance, the host image $f(x,y)$ is shifted to its central point and a centralized image $f_c(x_c,y_c)$ is determined. For this purpose, the following spatial coordinate transformation is used:

$$x_c = x_a - \bar{x}_a , y_c = y_a - \bar{y}_a \tag{3}$$

Let (x_l,y_l) denote the spatial coordinates of the scaled image $f_l(x_l,y_l)$ determined from the centralized image $f_c(x_c,y_c)$. Then, invariance to scaling can be obtained by using Eq. (4)

$$\begin{pmatrix} x_1 \\ y_1 \end{pmatrix} = \begin{pmatrix} \alpha & 0 \\ 0 & \delta \end{pmatrix} \begin{pmatrix} x_c \\ y_c \end{pmatrix} \tag{4}$$

where α and δ are computed from

$$\alpha = \pm \frac{1}{\sqrt{\mu_{20}^{(c)}}} \quad \delta = \pm \frac{1}{\sqrt{\mu_{02}^{(c)}}} \tag{5}$$

Out of the four possible solutions, the one for which $\mu_{50}^{(1)} > 0$ and $\mu_{05}^{(1)} > 0$ is chosen. Finally, let (x_n,y_n) denote the spatial coordinates of the normalized image $f_n(x_n,y_n)$ obtained from the scaled image $f_l(x_l,y_l)$. If the following spatial coordinate transformation is used, rotation invariance can be obtained

$$\begin{pmatrix} X_n \\ y_n \end{pmatrix} = \begin{pmatrix} \cos\phi & \sin\phi \\ -\sin\phi & \cos\phi \end{pmatrix} \begin{pmatrix} x_1 \\ y_1 \end{pmatrix} \tag{6}$$

where the rotation angle ϕ is computed from

$$\phi_1 = \tan^{-1} \left(-\frac{\mu_{30}^{(1)} + \mu_{12}^{(1)}}{\mu_{03}^{(1)} + \mu_{21}^{(1)}} \right), \phi_2 = \phi_1 + \pi \tag{7}$$

Out of the possible two solutions, the one for which $\mu_{03}^{(n)} + \mu_{21}^{(n)} > 0$ is chosen. In summary, applying the spatial transformations given in Equations (3), (4) and (6) results in the normalized image denoted by $f_n(x_n,y_n)$ in Fig. 2. Note that superscripts associated with the moments indicate the image from which the moment calculations are made.

B. 2D DT-CWT

For one dimensional (1-D) signals, a naive implementation of the CWT is to use a single analysis filter bank with complex coefficients satisfying perfect reconstruction (PR) properties. However designing such complex filters is a difficult task. Furthermore, complex filters that have PR properties amplify noise during the reconstruction stage. For 1-D, signals dual tree implementation of the CWT uses two analysis filter banks each with real coefficients. The two filter banks are designed such that the wavelet associated with one filter bank is an approximate Hilbert transform of the wavelet associated with the other filter bank. For 2-D signals, the two filter banks are applied to the rows followed by the columns of the data [7]. Hence, as shown in Figure 1, at each resolution there are six subbands as opposed to 2D DWT case, which has three subbands.

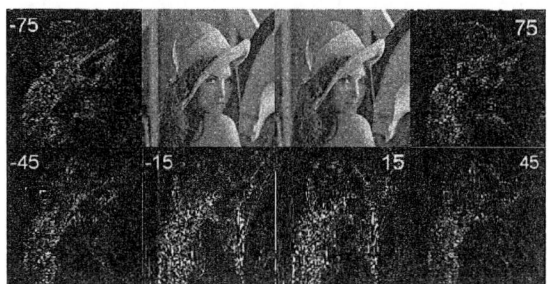

Fig. 1. Dual tree complex wavelet coefficients images

III. THE PROPOSED METHOD

A. Watermark Insertion

Figure 2 shows how a watermarked image is obtained. The steps involved in the watermark insertion process can be summarized as follows by assuming that the size of the cover image is 2Mx2N and the number of bits in the watermark signal is P:

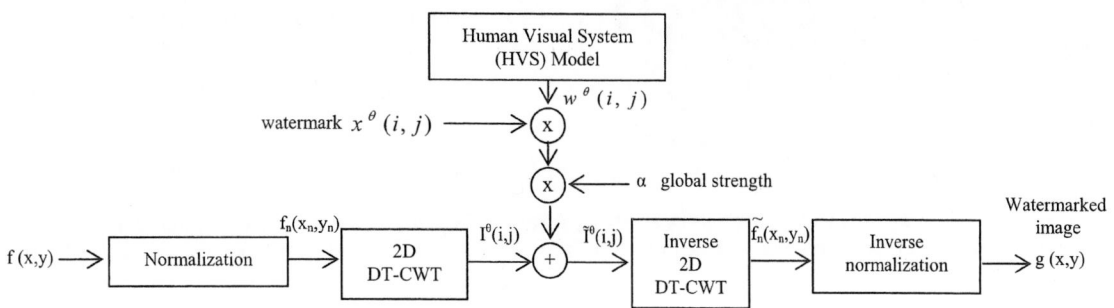

Fig. 1. The proposed digital image watermarking algorithm.

- Normalize the host image by using the procedure described in Section 2.1.
- Compute one level 2D DT-CWT of the normalized image $I^\theta(i,j)$.
- Generate 6P 2-D binary pseudo-random signals $x_l^\theta(i,j)$ whose support is MxN, where l=1,2,…,P, θ=0,1,2,3,4,5 (Note that θ shows the six subbands in 2D DT-CWT).
- Create a 2-D watermark signal $x^\theta(i,j)$ by using

$$x^\theta(i,j) = \sum_{l=1}^{P}(2m_l - 1)x_l^\theta(i,j)$$

where m_l is the lth bit (0 or 1) in the watermark signal.
- Obtain masking coefficients $w^\theta(i,j)$ by using Equation (8) that take the properties of the HVS into account.
- Multiply watermark signal $x^\theta(i,j)$ by $\alpha w^\theta(i,j)$, where α is a constant controlling global strength of the watermark signal, and than add it to the 2D DT-CWT coefficients of the normalized image to get $\tilde{I}^\theta(i,j)$.
- Compute inverse 2D DT-CWT of $\tilde{I}^\theta(i,j)$ to obtain the normalized watermarked image $\tilde{f}_n(x_n,y_n)$.
- Apply inverse normalization to $\tilde{f}_n(x_n,y_n)$ to get the final watermarked image g (x,y).

Watermark coefficients $x^\theta(i,j)$ must be scaled according to the human visual perception. A simple model to compute the Just Noticible Distortion (JND) is

$$w^\theta(i,j) = \sqrt{k^2 \cdot |I^\theta(i,j)|^2 + \gamma^2} \qquad (8)$$

where $|I^\theta(i,j)|^2$ is the average square magnitude of the CWT coefficients in a 3x3 neighbourhood centred at location (i,j). The constants k and γ for the first

resolution level and each subband θ were calculated in [8] and are given in Table 1.

TABLE I. Values of k, γ for the first resolution level and for each subband.

Subband	$\pm15^0$	$\pm75^0$	$\pm45^0$
k	1.8	1.8	1.35
γ	2.4	2.4	4

B. Watermark Detection

The watermark can be extracted by applying the steps summarized below:

- Apply the normalization procedure to the degraded watermarked image to obtain $g'(x_n,y_n)$
- Compute 2D DT-CWT of g'(x,y) $I'^\theta(i,j)$
- Using a correlator detector decode the watermark message bit-by-bit. Correlation between each binary 2D pseudo-random watermark signal $x_l^\theta(i,j)$ and $I'^\theta(i,j)$ is given by

$$\rho_l = \frac{1}{6MN}\sum_{\theta=0}^{5}\sum_{i=0}^{M-1}\sum_{j=0}^{N-1}I'^\theta(i,j)x_l^\theta(i,j) \qquad (9)$$

Then, lth bit of the watermark signal is decoded as

$$\hat{m}_l = \begin{cases} 1, & \rho_l > 0 \\ 0, & \text{otherwise} \end{cases} \qquad (10)$$

IV. SIMULATION RESULTS

The classical Lena image was chosen as the original image and the watermark was added by using the proposed method. Figure 3 shows the original image, watermarked image, and

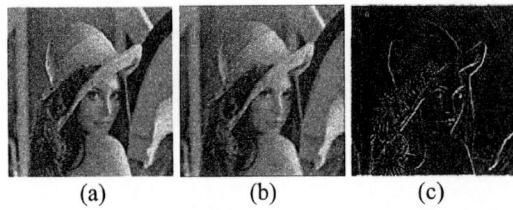

(a)   (b)   (c)

Fig. 2. (a) Original image, (b) watermarked image, (c) difference between the original and watermarked images.

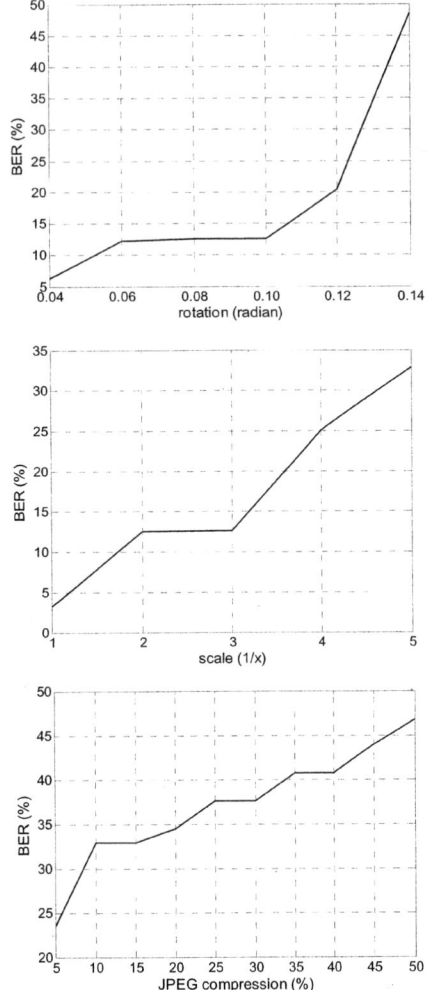

Fig. 3. BER curves for JPEG compression, rotation and scaling attacks.

the difference between the original and the watermarked images. Watermark is invisible and inserted mainly at high frequencies. JPEG compression, rotation and scaling were applied to the watermarked image. Then, the watermark was detected from the degraded watermarked images without

using the original image. Bit Error Rates (BER) defined as (number of incorrectly decoded bits / number of added bits) for each attack case are shown in Figure 4. As can be seen from the figure the proposed method provides promising results for these attacks.

V.   CONCLUSIONS

Most of the digital image watermarking algorithms proposed in the literature address the geometrical distortions and filtering artifacts separately. In other words, there do not exist efficient watermarking algorithms that work well for geometrical and filtering distortions simultaneously. In this study, a new digital image watermarking algorithm that is robust to geometrical distortions as well as linear and non-linear filtering was proposed. The method is based on the moment based image normalization, which provides immunity against geometrical distortions, and 2D DT-CWT, which provides robustness for various linear and non-linear filtering by exploiting the properties of the HVS. Simulation results show that the method provides promising robustness results for various kinds of image manipulations.

REFERENCES

[1] P. Moulin and R. Koetter, "Data-Hiding Codes" *Proceedings of the IEEE*, vol.93, pp. 2083-2126, 2005.

[2] J. J. K. O Ruanaid, T. Pun, "Rotation, scale and translation invariant spread spectrum digital image watermarking", *Signal Processing,* vol. 66, pp. 303-317, 1998.

[3] H. S. Kim and H. K. Lee, "Invariant image watermark using Zernike moments", *IEEE Transactions on Circuits and Systems For Video Technology*, vol. 13, pp. 766-775, 2003.

[4] M. Barni, F. Bartolini, and A. Piva, "Improved wavelet-based watermarking through pixel-wise masking", *IEEE Transactions on Image Processing*, vol. 10, pp. 783-791, 2002.

[5] P. Dong, , J. G. Brankov, N. P. Galatsanos, Y. Yang , F. Davoine, "Digital Watermarking Robust to Geometric Distortions", *IEEE Transactions on Image Processing,* Vol. 14, pp. 2140-2150, 2005.

[6] M. Alghoniemy and A. H. Tewfik, "Geometric invariance in image watermarking", *IEEE Transactions on Image Processing,* vol. 13, pp. 145-153, 2004.

[7] Selesnick, I. W., Baraniuk, R. G. ve Kingsbury, N. G., "The Dual-Tree Complex Wavelet Transform", *IEEE Signal Processing Magazine,* vol. 22 pp. 123-151, 2005.

[8] Loo, P. and Kingsbury, N., "Digital Watermarking Using Complex Wavelets", *Proc. IEEE International Conferance on Image Processing,* vol. 3, pp. 29-32, 2000.

Efficient Fractional Delay Hilbert Transform Filter in the Farrow Structure

Agnieszka Paruzel, Ewa Hermanowicz

Multimedia Systems Department, Faculty of Electronics, Telecommunications and Informatics
Gdansk University of Technology
Gdansk, Poland

Abstract—**In this paper the design and application of a Fractional Delay Hilbert Transform Filter (FDHTF) into an adaptive sub-sample delay estimation between two separated sinusoidal signals is considered. The FDHTF incorporates the functions of Hilbertian and variable fractional delay filtering of the incoming signal simultaneously, in one stage. In traditional approach each of these operations was performed separately. Obtained value of delay is utilized for steering the adaptive FDHTF. Efficiency of this filter was enhanced by using the Farrow structure.**

I. INTRODUCTION

Recently an increasing interest has been noticed for using the aggregated filter structure in various applications of digital signal processing (DSP), e.g. in wireless communication systems in adaptive sub-sample estimation [1], [2], [3]. Attempting to fulfill these needs we want to present a novel concept of Fractional Delay Hilbert Transform Filter (FDHTF) design. To enhance the performance of this filter we used a pair of fractional delay filters (FDFs) with identical structure coefficients. It was possible due to the usage of efficient solution, known as Farrow structure, due to real-time applications.

Processes with variable delay are very common in applications of signal processing. Systems for adaptive applications are usually very sensitive to the variations of the delay and require fast response and correction. The purpose of the system considered herein is to determine the sub-sample delay between two sinusoidal signals received from separated sources (receivers). Additional advantage of our designed FDHTF is the Lagrangian interpolation [4], [5] used to calculate Farrow structure's subfilters' coefficients, which allows achieving high accuracy of the delay estimate.

The organization of this paper is as follows: Section II starts with basic information about FDFs and method of designing them with the use of the Farrow structure. Section III is dedicated to present the design of the FDHTF. This method requires the usage of bi-phase decomposition of filter's impulse response. Resulting filter is composed of two FDFs from Section II. In Section IV obtained FDHTF is used in the adaptive subsample estimator, replacing a cascaded connection of Hilbert transform filter and FDF. Section V is devoted to the FDHTF performance evaluation.

II. DESIGN OF THE FDF USING THE FARROW STRUCTURE

The frequency response of an ideal FDF is defined as

$$D_\tau(e^{j\omega}) = \exp(-j\omega\tau) , \ |\omega| < \pi \qquad (1)$$

The impulse response of this filter is

$$d_\tau[n] = \frac{1}{2\pi}\int_{-\pi}^{\pi} D_\tau(e^{j\omega}) \, e^{j\omega n} d\omega = \mathrm{sinc}(n-\tau) \qquad (2)$$

where $n = 0, \pm 1, \dots$. In the case of an FIR approximation of length N, τ is defined as a sum of a transport delay of the system and introduced fractional delay α

$$\tau = (N-1)/2 + \alpha \qquad (3)$$

where the values of α are restricted to the interval $-0.5 \leq \alpha \leq 0.5$ in order to deal with the most accurate interpolation – central one.

The transfer function of the FIR FDF of length N in a direct form is defined as

$$D_{N\alpha}(z) = \sum_{n=0}^{N-1} d_{N\alpha}[n]z^{-n} \qquad (4)$$

where $d_{N\alpha}[n]$, $n = 0, 1, ..., N-1$ stands for the impulse response of the filter. The application of the Farrow structure demands expressing (4) in a form of a polynomial of a fractional delay α. The coefficients of the filter can be rewritten as

$$d_{N\alpha}[n] = \sum_{k=0}^{M} c_k[n]\alpha^k , \ n = 0, 1, ..., N-1 \qquad (5)$$

In the above equation $M+1$ stands for a number of subfilters in the Farrow structure. From (4) and (5) we have

$$D_{N\alpha}(z) = \sum_{k=0}^{M}\sum_{n=0}^{N-1} c_k[n]z^{-n}\alpha^k = \sum_{k=0}^{M} C_k(z)\alpha^k \qquad (6)$$

978-1-4244-1983-8/08/$25.00 ©2008 IEEE

where transfer functions of sub-filters can be written as

$$C_k(z) = \sum_{n=0}^{N-1} c_k[n] z^{-n} \qquad (7)$$

In the original Farrow work [6] they are implemented in direct form (see Fig. 1). This structure gives us flexibility to change the delay by manipulating only one coefficient, the fractional delay. Therefore this filter is also known as Variable Fractional Delay Filter – VFDF.

III. DESIGN OF THE FDHTF USING FDFs IN THE FARROW STRUCTURE

The frequency response of the ideal FDHTF with so-called generalized phase-response is defined as

$$H_\beta(e^{j\omega}) = \begin{cases} 2\exp(-j(\omega - \pi/2)\beta) & \omega \in (0,\pi) \\ 0 & \omega \in (-\pi,0) \end{cases} \qquad (8)$$

where β stands for the total delay value introduced by the filter. We introduced new symbol β in order to denote the total delay value of complex filter to distinguish it from τ - the total delay value of component FDFs: $\beta - \tau = ((2N-1)/2 + 2\alpha - 1/2) - ((N-1)/2 + \alpha) = (N-1)/2 + \alpha = \tau$

We can write the impulse response of this filter using a pair of FDFs

$$h_{2N,\beta}[n] = \begin{cases} (-1)^{n/2} d_{N,\beta/2}[n/2], & n=0,\ \pm 2,\ \dots \\ j(-1)^{(n-1)/2} d_{N,(\beta-1)/2}[(n-1)/2], & n=\pm 1,\ \pm 3,\ \dots \end{cases} \qquad (9)$$

The design of an FIR approximation of length $2N$ for this ideal FDHTF requires two VFDFs of length N each. The general scheme of implementing this FDHTF as an FIR filter in the Farrow structure is presented in Fig. 2. The total delay of this filter is

$$\beta = (2N-1)/2 + 2\alpha - 1/2 = N + 2\alpha - 1 \qquad (10)$$

To approximate the coefficients of a FDF we used the Lagrangian interpolation because of its easy and explicit formulas, very good response at low frequencies and the smoothness of magnitude response [4], [5]. For FIR Lagrange approximation, of length N, we have

$$d_{N,\beta}[n] = \prod_{\substack{k=0 \\ k \ne n}}^{N-1} \left.\frac{\alpha - k}{n - k}\right|_{\alpha = (\beta - N+1)/2} = \prod_{\substack{k=0 \\ k \ne n}}^{N-1} \frac{(\beta - N+1)/2 - k}{n - k} \qquad (11)$$

for $n = 0,1,\dots,N-1$. In order to apply the Farrow structure we decompose $d_{N,\beta}[n]$ from (11) using (5)-(7), to obtain the coefficients of the FDF in a direct form. In next step we "rotate" these coefficients by multiplying them by appropriate powers of (-1), as given below

$$\left\{ \tilde{c}_k[n] \right\}_0^{N-1} = (-1)^n \left\{ c_k[n] \right\}_{n=0}^{N-1} \qquad (12)$$

The resulting filter is called in Fig. 2 VFDR - variable fractional delayer "rotated".

Notice the difference in delays (0.5 [Sa] – sample intervals) in both branches aimed to properly interlace filter's coefficients.

The output of the FDHTF is a complex signal $y[n] = \mathrm{Re}\, y[n] + j\,\mathrm{Im}\, y[n]$

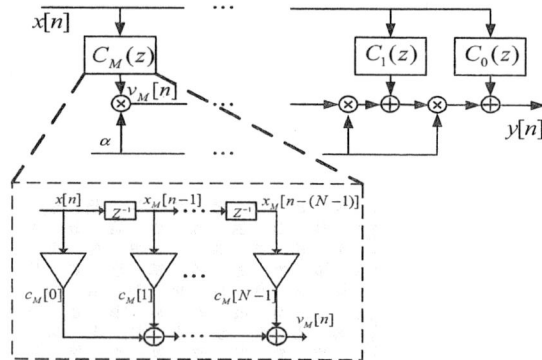

Figure 1. General idea of the Farrow structure (top part of scheme); lower part of scheme: construction of a given sub-filter of the Farrow structure, in our paper, as in original Farrow [6] work – FIR type filter in a direct form.

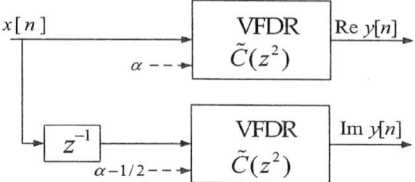

Figure 2. Block scheme of the FDHTF implementation.

The transfer function of the resulting FDHTF with fractional delay $d = 2\alpha - 1/2$ is

$$H_{\mathrm{FDHTF}}(z) = \sum_{k=0}^{N-1} \tilde{C}_k(z^2)\alpha^k + jz^{-1} \sum_{k=0}^{N-1} \tilde{C}_k(z^2)(\alpha - 1/2)^k \qquad (13)$$

Example 1.

Below we present the set of coefficients of the Farrow sub-filters of the Lagrangian FDHTF using VFDRs of length $N = 6$ each. Magnitude and group delay responses of this filter are presented in Figs. 3 and 4.

$$\{c_0[n]\}_0^5 = \left\{ \frac{3}{256}, \frac{25}{256}, \frac{75}{128}, -\frac{75}{128}, -\frac{25}{256}, -\frac{3}{256} \right\}$$

$$\{c_1[n]\}_0^5 = \left\{ -\frac{3}{640}, -\frac{25}{384}, -\frac{75}{64}, \frac{75}{64}, \frac{25}{384}, \frac{3}{640} \right\}$$

$$\{c_2[n]\}_0^5 = \left\{ -\frac{5}{96}, -\frac{13}{32}, \frac{17}{48}, \frac{17}{48}, \frac{13}{32}, \frac{5}{96} \right\}$$

$$\{c_3[n]\}_0^5 = \left\{ \frac{1}{48}, \frac{13}{48}, \frac{17}{24}, \frac{17}{24}, \frac{13}{48}, \frac{1}{48} \right\}$$

$$\{c_4[n]\}_0^5 = \left\{ \frac{1}{48}, \frac{1}{16}, \frac{1}{24}, -\frac{1}{24}, -\frac{1}{16}, -\frac{1}{48} \right\}$$

$$\{c_5[n]\}_0^5 = \left\{ -\frac{1}{120}, -\frac{1}{24}, -\frac{1}{12}, \frac{1}{12}, \frac{1}{24}, \frac{1}{120} \right\}$$

The advantage of this system over the traditional one [1], [2], [3] (Fig. 6) is smaller integer delay introduced and wider frequency response bandwidth [7].

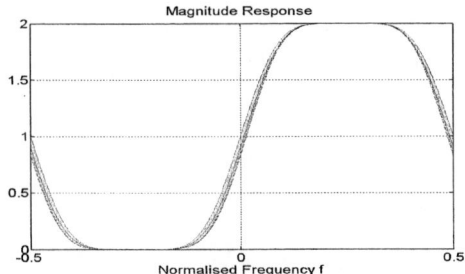

Figure 3. Magnitude responses of the FDHTF of length 12 for various values of d.

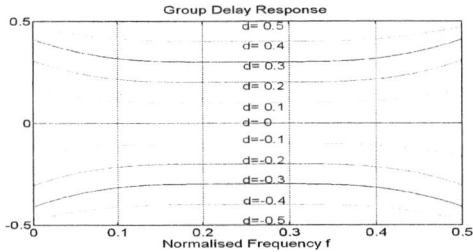

Figure 4. Group delay responses of the FDHTF of length 12.

IV. APPLICATION OF FDHTF INTO THE ADAPTIVE SUBSAMPLE ESTIMATION

In this Section the adaptive sub-sample delay estimation is considered. The block scheme for our system is presented in Fig. 5. We assumed two received sinusoidal signals, defined as

$$x_1[n] = \cos(2\pi f_0 n) \tag{14a}$$

$$x_2[n] = \cos(2\pi f_0 n + \varphi) = \cos(2\pi f_0 (n + \varphi_d)) \tag{14b}.$$

The main goal here is to determine the phase delay φ_d (in samples). We generate the analytic signals using the FDHTF, and extract their proper parts (the real part from upper branch and imaginary from the lower one)

$$\tilde{x}_1[n] = \text{Im}\{\text{FDHTF}(x_1[n])\} = \sin(2\pi f_0 n + \hat{\varphi}) \quad \text{and} \tag{15a}$$

$$\tilde{x}_2[n] = \text{Re}\{\text{FDHTF}(x_2[n])\} = \cos(2\pi f_0 n + \varphi) \tag{15b}$$

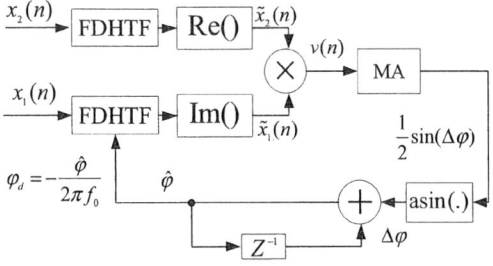

Figure 5. Adaptive sub-sample delay estimator with FDHTF. Here $d = -\varphi_d$.

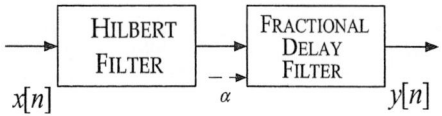

Figure 6. Cascaded connection of Hilbert transform filter and FDF, part of adaptive quadrature delay estimator block diagram from [1], [2] and [3].

Let us assume that $\hat{\varphi}$ is the estimated value of φ between the above-mentioned signals. Then we multiply signals from both branches

$$v(n) = \tilde{x}_1[n]\tilde{x}_2[n] = \sin(2\pi f_0 n + \hat{\varphi})\cos(2\pi f_0 n + \varphi) \tag{16}$$

After some computations using trigonometric formulae we can rewrite it as

$$v[n] = \left(\sin(4\pi f_0 n + \varphi + \hat{\varphi}) + \sin(\Delta\varphi)\right)/2 \tag{17}$$

where $\Delta\varphi = \hat{\varphi} - \varphi$. The next step is the use of the simple moving average (MA) filtering on this signal. In our experiments we employed Dolph-Chebyshev window of length 3 having 100 dB of relative sidelobe attenuation. With some approximation we can write that

$$MA(v[n]) \approx \sin(\Delta\varphi)/2 \tag{18}$$

We can calculate the value of $\Delta\varphi$ with the inverse *sin* function (*arcus sinus*). To obtain the current value of $\hat{\varphi}$ we use the adaptive algorithm

$$\hat{\varphi}[n] = \hat{\varphi}[n-1] + \Delta\varphi[n] \tag{19}$$

and the current (estimated) value of phase delay introduced into FDHTF is

$$\varphi_d[n] = -\hat{\varphi}[n]/(2\pi f_0) \tag{20}$$

Example 2.
To illustrate the behavior of the system described above we chose two sinusoidal signals with unit amplitudes and equal carrier frequencies (we assumed $f_0 = 0.24$ and $\varphi_d = 0.2$). Results of this research are shown in Fig. 7. In the upper part of this figure two incoming signals are presented. In the lower part the results of sub-sample delay estimation at the output of the system from Fig. 5 are shown. It can be noticed that after several iterations (the number of iterations increases as the distance between f_0 and central frequency of FDHTF, $f = 0.25$, increases) the difference φ_d between signals is estimated. In the upper figure we can notice that $x_1[n]$ and $x_2[n]$ begin to overlap as the φ_d reaches the correct value.

V. FDHTF PERFORMANCE EVALUATION

To evaluate the performance of the designed FDHTF we calculated the complex approximation error (CAE), which is defined as a difference between frequency responses of the ideal and designed FDHTF [8], [9]:

$$CAE[k] \triangleq H_d[k] - H_i[k] \tag{21}$$

where $H_d[k]$ and $H_i[k]$ for $k = 0, \pm 1, \ldots, \pm(K-1)/2$ are the sets of K equally spaced frequency points of the frequency responses of the designed and ideal filter, respectively. Note that the frequency response of this ideal complex filter was defined in (8). The results of CAE evaluation for the filter from Example 1 are shown in Fig. 8.

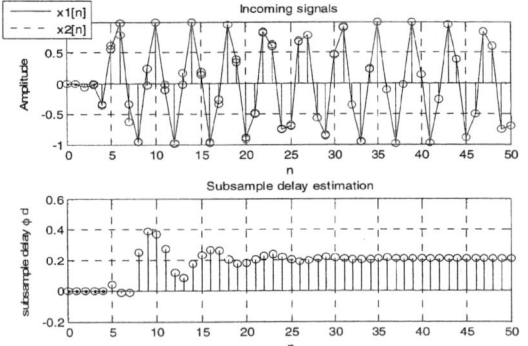

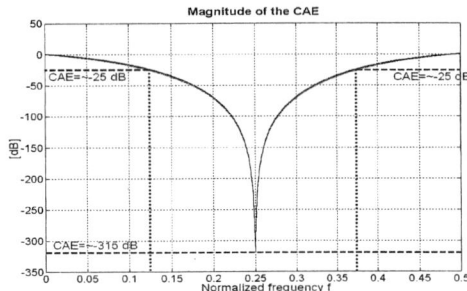

Figure 7.   Estimated value of delay between signals.

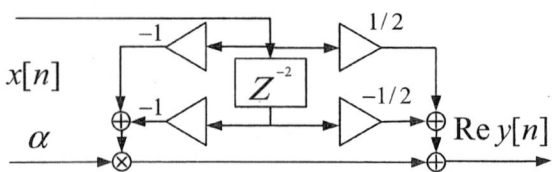

Figure 8.   CAE of FDHTF from Example 1.

It can be easily read out from that figure that the CAE characteristics of our filter are the best in the vicinity of normalized frequency $f = 0.25$. This corresponds with the maximally flat region on magnitude and delay characteristics from Figs. 3 and 4.

Generally, the filters which incorporate Farrow structure in its basic form have considerably high numerical complexity. But in set of coefficients from Example 1 we can notice the following (anty-) symmetry conditions in both FDF and FDHTF coefficients. We have for subfilters of even length

$$\tilde{c}_k[n] = -\tilde{c}_k[N-1-n] \text{ if } k = 0, 2, 4, \ldots \tag{22a}$$

$$\tilde{c}_k[n] = \tilde{c}_k[N-1-n] \text{ if } k = 1, 3, 5, \ldots \tag{22b}$$

and for subfilters of odd length

$$\tilde{c}_k[n] = \tilde{c}_k[N-1-n] \text{ if } k = 0, 2, 4, \ldots \tag{23a}$$

$$\tilde{c}_k[n] = -\tilde{c}_k[N-1-n] \text{ if } k = 1, 3, 5, \ldots \tag{23b}$$

Using relations (22a), (22b), (23a) and (23b) and sharing the delay elements among sub-filters we can save:

- Multipliers: $3N(M+1)/2$ for N even and
  $$3(N-1)(M+1)/2 \text{ for } N \text{ odd}$$

- Adders: $M+1$

- Delaying elements: $(2M+1)(N-1)$.

Figure 9.   VFDR for FDHTF of length 2 (shared delay elements within sub-filter).

In Fig. 9 a very simple example of a short VFDR ($N = 2$) for FDHTF of length $2N$ (cf. Fig. 2), where above savings are clearly visible.

VI.   CONCLUSIONS

In this paper we present the main results of our research on the design and application of the FDHTF – a novel one-stage filter. The usage of variable FDFs as compound sub-filters was proved to be well-chosen as it exhibits very high efficiency and accuracy in filtering the chosen signals.

We also proposed the application of FDHTF in an adaptive sub-sample delay estimation system. Farrow structure used in the FDHTF allowed for on-line tuning of the delay between two sinusoidal signals without redesigning the filter. We illustrated our work with the results of MATLAB simulations.

REFERENCES

[1] Maskell D.L., Woods G.S.: *Adaptive subsample delay estimation using a modified quadrature phase detector*, IEEE Trans. Circuits Syst. – II: Express Briefs, vol. 52, No. 10, Oct. 2005, pp. 669-674

[2] Maskell D.L., Woods G.S., Kerans A.: *A hardware efficient implementation of an adaptive subsample delay estimator*, Proc. ISCAS'2004, vol. III, pp. 317-320

[3] Maskell D.L. and Woods G.S.: *Adaptive subsample delay estimator using quadrature estimator*, Electr. Letters, vol. 40, No. 5, 4th March 2004, pp. 347-349

[4] Laakso T.I, Välimäki V., Karjalainen M., Laine U.K.: *Splitting the unit delay*, IEEE Signal Processing Magazine, January 1996, pp. 30-60

[5] Välimäki V.: *A New Filter Implementation Strategy for Lagrange Interpolation*, IEEE Int. Symposium on Circuits and Systems, 30 April- 4 May, 1995, vol. 1, pp. 361-364

[6] Farrow C.W.: *A continuously variable delay element*, Proc. IEEE Int. Symp. Circuits Syst., vol. 3, Espoo, Finland, June 7-9, 1988, pp. 2641-2645

[7] Hermanowicz E., Paruzel A.: *The design of complex Hilbert transform filter using the Farrow structure*, ICSES'06 Proceedings, Poznan, Poland, vol. 1 of 2, pp. 267 -270

[8] Hermanowicz E.: *Special Discrete-Time Filters and Applications*, Akademicka Oficyna Wydawnicza EXIT, Warszawa 2005, [in English]

[9] Blok M.: *Optimal Fractional Sample Delay Filter with Variable Delay*,
http://techonline.com/community/ed_resource/feature_article/20291

IMAGE REPRESENTATION AND ENHANCEMENT ON ELASTIC MANIFOLDS

Vadim Ratner   *Yehoshua Y. Zeevi*

Department of Electrical Engineering
Technion – Israel Institute of Technology, Haifa 32000
vad@techunix.technion.ac.il,   zeevi@ee.technion.ac.il

Abstract -**Enhancing detail and contours of objects in low resolution noisy natural images is still one of the most outstanding challenges in image processing. This study is devoted to a new approach to adaptive image enhancement, based on the application of a damped elastic deformation process to single images. The proposed approach results in a Telegraph-Diffusion (TeD) denoising-and-sharpening filtering scheme which enhances fine details in images. Three efficient numerical schemes are presented. Advantages of the algorithm are discussed with reference to computational results.**

Index Terms— Diffusion process, image enhancement, image edge analysis

I. INTRODUCTION

There is an inherent conflict in image processing between the desire to reduce the noise and yet sharpen the image. To cope with these conflicting demands, the development and application of more advanced and innovative techniques of image enhancement has become most desirable. Image processing techniques based on the application of partial differential equations (PDE) appear to be most promising in that they have been advanced to the stage that they are non isotropic and adaptive, and can therefore enhance selectively areas and features in an image-content-dependent manner [2], [4], [11].

This paper presents such a new scheme based on the novel Telegraph-Diffusion (TeD) image filter ([7]) that overcomes some shortcomings characteristic of the previously reported PDE-based schemes. It then implements this new framework of enhancement on images represented as high-dimensional manifolds [7].

II. ANISOTROPIC DIFFUSION

Previous research on PDE image processing ([1, 2, 6, 9]) has focused primarily on the diffusion equation:

$$(1) \quad -\nabla \cdot \left(k(\nabla u) \nabla u \right) + u_t = 0 .$$

Applying a linear diffusion process to an image is equivalent to filtering it by Gaussian lowpass kernel:

$$(2) \quad H_d(\lambda, t) = \exp\left(-k\lambda^2 t\right) ,$$

where λ is the spatial frequency and t is time (Fig. 1). Varying diffusivity, k, over an image in an adaptive manner allows feature dependant smoothing. This, in turn, facilitates selective denoising, while preserving important image features, e.g. strong smoothing of flat areas (where most fluctuations in image contrast are due to noise), and almost no smoothing around edges [6].

In [2], Gilboa et. al. extended the diffusion based processing to the ill-posed negative time regime. They have shown that when properly localized, forward-and-backward (FAB) diffusion results in edge enhancement.

Weickert ([9]) proposed a fully anisotropic diffusion filtering scheme, where the application of structure tensor instead of scalar coefficient k, allows control also over orientation of smoothing and not only its strength.

III. ANISOTROPIC TELEGRAPH DIFFUSION

Other physical processes besides diffusion have smoothing properties. One such process is damped elastic deformation, described by damped wave equation (telegraphers' eq.):

$$(3) \quad u_{tt} - \nabla \cdot \left(k(\nabla u) \nabla u \right) + c(\nabla u) u_t = 0 .$$

Similarly to the diffusion process, elastic deformation can also be applied to an image, resulting in Telegraph-Diffusion (TeD) filtering scheme [7]. Linear TeD is equivalent to applying the following filter:

$$(4) \quad H_{td}(\lambda, t) = \left(\frac{c + \sqrt{c^2 - 4k\lambda^2}}{2\sqrt{c^2 - 4k\lambda^2}} \right) \exp\left(\frac{-c + \sqrt{c^2 - 4k\lambda^2}}{2} t \right) + \left(\frac{-c + \sqrt{c^2 - 4k\lambda^2}}{2\sqrt{c^2 - 4k\lambda^2}} \right) \exp\left(\frac{-c - \sqrt{c^2 - 4k\lambda^2}}{2} t \right)$$

As is highlighted by Fig. 1, (4) defines a filter that is closer to an ideal lowpass than the one offered by the diffusion equation.

Increasing k and c ($k := \alpha k, c := \sqrt{\alpha} c$) is equivalent to increasing time ($t := \sqrt{\alpha} t$). From physical viewpoint this implies that adaptive coefficients control locally the pace of smoothing. From image processing viewpoint, this means

978-1-4244-1983-8/08/$25.00 ©2008 IEEE

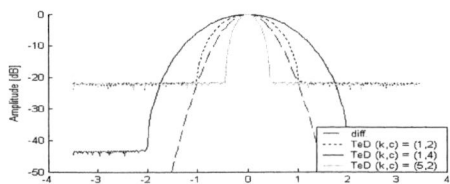

Figure 1: Comparison of the linear diffusion (dashed) and linear TeD spatial frequency transfer functions obtained at a fixed time

local control over lowpass width. In fact varying k alone achieves similar effect.

Allowing k to become negative around edges results in edge enhancement. This algorithm, which advances TeD into negative time regime, is called Forward-and-Backward (FAB) TeD.

IV. TENSOR TELEGRAPH-DIFFUSION

Although non-linear PDE-based methods allow spatially varying processing, they are still isotropic, i.e. display no directional dependency when viewed on small enough scale. When dealing with oriented structures, not only strength of smoothing needs to be controlled by image features, but also its direction. A method proposed by Weickert [9] achieves it by manipulating eigenvalues of a smoothed structure tensor (7), enhancing diffusion along edges and reducing, or even reversing it across them. This is achieved by replacing the $\nabla \cdot \left(k \nabla u \right)$ member in the diffusion and TeD equations by the following:

$$(5) \quad \nabla \cdot \left(D \nabla u \right).$$

In grayscale image, D is a 2x2 matrix which contains structural information about the image. It is defined as follows:

$$(6) \quad D = \left(\omega_1 ; \omega_2 \right) \begin{pmatrix} \lambda_1 & 0 \\ 0 & \lambda_2 \end{pmatrix} \begin{pmatrix} \omega_1 \\ \omega_2 \end{pmatrix} = \begin{pmatrix} a & b \\ b & c \end{pmatrix},$$

where ω_i are eigenvectors of the smoothed structure tensor J_ρ pointing in the direction of steepest descent and in the direction perpendicular to it:

$$(7) \quad J_\rho \left(\nabla u_\partial \nabla u_\partial^T \right) = \begin{pmatrix} j_{11} & j_{12} \\ j_{12} & j_{22} \end{pmatrix} = K_\rho * \left(\nabla u_\partial \nabla u_\partial^T \right)$$

$$u_\partial = K_\partial * u \left(\bullet, t \right)$$

$$K_\chi \left(x \right) = \frac{1}{2\pi\chi^2} \exp\left(-\frac{|x|^2}{2\partial^2} \right).$$

The values of λ_i are as follows:

$$(8) \quad \begin{aligned} \lambda_1 &= \alpha \\ \lambda_2 &= \alpha + \left(1 - \alpha \right) \exp\left(\frac{-1}{\left(j_{11} - j_{22} \right)^2 + 4 j_{12}} \right), \end{aligned}$$

where α is a small parameter controlling diffusion in the direction of ∇u (ω_1), i.e. across edges. λ_2 controls diffusion in direction of ω_2, i.e. orientation of strongest coherence.

It is also possible to use negative α and thereby enhance edges. Image dependant α is also useful, e.g. in stronger smoothing of flat areas.

V. NUMERIC SCHEMES

Discretization schemes and their convergence received little attention so far. Weickert et. al. in [11] and [10] presented three such schemes, which differ from the basic approach by increased rotational invariance and improved convergence rate. Their application to the TeD method is presented below.

A. Rotation-invariant tensor TeD

Weickert and Scharr ([11]) proved for tensor diffusion that replacing central difference spatial derivative:

$$\frac{1}{2dx} \begin{bmatrix} 0 & 0 & 0 \\ -1 & 0 & 1 \\ 0 & 0 & 0 \end{bmatrix}$$

by the following

$$\frac{1}{32dx} \begin{bmatrix} -3 & 0 & 3 \\ -10 & 0 & 10 \\ -3 & 0 & 3 \end{bmatrix}$$

improves rotation invariance and numerical stability of the algorithm. This is also true for TeD.

B. Backward Euler Semi-implicit TeD

To simplify matters we will discuss one-dimensional case. $u^{i,j}$ is the i-th element (spatial coordinate $x=i*dx$) of vector u at time $t=j*dt$. Basic discrete representation of (3) is as follows:

$$(9) \quad \frac{u^{x,t+1} - 2u^{x,t} + u^{x,t-1}}{\left(dt \right)^2} + c^{x,t} \frac{u^{x,t+1} - u^{x,t}}{dt} - \\ -\frac{k^{x,t} \left[u^{x+1,t} - u^{x,t} \right] - k^{x-1,t} \left[u^{x,t} - u^{x-1,t} \right]}{\left(dx \right)^2} = 0$$

It defines the following iterative update scheme of u:

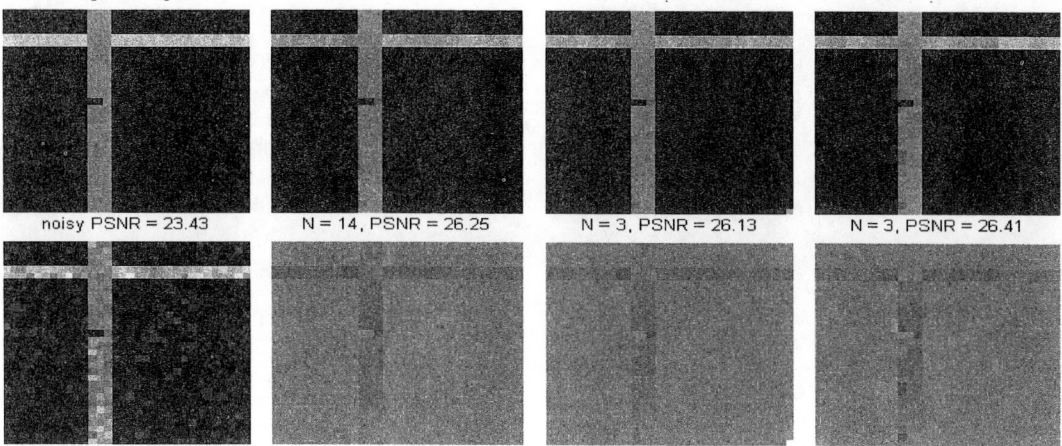

Figure 2: upper, left to right: original image, TeD denoising (14 iterations), semi-implicit raster scan TeD denoising (3 iterations), semi-implicit Hilbert TeD denoising (3 iterations). Lower, left to right: noisy image, respective error images.

$$(10) \quad \left(1+c^{x,t}dt\right)u^{x,t+1} = \left(2+cdt\right)u^{x,t} - u^{x,t-1} + \\ + \left(dt\right)^2 \frac{k^{x,t}\left[u^{x+1,t}-u^{x,t}\right]-k^{x-1,t}\left[u^{x,t}-u^{x-1,t}\right]}{\left(dx\right)^2} = 0$$

A scheme closely related to Backward Euler time discretization, was proposed for the diffusion equation by Weickert et. al. in [10], resulting in the following discrete equation:

$$(11) \quad \frac{u^{x,t+1}-2u^{x,t}+u^{x,t-1}}{\left(dt\right)^2} + c^{x,t}\frac{u^{x,t+1}-u^{x,t}}{dt} - \\ - \frac{k^{x,t}\left[u^{x+1,t+1}-u^{x,t+1}\right]-k^{x-1,t}\left[u^{x,t+1}-u^{x-1,t+1}\right]}{\left(dx\right)^2} = 0$$

This leads to semi-implicit update scheme:

$$(12) \quad \left(\left(1+cdt\right)I - dt^2 A\right)\underline{u}^{j+1} = \left(2+cdt\right)\underline{u}^j - \underline{u}^{j-1} \;,$$

where $\underline{u}^j$ is the input vector at time $j*dt$ and A is a tri-diagonal matrix, which makes $\left[\left(1+cdt\right)I + dt^2 A\right]$ easily invertible using Thomas algorithm. Although it requires additional calculations the scheme retains linear complexity, and is much more stable then the standard approach, allowing larger time step and fewer iterations (Fig. 2).

Generalization of the algorithm to higher dimensions is not trivial. The naïve approach of writing the Backward Euler scheme for 2D results in non-zero values outside the main three diagonals of A, requiring $O(n^2)$ operations. Solution proposed in [10] performs the update separately on each axis, resulting in a good approximation of the original scheme. The 2D input image is raster-scanned twice – along x and y axes during each iteration and the two resulting vectors are updated.

We propose another solution which involves update of a single vector during each iteration. The vector is obtained by Hilbert-sampling the input, rather then raster-scanning (Fig. 2, rightmost column).

C. Crank-Nicolson Semi-implicit TeD

Other time discretization schemes are possible. One of them, Crank-Nicolson is defined as follows:

$$(13) \quad \frac{u^{x,t+1}-2u^{x,t}+u^{x,t-1}}{\left(dt\right)^2} + c^{x,t}\frac{u^{x,t+1}-u^{x,t}}{dt} - \\ - \frac{k^{x,t}\left[u^{x+1,t+1}-u^{x,t+1}\right]-k^{x-1,t}\left[u^{x,t+1}-u^{x-1,t+1}\right]}{2\left(dx\right)^2} - \\ - \frac{k^{x,t}\left[u^{x+1,t}-u^{x,t}\right]-k^{x-1,t}\left[u^{x,t}-u^{x-1,t}\right]}{2\left(dx\right)^2} = 0$$

This leads to the following update scheme:

$$(14) \quad \left(\left(1+cdt\right)I - \frac{dt^2}{2}A\right)\underline{u}^{j+1} = \left(\left(2+cdt\right)I + \frac{dt^2}{2}A\right)\underline{u}^j - \underline{u}^{j-1}$$

Even at a first glance it appears that Crank-Nicolson scheme should allow larger time steps (dt^2 is replaced by $\frac{dt^2}{2}$).
More rigorous stability and convergence analysis will be presented elsewhere.

VI. ENHANCEMENT SCHEME

Based on the methods described above, an image enhancement method can be derived, which both denoises and sharpens low contrast noisy images. TeD denoising, coupled with FAB TED ([7]) and Tensor-TeD successfully preserve and enhance edges and other flow-like structures. An example of such image enhancement appears in fig. 3.

978-1-4244-1983-8/08/$25.00 ©2008 IEEE

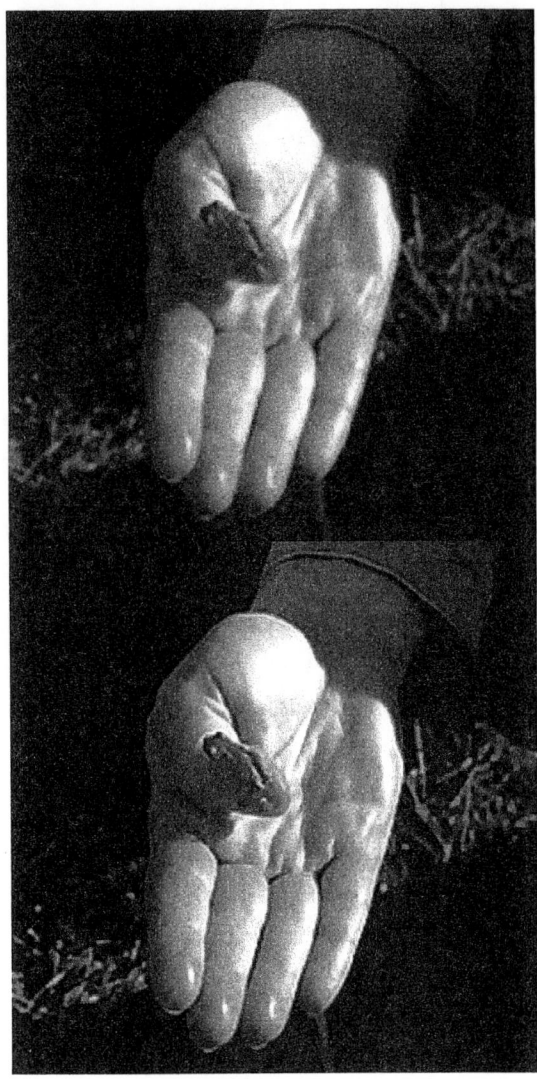

Figure 3: TeD enhancement: Original low resolution image (upper) and TED result (lower).

VII. CONCLUSION

The proposed TeD approach to image enhancement provides a broader and a richer framework for the development new powerful algorithms. However, further careful investigation has yet to verify that such new algorithms do not entail detrimental effects on edge location and/or noise-generated fine detail. The latter may prohibit certain applications in such fields of medical image processing, but will nevertheless permit many other applications. Since the performance of the proposed algorithms is controlled by several parameters, fine tuning for matching the algorithms to applications is possible.

VIII. ACKNOWLEDGEMENTS

We would like to thank Prof. Mitsuhiro Nakao for highlighting proper mathematical foundation for the presented algorithms. We especially thank Dr. Boris Ratner who contributed original ideas to this study. We are grateful to Veronica Lenarth (citrit1) from flickr.com for kindly permitting us the use of her images (fig. 3).

IX. REFERENCES

[1] L. Alvarez, P.L. Lions, and J.M. Morel, "Image Selective Smoothing and Edge Detection by Nonlinear Diffusion", *SIAM Journal on Num. Analysis*, vol. 29, no. 3, pp. 845-866, June 1992.

[2] G. Gilboa, N. Sochen, and Y.Y. Zeevi "Forward-and-Backward Diffusion Processes for Adaptive Image Enhancement and Denoising", *IEEE Trans. on Image Proc.*, vol. 11, no. 7 , July 2002.

[3] G. Gilboa, N. Sochen, and Y.Y. Zeevi, "Image Enhancement and Denoising by Complex Diffusion Process", *IEEE Trans. Pattern Analysis and Machine Intelligence,* vol. 26, no. 8, pp. 1020-1036, August 2004.

[4] R. Kimmel, R. Malladi and N. Sochen, "Images as Embedded Maps and Minimal Surfaces: Movies, Color, Texture, and Volumetric Medical Images", *Int'l Journal of Computer Vision* 39 vol. 2, pp. 111–129, 2000.

[5] M. Nakao, "Decay and global existence for nonlinear wave equations with localized dissipations in general exterior domains", *Operator Theory, Advances and Applications*, Vol.159, pp.213-299, 2007.

[6] P. Perona and J. Malik, "Scale-Space and Edge Detection Using Anisotropic Diffusion", *IEEE Trans. Pattern Analysis and Machine Intelligence*, vol. 12, no. 7, pp. 629-639, July 1990.

[7] V. Ratner, Y.Y. Zeevi, "Image Enhancement Using Elastic Manifolds", *ICIAP 2007 Proceedings*, September 2007.

[8] N. Sochen, and Y.Y. Zeevi, "Images as Manifolds Embedded in a Spatial-Feature Non-Euclidian Space", *EE-Technion report* No. 1181, November 1998.

[9] J. Weickert, "Coherence-Enhancing Diffusion of Color Images", *Int. J. Comp.Vision*, vol. 17, pp.199-210, 1999.

[10] J. Weickert, B. M. ter Haar Romeny, and Max A. Viergever, "Efficient and Reliable Schemes for Nonlinear Diffusion Filtering", *IEEE Trans on Image Processing*, vol. 7, No. 3, March 1998.

[11] J. Weickert, H. Scharr, "A Scheme for Coherence-Enhancing Diffusion Filtering with Optimized Rotation Invariance", *Journal of Visual Communication and Image Representation*, Vol. 13, pp. 103–118, 2002.

[12] E. Zauderer, *Partial Differential Equations of Applied Mathematics*, 2ed.,Wiley, New York, 1998.

A New Window Based on Exponential Function

Kemal Avci   Arif Nacaroğlu

Department of Electrical and Electronics Engineering
University of Gaziantep
Gaziantep, Turkey
avci@gantep.edu.tr   arif1@gantep.edu.tr

Abstract—**In this paper a new class of adjustable windows based on the exponential function is proposed. The proposed window has been derived in the same way of the derivation of Kaiser window, but it has the advantage of having no power series expansion in its time domain function. The spectrum design equations for the proposed window are established, and the spectral comparisons are performed with Kaiser and ultraspherical windows. Comparison with Kaiser window shows that the proposed window provides better sidelobe roll-off ratio, which is important for some applications, but worse ripple ratio for the same window length and mainlobe width. As for the comparison with ultraspherical window, for the same window length, mainlobe width and sidelobe roll-off ratio the proposed window presents better ripple ratio for the narrower mainlobe width and larger sidelobe roll-off ratio, but worse ripple ratio for the wider mainlobe width and smaller roll-off ratio.**

I.   INTRODUCTION

Window functions (or simply as windows) are widely used in digital signal processing for the applications in signal analysis and estimation, digital filter design and speech processing [1-2]. In literature many windows have been proposed [3-6]. They are known as suboptimal solutions, and the best window is depending on the applications.

Kaiser window is a well known flexible window and widely used for FIR filter design and spectrum analysis applications [2-3] since it achieves close approximation to the discrete prolate spheroidal functions that have maximum energy concentration in the mainlobe. With adjusting its two independent parameters, namely the window length and the shape parameter, it can control the spectral parameters main lobe width and ripple ratio for various applications.

Sidelobe roll-off ratio is another spectral parameter and important for some applications. For beamforming applications, the higher sidelobe roll-off ratio means that it can reject far end interferences better [6]. For filter design applications, it can reduce the far end attenuation for stopband energy. And for speech processing, it reduces the energy leak from one band to another [7].

Kaiser window has a better sidelobe roll-off characteristic than the other well known adjustable windows such as Dolph-Chebyshev [4] and Saramaki [5], which are special cases of ultraspherical window [6], but obtaining a window which performs higher sidelobe roll-off characteristic than Kaiser window will be useful.

In this paper, a new window based on the exponential function is proposed to provide higher sidelobe roll-off ratio than Kaiser window to be useful for some applications.

II.   DERIVATION OF THE PROPOSED WINDOW

A.   Spectral Characteristic of Windows

A window, $w(nT)$, with a length of N is a time domain function which is nonzero for $|n| \leq (N-1)/2$ and zero for otherwise. Windows are generally compared and classified in terms of their spectral characteristics. The frequency spectrum of $w(nT)$ can be found by

$$W(e^{jwT}) = |A(w)|e^{j\theta(w)} = w(0) + 2 \sum_{n=1}^{(N-1)/2} w(nT)\cos wnT \quad (1)$$

where T is the sample period. A typical window has a normalized amplitude spectrum in dB range as in Fig. 1.

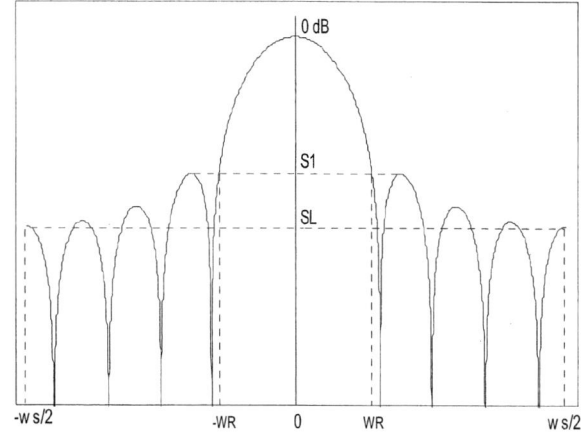

Figure 1. A typical window's normalized amplitude spectrum

978-1-4244-1983-8/08/$25.00 ©2008 IEEE

Normalized spectrum in Fig.1 can be obtained from

$$\left|W_N(e^{jwT})\right| = 20\log_{10}(\left|A(w)\right|/\left|A(w)\right|_{max}) \qquad (2)$$

The common spectral characteristic parameters to distinguish windows performance are the mainlobe width (w_M), ripple ratio (R) and sidelobe roll-off ratio (S). From Fig. 1, these parameters can be described as

w_M = Two times half mainlobe width = $2w_R$

R = Maximum sidelobe amplitude in dB - Mainlobe amplitude in dB = S_1

S = Maximum sidelobe amplitude in dB- Minimum sidelobe amplitude in dB = S_1-S_L

In the applications, it is desired for a window to have smaller ripple ratio and narrower mainlobe width. But, this requirement is contradictory [1].

B. Kaiser Window

In discrete time domain, Kaiser window is defined by [3]

$$w_k(n) = \begin{cases} \dfrac{I_0(\alpha_k\sqrt{1-\left(\frac{2n}{N-1}\right)^2})}{I_0(\alpha_k)} & |n| \leq \dfrac{N-1}{2} \\ 0 & otherwise \end{cases} \qquad (3)$$

where α_k is the adjustable shape parameter, and $I_0(x)$ is the modified Bessel function of the first kind of order zero and it is described by the power series expansion as

$$I_0(x) = 1 + \sum_{k=1}^{\infty}\left[\frac{1}{k!}\left(\frac{x}{2}\right)^k\right]^2 \qquad (4)$$

While an approximation closed formula for the Kaiser window spectrum is defined [1], the exact Kaiser spectrum can be obtained from (1). Note that T=1 is considered as the normalization for the rest of paper.

As known from the fixed windows while the window length, N, increases the mainlobe width decreases but ripple ratio remains generally constant. As for the shape parameter, α_k, its larger values result in a wider mainlobe width and a smaller ripple ratio.

C. Proposed Window

From Fig. 2, it can be seen that exp(x) and Io(x) have the same shape characteristic. Therefore, a new window can be proposed as

$$w_e(n) = \begin{cases} \dfrac{e^{\left(\alpha\sqrt{1-\left(\frac{2n}{N-1}\right)^2}\right)}}{e^{\alpha}} & |n| \leq \dfrac{N-1}{2} \\ 0 & otherwise \end{cases} \qquad (5)$$

As for the case in Kaiser window, the exact spectrum for the proposed window can be obtained from (1).

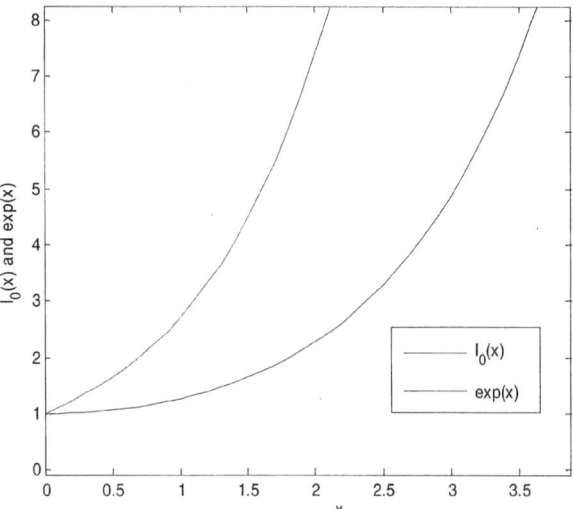

Figure 2. The functions Io(x) and e^x

Fig. 3 shows the frequency domain plots of proposed window for a fixed value of length N = 51. As in Kaiser window, α=0 corresponds to the rectangular window.

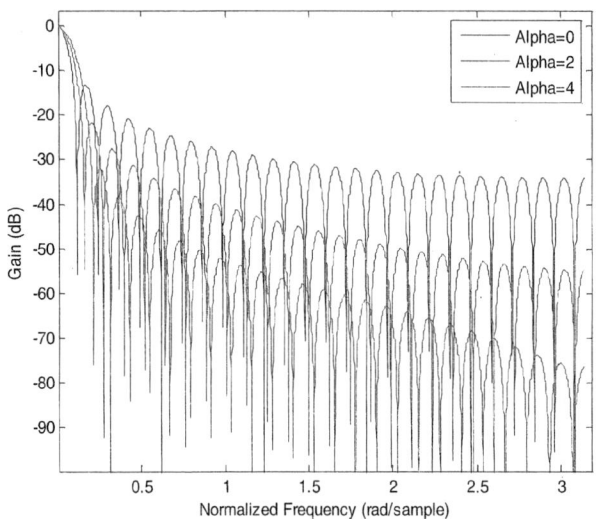

Figure 3. Proposed window spectrum in dB for α = 0, 2, and 4 and N=51

From Fig. 3 and TABLE I, it can be easily seen that as in the case for Kaiser window, when α increases the mainlobe width increases and ripple ratio decreases.

TABLE I. DATA FOR THE PROPOSED WINDOW SPECTRUM N=51

Window	N	α	w_R	R	S
Proposed-1	51	0	0.1	-13.25	20.9
Proposed-2	51	2	0.15	-21.73	32.95
Proposed-3	51	4	0.21	-31.84	44.54

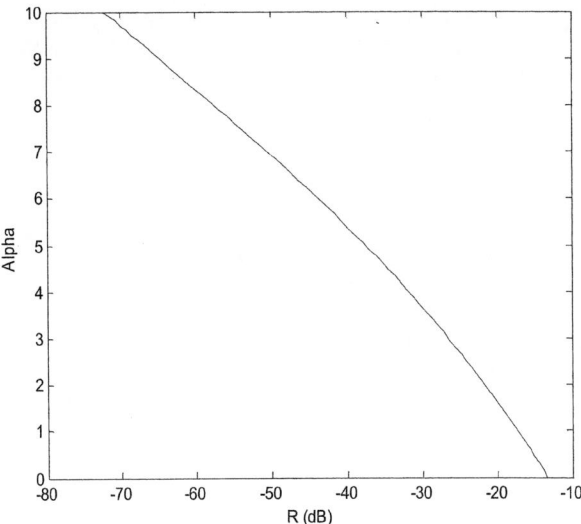

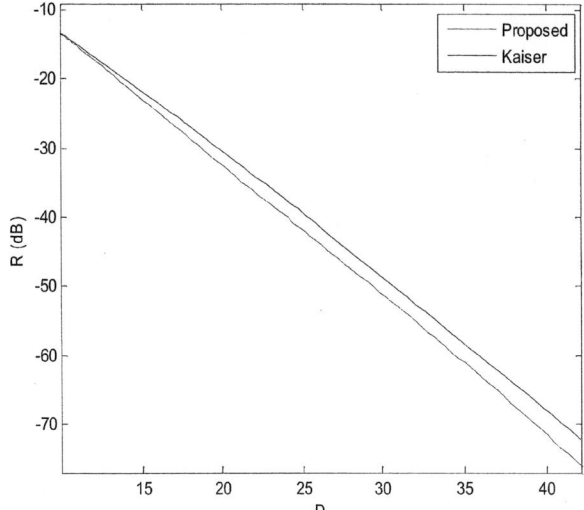

Figure 4. The relation between α and R for the proposed window

Figure 5. Comparison of the proposed and Kaiser windows in terms of the ripple ratio for N =51

Fig. 4 shows the relationship between the shape parameter and ripple ratio for the proposed window. For some applications such as the spectrum analysis [2], the design equations which give window parameters in terms of spectrum parameters are required. From Fig. 4, an approximate relationship for α in terms of R can be found by using polynomial curve fitting method for R≤-13.25 dB

$$\alpha = -1.552 \times 10^{-5} R^3 - 2.923 \times 10^{-3} R^2 - 0.3211 R - 3.763 \quad (6)$$

The second design equation is the relation between the window length, N, and the ripple ratio. To predict the window length N for a given quantities R and w_R, the normalized width D=$2w_R$(N-1) is used [6]. The relation between D and R for proposed window is given in Fig. 5.

By using quadratic polynomial curve fitting method, an approximate relation between D and R can be established as

$$D = -8.702 \times 10^{-4} R^2 - 0.6184 R + 1.972 \quad \text{for } R \le -13.25 \quad (7)$$

An integer value of the window length N can be predicted from

$$N \ge \frac{D}{2w_R} + 1 \quad (8)$$

Using (6) through (8), the proposed window can be designed for satisfying the prescribed values of ripple ratio and mainlobe width.

III. SPECTRUM COMPARISIONS

A. Comparison with Kaiser Window

To be able to make a comparison between proposed window and Kaiser window, it is necessary to plot their

spectrum and compare their spectral characteristic parameters in terms of ripple ratio, mainlobe width and sidelobe roll-off ratio.

Fig. 5 shows the comparison of the proposed and Kaiser windows in terms of ripple ratio versus normalized width for N=51. It can be observed that Kaiser window performs smaller ripple ratio for the same mainlobe width.

As for the comparison between the proposed and Kaiser windows in terms of sidelobe roll-off ratio, the simulation result is given in Fig. 6. It can be seen that the proposed window performs higher sidelobe roll-off ratio than Kaiser window for the same mainlobe width.

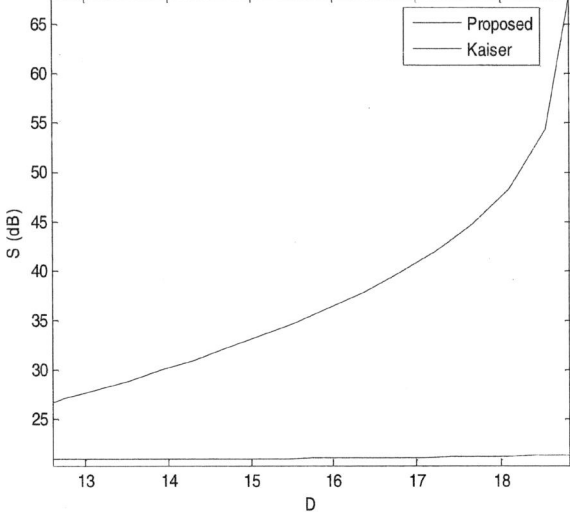

Figure 6. Comparison of the proposed and Kaiser windows in terms of the sidelobe roll-off ratio for N=51

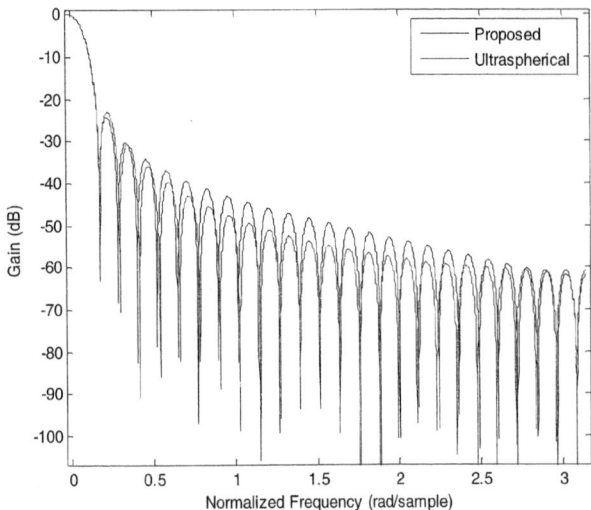

Figure 7. Comparison of the proposed and ultraspherical windows for narrower mainlobe width and larger sidelobe roll-off ratio for N=51

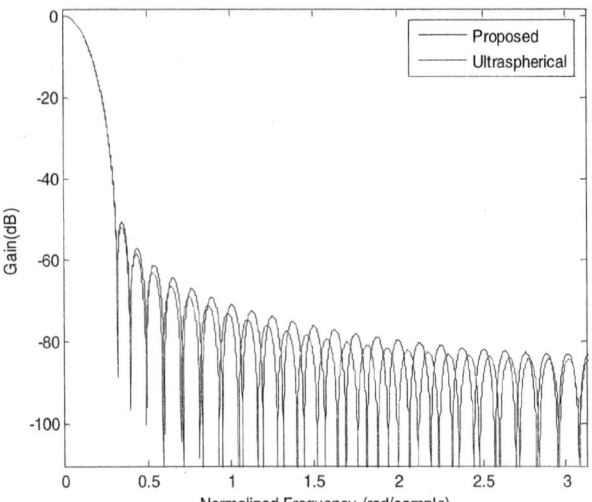

Figure 8. Comparison of the proposed and ultraspherical windows for wider mainlobe width and smaller sidelobe roll-off ratio for N=51

TABLE II. DATA FOR THE FIRST COMPARISION EXAMPLE BETWEEN THE PROPOSED AND ULTRASPHERICAL WINDOWS

Window	N	w_R	S	R
Proposed	51	0.164	37.81	-24.1
Ultraspherical	51	0.164	37.81	-23.02

B. Comparison with Ultraspherical Window

Two examples are given for the comparison between proposed and ultraspherical windows. The first comparison example is performed for the narrower mainlobe width and larger sidelobe roll-off ratio. The simulation result given in Fig. 7 and TABLE II shows that the proposed window provides a better ripple ratio than ultraspherical window for the same window length, mainlobe width and sidelobe roll-off ratio. The ultraspherical window parameters for this example are μ=1.99997 and xμ=1.00039.

The second comparison example is given for the wider mainlobe width and smaller sidelobe roll-off ratio. The simulation result given in Fig. 8 and TABLE III shows that ultraspherical window provides a better ripple ratio than the proposed window in this case. The ultraspherical window parameters for this example are μ=1.6662 and xμ=1.00973.

From Fig. 7 and Fig. 8, the ripples between the maximum and the minimum sidelobe amplitudes can be also seen to be higher for the proposed window.

TABLE III. DATA FOR THE SECOND COMPARISION EXAMPLE BETWEEN THE PROPOSED AND ULTRASPHERICAL WINDOWS

Window	N	w_R	S	R
Proposed	51	0.31	32.48	-50.53
Ultraspherical	51	0.31	32.48	-51.75

IV. CONCLUSION

In this paper, we propose a new 2-parameter window family based on the exponential function. After giving the definition and finding the spectrum design equations for the proposed window, the spectral comparisons are performed with Kaiser and ultraspherical windows. Comparison with Kaiser window shows that the proposed window provides better sidelobe roll-off ratio, but worse ripple ratio for the same window length and mainlobe width. As for the comparison with ultraspherical window, by fixing the window length, mainlobe width and sidelobe roll-off ratio, the proposed window presents better ripple ratio for the narrower mainlobe width and larger sidelobe roll-off ratio, but worse ripple ratio for the wider mainlobe width and smaller roll-off ratio.

REFERENCES

[1] A. Antoniou, Digital signal processing: Signal, systems, and filters, McGraw-Hill, 2005.

[2] J.F.Kaiser and R.W.Schafer, "On the use of the Io-sinh window for spectrum analysis" IEEE Trans. Acoustics,Speech, and Signal Processing, vol.28, no.1, pp. 105-107, 1980.

[3] J. F. Kaiser, "Nonrecursive digital filter design using I_0-sinh window function" in Proc. IEEE Int. Symp. Circuits and Systems (ISCAS'74), San Francisco, Calif, USA, pp.20-23, April 1974.

[4] C. L. Dolph, "A Current Distribution for Broadside Arrays Which Optimizes the Relationship Between Beamwidth and Side-lobe Level" Proc. IRE, vol.34, pp.335-348, June 1946.

[5] T. Saramaki, "A class of window functions with nearly minimum sidelobe energy for designing FIR filters" in Proc. IEEE Int. Symp. Circuits and systems (ISCAS'89), Portland, Ore, USA, vol.1, pp. 359-362, 1989.

[6] S. W. A. Bergen and A. Antoniou, "Design of ultraspherical window functions with prescribed spectral characteristics" EURASIP Journal on Applied Signal Processing, no.13, pp. 2053-2065, 2004.

[7] A. Jain, R. Saxena and S. C. Saxena, "A simple alias-free QMF system with near-perfect reconstruction" J. Indian Ins. Sci., Jan-Feb, no.12, pp. 1-10, 2005.

A model for piezoresistance in torsional MEMS springs

Silvia Lenci, Andrea Nannini, Francesco Pieri

Dipartimento di Ingegneria dell'Informazione
Università di Pisa
Via G. Caruso, 16, I-56122 Pisa, Italy
e-mail: silvia.lenci@iet.unipi.it; phone: +390502217640

Abstract- **A piezoresistive readout method for MEMS torsional resonators is presented. To this purpose, a new theoretical model describing the piezoresistive effects on monocrystalline and polycrystalline beams, used as suspended resistors, is investigated. A square law, linking the angular displacement θ to the resistance variation ΔR, is derived. The theory is confirmed by ANSYS FE (finite element) simulations. When adding a feedback to an actuator, a piezoresistive sensing part can read the output signal without increasing the process complexity and represents an alternative to capacitive sensing. A measurement configuration, based on a Wheatstone bridge, as well as an implementation of this measurement scheme to silicon-germanium (SiGe) resonators are also proposed.**

I. INTRODUCTION

Parallel-plate MEMS structures can be electrostatically driven in a closed-loop configuration by a dynamic voltage suitable for full-gap operation [1, 2]. However, closed-loop operation requires measurement of the instantaneous actuator displacement, which is not always readily available, and typically requires redesign of the device to include an additional sensing capacitor. Despite the known limitations of piezoresistive sensing (in terms of materials, temperature dependence, etc.), piezoresistance reading of the displacement could be easily implemented without significant redesign of the electrostatic actuator. In this work, piezoresistive sensing on the suspension springs (always present in the typical capacitive MEMS) is proposed as an alternative method to detect the MEMS displacement. Specifically, we propose a general method for reading the angular displacement of torsional microresonators. The sensing operation consists in measuring the resistance variation of the resonator springs, used also as suspended piezoresistors. Both monocrystalline and polycrystalline materials are considered. A square-law relationship between the angular displacement θ and the resistance variation ΔR of a torsional spring is analytically found. The resulting theory, which is general, is validated by FE simulations. An experimental verification on microresonators based on a SiGe technology [3, 4] is planned. Temperature compensation is to be provided by putting four microresonators in a Wheatstone bridge. The expected output signal of this configuration is finally presented.

II. THEORY: PIEZORESISTIVITY OF A TORSIONAL BEAM

Piezoresistivity is a phenomenon that changes the bulk resistivity of a material under applied stresses. The piezoresistive matrix links the change in resistivity to the stresses. In a cubic crystal like silicon and silicon-germanium, the matrix takes the form [5]:

$$\frac{1}{\rho}\begin{pmatrix} \Delta\rho_1 \\ \Delta\rho_2 \\ \Delta\rho_3 \\ \Delta\rho_4 \\ \Delta\rho_5 \\ \Delta\rho_6 \end{pmatrix} = \begin{pmatrix} \pi_{11} & \pi_{12} & \pi_{12} & 0 & 0 & 0 \\ \pi_{12} & \pi_{11} & \pi_{12} & 0 & 0 & 0 \\ \pi_{12} & \pi_{12} & \pi_{11} & 0 & 0 & 0 \\ 0 & 0 & 0 & \pi_{44} & 0 & 0 \\ 0 & 0 & 0 & 0 & \pi_{44} & 0 \\ 0 & 0 & 0 & 0 & 0 & \pi_{44} \end{pmatrix} \begin{pmatrix} \tau_{xx} \\ \tau_{yy} \\ \tau_{zz} \\ \tau_{yz} \\ \tau_{xz} \\ \tau_{xy} \end{pmatrix} \quad (1)$$

where $\pi_{11,12,44}$ are the piezoresistive coefficients, τ_{ik} are the normal ($i=k$) and shear ($i \neq k$) stresses ($i, k = x, y, z$), while ρ and $\Delta\rho_{1..6}$ are the zero-stress resistivity and the resistivity variations. The relationship between the electrical field and the current density in the piezoresistive solid is then [5]:

$$\begin{pmatrix} \varepsilon_x \\ \varepsilon_y \\ \varepsilon_z \end{pmatrix} = \begin{pmatrix} \rho+\Delta\rho_1 & \Delta\rho_6 & \Delta\rho_5 \\ \Delta\rho_6 & \rho+\Delta\rho_2 & \Delta\rho_4 \\ \Delta\rho_5 & \Delta\rho_4 & \rho+\Delta\rho_3 \end{pmatrix} \cdot \begin{pmatrix} j_x \\ j_y \\ j_z \end{pmatrix}, \quad (2)$$

where $\varepsilon_{x,y,z}$ and $j_{x,y,z}$ are the electric field and the current density along the three cartesian coordinates, respectively.

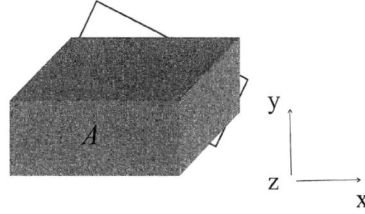

Figure 1. Section *A* of a rectangular piezoresistor under torsional load, and coordinate convention used in the text.

A. Resistance variation as a function of the induced stresses

Consider a semiconductor bar (length l and rectangular section of area A) under a torsional load. The generic section of the solid is illustrated in fig. 1. When applying a current density j_z along the direction perpendicular to the section, transverse current densities $j_x(x,y,z)$ and $j_y(x,y,z)$ are created because of the piezoresistivity-induced variation of ρ. j_z is assumed to be constant in the volume, while the transverse current densities are functions of the space coordinates. The resistor can be divided, along the length l (z-axis), in a series of infinitesimal resistors r_i, each of them having constant section A and length Δz. The conductance of each resistor can be expressed as:

$$\frac{1}{r_i} = \frac{1}{r_{i0} + \Delta r_i} \cong \frac{1}{r_{i0}} - \frac{\Delta r_i}{r_{i0}^2}, \qquad (3)$$

where r_{i0} and Δr_i are, respectively, the zero-stress resistance and the resistance variation of the infinitesimal resistor and $r_{i0} = \rho \Delta z / A$. Moreover, we have

$$\frac{1}{r_i} = \frac{1}{\Delta z} \int_0^t \int_0^h \frac{j_z}{\varepsilon_z(x,y,z_i)}\, dx dy, \qquad (4)$$

where $z_{i+1} = z_i + \Delta z$ and t and h are the width and the height of the section ($A = th$). We assume the reasonable hypothesis that the resistivity variations and both j_x and j_y are only functions of (x,y) and (4) becomes

$$\frac{1}{r_i} = \frac{1}{\Delta z} \int_0^t \int_0^h \frac{dx dy}{\rho + \Delta\rho_5(x,y)\dfrac{j_x(x,y)}{j_z} + \Delta\rho_4(x,y)\dfrac{j_y(x,y)}{j_z} + \Delta\rho_3(x,y)}. \qquad (5)$$

It is reasonable to assume that the first term in the denominator of the integral argument in (5) is much greater than the others, both because $\Delta\rho_{3,4,5} \ll \rho$ and $|j_x|, |j_y| \ll |j_z|$. Approximating the integral argument with its first-order power series, we obtain

$$\frac{1}{r_i} \cong \frac{A}{\rho \Delta z} - \frac{1}{\rho \Delta z}\left(I_{\rho5} + I_{\rho4} + I_{\rho3}\right) = \frac{1}{r_{i0}} - \frac{\Delta r_i}{r_{i0}^2}, \qquad (6)$$

where

$$I_{\rho5} = \frac{\pi_{44}}{j_z} \int_0^t \int_0^h \tau_{xz}(x,y) j_x(x,y)\, dx dy, \qquad (7)$$

$$I_{\rho4} = \frac{\pi_{44}}{j_z} \int_0^t \int_0^h \tau_{yz}(x,y) j_y(x,y)\, dx dy, \qquad (8)$$

$$I_{\rho3} = \int_0^t \int_0^h \left[\pi_{11}\tau_{zz}(x,y) + \pi_{12}\left(\tau_{xx}(x,y) + \tau_{yy}(x,y)\right)\right] dx dy. \qquad (9)$$

Due to the symmetry of the bar, it is reasonable to assume that the normal stresses are antisymmetric over the surface A. The hypothesis is confirmed by ANSYS FE simulations. As a consequence of the stress symmetry, the integral in (9) is equal to zero.

Being R_0 the zero-stress resistance of the semiconductor bar (length l and section A) and ΔR the resistance variation due to torsion, we can write (3) as

$$\frac{1}{R} \cong \frac{1}{R_0} - \frac{\Delta R}{R_0^2}, \qquad (10)$$

where $R_0 = \rho l / A$. Thus, the resistance variation results to be

$$\Delta R \cong \frac{\rho l}{A^2}\left(I_{\rho5} + I_{\rho4}\right) = \frac{R_0}{A}\left(I_{\rho5} + I_{\rho4}\right). \qquad (11)$$

An expression for the transverse current densities, which are generated in a beam under torsion, is derived in [6] and results to be proportional to the shear stresses as

$$j_k(x,y) \cong -\frac{\pi_{44}\tau_{kz}(x,y)}{\rho}\varepsilon_z, \qquad (12)$$

where $k = x, y$. Equations (12) are substituted into (7) and (8) and the resulting integrals into (11). ε_z is considered to be constant, since the resistance variations are $\ll\rho$. Finally, ΔR is found to be

$$\Delta R \cong -R_0 \pi_{44}^2 \left(\overline{\tau_{xz}^2} + \overline{\tau_{yz}^2}\right), \qquad (13)$$

where $\overline{\tau_{kz}^2} = \dfrac{1}{A}\displaystyle\int_0^t \int_0^h \left(\tau_{kz}^2(x,y)\right) dx dy$ and $k = x, y$. The resistance change of a monocrystalline beam under torsional load depends on the products between the current densities and the shear stresses. Since such currents are linear with the stresses, ΔR results to be a square function of the same stresses.

B. Application to a Torsional Microresonator

The derived theory is applied to an electrostatically-actuated torsional microresonator (fig. 2). A constant voltage and torque are considered and the resistance variation ΔR is calculated as a function of the torsion angle θ.

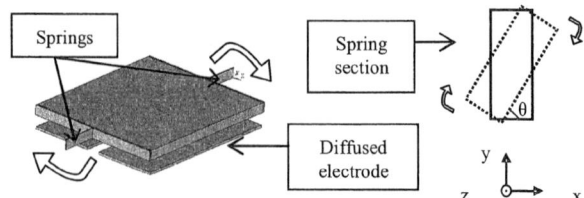

Figure 2. Torsional resonator scheme

Each spring is a rectangular-section beam with piezoresistive properties. The hypothesis that only the springs contribute to ΔR is assumed. Equation (13) shows that the resistance variation is a square function of the shear stresses. Since it is reasonable to assume such stresses to be linear with θ, a square relationship between ΔR and θ is expected:

$$\Delta R = \alpha \theta^2 . \qquad (14)$$

While relationship (14) holds for any beam section, direct computation of α is not straightforward unless a closed-form expression is available for the stresses which can be used to compute the averages in (13). For this reason, the square-law dependency in Eq. (14) was verified by ANSYS FE simulations.

III. SIMULATIONS AND RESULTS

A single spring, consisting of coupled-field piezoresistive elements (solid 226), is modeled. Default solver options are used. While the main interest of this study is aimed at the design of a polycrystalline SiGe resonator, the piezoresistivity of SiGe is still a subject of study [7, 8] and no reliable values of the piezoresistive coefficients exist yet. Thus, the theory of section II is validated by simulations using the p-doped, monocrystalline silicon piezoresistive coefficients [5]: $\pi_{11} = -102.2 \cdot 10^{-11}$, $\pi_{12} = 53.4 \cdot 10^{-11}$ and $\pi_{44} = -13.6 \cdot 10^{-11}$ Pa^{-1}. The torsional spring is a 0.6 μm thick, 4 μm high, and 25 μm long parallelepiped. A Young's modulus E = 146 GPa and a Poisson ratio ν = 0.23 are used. One end section of the beam is clamped, while the other one is twisted by moment values of 20.2, 30.3, 40.4 and 80.8 μN μm (see fig. 3 and table 1).

A. Monocrystalline springs simulations

Table 1 shows the results; in terms of ΔR and θ. The simulated ΔR values are fitted against a power relationship (see fig. 4) and show a square-law behavior, as expected from (14): $\Delta R = \alpha \theta^2$. From the fitting, $\alpha = -6.84$ Ω/rad^2 is found.

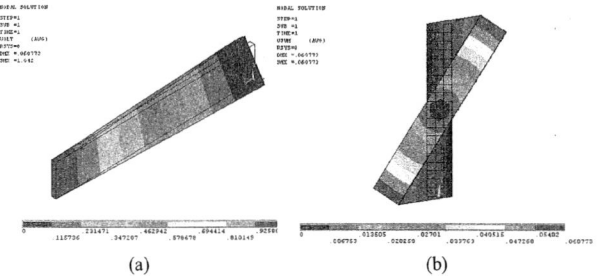

(a)   (b)

Figure 3. Monocrystalline resonator spring under torsion. ANSYS plot of the voltage drop (a) and total displacement vector (b) for a moment $M = 20.2 \cdot 10^{-12}$ N m. A static current of 10 mA is applied and the voltage change allows the evaluation of ΔR.

TABLE I. RESULTS

M [μN·μm]	20.2	30.3	40.4	80.8
ΔR [mΩ]	-6.150	-13.86	-24.67	-99.85
θ [rad]	0.0301	0.0451	0.0601	0.120

Simulations for a monocrystalline spring.

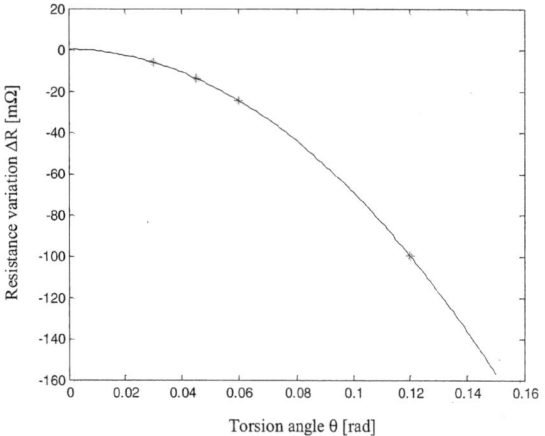

Figure 4. Resistance variation as a function of torsion angle.

B. Polycrystalline springs simulations

A polycrystalline spring is also modeled by dividing the resistor into small volumes. A uniform random crystal orientation is assigned to each sub-volume, which represents a grain. To simplify the computational approach, the volume is divided only along the spring axis, resulting in column grains spanning the whole height and width of the bar. Montecarlo simulations are performed to evaluate the average values of the resistance variation and of α. Also for a polycrystalline material the relationship between ΔR and θ shows a square-law behavior. For a typical grain dimension (~1 μm), mean values of $\langle \alpha \rangle \sim 10$ Ω/rad^2 are found.

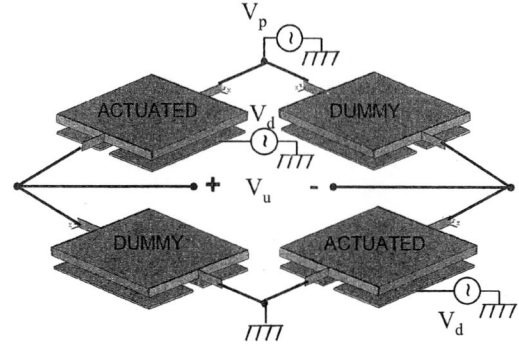

Figure 5. Measurement setup: Wheatstone bridge configuration consisting of two actuated and two "dummy" resonators. A bias supply V_p and an actuation voltage V_d are provided.

978-1-4244-1983-8/08/$25.00 ©2008 IEEE

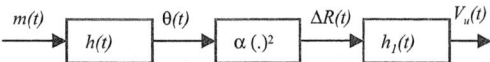

Figure 6. Electromechanical system formed by 4 Wheatstone bridge-connected torsional resonators. $m(t)$ is the torsional moment.

IV. PROPOSED MEASUREMENT CONFIGURATION

Although a static moment is used for the simulation in order to find the square relationship (14), ΔR will be measured by driving the resonators on the first resonance frequency ω_0 (torsion about the springs axis). Since piezoresistive measurements are affected by temperature variation, a measurement configuration as Wheatstone bridge is proposed (fig. 5), which requires, as a disadvantage, the use of dummy resonators. The output voltage V_u is proportional to ΔR, and thus contains information about θ. The system results in a piezoresistive MEMS sensor of torsion angle, according to the scheme illustrated in fig. 6, where $m(t)$ is the torsional moment. The angle is linked to the moment by the transfer function

$$H(j\omega) = \frac{\Theta(j\omega)}{M(j\omega)} = \frac{1}{k_\theta} \cdot \frac{1}{1 - \left(\dfrac{\omega}{\omega_0}\right)^2 + j\dfrac{\omega}{\omega_0 Q}}, \qquad (15)$$

where k_θ is the elastic constant of the torsional spring and Q is the quality factor. When the system is driven on resonance ω_0, assuming $Q = 100$, the moment is proportional to θ as $m(\theta) = \dfrac{k_\theta}{Q}\theta$. Considering small angles, it can be shown that [6]

$$m(\theta) = \frac{1}{2}\beta\left(\frac{V_p}{2} - V_d\right)^2, \qquad (16)$$

where $\beta = \left.\dfrac{\partial C}{\partial \theta}\right|_{\theta=0}$ and C is the capacitance formed by the resonator and the underlying diffused electrode. The output voltage can be expressed as

$$V_u(t) \cong \beta^2 \frac{\alpha}{R_0}\left(\frac{Q}{k_\theta}\right)^2 \frac{V_p(t)}{4}\left(\frac{V_p(t)}{2} - V_d(t)\right)^4. \qquad (17)$$

The output voltage can be driven by choosing a suitable configuration for $V_p(t)$ and $V_d(t)$. The following input voltages, where the bias contains an intermediate frequency ω_1, are presented [6]:

$$\begin{aligned} V_p(t) &= V_{p0} + V_{p1}\cos(\omega_1 t) \\ V_d(t) &= V_{d1}\cos(\omega_0 t). \end{aligned} \qquad (18)$$

When applying the above configuration, the output voltage contains a DC part and 7 harmonic components, at the following frequencies: ω_0, $2\omega_0$, ω_1, $\omega_1 + \omega_0$, $\omega_1 - \omega_0$, $2\omega_0 + \omega_1$ and $2\omega_0 - \omega_1$. As an advantage, the intermediate

frequency allows tuning the output far from the resonance frequencies of the system. Experimental results are needed: polycrystalline SiGe resonators, with the same geometries as those discussed in section III, have been processed at IMEC and will be measured in DC and AC. Further input configurations need to be performed, in order to reduce possible matching errors that affect the resonators, changing the zero-stress resistance values.

V. CONCLUSIONS

A theoretical model for the piezoresistive detection of angular displacements of MEMS torsional microresonators is discussed. Theory shows a square relationship between the angle and the resistance variation, confirmed by ANSYS FE simulations on the piezoresistivity of rectangular-section torsional springs. Although this relationship implies a more sensitive measurement method with respect to a linear law, the advantage of direct (i.e. without modifications to the design of the device) inline measurement of the deflection of a MEMS could make this approach feasible for closed-loop control of capacitive MEMS. Moreover, the method provides a sensing part without adding any process step. A measurement method is also proposed. Experimental measurements on SiGe resonators to validate the proposed model, as well as effect of resonator mismatch on the output signal are planned.

ACKNOWLEDGMENT

This work has been partly financed by STMicroelectronics. We are grateful to IMEC (Leuven, Belgium) for the process on the SiGe microresonators. A special acknowledgment goes to Antonio Molfese for his kind help.

REFERENCES

[1] L.A. Rocha, E. Cretu, R.F. Wolffenbuttel, "Using dynamic voltage drive in a parallel-plate electrostatic actuator for full-gap travel range and positioning", Journal of Microelectromechanical Systems, vol.15, pp. 69-83, 2006.

[2] B. Borovic, A. Q. Liu, D. Popal, H. Cai, F. L. Lewis, "Open-loop versus closed-loop control of MEMS devices: choices and issues", Journal of Micromechanics and Microengineering, vol. **15, p.** 1917-1924, 2005.

[3] A. Mehta, M. Gromova, P. Czarnecki, K. Baert, A. Witvrouw, "Optimisation of PECVD poly-SiGe layers for MEMS postprocessing on top of CMOS", Digest Tech. Paper Transducers '05, Vol. 2, 1326-1329, 2005.

[4] M. Gromova et al., "Highly reliable and extremely stable SiGe micro-mirrors", Proc. MEMS '07, pp. 759-762. 2007.

[5] S.M. Sze, Semiconductor Sensors, John Wiley & Sons.

[6] S. Lenci, Master DegreeThesis, A.A 2005- 2006.

[7] J. Richter et al., "Piezoresistance of Silicon and Strained $Si_{0.9}Ge_{0.1}$", Sensors and Actuators A, Vol. 123-124, pp. 388-396, 2005.

[8] S. Lenci et al., "Determination of the piezoresistivity of microcrystalline silicon-germanium and application to a pressure sensor", Proc. MEMS '08, pp. 427-430, 2008.

978-1-4244-1983-8/08/$25.00 ©2008 IEEE

Development of a Bond Graph Based Model of Computation for SystemC-AMS

Torsten Maehne, Alain Vachoux, and Yusuf Leblebici

Laboratoire de Systèmes Microélectroniques (LSM), École Polytechnique Fédérale de Lausanne (EPFL)
EPFL/STI/IEL/LSM, Bâtiment ELD, Station 11, CH-1015 Lausanne, Switzerland
Phone: +41(21)69-36922, Fax: +41(21)69-36959, WWW: http://lsm.epfl.ch/, E-Mail: torsten.maehne@epfl.ch

Abstract—The modelling and simulation capabilities of SystemC-AMS concerning conservative continuous time systems involving the interaction of several physical domains and with digital control components are currently limited. Bond graphs unify the description of multi-domain systems by modelling the energy flow between the electrical and non-electrical components. They integrate well with block diagrams describing the signal processing part of a system. The goal of this work is to integrate the bond graph formalism as a new Model of Computation (MoC) into the SystemC-AMS prototype.

I. INTRODUCTION

The advances in processing technologies supports since several years the trend towards more and more feature rich and heterogeneous Systems-on-Chips (SoCs) or Systems-in-Packages (SiPs). The increasing complexity of the implemented functionalities, the diversity of possible implementations (e.g., digital/analog/RF hardware, software, Micro-Electro-Mechanical System (MEMS)), and the rapid evolution of the needs (e.g., time to market, product diversity, compliance to standards) ask, on the one hand, for early partitioning decisions that meet design constraints and, on the other hand, for easy reuse and retargeting. To cope with the resulting design complexity, application domain dependent modelling and abstractions concepts need to be used in parallel throughout the design flow.

SystemC [1] is a C++ library, which allows to model complex digital hardware/software systems by mapping them on communicating processes, which are executed and synchronized by a *Discrete-Event* (DE) MoC based simulation kernel. There have been several attempts to extend SystemC to support the design of heterogeneous systems as described earlier. *SystemC-A* [2] and *SystemC-AMS* [3] use a SPICE-like approach to model energy conserving systems using generalized networks, which limits possible improvements of the simulation performance. *SystemC-AMS* also implements the *Synchronous Data Flow* (SDF) MoC to model signal processing dominated continuous time behavior. A synchronization layer handles the communication between both continuous time MoCs and the discrete event MoC of the SystemC simulation kernel. This is an advantage over *SystemC-A*, which requires modifications to the standard SystemC kernel thus hampering the integration of other extensions. *SystemC-WMS* [4] uses another approach based on the Wave Digital

Filter (WDF) theory by describing the conservative components through a scattering matrix and their interaction through incident and reflected energy waves. The implementation is entirely based on the hierarchical channels of SystemC. This implies the scheduling of many discrete events per analog solution point and thus limits the simulation performance.

This work aims for improving the modelling and simulation capabilities of SystemC-AMS regarding conservative continuous time components from different physical domains and their interaction with discrete time (digital) control components by implementing a new MoC based on the *bond graph formalism*. Section II introduces this formalism and Section III describes the resulting requirements and first results of its integration as a new MoC into SystemC-AMS.

II. BOND GRAPH FORMALISM

The bond graph formalism [5] unifies the description of multi-domain systems and is mostly used in mechanical engineering, mechatronics, and control theory. Each domain-specific system model (e.g., electrical circuit, mechanical multi-body system, fluidic or thermal networks) can be transformed into a bond graph representing the energy flow between generalized elements of a multi-domain system. The energy link between the ports of two elements El_i and El_k is represented with an half-arrow shaped *bond*:

$$El_i \xrightarrow[f]{e} El_k$$

Associated to each bond are an *effort e* and a *flow f variable*. They are called *power variables* because their product is the power P. By definition, the half-arrow points into the direction, in which the power flows for positive e and f. The description of dynamic systems requires also to introduce two *energy variables*, called *(generalized) momentum p(t)* and *(generalized) displacement q(t)*, which are defined as the time integral of an effort and flow, respectively:

$$p(t) = \int^t e(t)\,dt \qquad q(t) = \int^t f(t)\,dt \tag{1}$$

The energy E can be expressed as a function of time or of one of the energy variables:

$$E(t) = \int^t P(t)\,dt \quad E(q) = \int^q e(q)\,dq \quad E(p) = \int^p f(p)\,dp \tag{2}$$

This work has been funded by the Hasler Stiftung under project № 2161.

978-1-4244-1983-8/08/$25.00 ©2008 IEEE

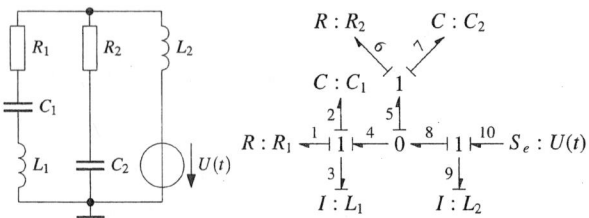

Fig. 1.   Interpretation of a bond as a bilateral signal flow

The definition of the *generalized* power and energy variables is independent of a particular physical domain. The link to the physical domain is kept through the units attached to the variables and parameters of the generalized components. Effort and flow can, e.g., represent voltage/current, force/velocity, or pressure/volume flow rate in electrical, mechanical, or hydraulic systems, respectively.

The energy exchange through the bond between two elements causes the effort and flow variables to act in opposite directions. This can be used to determine the computational direction, which is indicated by a perpendicular stroke at one end of the bond. This *causal stroke* states that at this side the effort variable e is known (it acts as an input) and f can be calculated as a function $f := \Phi_k^{-1}(e)$. Consequently, the flow f is known on the other side of the bond and acts as an input to a function to calculate the effort: $e := \Phi_i(f)$ (Fig. 1). The equations describing the component behavior impose a *required*, *preferred* (e.g., due to numerical reasons), or *free* causality (*effort-in* or *effort-out*) on the element ports. The resulting constraints need to be propagated to all related ports. Methods like the Sequential Causality Assignment Procedure (SCAP) [5] exist to systematically define the causality of a bond graph. This and the compact visualization of the energy flow and computational structure are their main advantages. The assigned causalities allow to sort the element equations in the right order for an efficient model execution and to do some further formal checks on the model: the number of states and non-states in the system, the presence of algebraic loops during model execution, or if it is an ill-posed model.

Three generalized 1-port elements represent the resistive R, inertial I, and capacitive C elements independent of the considered physical domain. Energy sources are modelled as effort source S_e and flow source S_f elements. Quantity transformations (also across domain boundaries) are represented through the transformer TF and gyrator GY 2-port elements, which ensure the conservation of energy. Multi-port junction elements represent explicitly the connexion of elements, which are exposed to a common effort (0-junction) or a common flow (1-junction). All elements may have non-linear characteristic equations. TF, GY, S_e, and S_f can be modulated by an external signal. As an example, Fig. 2 shows a simple electrical circuit, its equivalent bond graph annotated with causality, and the derived computational structure. The bond graph is constructed from the circuit by using the equivalences between the electrical and bond graph elements and 0-/1-junctions to represent the parallel/serial connections of the circuit, respectively. The computational structure results from the dependencies between the quantities due to the element equations, the bond graph structure, and the assigned causality.

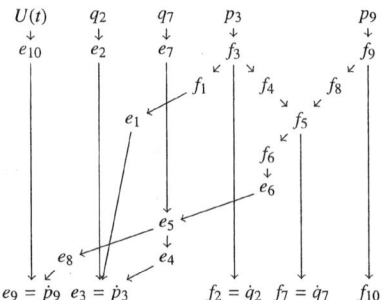

(a) Simple electrical circuit   (b) Bond graph with annotated causality

(c) Dependency graph representing the computational structure of the bond graph

Fig. 2.   Example of the transformation of an electrical circuit to an equivalent bond graph annotated with causality and then derived computational structure

The causality assignment allows also for a natural integration of bond graphs with signal flow graphs. An example is given in Fig. 3 in the form of a car wheel model of an electronically controlled suspension system incorporating a damper and a load-leveller [5]. The system is once modelled in the "classic" domain-specific way and once using bond graphs for the energy conserving part and block diagrams for the signal processing part. The blocks in the signal flow graph can take power or energy variables as input and can modulate the sources or element parameters of the bond graph.

Complex systems such as Analog and Mixed-Signal (AMS)-SoCs require a hierarchical description of their structure. This can be achieved by using the more abstract *word bond graphs* [5], which add the usage of "macros" to the classic bond graph modelling to represent a multi-port component of the overall system. Such systems are also often characterized through the presence of digital units, which control the continuous time parts through feedback and switching of energy flows due to digital signals. The switching can be modelled using idealized controlled junctions. Their presence indicate a *hybrid bond graph* [6]. An active junction constrains the causality of attached bonds like a normal junction. When it gets inactive, it imposes a zero on the power variable, which is common to all attached bonds, leading to a causality change. In electrical terms this means that a parallel connexion modelled by a 0-junction is "short-circuited" and a series connexion modelled by a 1-junction is "left open". The causality change needs to be propagated into the graph and requires a solver reinitialization imposing a simulation performance penalty.

Specialized bond graph tools such as MTT, 20-sim, and En-

978-1-4244-1983-8/08/$25.00 ©2008 IEEE

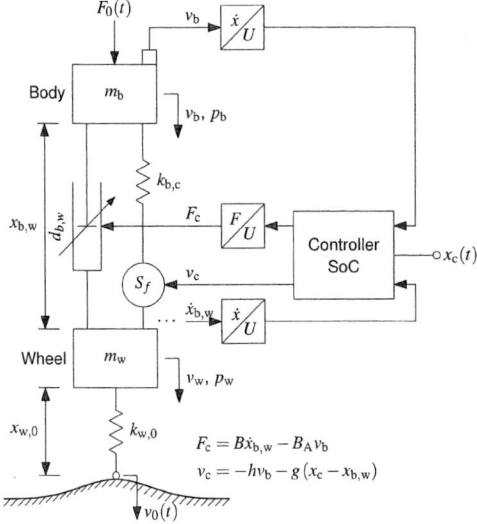

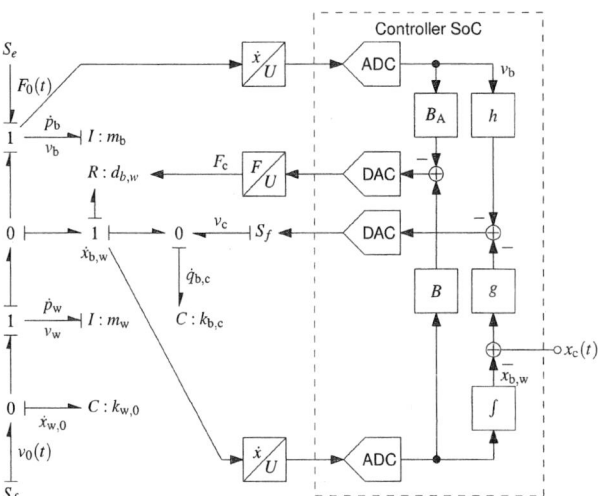

(a) "Classic" domain-specific system representation

(b) Representation coupling a bond graph for the energy conservation part with a block diagram for the signal processing part

Fig. 3. Car wheel model of an electronically controlled suspension system incorporating a semi-active damper and a fast load-leveller [5]

port (reviews available on, e.g., http://www.bondgraphs.com/) are not popular in the microelectronics community. There are also efforts to integrate bond graph support in widely used mathematical tools such as MATLAB/Simulink (e.g., Bond-Lab) or Mathematica (e.g., Bond graph tool box), which usage is limited to very early design stages. Another approach is to use the capabilities of an AMS-HDL to represent bond graphs. BondLib [7] implements causal/a-causal bond graphs support for Modelica/Dymola. However, Modelica's capabilities are weak on the discrete event side, which is one reason why it is not suitable for AMS-SoC design. [8] shows that bond graphs can be represented in VHDL-AMS—however, not very efficiently as equations are duplicated due to their mapping to

generalized networks. Those impose Kirchhoff's voltage and current laws, which are already implicitly considered through the 0- and 1-junctions. Causality cannot be assigned to the bonds and thus not be benefited from. All three aspects have a negative impact on the simulation performance of the model.

III. SystemC-AMS Bond Graph Extension

The previous section showed that the bond graph formalism has some properties, which make its application to AMS-SoC design attractive and would complement well the already available MoCs in SystemC-AMS. This motivates the development of a bond graph MoC. These properties are:

- The unified description of conservative multi-domain systems with dimension units attached to variables/parameters keeps the link to the related physical domain. A dimensional analysis can thus ensure proper model assembly. This is currently not possible in SystemC-AMS since all non-electrical conservative behavior needs to be mapped to electrical primitives of the Electrical Linear Network (ELN) MoC discarding the dimension units.
- The causality analysis of a bond graph allows the ordering of the describing equations for an optimized (procedural) execution (similar to the static scheduling of the Synchronous Data Flow (SDF) MoC). This alternative to the set-up and solution of a global Differential Algebraic Equation (DAE) system for a generalized network description promises a higher simulation performance.
- The semiformal checks (number of states/non-states, algebraic loops, ill-posed model) possible after causality analysis offer insight into the model's physical and computational structure to the designer.
- Word bond graphs permit hierarchical modelling to tackle complexity and allow reuse.
- Bond graphs integrate well with signal-flow models (SDF MoC) and allow switching of energy flows due to external signals (DE MoC).

To retain these advantages and implementing them in a way coherent to the philosophy and modular architecture of SystemC-AMS are the principal design requirements to the bond graph MoC [9]. The MoC is in an early development stage. This includes the definition of how the layered architecture of SystemC-AMS is going to be extended (Fig. 4) and the start of the design and implementation of the necessary class hierarchy. The *module layer* provides the user interface to the library. Templated `sca_bg_port` and `sca_bg_bond` classes will allow to interconnect instances derived from `sca_bg_module`, which is the base class to implement the bond graph primitives. The template nature of the port and bond classes allows them to be typed to not only carry **double** quantities but properly typed quantities with a dimension unit. It is planned to use the *quantitative units library*, which is part of the *Boost project* (http://www.boost.org/) and implements units not only as "syntactical sugar" but to use them for compile time dimensional analysis. Fig. 5 shows how this translates into a proposed syntax for the structural description of a simple bond graph with an `sc_module`. The *view*

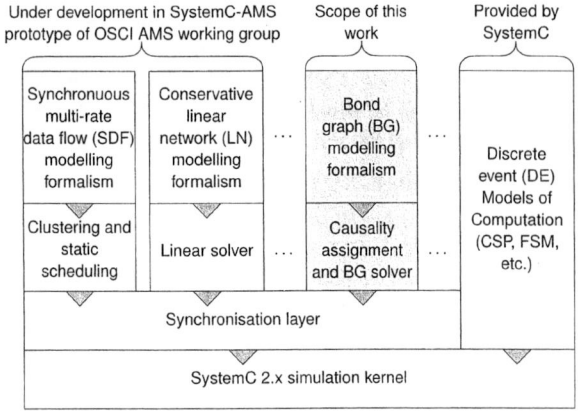

Fig. 4. Architecture of SystemC-AMS showing the MoCs provided by the SystemC kernel and the AMS extension as well as the synchronization layer

layer implemented by `sca_bg_view` will cluster the interconnected `sca_bg_modules`. The therefore necessary flattening of the hierarchy is handled by the SystemC kernel. The causality analysis, formal checks, and ordering of the equations of the bond graph primitives is then done per cluster. The *solver layer* executes the ordered equations, which will be handled by an `sca_bg_solver` object per cluster. The scheduling and *synchronization layer* executes the `sca_bg_solvers` as part of the surrounding SDF cluster. It also notifies the solver object of switching events causing an energy flow commutation. This will toggle causality reassignment and regeneration of the computation model at runtime. Their efficient implementation is still topic of active research [6].

IV. CONCLUSIONS AND OUTLOOK

The reviewed current state of AMS extensions to SystemC shows a gap between low-level SPICE-like modelling of conservative behavior and high-level signal-flow modelling of non-conservative behavior. The presented work addresses this issue through the development of a bond graph MoC for SystemC-AMS to improve its modelling and simulation capabilities in the field of energy conserving multi-domain components and their interaction with digital control components. The design and implementation of the bond graph MoC for SystemC-AMS are under way. A first definition of its architecture and a preliminary syntax for describing bond graphs were presented. In a first step, the "classic" bond graph primitives will be implemented. Future working directions are an optimized implementation of hybrid bond graphs controlled by discrete events, the optimization of the computational structure to reduce the number of algebraic loops, and further verification of the models using the results from causality analysis.

REFERENCES

[1] *IEEE Standard 1666-2005, SystemC Language Reference Manual.*

[2] H. Al-Junaid, T. Kazmierski, and L. Wang, "SystemC-A modeling of an automotive seating vibration isolation system," in *Forum on Specification and Design Languages (FDL) 2006.*

$$S_e : e(t) \xrightarrow{\;1\;} \begin{array}{c} R : R_1 \\ {\scriptstyle 2}\big| \\ 1 \xrightarrow{\;3\;} I : I_1 \\ {\scriptstyle 4}\big| \\ C : C_1 \end{array}$$

(a) Bond graph of `simple_bg`

```
#include "sca_bond_graph"

using namespace sca_bg;
using namespace sca_bg::domains;
using namespace boost::units;
using namespace boost::units::SI;

struct simple_bg : public sc_module {
  // bond graph elements
  sca_R<electrical> R1;
  sca_C<electrical> C1;
  sca_I<electrical> I1;
  // sine source derived from sca_Se
  Se_sine<electrical> Se1;
  // junctions and bonds
  sca_1<electrical> CFJ1;
  sca_bond<electrical> b1, b2, b3, b4;

  // set element parameters
  simple_bg(const sc_module& name)
  : Se1("Se1", 0.0 * volt, 1.5 * volt,
        50.0 * hertz, 0.0 * radian),
    R1("R1", 1.0e3 * ohm),
    C1("C1", 10.0e-9 * coulomb / volt),
    I1("I1", 1.0 * micro * henry),
    CFJ1("CFJ1"),
    b1("b1"), b2("b2"), b3("b3"), b4("b4")
  {
    // connectivity
    b1.bind(Se1.p, CFJ1.p);
    b2.bind(CFJ1.p, R1.p);
    b3.bind(CFJ1.p, C1.p);
    b4.bind(CFJ1.p, I1.p);
  }
};
```

(b) Proposed syntax

Fig. 5. Structural description of a simple bond graph with an `sc_module`

[3] A. Vachoux, C. Grimm, and K. Einwich, "Extending SystemC to support mixed discrete-continuous system modeling and simulation," in *IEEE International Symposium on Circuits and Systems (ISCAS) 2005.*

[4] S. Orcioni, G. Biagetti, and M. Conti, "SystemC-WMS: A wave mixed-signal simulator," in *8th International Forum on Specification & Design Languages (FDL) 2005.*

[5] D. C. Karnopp, D. L. Margolis, and R. C. Rosenberg, *System Dynamics: Modeling and Simulation of Mechatronic Systems*, 4th ed. Wiley, 2006.

[6] C. D. Beers, E.-J. Manders, G. Biswas, and P. J. Mosterman, "Building efficient simulations from hybrid bond graph models," in *2nd IFAC Conference on Analysis and Design of Hybrid Systems 2006.*

[7] F. E. Cellier and R. T. McBride, "Object-oriented modeling of complex physical systems using the Dymola bond-graph library," in *6th SCS International Conference on Bond Graph Modeling and Simulation (ICBGM) 2003.*

[8] F. Pêcheux, B. Allard, C. Lallement, A. Vachoux, and H. Morel, "Modeling and simulation of multi-discipline systems using bond graphs and VHDL-AMS," in *International Conference on Bond Graph Modeling and Simulation (ICBGM) 2005.*

[9] T. Maehne and A. Vachoux, "Proposal for a bond graph based model of computation in SystemC-AMS," in *Forum on specification & Design Languages (FDL) 2007.*

978-1-4244-1983-8/08/$25.00 ©2008 IEEE

Efficient and Refined Modeling of Wireless Sensor Network Nodes Using SystemC-AMS

Michel Vasilevski, Nicolas Beilleau, Hassan Aboushady, Francois Pecheux
Laboratoire UPMC/LIP6/SOC, Paris, France
{michel.vasilevski,nicolas.beilleau,hassan.aboushady,francois.pecheux}@lip6.fr

Abstract—**The paper introduces a model of wireless sensor network (WSN) nodes with SystemC-AMS, a C++ based language for system-level description, then presents model refinements and simulation improvements of the system. Firstly the paper details the components of a WSN node : a sensor, an ADC, a microcontroller and a RF transceiver with some introduced impairements. Secondly, it presents the implementation of RF components refinements, for a designer closer model specifications, such as power gain, noise figure, IIP3 specifications. Finally, the third part shows the simulation improvements introduced by a baseband equivalent implementation made easier with a C++ based language.**

I. INTRODUCTION

Wireless sensor network design has become an issue of reflection since a large amount of applications is developed. Modeling such systems implies to deal with heterogeneous [1] simulations in terms of signal type (digital and analog) as well as operating frequencies. An expected purpose for modeling wireless sensor network systems is to use the same language to simulate the architectural description of the system for components specifications validation, and to run several transmissions for protocol communications. Thus, architecture designers along with network designers would work in cooperation. When some languages like VHDL-AMS [2] [3] or Verilog-AMS [3] show rapidly their limits regarding simulation performance and interoperability, it is interesting to experiment such complex architecture with SystemC-AMS [4] [5] [6] to reveal its advantages.

SystemC-AMS is an extension of SystemC [7], an open source C++ based language. It introduces two models of computation, SystemC-AMS synchronous dataflow (SDF) and SystemC-AMS conservative. In addition of SystemC event-driven model of computation (for digital description), we used such SystemC-AMS views to model WSN system. The conservative view uses linear networks to make a descritption, the sensor is described in this way. The synchronous dataflow view embeds continuous-time modules into dataflow clusters. Each cluster contains modules that exchange a dataflow. It is a set of timed values that defines the signal at a certain time. When scheduled by the SystemC simulation kernel, a dataflow cluster runs at a constant simulation time step. Each module of a cluster can specify a different time step when produced sample number is different of consumed sample number. It permits to make a heterogeneous frequency operating simulation, computing an adapted number of samples according to low or high varying dataflow. Hence, SDF is specially suited for communication systems like WSN with strong oversampling. ADC and RF parts are described with SDF models. Microcontroller is described with SystemC event-driven model of computation. For more details about language implementation, some code listings of certain modules are given in [8].

II. WIRELESS SENSOR NETWORK SYSTEM AND ITS DIRECT SYSTEMC-AMS IMPLEMENTATION

The WSN system consist of two nodes (*N1*, *N2*) that exchange information through a 2.4 GHz communication channel. As shown in figure 1, each node contains a sensor, an ADC, a microcontroller and a RF transceiver.

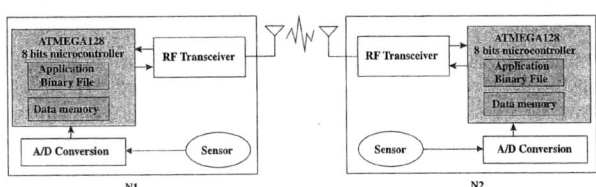

Fig. 1. The WSN, consisting of two nodes N1 and N2.

The behavior of the WSN is the following: *N1* acquires an analog measure from its sensor and uses an ADC to convert it into its 8-bits digital equivalent. The *N1* ATMEGA128 microcontroller reads the 8-bits value on its ports, continuously executes instructions to serialize the read data, and propagates the corresponding bitstream on a 1-bit port configured as output. The bitstream is emitted by the *N1* RF transmitter, and is received by the *N2* RF receiver. The received bitstream is not propagated to the *N2* microcontroller, because no synchronization protocol has been implemented yet. In the same time, the *N2* mote performs exactly the same work and propagates a sensor measure to *N1*. Simulation results given below take into account this full duplex behavior.

A. Sensor

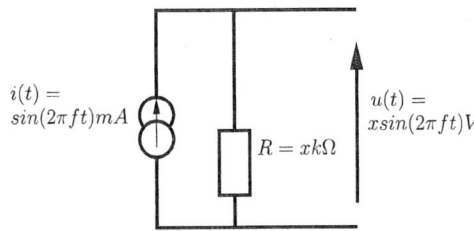

Fig. 2. Simple electrical model of a sensor.

The sensor is described in figure 2, it is a simplified model that translate analog physical variations (like temperature, pressure, wetness) to a voltage value thanks to a load resistor. The analog physical variations are modeled with a sinusoidal current source that covers a specified amplitude. The current source and load resistor are modeled using the conservative view offered by SystemC-AMS.

B. A/D converter

The ADC contains a second order sigma-delta 1-bit modulator with delayed return-to-zero feedback [9] and a decimator using a third order FIR2 filter [10] that can be parameterized to generate a n-bit word, as shown in the theoretical figure 3. Oversampling rate and

978-1-4244-1983-8/08/$25.00 ©2008 IEEE

number of bits produced are parameters that can be fixed following particular specifications.

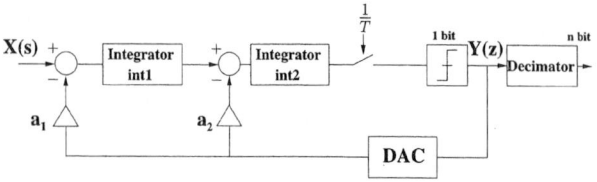

Fig. 3.   Second order sigma-delta continuous-time modulator and decimator.

C. ATMEL ATMEGA128 Microcontroller

The microcontroller is ATMEGA128 [11], it is an AVR family device from ATMEL. It is a RISC microcontroller with 16-bit wide instructions and a flash program memory of 128 Kbytes. In this first approach, we used C language compiled binary file and a simplified behavior, for embedded application. Evolution of project will be oriented to TinyOS solution, an embedded operating system for wireless sensor networks [12].

D. 2.4 Ghz QPSK RF transceiver

The RF transceiver uses a QPSK (Quadrature Phase Shift Keying) transmission, as explained in [13] and shown in figures 4 and 5 with a f_c =2.4 GHz carrier frequency and a f_b =2.4 MHz data frequency. An AWGN (Additive White Gaussian Noise) noisy channel is introduced for calculating the fundamental RF characteristic BER (Bit Error Rate) with respect to SNR (Signal-to-Noise Ratio).

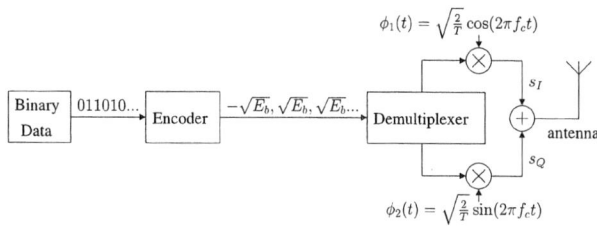

Fig. 4.   QPSK RF transmitter.

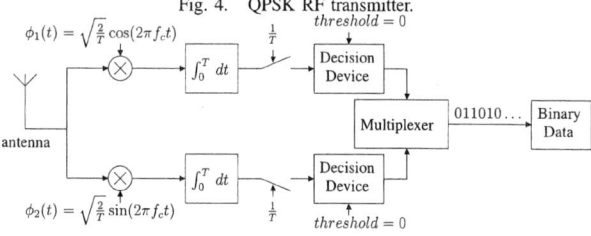

Fig. 5.   QPSK RF receiver.

E. Simulated platform

Simulation results can be displayed with gnuplot. All the tools (SystemC, SystemC-AMS, gnuplot) needed to obtain simulation results are totally open source. ADC oversampling rate has been set to 64, and decimator has been configured to 8 bits. Gain values of $\Sigma\Delta$ feedback loop are specified from amplitude histogram analysis, and are respectively set to 2 and 7/6. We used a 2.4 MHz clock frequency for microcontroller, so bit rate for RF transmission is 2.4 Mbps. Carrier frequency is 2.4 GHz.

In ADC part, we simulated a variable sine amplitude input, to compute SNR characteristics of ADC and we compared to similar matlab/simulink model result.

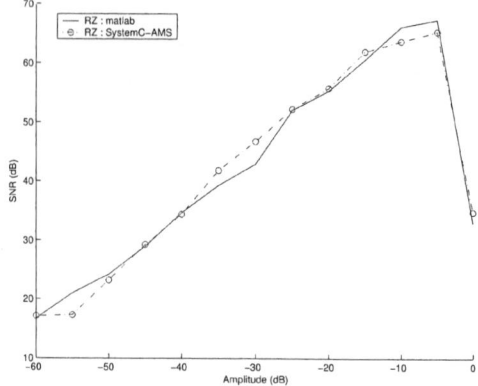

Fig. 6.   SNR analysis for ADC in relation to input amplitude

In the RF part, we performed a bit error rate (BER) analysis. Figure 7 shows the match between simulation and a theorical BER from AWGN characteristics.

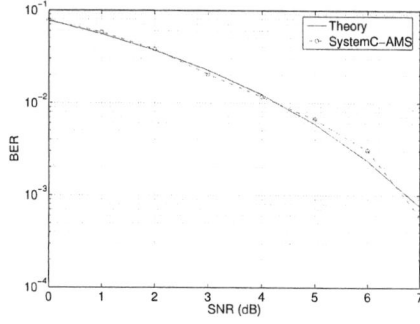

Fig. 7.   Bit error rate for QPSK transmission through an AWGN channel.

Other simulations about impairements permited to display constellations according to gain mismatch, frequency offset, phase mismatch, DC offset, they are presented in figure 8.

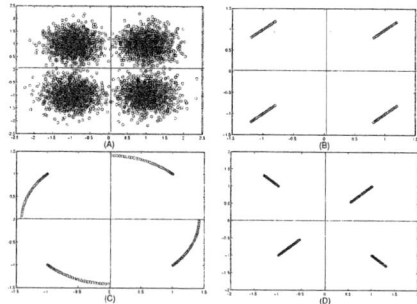

Fig. 8.   Bit error rate for QPSK transmission through an AWGN channel.

Table I presents simulation times with respect to Matlab. We specified an exact equivalent configuration and used the same number of samples. Simulation of communication between 2 motes cannot be performed with Matlab, because of the complexity of microcontroller modeling. This problem reveals one true advantage of the SystemC-AMS simulation, we are able to simulate both digital and analog models simultaneously.

978-1-4244-1983-8/08/$25.00 ©2008 IEEE

TABLE I
SIMULATION RESULTS.

	Configuration	Simulation	Matlab	SystemC AMS
ADC	OSR=64 8 bits	1 ms 16*1024 pts	1.60s	0.93s
RF	2.4 GHz carrier freq.	416.67 μs 10^3 pts for digital part 10^7 pts for RF part	2m30.74s	54.36s
2-mote WSN	Same settings	416.67 μs	–	3m1.65s

III. RF MODEL REFINEMENT IN SYSTEMC-AMS

The quality of the transmission scheme relies on the knowledge of the RF designer and on the refinement of the main specifications of the RF modules. Considering the theoretical model of the transceiver detailed earlier, this section, describes how to take RF specifications into account and how to express them with SystemC-AMS. The Low Noise Amplifier is used as an example.

A. LNA RF specifications

Based on the models used in [14] and illustrated in figure 9, the input parameters of the LNA are identified as the power available gain, the input and output impedances, the Noise Figure (NF) and the 3rd order Input Intercept Point (IIP3).

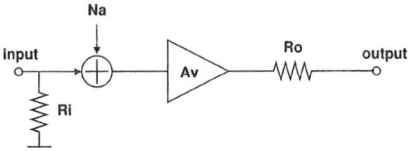

Fig. 9.   RF block model

In a first step, the thermal noise of the amplifier is added to the signal. Thermal noise is given by the specified noise figure (NF):

$$N_a = 4KT(NF - 1) \tag{1}$$

Then the gain and non-linearities are added with a polynomial representation:

$$V_{out} = 0 + A_1 * V_{in} + A_3 * V_{in}^3 \tag{2}$$

where A_1 is extracted from the power available gain, input resistance (R_i) and next block input resistance (R_l) with the following equation:

$$A_1 = \sqrt{\frac{G_p R_l}{R_i}} \tag{3}$$

and A_3 is derived from the IIP3 parameter:

$$A_3 = \frac{4 * A_v}{3 * IIP3^2} \tag{4}$$

B. SystemC-AMS implementation and simulations results

Considering these designer specifications, the implementation in SystemC-AMS of the LNA is straightforward: The LNA module has SDF input and output ports (line 4 and 5 in listing 1) that carry *double* sample values. The **init()** function (line 9 to 19) is called once during model elaboration and computes the RF coefficients used throughout simulation. The **sig_proc()** function (line 21 to 31) contains the actual behavior of the LNA module. The LNA module output is connected to two mixers (I and Q) for demodulation, each mixer have a load resistance. That's why we computed the equivalent resistance of such two parallel resistors for global load resistance.

Listing 1.   LNA implementaion

```
1  ...
2  SCA_SDF_MODULE (lna)
3  {
4    sca_sdf_in < double >in;
5    sca_sdf_out < double >out;
6    double gain_power, a1, a3, AIP3, sigma;
7    double rin, *rloadI, *rloadQ;
8
9    void init(sc_time ts, double gain_power_db,
10            double iip3, double nf,double rin,
11            double *rloadI, double *rloadQ){
12      double f  = pow(10,nf/10), N0 = 4*(f-1)*K*T*50;
13      double fs = 1/ts.to_seconds();
14      this->sigma=sqrt(N0*fs/2);
15      srand (time(NULL)); //randomize
16      this->rin=rin; this->rloadI=rloadI;
17      this->rloadQ=rloadQ; this->AIP3=undbm(iip3);
18      this->gain_power=pow(10,gain_power_db/10);
19    }
20
21    void sig_proc () {
22      double rload=
23        (*rloadI)*(*rloadQ)/((*rloadI)+(*rloadQ));
24      this->a1 =
25        sqrt(gain_power*rload/rin);
26      this->a3 =
27        a1/(3*pow(AIP3,2)/4);
28      double input = in.read()+sigma*randn();
29      out.write (a1*input-a3*pow(input,3));
30      ...
31    }
32    SCA_CTOR (lna) {}
33  };
```

Figures 10 present simulation results displayed with a simple plot program, like Gnuplot, Scilab or Matlab. The points to be plotted are computed thanks to a C++ FFT free library integrated in SytemC-AMS "signal analyser" module.

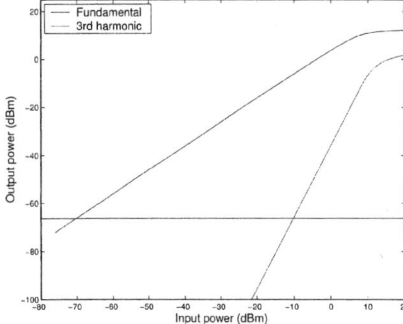

Fig. 10.   Simulations results of the LNA illustrating the non-linearities and the thermal noise implementations.

IV. BASEBAND MODELS IN SYSTEMC-AMS

The standard WSN simulation used so far shows that most of the time is spent in the simulation of the RF part, with exactly 24 billion samples generated for 1 second of simulation time. To prevent this simulation time to become too prohibitive, and hence to validate and optimize parts of the WSN, a common technique [15] [16] is to abstract the signal carrier frequency ω oscillation. Considering a signal represented by:

$$x(t) = DC + I_1 \cos(\omega t) + I_2 \cos(2\omega t) + I_3 \cos(3\omega t)$$
$$+ Q_1 \sin(\omega t) + Q_2 \sin(2\omega t) + Q_3 \sin(3\omega t) \quad (5)$$

In the baseband equivalent transmission scheme, the only data actually transmitted over the RF channel are the 7 coefficients of equation 5 that represent signal harmonics. The target of such representation is to plan the effect of each module behaviour to the signal harmonics, including impairements. The shift from scalar representation (*double* values) to vector (DC, I1, I2, I3, Q1, Q2, Q3 list) can be simply done with SystemC-AMS, by taking advantage of the C++ power. This power is to permit operators overloading : defining the operators function according to its operands type. Thus, rather than computing each modules behaviour in baseband equivalent, the only modification to be done is to template the **sca_sdf_in** and **sca_sdf_out** module ports with *BB* instead of *double*. As shown in listing 2, a class called BB has been defined implementing the vector and related operators.

Listing 2. Baseband equivalent implementation

```
1  class BB{
2   public:
3     double DC,I1,I2,I3,Q1,Q2,Q3,w;
4     ...
5     BB operator* (double x) const{
6       BB z(DC*x,I1*x,I2*x,I3*x,Q1*x,Q2*x,Q3*x,w);
7       return z;
8     }
9     BB operator* (BB x) const{
10      BB z(
11  DC*x.DC+I1*x.I1/2+I2*x.I2/2+I3*x.I3/2
12            +Q1*x.Q1/2+Q2*x.Q2/2+Q3*x.Q3/2,
13  ...
14  Q3*x.DC+Q2*x.I1/2+Q1*x.I2/2
15            +I2*x.Q1/2+I1*x.Q2/2+DC*x.Q3,
16      w);
17      return z;
18    }
19    BB operator+ (BB x) const{
20      BB z(
21          DC+x.DC,
22          I1+x.I1, I2+x.I2, I3+x.I3,
23          Q1+x.Q1, Q2+x.Q2, Q3+x.Q3,
24          w
25          );
26      return z;
27    }
28  };
```

Data flow time variation is significantly pull down, thereby it is possible to reduce sample simulation period and observe improved simulation timing results. The simulation results in table II correspond to sections III and IV. As expected, simulation time has decreased by several order of magnitude.

TABLE II
SIMULATION RESULTS.

Simulation	SC-AMS classical simulation with refinements	SC-AMS BB equivalent RF simulation
1000 bits transmission	1m2.958s	0m0.036s
DC offset -1e5:5e3:1e5	0m19.916s	0m0.018s
Freq. offset 0:20:1e3	0m24.918s	0m0.022s
Phase mismatch $0:\frac{\pi}{360}:\frac{\pi}{4}$	0m44.407s	0m0.031s

V. CONCLUSION

The advantages of SystemC-AMS are the capacity of interoperability and multi-frequency simulations. Moreover, a C++ based language opens the possibility to use a large choice of existent libraries or to request software programmers that don't have to learn a description language. Such asset have been exploited when including an FFT analyser to verify IIP3 specifications. Introducing baseband equivalent modeling allows to implement protocol communications such as Zigbee, adapted for wireless sensor networks. In fact, the simulation timing results are clearly improved.

REFERENCES

[1] P. Schwarz, "Physically Oriented Modeling of Heterogeneous Systems," *3rd IMACS Symposium of Mathematical Modelling (MATHMOD), Wien, 2-4 Feb. 2000, pp. 309-318 (vol1)*.

[2] J. Ravatin, J. Oudinot, S. Scotti, A. Le-clercq, and J. Lebrun, "Full transceiver circuit simulation using VHDL-AMS," *Microwave Engineering*, May 2002.

[3] F. Pecheux, C. Lallement, and A. Vachoux, "VHDL-AMS and Verilog-AMS as Alternative Hardware Description Languages for Efficient Modeling of Multi-Discipline Systems," *IEEE Transactions on Computer-Aided Design of Integrated Circuits and Systems(TCAD)*, Feb. 2005.

[4] "SystemC-AMS," http://www.systemc-ams.org.

[5] E. Markert, M. Dienel, G. Herrmann, D. Müller, and U. Heinkel, "Modeling of a new 2D Acceleration Sensor Array using SystemC-AMS," *Internationnal MEMS Conference (IMEMS)*, May 2006.

[6] A. Vachoux, C. Grimm, and K. Einwich, "Analog and Mixed Signal Modelling with SystemC-AMS," *IEEE International Symposium on Circuits and Systems (ISCAS)*, May 2003.

[7] "SystemC," http://www.systemc.org.

[8] M. Vasilevski, F. Pecheux, H. Aboushady, and L. de Lamarre, "Modeling Heterogeneous Systems Using SystemC-AMS, Case Study: A Wireless Sensor Network Node," *IEEE International Behavioral Modeling and Simulation Conference (BMAS)*, Sep. 2007.

[9] H. Aboushady, F. Montaudon, F. Paillardet, and M. M. Louerat, "A 5mW, 100kHz Bandwidth, Current-Mode Continuous-Time Sigma-Delta Modulator with 84dB Dynamic Range," *IEEE European Solid-State Circuits Conference (ESSCIRC) Florence,Italy*, 2002.

[10] H. Aboushady, Y. Dumonteix, M. Louerat, and H. Mehrez, "Efficient Polyphase Decomposition of Comb Decimation Filters in Sigma-Delta Analog-to-Digital Converters," *IEEE Trandactions on Circuits and Systems-II (TCASII)*, Oct. 2001.

[11] "8-bit AVR Microcontroller with 128K bytes in-System programmable flash ATmega128 Datasheet," Oct. 2006, http://www.atmel.com/dyn/resources/prod%5Fdocuments/doc2467.pdf.

[12] "An open-source operating system designed for wireless embedded sensor networks." http://www.tinyos.net/.

[13] S. Haykin, *communication systems, 3rd ed.* Wiley.

[14] D. Leenaerts, J. van der Tang, and C. Vaucher, *Circuit design for RF transceivers.* Kluwer Academic Publishers.

[15] K. Kundert, "Introduction to RF Simulation and its Application," *Journal of Solid-State Circuits (JSSC), vol.34, no. 9*, Sep. 1999.

[16] D. G.-W. Yee, "A design methodology for highly-integrated low-power receivers for wireless communications," Ph.D. dissertation, University of California, Berkeley, 2001.

978-1-4244-1983-8/08/$25.00 ©2008 IEEE

Methodology for Placing Localized Guard Rings to Reduce Substrate Noise in Mixed-Signal Circuits

Emre Salman and Eby G. Friedman

Department of Electrical and Computer Engineering
University of Rochester
Rochester, New York 14627
[salman, friedman]@ece.rochester.edu

Abstract—A methodology is proposed to improve the efficacy of placing guard rings to reduce substrate coupling noise in mixed-signal circuits. The methodology is based on a *localized guard ring* structure within an aggressor circuit by redesigning the standard cells in a library. Specifically, a noise aware library is generated where each standard cell contains a dedicated substrate contact. This library is used within the aggressor block. Dedicated contacts within the cells generate a *localized guard ring* structure within each aggressor block. The proposed methodology achieves enhanced isolation as compared to conventional guard rings by minimizing the number of vertical current paths within the substrate.

I. INTRODUCTION

The ever increasing demand to integrate a variety of functions on the same monolithic substrate, commonly referred to as a system-on-chip (SoC), places stringent signal integrity constraints. The decreasing physical distance between the noisy digital circuits and the sensitive analog/RF circuits exacerbates this issue. A common coupling medium is the monolithic substrate, forming a conductive path between the switching digital circuits and the sensitive analog/RF circuits [1].

Three primary mechanisms exist for injecting noise into the substrate: coupling from the noisy digital ground and power rails, coupling from the junction capacitance of the devices during switching, and impact ionization [2]. The injected noise propagates through the substrate, reaching the boundary of the sensitive circuit. The substrate noise can affect the sensitive circuit by modifying the threshold voltage of the devices through the body effect or capacitively coupling into the power/ground and signal lines. Significant performance degradation and functional failure due to substrate noise have been demonstrated [3], [4], [5].

A common technique to reduce substrate coupling noise is to modify the transfer function of the substrate medium by placing guard rings around the sensitive or aggressor blocks [4]. These guard rings consist of substrate contacts (p+ diffusion areas for p- substrates) connected to a dedicated

This research is supported in part by the Semiconductor Research Corporation under Contract No. 2004-TJ-1207, the National Science Foundation under Contract No. CCF-0541206, grants from the New York State Office of Science, Technology & Academic Research to the Center for Advanced Technology in Electronic Imaging Systems, and by grants from Intel Corporation, Eastman Kodak Company, and Freescale Semiconductor Corporation.

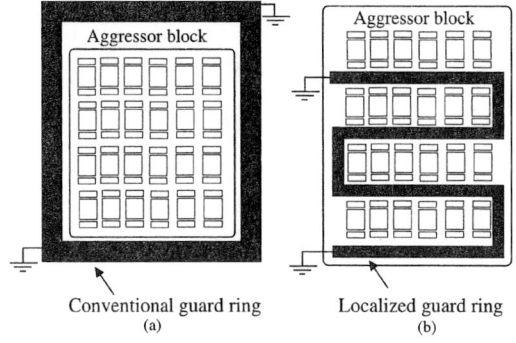

Fig. 1. Simplified representations of a (a) conventional guard ring around an aggressor block, (b) proposed localized guard ring within an aggressor block.

ground pad. The guard ring acts as a low impedance path, filtering the current noise within the substrate. The efficacy of conventional guard rings, however, is limited due to the vertical current propagation paths throughout the substrate. A portion of the noise current can flow deeper into the substrate, thereby bypassing the guard ring, making the isolation less effective. This inefficiency is significant, particularly in large aggressor blocks, since the noise current is more likely to spread throughout the substrate until the current reaches the guard ring surrounding the block.

A methodology is proposed in this paper to improve the efficiency of guard ring structures by generating a localized ring within the aggressor block rather than placing the ring around the block, as illustrated in Fig. 1. This localized structure is achieved by designing a noise aware cell library where each cell contains a dedicated substrate contact that adds to the local guard ring. The proposed methodology reduces the substrate noise by 72%, on average, as compared to a conventional guard ring. Furthermore, the methodology can be automated within existing digital standard cell design flows.

The rest of the paper is organized as follows. Conventional guard ring placement and associated limitations are reviewed in Section II. The proposed methodology is described in Section III. Simulation results are presented in Section IV. Finally, some conclusions are drawn in Section V.

978-1-4244-1983-8/08/$25.00 ©2008 IEEE

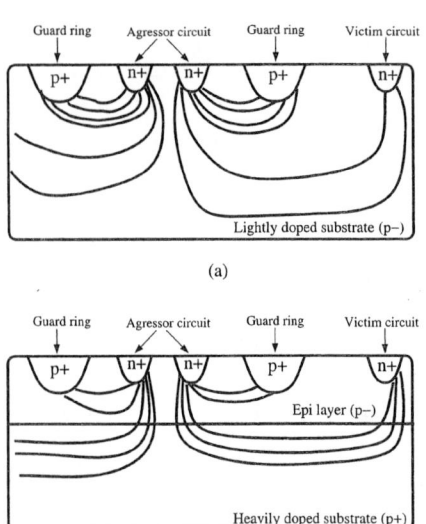

(a)

(b)

Fig. 2. Representative current flow within the substrate among a guard ring, aggressor, and victim circuit: (a) Lightly doped (bulk type) substrate, (b) Heavily doped (epi type) substrate.

II. CONVENTIONAL GUARD RING PLACEMENT

A guard ring is traditionally placed around a victim or an aggressor block to filter the noise current within the substrate. An order of magnitude and nine decibel reduction in the substrate noise has been shown, respectively, in [4] and [6]. The existing vertical current paths, however, limit the efficiency of conventional guard rings [4], [7], as illustrated in Fig. 2.

The flow of the current within a lightly doped substrate is illustrated in Fig. 2(a). The p+ contacts of the guard ring filter a significant portion of the substrate current generated by the aggressor circuit. The vertical current paths, however, bypass the guard rings, reaching the victim circuit. This limitation of conventional guard rings is more significant in heavily doped or epi type substrates since the bulk can be modeled as a single equipotential node, as illustrated in Fig. 2(b). The guard rings should therefore be placed as close as possible to an aggressor or victim block to enhance the isolation. A brief guideline is provided in [8] for placing and biasing guard rings.

The dependence of noise isolation on the width of the guard ring is investigated in [7]. The isolation is shown to be a weak function of the guard ring width. Increasing the width is therefore an ineffective technique for enhancing isolation. An alternative methodology is proposed in this paper to improve the isolation based on a localized guard ring rather than increasing the width, as described in the following section.

III. PROPOSED LOCALIZED GUARD RING METHODOLOGY

The design of a standard cell library with dedicated substrate contacts is explained in Section III-A. An analysis of the

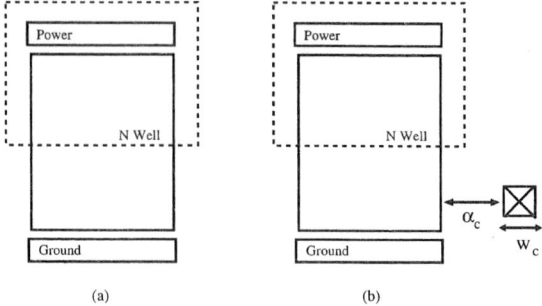

(a)   (b)

Fig. 3. Standard cell: (a) Conventional cell, (b) Noise aware cell with a dedicated substrate contact for the local guard ring. α_c is the minimum distance between the contact and the diffusion, and w_c is the width of the contact.

substrate noise reduction mechanism by generating a localized guard ring using these standard cells is described in Section III-B.

A. Standard Cell with a Dedicated Substrate Contact

In the design of a digital integrated circuit, placing the substrate contacts is usually accomplished after the place-and-route phase of the design flow is completed. The latch-up design rules determine the minimum distance among the contacts.

A standard cell design approach is proposed in this paper where each cell in the library has a dedicated substrate contact used to generate a localized guard ring. This dedicated substrate contact is placed in close proximity to the cell, specified by technology based design rules. Conventional and *noise aware* standard cells are illustrated in Fig. 3. Note that these noise aware cells are in addition to existing conventional cells in the library. The choice between a conventional and noise aware cell depends upon several factors such as the switching activity of the digital blocks and the physical distance between the digital and sensitive analog blocks.

The physical representation of an aggressor device with a dedicated substrate contact is shown in Fig. 4. C_1 represents an existing substrate contact placed according to latch-up design rules and C_2 represents the dedicated substrate contact of the cell within the local guard ring. Note that C_1 is connected to the ground network of the digital circuit; a separate ground network, however, is necessary for the dedicated contacts to isolate these contacts from the noisy ground network. This isolation is required for the local guard ring to filter noise from the substrate rather than inject additional noise into the substrate. Noise reduction is achieved through the low impedance path between C_1 and C_2, and between the bulk of the aggressor device and C_2. The injected noise from the noisy contact C_1 and the bulk is filtered through C_2 rather than propagated into the substrate.

B. Analysis of Noise Reduction Mechanism

A localized guard ring significantly reduces the noise current propagating through the substrate. The current injected into the substrate is more effectively filtered by the local ring as compared to a conventional ring due to the decreased

978-1-4244-1983-8/08/$25.00 ©2008 IEEE

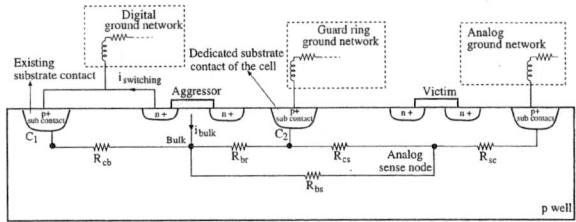

Fig. 4. The effect of the local guard ring on reducing substrate noise. The noise injected from the noisy contact C_1 and bulk of the aggressor device is filtered through the dedicated contact C_2 of the guard ring rather than propagated into the substrate.

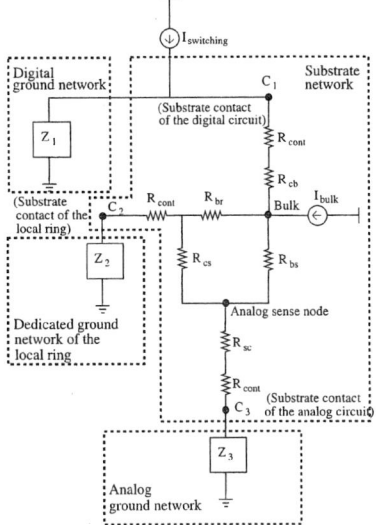

Fig. 5. Equivalent circuit model to analyze the effect of the localized guard ring consisting of the ground network of the digital circuit (Z_1), ground network of the local guard ring (Z_2), and ground network of the analog circuit (Z_3). The substrate network is represented by the equivalent resistance among the substrate contacts and the analog sense node.

substrate resistance between the noise source (the substrate contact and bulk of the aggressor circuit) and the noise filter (the substrate contact of the ring). The number of vertical current paths that bypass the ring are therefore decreased.

An equivalent circuit model to analyze the significance of a localized guard ring is shown in Fig. 5. Z_1, Z_2, and Z_3 represent, respectively, the parasitic impedance of the ground network of the digital (aggressor) circuit, dedicated ground network of the local guard ring, and ground network of the analog (victim) circuit. $I_{switching}$ and I_{bulk} represent, respectively, the switching current of the aggressor circuit and the bulk current injected into the substrate through the source/drain junctions. R_{cont} is the resistance of the substrate contact. A physical perspective of the substrate resistances is provided in Fig. 4.

The noise voltage at the analog sense node is determined from the superposition of the noise due to I_{swi} and I_{bulk}. Assuming the sense node is sufficiently far from the aggressor circuit to ensure that $R_{br} \ll R_{bs} \approx R_{cs}$, and the contact resistance and ground network impedance are much smaller

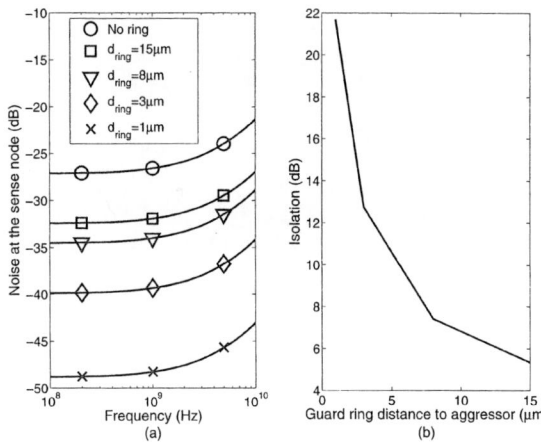

Fig. 6. Analytic model illustrating the significance of the distance of the guard ring from the aggressor: (a) Noise voltage at the sense node with respect to frequency, (b) Isolation as a function of distance of the guard ring from the aggressor.

than the substrate impedance, the noise current caused by the switching current I_{sense}^{swi}, bulk current I_{sense}^{bulk}, and noise voltage at the sense node V_{sense} can be approximated, respectively, as

$$I_{sense}^{swi}(\omega) \approx \frac{I_{swi}(\omega)Z_1(\omega)R_{br}}{(R_{cb}+R_{br})(R_{br}+R_{bs}+R_{sc})}, \quad (1)$$

$$I_{sense}^{bulk}(\omega) \approx \frac{I_{bulk}(\omega)(R_{br}\backslash\backslash R_{cb})}{(R_{br}\backslash\backslash R_{cb})+R_{bs}+R_{sc}}, \quad (2)$$

$$V_{sense}(\omega) \approx [I_{sense}^{swi}(\omega)+I_{sense}^{bulk}(\omega)](R_{sc}+R_{cont}+Z_3). \quad (3)$$

Assuming $Z_1 = Z_2 = Z_3$, the noise voltage predicted by (3) is illustrated as a function of frequency in Fig. 6(a) at different locations of the guard ring, i.e., the distance d between the ring and aggressor circuit is varied. The isolation as a function of distance at 1 GHz is shown in Fig. 6(b), where the isolation is

$$\Delta V_{noise}(dB) = V_{sense}^{without\ ring}(dB) - V_{sense}^{with\ local\ ring}(dB). \quad (4)$$

Note that the substrate resistances have been extracted using SubstrateStorm [9] for a 90 nm CMOS technology with a bulk type substrate. As shown in Fig. 6, a localized guard ring ($d = 1\ \mu m$) achieves an additional isolation of 17 dB as compared to a conventional guard ring ($d = 15\ \mu m$), demonstrating the importance of the location of the guard ring. These results have been validated by an industrial circuit, as described in the next section.

IV. SIMULATION RESULTS

The proposed guard ring placement methodology has been evaluated on an aggressor digital core located close to a sensitive block in an industrial transceiver circuit designed in a 90 nm CMOS technology with a bulk type substrate. The layout of the aggressor circuit and the sense nodes where the substrate noise is observed are shown in Fig. 7. Three different versions of the circuit have been investigated. The

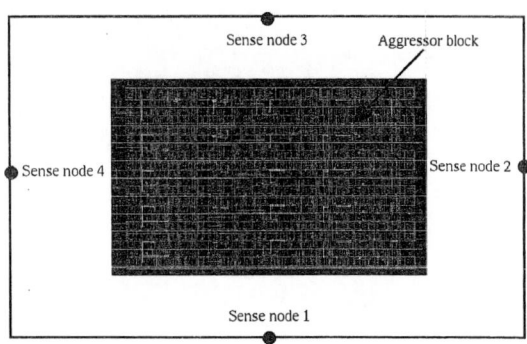

Fig. 7. Layout of the aggressor digital core located close to a sensitive block in an industrial transceiver circuit. The substrate noise is observed at the four sense nodes located on each side of the block.

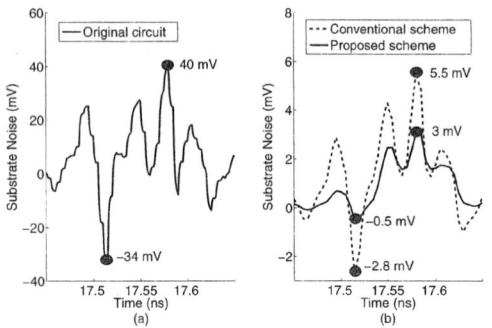

Fig. 8. Substrate noise voltages observed at sense node 2: (a) The original circuit, (b) Conventional and proposed guard ring schemes.

first circuit does not have a guard ring structure. The second circuit utilizes a conventional scheme where the guard ring is placed around the aggressor core. The third circuit utilizes the proposed methodology where the standard cells are replaced with noise aware cells and dedicated contacts are used to generate the localized guard ring scheme.

The layout and substrate impedances of the three circuits have been extracted using Assura and SubstrateStorm [9]. The substrate noise voltage is examined at the sense nodes, using Spectre. In addition to the extracted parasitic impedances of the on-chip ground distribution network, the bond wire package exhibits a parasitic impedance of 1 nH and 0.2 Ω.

The substrate noise voltage waveforms observed at sense node 2 are shown in Fig 8. The peak-to-peak substrate noise voltage for the original circuit is 74 mV, as illustrated in Fig. 8(a). The conventional guard ring scheme achieves an 89% improvement, reducing the peak-to-peak substrate noise to 8.3 mV. The proposed guard ring placement methodology achieves an additional 58% reduction with a peak-to-peak noise of 3.5 mV. The peak-to-peak noise voltage for conventional and proposed guard ring placement schemes are listed in Table I for the four sense nodes. On average, the proposed methodology reduces the peak-to-peak substrate noise voltage by 72% as compared to a conventional scheme. Note that the improvement obtained by the localized guard ring placement methodology will be greater in larger aggressor blocks and

TABLE I

COMPARISON OF PEAK-TO-PEAK SUBSTRATE NOISE VOLTAGE FOR CONVENTIONAL AND PROPOSED GUARD RING SCHEMES.

Sense location	Peak-to-peak substrate noise (mV)		Reduction in substrate noise
	Conventional	Proposed	
Sense node 1	15.1	1.5	90%
Sense node 2	8.3	3.5	58%
Sense node 3	21.5	6.5	70%
Sense node 4	5.7	1.7	70%

epi type substrates due to the additional vertical current paths within the substrate.

Two primary drawbacks exist for the proposed technique. One drawback is the increased area due to the additional substrate contacts since a contact is required for each cell in the aggressor block to generate the local ring. This increased area, however, is not significant since these modified cells are only used in the primary noise generating blocks within a circuit. The second drawback is the use of a metal layer to route the ring within the aggressor block. This metal layer separates the ground network of the ring from the ground network of the aggressor circuit.

V. CONCLUSIONS

A methodology is proposed for improving the efficiency of guard ring structures around an aggressor circuit at the expense of area and a metal layer. The guard ring is localized by redesigning each cell within a standard cell library with a dedicated substrate contact. These dedicated contacts generate a localized guard ring. The proposed methodology achieves enhanced isolation as compared to a conventional guard ring by decreasing the number of vertical current paths within the substrate. Furthermore, this methodology is amenable to automation since it is standard cell based.

REFERENCES

[1] A. A. Kusha, M. Nagata, N. K. Verghese, and D. J. Allstot, "Substrate Noise Coupling in SoC Design: Modeling, Avoidance, and Validation," *Proceedings of the IEEE*, Vol. 94, No. 12, pp. 2109–2138, December 2006.

[2] J. Briaire and K. S. Krisch, "Principles of Substrate Crosstalk Generation in CMOS Circuits," *IEEE Transactions on Computer-Aided Design of Integrated Circuits and Systems*, Vol. 19, No. 6, pp. 645–653, June 2000.

[3] C. Soens *et al.*, "Performance Degradation of LC-tank VCOs by Impact of Digital Switching Noise in Lightly Doped Substrates," *IEEE Journal of Solid-State Circuits*, Vol. 40, No. 7, pp. 1472–1481, July 2005.

[4] D. K. Su, M. J. Loinaz, S. Masui, and B. A. Wooley, "Experimental Results and Modeling Techniques for Substrate Noise in Mixed-Signal Integrated Circuits," *IEEE Journal of Solid-State Circuits*, Vol. 28, No. 4, pp. 420–430, April 1993.

[5] N. K. Verghese, D. J. Allstot, and M. A. Wolfe, "Verification Techniques for Substrate Coupling and Their Application to Mixed-Signal IC Design," *IEEE Journal of Solid-State Circuits*, Vol. 31, No. 3, pp. 354–365, March 1996.

[6] S. Hazenboom, T. Fiez, and K. Mayaram, "Digital Noise Coupling Mechanisms in a 2.4GHz LNA for Heavily and Lightly Doped CMOS Substrates," *Proceedings of the IEEE Custom Integrated Circuits Conference*, pp. 367–370, May 2004.

[7] R. Gharpurey and R. G. Meyer, "Modeling and Analysis of Substrate Coupling in Integrated Circuits," *IEEE Journal of Solid-State Circuits*, Vol. 31, No. 3, pp. 344–352, March 1996.

[8] M. Ingels and M. S. Steyaert, "Design Strategies and Decoupling Techniques for Reducing the Effects of Electrical Interference in Mixed-Mode IC's," *IEEE Journal of Solid-State Circuits*, Vol. 32, No. 7, pp. 1136–1141, July 1997.

[9] "Assura RCX^{TM}, SubstrateStorm^{TM}, Spectre^{TM} tools.," [Online]. Available: http://www.cadence.com.

Double-Ended Tuning Fork Resonator in 0.35um CMOS Technology for RF Applications

J.L. López, F. Torres, G. Murillo, J.Giner, J.Teva, J.Verd, A.Uranga, G. Abadal and N.Barniol

Dept. Enginyeria Electrònica
Universitat Autónoma de Barcelona (UAB)
08193 Bellaterra (Barcelona), Spain
joanlluis.lopez@uab.es

Abstract— A double-ended tuning fork (DETF) fabricated in a 0.35um commercial CMOS technology is presented. Resonator performance for the application of this device in a RF front-end is measured using electrical test. DEFT offers a higher isolation between ports than clamped –clamped beams and the possibility to create a band-pass for frequency filtering or mixing using a single resonator. Discrepancies between expected and obtained results are studied using FEM mechanical simulations.

I. INTRODUCTION

The milestones of today's RF front-ends are the reduction of area and reduced power consumption, with special focus on portable systems. The integration of Micro-Electro-Mechanical Systems (MEMS) in RF systems is a promising way to get rid of these challenges [1].

Several MEMS resonators as mixers and filters (also named mixlers) in standard and CMOS-compatible technologies have been presented [2-5]. For electrostatically driven and capacitively readout MEMS, the exerted force to the resonator presents a square power dependence with the voltage difference between the excitation driver and the resonator [2].Thanks to this quadratic behaviour, frequency mixing can be performed. MEMS resonators are also frequency selective devices: their behaviour depends on the frequency of the applied force, their maximum displacement is obtained for frequencies near the natural resonance frequency of each oscillation mode. This resonator displacement generates a current on the read-out electrode, and therefore there is a maximum current for frequencies near the natural resonance of the structure (i.e. the resonator mechanically filters the input signal).

In this paper we present a MEMS resonator fabricated using a standard CMOS technology and capable to be used as a mixer with a programmable bandwidth.

II. DESIGN AND FABRICATION

A. Double-ended tuning fork

The selected MEMS resonator topology is the double ended tuning fork (Fig. 1). One of the motivations for the use of this resonator is to increase the distance between the input and readout drivers (and therefore increase the isolation between these ports). This resonator exhibits two lateral resonance modes whose frequency distance is determined by the geometry of the resonator providing a finite bandwidth whose value depends on geometrical design variables, and using a sole resonator, like in [6]. In particular, key parameters are the decoupling area width (Wda), support beam length (La) and the distance between the tines of the resonator (2·d). Fig. 2 shows the two lateral oscillation modes of interest and the dependence of the frequency shift in terms of the distance between tines, for a given support beam length (La=0.5um) and a decoupling area width of 0.5um.

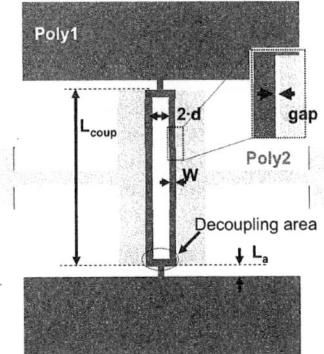

Figure 1. Double ended tuning fork. Design variables are shown.

This work is funded by MEC (TEC2006-03698/MIC)

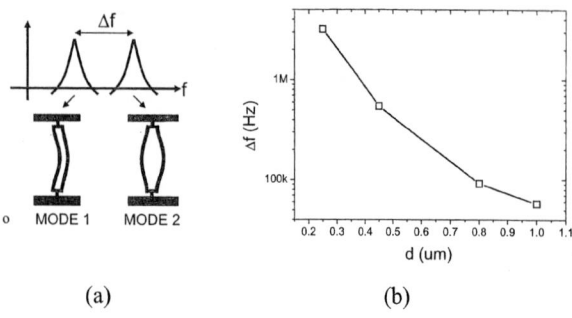

(a)   (b)

Figure 2.   First two lateral modes of oscillation (a). Frequency shift vs. distance between tines (b).

The device was designed to have an oscillation frequency around 40MHz for the first mode and a frequency shift between oscillation modes of 500kHz. The dimensions of the designed resonator are: Lcoup=9, La=0.5um, d=0.24um and W=0.4um

B.   MEMS Fabrication Process

The technology selected for MEMS fabrication is the AMS standard CMOS 0.35um (2 poly, 4 metal). The resonator is fabricated in conducting layers along the CMOS process. Process pad window are used to prevent the passivation deposition above the resonator. This window allows the etchant to release the structure while the other parts of the chip are protected by the passivation layer. The post-process required to etch the silicon oxide is performed using a wet etching with an HF-based solution that does not affect conductor layers [7]. The structure is defined in the Poly1 layer whereas the drivers are drawn on Poly2, in order to obtain a very small transduction gap (40nm) [8].

III.   RESULTS

Fig. 3 shows a SEM image of the DEFT released structure. The device was submitted to different electrical tests to evaluate its performance. The first test is the S21 measurement of the device for different applied DC voltage to the resonator, (Fig.4). This initial test is performed on air using Cascade Microtech Infinity GSGSG probes mounted on a probe station.

Fig. 4 shows the presence of two resonant modes, the first one located around 42.5MHz and the second one around 44.5MHz. Whereas the first mode shows the resonance-anti-resonance typical response, the second mode is flipped, due to the out of phase relationship between the parasitic current and the motional current. The difference of 2MHz (instead of the 500kHz designed) will be further discussed in section IV.

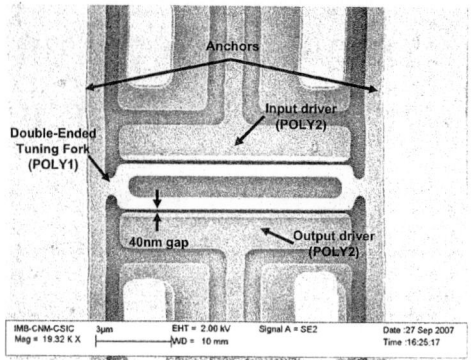

Figure 3.   SEM image of the released DETF

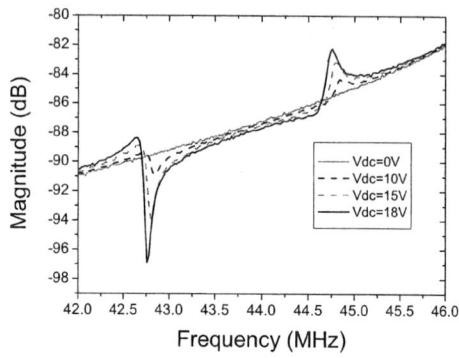

Figure 4.   S21 magnitude measurement for different applied voltages (from 0 to 18V). The two lateral oscillator modes are shown

Fabricated device exhibits also the spring softening effect (the peak frequency is reduced by increasing the applied DC voltage). Fig. 4 shows a typical antiresonance peak found in electrical characterization, due to the existence a parasitic capacitor. This parasitic effect causes a degradation of the measured Q, even in the case of the DETF (where this capacitor is reduced by increasing the distance between drivers).

Measurements of the resonator in vacuum are performed to decrease the air squeezing effect. The device is wired-bonded and introduced in a custom-made vacuum chamber. The sample is then characterized by means of the mixing technique using the test setup depicted in Fig.5. In this test setup, the DETF is biased with a DC plus an LO signal (by means of a Bias-T). The LO frequency is kept constant to 50MHz. The input signal is swept using a central frequency of f0+50MHz (where f0 is the oscillation frequency of the resonator). The resonator performs a down-conversion, and therefore the output signal is centred at f0. With this technique the functionality of the resonator as a mixer-filter

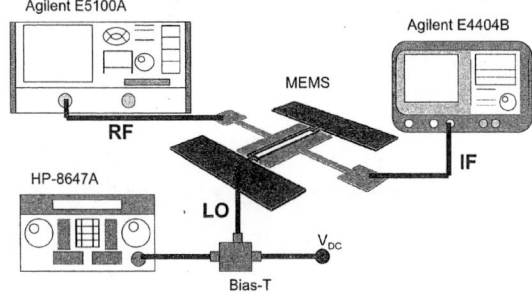

Figure 5. Mixing measurement setup. The network analyzer is used to perform input frequency sweep. The spectrum analyzer is used with MAXHOLD function

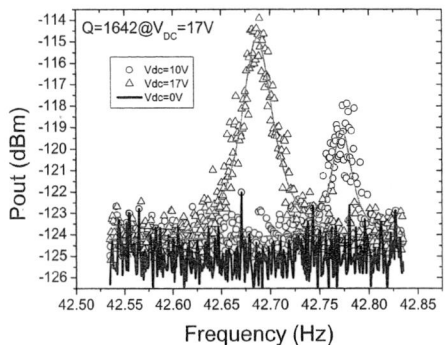

Figure 6. Mixing curve for the first resonant mode at different DC voltages.

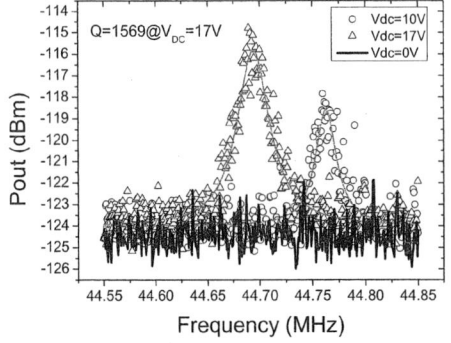

Figure 7. Mixing curve for the second resonant mode at different DC voltages.

is demonstrated, Moreover, the effect of the parasiticcapacitor is removed because excitation and detection are performed at different frequencies, and therefore only resonance peak is obtained.

Fig.6 and Fig.7 shows the obtained mixing curves for the first resonant mode and the second one, respectively. The applied input powers are PRF=15dBm and PLO=10dBm, whereas the pressure is of 4.9E-3 mBar. As can be observed, there is no antirresonance peak, due to the use of mixing.

No peak is observed for DC=0, which means that the mixer is turned OFF. From these peaks the quality factor of the resonator in each mode can be calculated. Quality factors of each mode are: QM1=1642 and QM2=1569, for lateral mode 1 and mode 2 respectively. These values are lower than the expected Q specially for the case of the second mode of oscillation, that is expected to be much higher than the Q of the first mode, due to the moment of inertia cancellation on the decoupling area [9]. This effect, as well as the deviation of the shift of the resonance frequencies will be described on next section.

The output power at input frequencies (fRF and fLO) is also measured to evaluate the isolation improvement inherent to this resonator when compared to clamped-clamped beams MEMS mixlers. An improvement of 10dB in RF to IF ports isolation is obtained based on previous results [5].

An important drawback of MEMS resonators as mixers is the power conversion gain measured. The main reason of the low values is the poor matching of the structures with the RF test setup (adapted to 50Ω). However several methods to improve this power conversion gain have been presented like CMOS circuit amplification [4] and parametric amplification [10], which will be tested soon.

IV. DISCUSSION AND CONCLUSIONS

The measured device differs from the designed in two main aspects: one of them is the fact that the frequency difference between the two oscillation modes is four times the expected (2MHz instead of the designed 500kHz). The other one is the reduced Q factor found in the anti-symmetric resonant mode. Therefore additional FEM simulations (using Coventor) where performed to study in deep the cause of these errors.

To obtain more accurate results in simulation, the true dimensions of the resonators are obtained from SEM images. These measurements give resonator dimensions of: La=0.4um, d=0.4um and Wda=0.5um. Simulations of this device with experimental dimensions give a distance between oscillation modes of 2MHz, an excellent agreement with experimental results. Therefore for future designs, the deviation between the designed and obtained dimensions must be considered.

On the degradation of Q factor in the second lateral oscillation mode, Coventor simulations show that, for the designed resonator, modal displacement is different for each tine and oscillation node (Fig.8). In this figure the dark areas

978-1-4244-1983-8/08/$25.00 ©2008 IEEE

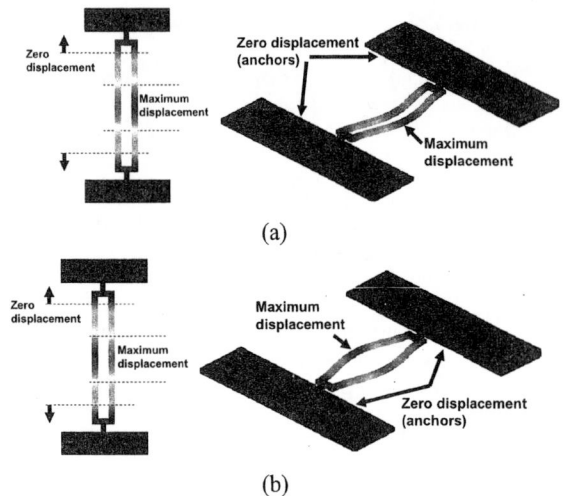

(a)

(b)

Figure 8.   Modal displacement simulations of the resonator with experimental dimensions (a) First lateral oscillation mode and (b) Second lateral oscillation mode

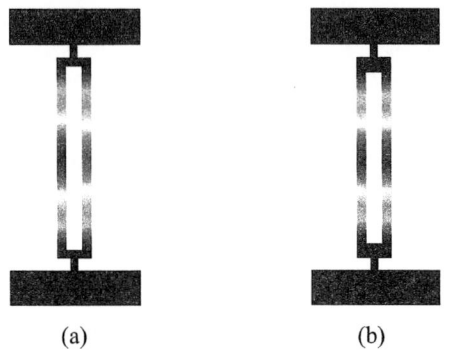

(a)   (b)

Figure 9.   Second lateral mode displacement simulation for a resonator with measured decoupling area (a), and with a wider decoupling area (b). The displacement of the second resonator shows less mismatch between tines.

located in the middle of the tuning fork represents the maximum displacement. It can be observed that the shading in each one of the tines is different, showing an unbalance of the displacement. The origin of unbalance of the movement is the poor anchoring of the tines with the decoupling region. Fig.9 shows the improvement of this unbalance for the second oscillation mode by increasing the decoupling regionwidth. Additional attention has to be paid for the design of this decoupling area in the design process.

The double ended tuning fork fabricated in CMOS technology shows an improved isolation between ports in mixer+filter application. Some problems encountered in the behavior of the fabricated and tested resonator are discussed and advises for the improvement of these results are given. The DETF is a very interesting structure for mixer+filter due

to the possibility to use the two lateral modes to widen the filter band-pass, using a unique resonator.

REFERENCES

[1]   C. T. C. Nguyen, "Integrated Micromechanical Circuits for RF Front Ends," presented at Solid-State Circuits Conference, 2006. ESSCIRC 2006. Proceedings of the 32nd European, 2006.

[2]   W. Ark-Chew and C. T. C. Nguyen, "Micromechanical mixer-filters ("mixlers")," *Microelectromechanical Systems, Journal of*, vol. 13, pp. 100-112, 2004.

[3]   G. K. Fedder, "CMOS-MEMS resonant mixer-filters," presented at Electron Devices Meeting, 2005. IEDM Technical Digest. IEEE International, 2005.

[4]   A. Uranga, J. Verd, J. L. Lopez, J. Teva, G. Abadal, F. Torres, J. Esteve, F. Perez-Murano, and N. Barniol, "Fully integrated MIXLER based on VHF CMOS-MEMS clamped-clamped beam resonator," *Electronics Letters*, vol. 43, pp. 452-454, 2007.

[5]   J. L. Lopez, J. Teva, A. Uranga, F. Torres, J. Verd, G. Abadal, N. Barniol, J. Esteve, and F. Perez-Murano, "Mixing in a 220MHz CMOS-MEMS," presented at Circuits and Systems, 2007. ISCAS 2007. IEEE International Symposium on, 2007.

[6]   J. Yan, A. A. Seshia, K. L. Phan, P. G. Steeneken, and J. T. M. van Beek, "Narrow Bandwidth Single-Resonator MEMS Tuning Fork Filter," presented at Frequency Control Symposium, 2007 Joint with the 21st European Frequency and Time Forum. IEEE International, 2007.

[7]   J. Verd, A. Uranga, J. Teva, J. L. Lopez, F. Torres, J. Esteve, G. Abadal, F. Perez-Murano, and N. Barniol, "Integrated CMOS-MEMS with on-chip readout electronics for high-frequency applications," *Electron Device Letters, IEEE*, vol. 27, pp. 495-497, 2006.

[8]   J. Teva, G. Abadal, A. Uranga, J. Verd, F. Torres, J. L. Lopez, J. Esteve, F. Perez-Murano, and N. Barniol, "VHF CMOS-MEMS resonator monolithically integrated in a standard 0.35um CMOS technology," presented at Micro Electro Mechanical Systems, 2007. MEMS. IEEE 20th International Conference on, 2007.

[9]   S. P. Beeby, G. Ensell, and N. M. White, "Microengineered silicon double-ended tuning fork resonators," *Engineering Science and Education Journal*, vol. 9, pp. 265-271, 2000.

[10] M. Koskenvuori and I. Tittonen, "Improvement of the Conversion Performance of a Resonating Multimode Microelectromechanical Mixer-Filter Through Parametric Amplification," *Electron Device Letters, IEEE*, vol. 28, pp. 970-972, 2007.

LED Integrated Miniaturized Polymer MEMS Display

Y. Daghan Gokdel, Baykal Sarioglu, Arda D. Yalcinkaya

Electrical and Electronics Engineering Department
College of Engineering, Bogazici University
Bebek, TR-34342 Istanbul Turkey
Email: daghan.gokdel@boun.edu.tr

Abstract—**This study presents the process development and first experimental results of LED integrated polymer microelectromechanical systems (MEMS) for two-dimensional displays. Proposed integrated system is very cheap due to the fabrication simplicity and the choice of polymer material as the structural layer, and it is a strong candidate to show similar performance when compared with the existing flat panel displays. A process technology that enables the integration of the LED dies and the PCB-based MEMS structure is presented. The very first version of the MEMS structure is actuated by consuming 11.5 mW and resulting displacement is measured to be 9 mm. Along with the LED die size of $250~\mu$m $\times 250~\mu$m, 36 pixels of resolution can be achieved.**

I. INTRODUCTION

There are a number of references to semiconductor LED based dot matrix displays, which today is known to be a somewhat mature technology [1], [2], [3], [4], [5]. These conventional solutions incorporate modulation of 2D LED/OLED matrix with a driver electronic circuitry. For such a display, the resolution depends on the number of LEDs in the 2D matrix, which becomes excessively large in terms of device number and expensive for high-resolution systems. Conventional realization techniques for 2D dot-matrix displays not only suffer from the large number of LEDs used in the system, but also (1) the low yield due to difficulties in high-density semiconductor LED fabrication, (2) reliability problems (dead pixels, LED-to-LED variations in luminance-current curves etc.) and (3) high power consumption of 2D LED matrix. In this paper a novel method for combination of LED light sources and a movable MEMS platform is introduced. Theoretical and experimental work including the characterization of first generation of MEMS structures are presented.

II. DEVICE OPERATION

Fig.1 shows the schematic of the device that uses a novel method for realization of 2D displays by integrating the 1D array of light sources (LEDs) with the MEMS actuators. Operation of the device relies on the modulation of the light sources in one axis (called fast scan axis, horizontal axis in Fig.1) and the resonant movement of the LED integrated MEMS platform in the axis perpendicular to the fast scan axis (called slow-scan axis, vertical axis in Fig.1). By simultaneously controlling the MEMS platform movement and LED *on* periods, it is possible to generate a 2D image where, *virtual* pixels are

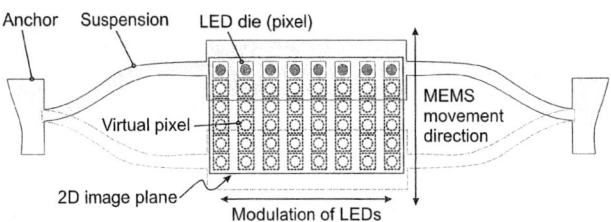

Fig. 1. Structure of MEMS based LED display.

utilized. This method has a number of novelties over the conventional techniques, such as (1) usage of only one-row of LED array leads to severe reduction in the cost of light-source devices (2) due to reduced number of light sources, reliability is expected to increase significantly having much smaller number of dead pixels, lower amount of LED-to-LED variation (depending on the resolution this enhancement is expected to be on the order of 50 to 100 times) (3) power consumption of the 2D display proposed here is the sum of the powers consumed by the 1D LED array and the MEMS actuator, which is much smaller than the power consumption of 2D LED matrix displays (4) the present display represents a technologically new class (5) Present method can use 1D array passive addressing simplicity. Integration of polymer MEMS actuators with LEDs results in medium resolution miniaturized 2D displays. Such a component can be used in consumer electronics products (mobile phones, MP3 players, digital cameras, hand-held PDAs etc.), reperesenting an alternative class of diplays as a reliable, robust, low-power and most importantly very cheap system.

The display basically consists of two main section: a MEMS actuator and an array of LEDs electronically modulated by a LED driver. The MEMS device uses electromagnetic actuation (Lorentz type) principle where a permanent magnet and an electro-coil energized by an electrical current are employed.

A. Mechanical structure

The proposed MEMS architecture is a variation of what is sketched in Fig. 1, and it generates out-of-plane angular movement as depicted in Fig 2, where the structure is vibrating at its fundamental mode. The natural frequency of this structure can be modeled by using cantilever beam with the end-mass as following:

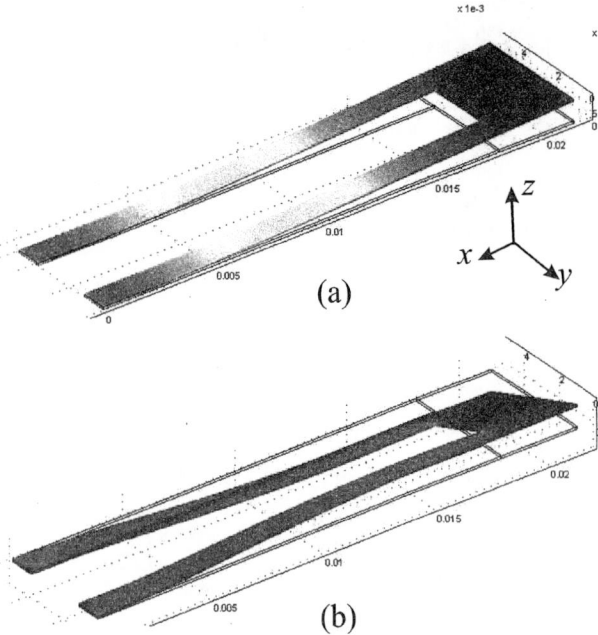

(a)

(b)

Fig. 2. Finite Element Simulation of the device (a) fundamental vibration mode at 22 Hz (out of plane displacement), (b) adjacent mode at 193 Hz.

$$f = \frac{1}{2\pi}\sqrt{\frac{3EI}{(M_f + M_s)L^3}} \qquad (1)$$

where E is the Young's Modulus of the structural material, I is the inertia, L is the length of the cantilever beam, M_f is the mass of flexure and M_s is the effective mass (including the permanent magnet) of the device. In order to get the targeted out of plane movement, flexure width is chosen much larger than flexure height. This choice assure that first mode of the structure makes the vertical movement in $z-$axis rather than an oscillation on $y-$direction. Making this difference higher also ameliorates the mode separation of the structure. Natural frequency and deflection of the proposed design are calculated as $f_0 = 22$ Hz and $d = 8.9$ mm respectively. The results are also confirmed by finite element simulations shown in Fig. 2. Mechanical vibration modes of the device is carefully designated to have sufficient frequency separation between the adjacent modes. The device is designed to have its fundamental mode to comply with the slow-scan motion of the display system shown in Fig. 1. Since the resonance mode operation is utilized for slow-scan operation, the quality factor of the mechanical structure becomes important. In order to increase the quality factor, the effective device area is designed to be as small as possible to reduce the fluid damping.

B. Electromagnetic Actuator

The actuator of the device is based on the Lorentz force generation due to the interaction between the time-varying electrical current and the magnetic field generated by the permanent magnet. By ignoring the fringe fields, the actuation

force, $F(\omega)$, exerted on the movable device can be approximated as [6]

$$F(\omega) = B \cdot i(\omega) \cdot L \cos\theta \qquad (2)$$

where B is the DC field, $i(\omega)$ is the AC current flowing in the electro-coil, L is the effective length of the current carrying conductor and θ is the deflection angle of the actuator tip. When the frequency of the force given with equation 2, enters into the pass band of the mechanical resonator, the displacement of the device is multiplied by the quality factor. This useful property of the mechanical structure helps to reduce the power consumption of the resonant type actuator.

III. PROCESS DEVELOPMENT AND FABRICATION

The fabrication process is outlined in Fig. 3. The starting material consists of 100 μm thick fire-resistant 4 material (FR4) with a 30 μm copper layer on top of it. The material is coated and patterned with a 4 μm thick Shipley 1828 photoresist mould as it is depicted in Fig.3(a). After that, the copper layer is etched electrochemically in HCl solution and the mould is stripped. DC electrochemical etch method is preferred for copper removal. Subsequently, a 100 μm thick SU8 photoresist mould is defined on the patterned copper layer in order to facilitate the placement of LED dies. The SU8 process and the resulting SU8 holes that have 100 μm depth and 250 μm width can be seen in Fig. 3(b) and Fig. 4.

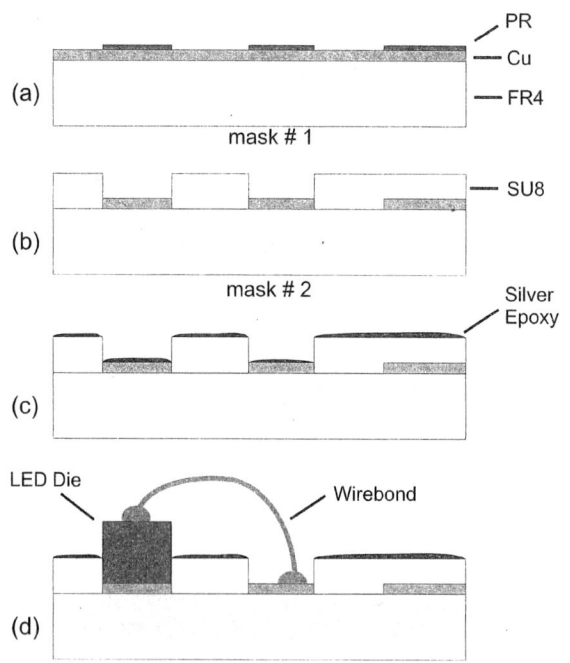

Fig. 3. Process sequence for LED integrated MEMS Display

The vicinity and the inside of SU8 holes are coated with silver epoxy to ensure both the electrical connectivity and the

978-1-4244-1983-8/08/$25.00 ©2008 IEEE

adhesion of LED dies to the copper paths that are located at the bottom as it is shown in figure 3(c). In this process, silver epoxy is preferred to ordinary solder due to its superior wetting and removal properties. Afterwards, LED dies are carefully placed into the SU8 holes and left to dry for 12 hours. Finally, as it is depicted in Fig. 3(d), LED dies are wire bonded to the cathode pads to complete electrical connectivity.

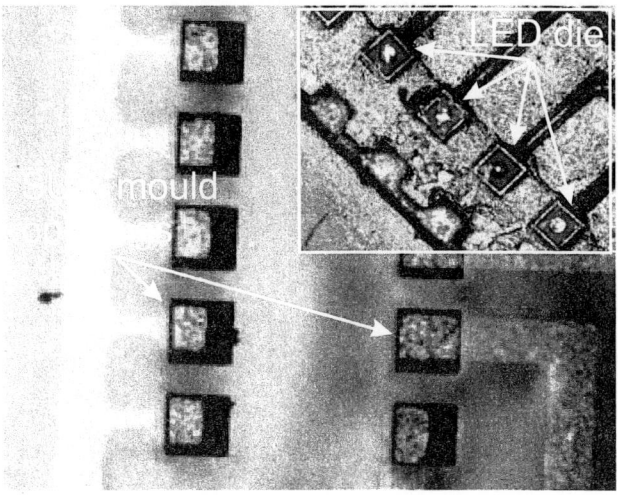

Fig. 4.   LED dies integrated with MEMS structure

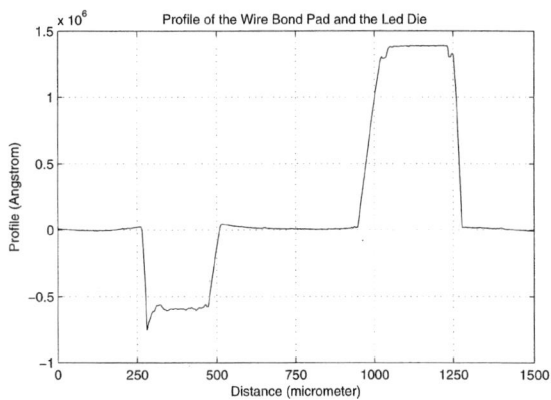

Fig. 5.   Profile of the structure after LED die placement

In Figure 4 a micrograph of the final structure is shown. LED dies are placed with approximately 100 μm spacing. The picture depicts a part of LED integrated MEMS structure without any wire bonding. Afterwards, the integrated structure is cut by VersaLaser cutter VLS2.30 in order to give MEMS device its final shape. A stylus-type surface profiler is used to characterize the topography of the resulting structure. The profile can be seen in Fig. 5, where in between 250 μm and 500 μm of the horizontal axis, an SU8 opening in the form of a valley (cathode pad) is visible which will then be used for wire bonding. The LED die is placed between 1000 μm and 1250 μm of the same graph, introducing a peak to the surface

topography.

IV. EXPERIMENTAL RESULTS

Fabricated device is tested to extract the slow-scan motion behavior. The device is actuated by a mechanically fixed electro-coil to generate time-varying forces on the movable MEMS device. A small piece of neodymium magnet is attached to the tip of the device and the electro-coil energizing current is swept in frequency. Resulting displacement is measured by a laser Doppler vibrometer (Polytec OFV-2500) at each frequency value. Figure 6 shows the mechanical (force-to-displacement) transfer function of the actuator measured in ambient air, for two distinct drive levels: 816 μW and 3.67 mW, respectively. As can be seen from the figure, the resonance characteristic shows a peak around 25 Hz where the fundamental mode of the actuator is observed. In this mode, the actuator generates an out-of-plane peak-to-peak deflection of 1 mm to 9 mm depending on the drive signal level. When combined with the LED dimensions of 250 μm $\times$ 250 μm, the achievable slow-scan axis resolution is approximately 36 pixels. The characterization of this first demonstrator clearly shows the potential of the presented technology: with more optimized designs better resolutions can be achieved.

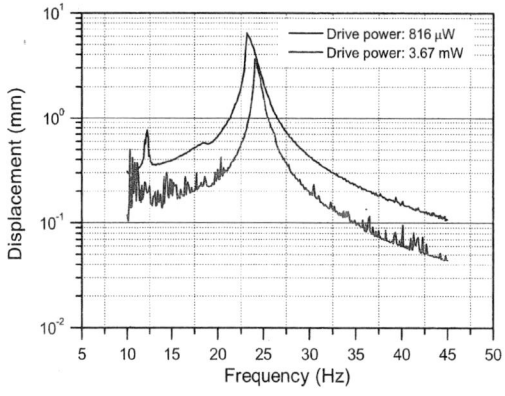

Fig. 6.   Resonance behavior of the MEMS actuator.

The mechanical quality factor of the device is approximately 30, which is an expected value [7], due to the low elastic modulus of the structural material. Yet, this value of the quality factor is sufficient to form displays with medium resolution on the order of 100×100. Fig. 7 plots the peak displacements and corresponding resonance frequencies of the actuator for different drive levels. As can be seen from this figure, there is a parabolic relationship between the drive power (RMS) and the obtained deflection (thus a linear relationship between the drive current and the displacement, as expected from equation 2). As the drive level is increased, electrical spring softening effect is observed [7], which results in downward shift in the resonance frequency.

978-1-4244-1983-8/08/$25.00 ©2008 IEEE   95

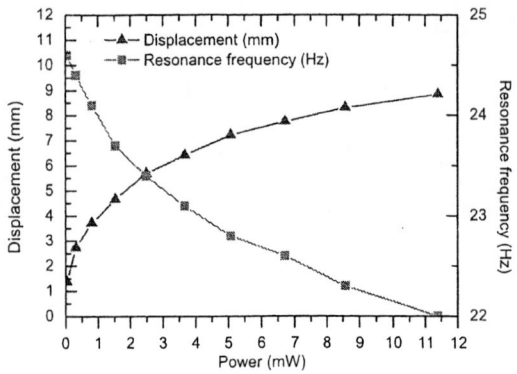

Fig. 7. Drive dependency of the mechanical displacement and the resonance frequeny.

[6] A.D. Yalcinkaya, H. Urey, T. Montague, D. Brown, and R. Sprague "Two-axis Electromagnetic Microscanner for High Reslolution Displays", *IEEE Journal of Microelectromechanical Systems*, Vol. 15, No. 4, pp. 786-794, 2006.

[7] A.D. Yalcinkaya, O. Ergeneman and H. Urey "Polymer Magnetic Scanner for Bardcode Reader Applications", *Sensors and Actuators A*, vol 135, Issue 1, pp. 236-243, 2007.

[8] H. Urey, S. Holmstrom and A.D. Yalcinkaya, "Electromagnetically Actuated FR4 Scanners" *IEEE Photonics Technology Letters* Vol.20, Issue 1, pp. 30-323, 2008.

[9] J E Rogers, R Ramadoss, P M Ozmun and R N Dean, "A microelectromechanical accelerometer fabricated using printed circuit processing techniques", *J. Micromechanics and Microengineering*, Vol.18, 2008.

V. CONCLUSION

Initial design and implementation efforts of the first generation moving LED display are prèsented. A fabrication technique that allows hybrid integration of the LED chips and the FR4 MEMS platform is used and is proven to be successful. Small LED die sizes (250 μm $\times$ 250 μm) allow to pack large number of pixels in one row and the displacements obtained from the initial designs clearly show the potential of the proposed system. Current performance of the device accommodates approximately 40 pixels in the slow scan axis, which can be improved by an order of magnitude with the optimized design. Depending on the drive level, the power consumption of the actuator is about 12 mW (RMS) which results in 10 mm peak-to-peak displacement of the device.

ACKNOWLEDGMENTS

This work is supported by the Turkish Research and Scientific Council (TUBITAK) project (project 107E053) and by the Bogazici University, Scientific Research Project (BAP 07A202). Authors would like to thank Umit Tumkaya of Aselsan Microelectronics Corp., Ankara, Turkey for his help in wire bonding.

REFERENCES

[1] J.H. Baek et al. "A Current-Mode Display Driver IC Using Sample-and-Hold Scheme for QVGA Full-Color AMOLED Displays" *IEEE J. Solid-State Circuits*, Vol. 41, No. 12, pp. 2974 - 2982, 2006.

[2] M. Mizukami, N. Hirohata, T. Iseki, K. Ohtawara, T. Tada, S. Yagyu, T. Abe, T. Suzuki, Y. Fujisaki, Y. Inoue, S. Tokito, and T. Kurita "Flexible AM OLED Panel Driven by Bottom-Contact OTFTs" *IEEE Electron Device Letters*, V 27, pp. 249-251, 2006.

[3] S.J. Ashtiani and A.Nathan, "A Driving Scheme for Active-Matrix Organic Light-Emitting Diode Displays Based on Feedback", *IEEE/OSA J. Disp. Tech.*, Vol.2, No.3, pp. 258-264, 2006.

[4] S-W Wen, M-T Lee and C.H. Chen "Recent development of blue fluorescent OLED materials and devices", *IEEE/OSA J. Disp. Tech.*, Vol.1, No.1, pp. 90-99, 2005.

[5] Y.Kijima, N. Asai, N. Kishii, S. Tamura "RGB Luminescence from Passive-Matrix Organic LED's" *IEEE Trans. Electron Devices*, Vol. 44, No. 8, 1997.

A MEMS Microphone Interface with Force-Balancing and Charge-Control

S. A. Jawed[1], D. Cattin[2], N. Massari[1], M. Gottardi[1]

[1] IRST – Fondazione Bruno Kessler, Trento, Italy.
[2] University of Trento, Trento, Italy.
{jawed, massari, gottardi}@fbk.eu, cattin@disi.unitn.it

A. Baschirotto[3]

[3]University of Milano – Bicocca, Milano, Italy.
andrea.baschirotto@unimib.it

Abstract—This paper presents a low-power CMOS interface for a MEMS capacitive sensor. The interface has embedded force-balancing capability which improves the linearity of the readout for higher sound pressures. The interface also features a bias-charge control functionality to enhance the sensitivity of the microphone. The interface employs boot-strapped pre-amplifier for active parasitic capacitance compensation, followed by a 3rd-order sigma-delta modulator. The interface is designed in 0.35um CMOS technology and its brief simulation results in Cadence-Spectre are presented.

I. INTRODUCTION

MEMS capacitive sensors, such as accelerometers and gyroscopes, have proved viable for high-density low-power applications. Silicon Microphones are one of the rapidly maturing areas of MEMS sensors owing to their improved aspects over the conventional Electret-Condenser-Microphones (ECM) [1]. The applications for a microphone sensor extend from industrial, medical to consumer markets, making it a pervasive and ubiquitous sensor. Silicon microphones exhibit improved aspects over ECMs such as compatibility with standard fabrication and assembly procedures, reduced size, low-power consumption, higher immunity to mechanical shocks and low temperature coefficient. This not only makes the silicon microphones a viable candidate for those applications where conventional ECMs are applied, but it opens-up the possibility of considering them in novel applications such as microphone sensor-arrays for enhanced acoustic performance[2].

However, interfacing to MEMS microphones gives rise to a challenging necessity of addressing the low-power and high-performance aspects of the interface at the same time. Like other MEMS micro-machined sensors, MEMS microphones suffer from sensitivity (ΔCapacitance / ΔPressure) deterioration owing to the inevitable electrode-to-substrate parasitic capacitances. Since Microphones have a relatively larger operational-bandwidth (50Hz – 20kHz), sensitivity is sometimes traded-off by employing specific mechanical structures [3], to achieve a flat response in the whole audio band. The large frequency band causes the parasitic capacitances to suffer from dielectric-dispersion, which can ultimately cause distortion in the readout.

MEMS microphone is a second-order electro-mechanical system and its response non-linearly depends on the movement of the mobile electrode. The sound pressure of 10Pa (114 dB-SPL) is considered as the acoustic overload threshold. For most of the relevant applications, a total harmonic distortion < 10% (sensor + readout) can be tolerated around the overload. However, depending on the sensor, it is possible that a moderate incident acoustic pressure causes significant distortion because of exaggerated movement of the mobile electrode. In such cases, force-balancing [4,5] of the electro-mechanical structure must be employed to control the motion of the mobile electrode to fulfill the linearity specs of the system.

The electrostatic force inside the MEMS depends quadratically on the gap between the two electrodes, while the mechanical restoration force of the spring has linear dependence on this inter-electrode gap. There exists a critical bias voltage where the electrostatic force exceeds the mechanical restoration force and the moving electrodes snaps on to the fixed one, this well-known phenomenon is called pull-in. The pull-in limits the travel-range of the moving electrode to one-third (1/3) of the inter-electrode gap [6]. This limitation in the travel-range constraints the sensitivity, since the magnitude of the capacitive variations inside the sensor depends on the total bias charge stored. Several approaches have been proposed to extend the travel-range beyond the pull-in, including current-biasing[6] and series capacitance[7].

This paper presents a low-power CMOS interface with embedded force-feedback and bias charge-control capability for a MEMS capacitive microphone, as shown in figure 1. Force-feedback and charge-control together increase the sensitivity and improve the linearity for higher acoustic pressures. The interface employs a boot-strapped voltage pre-amplifier, which interfaces directly to the sensor, to implement active parasitic compensation [12]. The pre-amplifier feeds a 3rd-order sigma-delta modulator (SDM), which shares a single opamp between second and the third integrator to reduce power consumption. Section 2 discusses implementation of force-feedback in the interface. Section 3 gives details of the bias charge-control and section 4 presents brief simulation results in Cadence-Spectre.

This work was funded by Provincia Autonoma di Trento Fondo-Unico under the project "Highly Configurable Distributed Microphone – MIDALCO", 2004-2008.)

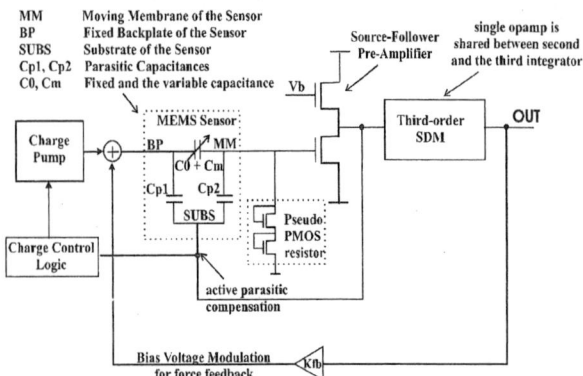

MM   Moving Membrane of the Sensor
BP   Fixed Backplate of the Sensor
SUBS   Substrate of the Sensor
Cp1, Cp2   Parasitic Capacitances
C0, Cm   Fixed and the variable capacitance

Figure 1.   Major Blocks of the Interface

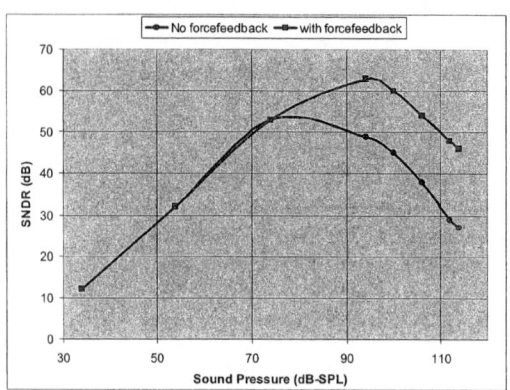

Figure 2.   Simulated SNDR of the MEMS Microphone, with and without force-feedback

II. FORCE-FEEDBACK LOOP

A force feedback loop counter balances the incident actuation by generating an opposite electrostatic force, which keeps the moving electrode almost stationary. Therefore, the non-linear terms that are associated with the large movements of the electrodes become negligible. Similarly, instead of making the electrode's movement very small, the movement can be reduced to such a point where the inherent distortion of the sensor is within the required specs. This relaxes the gain requirements for the force-feedback. i.e. k_{fb}.

The non-linear dependence of the electrostatic force, on the amplitude of bias voltage, limits the use of an analog force-feedback scheme [4,8]. Instead, a time-referenced feedback signal, such as pulse-density or pulse-width modulated signal, can avoid the amplitude related non-linearity. This is because the amplitude of the feedback pulse is always the same and the amplitude information is kept in time. This enables sigma-delta modulation to become the most common scheme in force-feedback loops[4,8].

A behavioral model of a MEMS microphone, based on electro-mechanical analogy, is simulated in Simulink. The parameters of the microphone correspond to IRST-Microphone [3], such as area of moving electrode, inter-electrode gap, spring-constant etc. The pre-amplifier is replaced by a Capacitance-to-Voltage gain, while the SDM is implemented using $\Sigma\Delta$-toolbox [9]. A force-feedback pulse of ±1V is applied to the fixed backplate of the microphone to modulate its bias voltage. The movement of the electrode is reduced from 2.1×10^{-7} to 4.8×10^{-8} for Pin=10Pa (114dB-SPL). Figure 2 shows that the SNDR (in the audio band) at the digital output is significantly improved for higher sound pressures, i.e. for Pin > 1Pa (94 dB-SPL).

The interface employs a 3rd-order feed-forward distributed-feedback SDM, which shares a single opamp between its second and third integrator [14]. This reduces the power consumption of the SDM.

The force-feedback in the CMOS interface is implemented by modulating the charge-pump's output voltage. The charge-pump biases the sensor by charging the backplate (BP) node, as shown in figure 3.. The charge-pump is implemented using four cascaded stages based on static charge-transfer-switches (CTS) [10]. The periodic boosting signals of the last stage, are adjusted in accordance with the output of the SDM to modulate the bias voltage. The pre-amplifier of the interface bootstraps the parasitic capacitances by tying its output to the substrate of the sensor, as shown in figure 3. This effectively removes Cp1 from the signal path and also bootstraps Cp2. A bootstrapped Cp2 offers much less loading to the charge-pump which has to modulate the bias voltage with the SDM's output. The dc-bias at the input of the ac-coupled pre-amplifier is established using pseudo-PMOS resistors [11].

MEMS microphone, mainly because of its vertical structure, is a single-ended electro-mechanical system. Other MEMS sensors, such as accelerometers, facilitate the application of the force feedback due to their differential structure. This is because the feedback pulse applied to the sensor for force-balancing is cancelled as common-mode noise before reaching the readout interface. For the single-ended microphone sensor, the applied force feedback pulse goes through the fixed bias capacitance (C_0) between the two electrodes and reaches the output of the pre-amplifier.

To convert the output of the pre-amplifier into a pseudo-differential output, the interface uses a dummy branch with a similar capacitive structure like the sensor. The dummy structure is bootstrapped just like the sensor and is also fed with the force-feedback pulse. However, it is obvious that a perfect matching between MEMS capacitances and poly/MIM capacitances in the dummy branch cannot be achieved. Consequently, the mismatch between the two branches will translate into a variable offset at input of the digital SDM. Figure 3 shows the scheme that is used in the interface to minimize the mismatch between the two branches[13].

A fully-differential switched-capacitor (SC) high-pass filter is used that senses the difference between the amplitudes of the force-feedback pulses coming out from both branches. This difference, along with the actual feedback pulse, can be used to estimate the amount of

978-1-4244-1983-8/08/$25.00 ©2008 IEEE

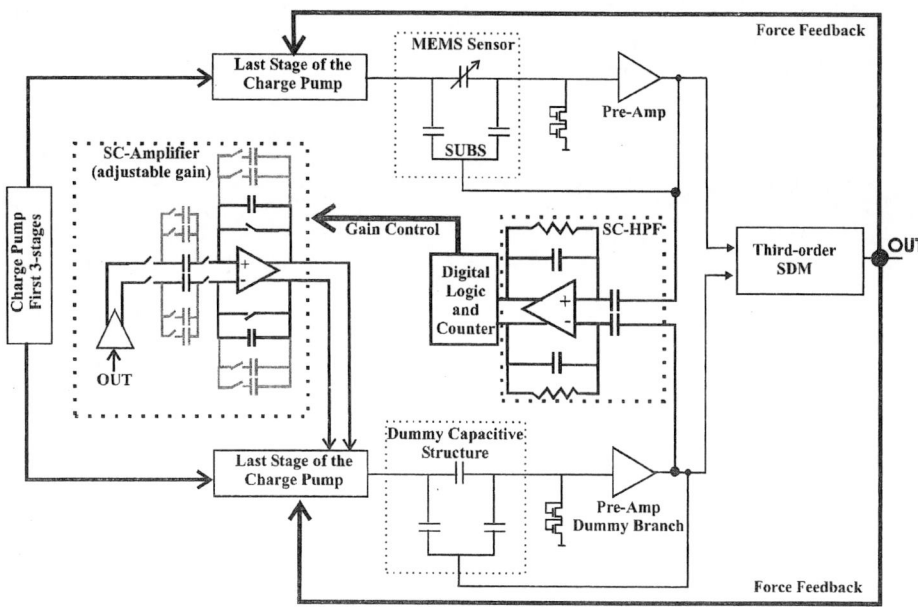

Figure 3.   Interface with the dummy-capacitive branch and the logic to minimize the mismatch between MEMS and the dummy structure

mismatch between the effective capacitances of the two branches. By using a SC-amplifier with adjustable gain, the amplitude of the force-feedback pulse to the dummy branch can be adjusted, hence minimizing the mismatch at the output of the pseudo-different pre-amplifier.

The amplitude of the dummy-branch's pulse is controlled by adjusting a 5-bit counter every time a pulse is applied. Depending on the polarity of the difference, the counter is either incremented or decremented. The value of the counter increases or decreases the integrating or the sampling capacitances in the SC-pulse generation that generates the force-feedback pulse for the dummy branch. This modifies the amplitude of the force-feedback pulse for the dummy branch according to the value of the counter.

In this approach, the two branches must have a separate last stage of the charge-pump, since the output of the charge-pump for the dummy branch is modified. To minimize the power consumption, the first-three stages of the charge-pump can be shared between both branches. This is shown in figure 3.

III.   CHARGE CONTROL

Because of the pull-in under voltage bias, the travel range of a MEMS moving electrode is limited to the 1/3 of the initial inter-electrode gap. A typical voltage biasing biases the sensor around 70% of the pull-in voltage, to leave certain room for the electrode's movement. A high-value resistor is also inserted between the charge-pump and the sensor. This resistor limits the amount of charge that can flow from the voltage source to the sensor and reduces the risk of pull-in if the mobile electrode moves excessively. However, very low-frequency signal can still cause large amount of charge-

accumulation and can ultimately cause pull-in for large electrode movements.

In the previous section, we observed a reduction in the movement of the mobile electrode to 4.8×10^{-8} for 10Pa. If the initial inter-electrode gap for a MEMS microphone [3] is 1.6×10^{-6}m, this implies that the maximum travel range is 0.533×10^{-6}m. If the dc-biasing at 70% of the pull-in, reduces the inter-electrode gap, because of the spring-softening, to 1.4×10^{-6}m, it leaves 0.333×10^{-6}m as the maximum available travel range. Without the force-feedback, the electrode movement under 10Pa, is 0.21×10^{-6}, which highlights that there is sufficient safety-margin available without the force-feedback. Obviously, here we are intentionally overlooking the fact that reduction of movement brought improvement in linearity for higher acoustic pressure. Nevertheless, this safety margin is achieved at the cost of sensitivity.

The force-feedback reduced the movement of the electrode so we require less travel range now for the movement of electrode. This extra room can be used for a stronger dc-biasing , which will increase the bias charge, and will further decrease the initial inter-electrode gap. However, increase in the bias charge will result in the improvement of sensitivity.

Following this scheme, the sensor is biased up around 90% of the pull-in voltage. Then a position-sense circuit keeps track of the movement of the mobile electrode. If the electrode crosses a certain threshold, the sense circuit controls the bias of the sensor to adjust the position of the electrode. The range is selected to be sufficient enough for accommodating the electrode's movement under the maximum incident pressure. In this way, the charge on the sensor is controlled and an infinite charge accumulation is never allowed to occur, which avoids the pull-in.

978-1-4244-1983-8/08/$25.00 ©2008 IEEE   99

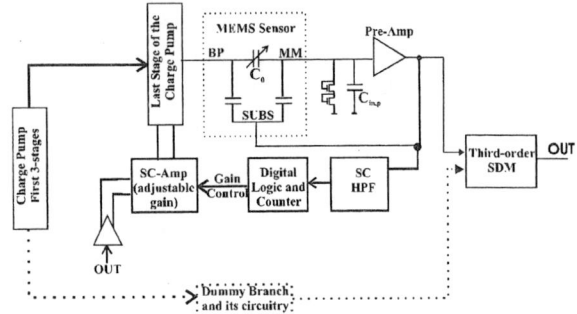

Figure 4. The charge-control logic for the MEMS

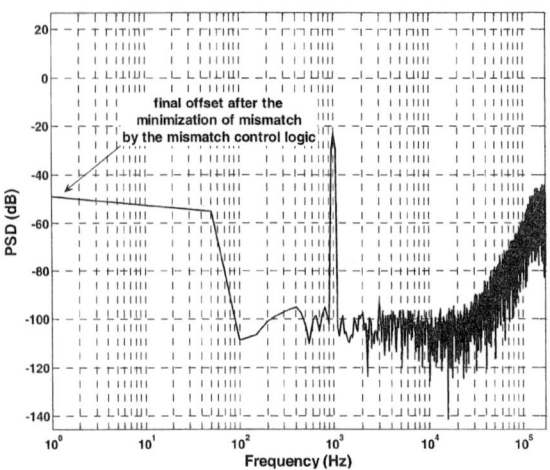

Figure 5. The PSD of the digital output of the interface, after the mismatch-minimization logic has settled

The position sense circuit, as shown in figure 4. makes use of the fact that the applied force-feedback pulse, which is applied at the backplate of the sensor, reaches the output of the pre-amplifier as eq(1) suggests:

$$C_0 = \frac{V_{MM}}{V_{fb,pulse} - V_{MM}} C_{in,p} \tag{1}$$

where C_0 is the fixed bias capacitance of the sensor and $C_{in,p}$ is the interconnect and pre-amplifier input parasitic capacitance at the node MM. X_0 (inter-electrode gap) is inversely proportional to C_0.

Eq. (1) suggests that, within a certain range, the inter-electrode gap can be linearly estimated by observing the value of C_0. The high-pass SC filter is needed to make the charge-control sensitive to only high-frequency feedback pulses and not to the signals within acoustic band. Otherwise, the negative charge feedback control would deteriorate the sensitivity of the readout.

Similar to the mismatch minimization logic, a counter and SC-amplifier with adjustable gain is used. But here it is used to control the amplitude of the periodic boost signal of the last-stage of the charge-pump.

IV. SIMULATION RESULTS

The interface is designed in 0.35um CMOS technology and is simulated in Cadence-Spectre. A maximum mismatch of 30% between the capacitances of the two branches is simulated. Figure 5 shows that the final offset is -50dB after the mismatch-minimization logic has settled.. An increase of ~6dB in the sensitivity is achieved by biasing the sensor around 90% of the pull-in voltage. The simulated digital output achieves 64 dBA SNDR for Pin=1Pa in audio band.

V. CONCLUSIONS

A CMOS interface for a MEMS capacitive microphone was presented. The interface features embedded force-balancing capability to improve the linearity of the readout for high input acoustic pressures. The interface also incorporates bias-charge control, which improves the sensitivity of the readout. Brief simulation results in Cadence-Spectre were presented.

VI. REFERENCES

[1] Report from Yole Development, "SIMM'05 : Silicon Microphone Market 2005 – From Si microphone to acoustic modules", Sept. 2005.

[2] "Midalco – A highly configurable distributed microphone" IRST – Fondazione Bruno Kessler, 2004.

[3] B. Margesin, A. Faes, F. Giacomozzi, A. Bagolini, M. Zen, "Fabrication of Piston-type Condenser Microphone with structured polysilicon diaphragm", Proceedings of 8th Italian Conference on Sensors and Microsystems, 2003.

[4] W. Yun, R. T. Howe P.R. Gray "Surface micromachined, digitally force-Balanced accelerometer with integrated CMOS detection circuitry" Solid-State Sensor and Actuator Workshop, 22-25 Jun 1992, USA.

[5] D. Cattin, A. Faes, B. Margesin, R. Oboe "Modelling and Control of IRST MEMS microphone", 9th International Workshop on Advanced Motion Control, March 27-29, 2006 Istanbul, Turkey.

[6] R. Nadal-Guardia, A. Dehe, R. Aigner, L.M. Castaner, "Current Drive Methods to Extend the Range of Travel of Electrostatic Microactuators Beyond the Voltage Pull-In Point", Journal of MicroElectroMechanical Systems, vol. 11, no. 3, June 2002.

[7] Joseph I. Seeger, Bernhard E. Boser, "Dynamics and Control of Parallel-Plate Actuators Beyond the Electrostatic Instability", the 10th International Conference on Solid-State Sensors and Actuators, Sendai, Japan, June 7-9, 1999.

[8] Y. Dong et al., "Force feedback linearization for higher-order electromechanical sigma-delta modulators", J. Micromech. Microeng. , 2006.

[9] A. Fornasari, P. Malcovati, F. Maloberti, "Improved Modeling of ssigma-delta modulator non-idealities in Simulink", ISCAS'05, May 23-26, 2005.

[10] J. Wu, K. Chang, "MOS Charge Pumps for Low-Voltage Operations", IEEE J. Solid-State Circuits, vol. 33, no. 4, April 1998.

[11] R. R. Harris, C. Charles, "A Low-Power Low-Noise CMOS Amplifier for Neural Recording Applications", IEEE Journal of Solid-State Circuits, June 2003.

[12] S. A. Jawed, M. Gottardi, N. Massari, A. Baschirotto, "A low-voltage bootstrapping technique for capacitive MEMS sensors interface", accepted in IEEE-IMTC07, May 1-3 2007, Poland.

[13] G. Amendola, G.N. Lu, L. Babadjian, "Signal-Processing Electronics for a Capacitive Micro-Sensor", Analog Integrated Circuits and Signal Processing, Kluwer Academic Publishers, 2001.

[14] S. A. Jawed et al., "A Low-Power Interface for the Readout and Motion-Control of a MEMS capacitive Sensor", AMC'08, Trento, Italy, March 26-28, 2008.

978-1-4244-1983-8/08/$25.00 ©2008 IEEE

MOEMS Thermal Imaging Camera

M. Fatih Toy, Onur Ferhanoglu, Hamdi Torun*, F. Levent Degertekin*, Hakan Urey

Department of Electrical Engineering, Koç University, Istanbul-Turkey
*Georgia Institute of Technology, Atlanta-USA

Abstract— **A novel thermo-mechanical Infrared (IR) imaging array with integrated diffraction gratings for optical readout was fabricated and tested. Parylene was used as a structural material for its low thermal conductivity and high thermal expansion properties. Tests were performed using an IR blackbody target and first order diffracted light was imaged on a CCD camera to monitor the entire array. Results show that it is possible to achieve <100mK NETD using a 12 bit CCD camera.**

I. INTRODUCTION

Thermal imaging has been utilized in many temperature monitoring applications such as surveillance, defense, biomedical imaging, monitoring circuits /machinery, and rescue. Cooled and uncooled thermal imaging methods are the two main approaches to thermal imaging. Uncooled thermal imaging has the advantages of low cost and fabrication simplicity with poorer performance compared to cooled approach. Operational principle of uncooled thermal imagers is based on the detection of an output, either electrical or mechanical, that is modulated with the absorbed IR heat. Micro Electro-mechanical System (MEMS) based uncooled thermomechanical detectors are used to monitor optical or electrical outputs caused by mechanical bending or rotation of a bimaterial structure due to absorbed IR radiation. In the group of MEMS based uncooled thermomechanical detectors, optical readout methods are superior to other methods in the sense of sensitivity [1] and elimination of electrical interconnects.

In this work; design, fabrication and performance of an uncooled thermomechanical detector is explained. Noise performance of detector is observed with an integrated diffraction grating underneath each element, monitoring 1st diffraction orders using a CCD camera. Furthermore, the effect of material choice on the performance is explored. Parylene is one of the most suitable structural materials with its high coefficient of thermal expansion (CTE) and thermal resistivity.

II. THEORY

Bimaterial thermal detector performance is composed from a group of parameters such as; IR absorbance, stiffness, thermal isolation, bending or rotation per temperature increase.

This work is partly sponsored by Aselsan.

A. Noise Analysis

The performance of uncooled bimaterial IR imager arrays is limited by temperature fluctuation noise, background fluctuation noise, thermo-mechanical noise. Datskos et al developed analytical measures for all three components of NETD [1]:

Thermal fluctuations are caused due to the heat exchange between the pixel and its environment, and the NETD due to thermal fluctuations can be expressed as [2]:

$$\text{NETD}_{\text{TF}} = \frac{8f^2 T_D \sqrt{k_B GB}}{\eta \tau_0 A_d (dP/dT)_{\lambda_1 - \lambda_2}} \quad (1)$$

where f is the f-number of the IR optics, T_D is detector temperature, k_B is Boltzmann's constant, G is the thermal conductance, B is measurement bandwidth, η is the absorbance of the detector, τ_0 is the transmission of optics, A_d is the surface are of the detector (area of the absorbing region) and dP/dT is the change of power with respect to temperature within the given wavelength range.

The fundamental limit of the thermal fluctuation noise arises due to the radiative heat exchange between the pixel and surroundings, so called the background fluctuation noise. NETD due to background calculations is expressed as [2]:

$$\text{NETD}_{\text{BF}} = \frac{8f^2}{\eta \tau_0 (dP/dT)_{\lambda_1 - \lambda_2}} \sqrt{\frac{2k_B \sigma_T B (T_D^5 + T_B^5)}{A_d}} \quad (2)$$

where σ_T is Stephan-Boltzmann constant and, T_B is background temperature.

Thermo-mechanical noise is caused due to the continuous exchange of the mechanical energy accumulated in the device and the thermal energy of the environment and the NETD due to thermo-mechanical fluctuations can be expressed as [2]:

$$NETD_{TM} = \frac{8f^2 G}{\eta \tau_0 A_d \left(\frac{dP}{dT}\right) \lambda_1 - \lambda_2 \left(\frac{dz}{dT}\right)} \sqrt{\frac{k_B T_D B}{kQ\omega_0}} \quad (3)$$

where dz/dT is the mechanical responsivity of the detector to unit temperature increase on the detector, k, Q, w_0 are spring constant, quality factor and resonant frequency of the structure respectively.

Moreover, NETD due to optical readout [3] is expressed as:

$$NETD_{RO} = \frac{dz/dT}{T_d/T_t} \frac{n_{CCD}}{S_{CCD}} \quad (4)$$

where T_d / T_t is the value of temperature change on the detector caused by unit temperature change at the target, n_{CCD} is the noise of the CCD camera, and S_{CCD} is the sensitivity of the detector as the ratio of readout intensity variation in the CCD units per unit deflection of the detector. CCD sensitivity and noise values vary with the selection of camera, so CCD device is considered to have 12 bits of dynamic range for theoretical NETD calculations. Overall NETD is calculated using:

$$NETD = \sqrt{NETD_{TF}^2 + NETD_{BF}^2 + NETD_{TM}^2 + NETD_{RO}^2} \quad (7)$$

Fig. 1 shows different detector designs, and calculated NETD for each design is given in Table 1.

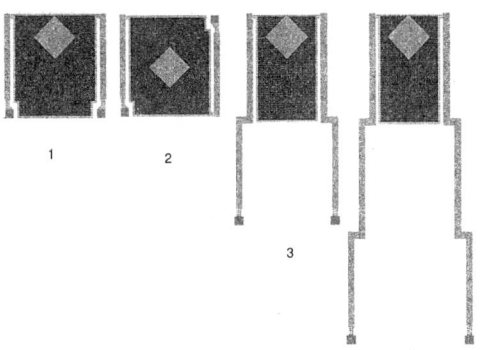

Figure 1. Various detector designs with different performances, performance comparison is given in Table 1

TABLE I.   THEORETICAL NETD VALUES FOR DIFFERENT DETECTOR DESIGNS

Design	NETD$_{TF}$ (mK)	NETD$_{BF}$ (mK)	NETD$_{TM}$ (mK)	NETD$_{RO}$ (mK)	NETD$_{TOTAL}$ (mK)
1	2.7	2.9	6.7	16.1	17.9
2	18.7	2.9	8.3	175	176
3	2.6	3.2	8.3	4.5	10.3
4	1.9	3.2	8.3	1.9	9.2

III. FABRICATION

A. Fabrication Steps

The devices were fabricated at the Microelectronics Research Center at Georgia Institute of Technology. Quartz wafers were used to serve as a transparent medium for the optical readout.

The gratings were evaporated and patterned on the substrate. Photoresist (PR) was used as a sacrificial layer to serve as a quarter wavelength gap in between the bottom metal and IR absorber to enhance absorption. Parylene dimmer was evaporated on top of the sacrificial layer as a structural material. Titanium was sputtered and patterned to serve as the secondary bimaterial pair. A thin layer of titanium was sputtered on top of the devices for IR absorption. Final step of the fabrication is given in Fig. 2. Microscope photograph of the fabricated array is shown in Fig. 3.

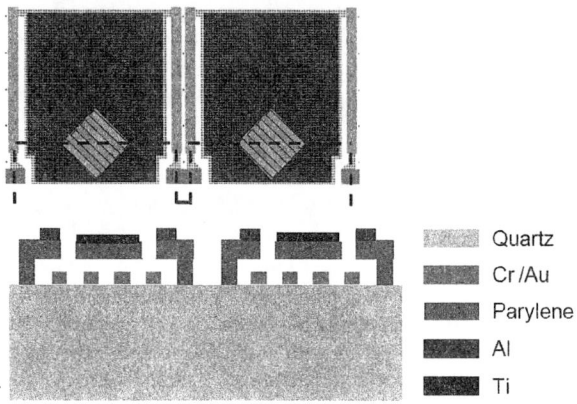

Quartz
Cr /Au
Parylene
Al
Ti

Figure 2. Final step of detector fabrication

Figure 3. Microscope photograph of fabricated array

IV. EXPERIMENTAL SETUP

Fabricated detector array is enclosed into a vacuum package which has an infrared window and a visible readout window. A He-Ne laser as the readout source is expanded to match with the size of the detector array and directed to the detector substrate through the visible window. Diffracted

978-1-4244-1983-8/08/$25.00 ©2008 IEEE

light returning from the detector array is band pass filtered to only transmit 1st diffraction order from every detector element and imaged on to an 8 bit CCD camera operating with a shutter speed of 33ms and a frame rate of 15 fps. A shutter is placed in front of a heater with high emissivity tape staring at the detectors through the infrared window. Fig. 4 shows the optical readout setup.

Figure 4. Photograph of the optical readout setup

V. RESULTS AND DISCUSSION

CCD camera at the readout setup is used to monitor the 1st diffraction order intensity modulation from detector elements. Diffracted light from each detector element is imaged on to a group of CCD pixels. Fig. 5 shows a snapshot from the CCD camera output.

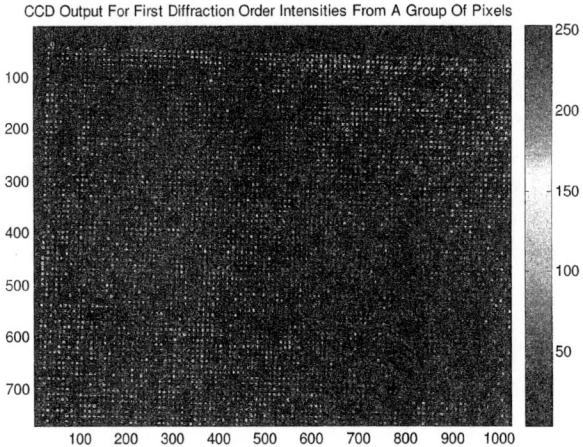

Figure 5. 1st Diffracted order image of a group of detector elements from the CCD camera.

With the off chip pixel binning for the CCD pixels receiving light from the same detector element, dynamic range of the CCD camera is enhanced to 10 bits. In order to measure experimental NETD, temperature of the heater is increased to 155°C. Then the heater is blocked by a shutter at the room temperature (25°C). Binned output from the CCD camera for a selected detector element is recorded during the experiment. Fig. 6 shows a sample CCD output for a selected pixel for this target temperature change.

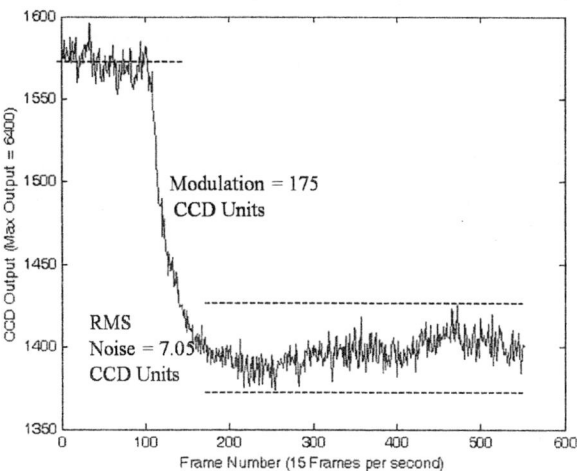

Figure 6. CCD Output modulation by the change of target temperature for a selected detector element.

From the data in Fig. 6, SNR can be found as 24.8. The increase in the IR target temperature is 130°K, therefore the NETD of this pixel is calculated as 5.24°K using ~f/1.5 optics for the light collection system. The reason for this mismatch between theoretical and experimental results was further explored, and it was realized that the self modulation of the readout source and the leak of the vacuum package are the dominating noise sources. These noise sources were not studied in the previous chapters, because the self modulation of the readout source, which appears as a cyclic modulation on the Fig. 6 can be completely eliminated with the application of a simple correlated double sampling method. Furthermore, the leak from the vacuum package causes the drift of output at Fig. 6. Permanent sealing of the vacuum package can stop the leak and this drift behavior. With the corrections for these two factors, experimental NETD is recalculated as 2.3°K.

VI. CONCLUSION AND FUTURE WORK

Parylene based thermal imaging array was designed, fabricated and tested. Experiments reveal that 2.3 °K is achieved for the proposed design, which is close to theoretical findings for the same test conditions. Incorporation of a 12bit CCD camera into experimental setup using f/1 optics should allow achieving an NETD of <100°mK, and further improvements in the experimental setup are planned as the future work.

ACKNOWLEDGMENT

This work is partly sponsored by ASELSAN (Turkey). We'd like to acknowledge the support from NSF International collaboration grant (award number: 0423403) and the MiRC staff at Georgia Tech.

REFERENCES

[1] W. Lee, N.A. Hall and F.L. Degertekin, "A grating-assisted resonant-cavity-enhanced optical displacement detection method for micromachined sensors", Appl .Phys. Lett. 3032 , 2004

[2] P.G. Datskos, N.V.Lavrik, S.Rajic, "Performance of uncooled microcantilever thermal detectors", Review of Scientific Instruments, vol. 75, no. 4, 2004.

[3] Y. Zhao, "Optomechanical Uncooled Infrared Imaging System", Dissertation for the Degree of Doctor of Philosophy, University of California, Berkeley, Fall 2002.

An *Asynchrobatic,* radix-four, carry look-ahead adder

David J. Willingham and Izzet Kale

Applied DSP and VLSI Research Group,
University of Westminster,
London, Great Britain.
D.Willingham@wmin.ac.uk & kalei@wmin.ac.uk

Abstract—A low-power, Asynchrobatic (asynchronous, quasi-adiabatic), sixteen-bit, radix-four, parallel-prefix adder circuit is presented. The results show that it is an efficient, low power design, and that as would be expected with an asynchronous design, its performance is determined by its operating conditions. On a 0.35μm CMOS process, under "typical" process conditions, operating at an effective frequency of 22MHz, an addition can be performed using 69pW, with 48.3pW used by the control logic and 20.7pW by the data-path.

I. INTRODUCTION

Asynchrobatic logic is a low-power design methodology that combines an asynchronous stepwise charging controller with a quasi-adiabatic data-path. In the authors' previous work [1], it has been shown that it is possible to implement simple inverter or buffer chains using this design methodology. This work extends that initial presentation and demonstrates that more complex data-path structures can be implemented using this novel low-power technology. To that end, this paper presents the design and simulation evaluation of a sixteen-bit, radix-four, carry look-ahead adder. In section the background of *Asynchrobatic* logic is presented. Sections III concentrates on the design and testing of the adder. The results are presented in Section IV.

II. ASYNCHROBATIC LOGIC

Asynchrobatic logic uses an asynchronous Step-Wise Charging (SWC) controller to drive what are in effect the local power-clock signals of dual-rail adiabatic logic families including Efficient Charge Recovery Logic (ECRL) [2] (also known as 2n-2p logic), 2n-2n2p logic [3] or Positive Feedback Adiabatic Logic (PFAL) [4]. This allows a data-path constructed of *Asynchrobatic* processing pipelines to be created. The asynchronous controller uses a Muller C--element [5] to drive a generator which creates a series of pulses. The duration of the pulses is controlled by N- and P-bias voltages, which are used to control a series of current-starved invertors. These pulses are routed to a SWC circuit [6] which progressively connects the local power-clock signals from V_{ss} to V_{dd} via a series of tank capacitors. Once the local power-clock is connected to V_{dd}, the handshake signals can be sent to the previous and subsequent stages. Once the next stage has completed its processing, the order

of the pulses is reversed to recover the charge to the tank capacitors. In the *Asynchrobatic* design style, the use of four-phase asynchronous signaling perfectly complements the four charging and discharging phases of the previously mentioned adiabatic logic families. Fig. 1 shows an ECRL buffer, Fig. 2 a 2n-2n2p buffer and Fig. 3 a PFAL buffer. These could be converted to inverter configurations by simply swapping the A_H and A_L labels. Fig. 4 shows the asynchronous Muller C-element controller and Fig. 5 shows the SWC circuit.

III. ADDER DESIGN

The adder style chosen for this demonstration was the parallel prefix structure [7]. However, because of the nature of the *Asynchrobatic* pipeline, it was decided to use a radix-four structure rather than the more common radix-two structure, as this reduces the number of stages in the *Asynchrobatic* pipeline, thus making the design more efficient. For this demonstration circuit, a Skylansky adder [8] was used. For adders larger than 16-bits wide, it is likely that fan-out will become a problem, if a Skylansky adder is used. However, due to the dual-rail nature of *Asynchrobatic* logic, the amount of wiring would become problematic if the Kogge-Stone structure was used. Therefore for wider adders, it is suggested that a novel, higher-radix extension of Knowles adders [9] is used. The use of Higher-Radix Knowles Adders (HRKA) would allow a designer to trade-off the capacitive load from the fan-out against the wiring flux, which due to the dual-rail nature of the design is something that could become problematic in wider designs.

The radix-four adder consists of an input stage of half-adders which create the Generate and Propagate signals, two stages of Look-ahead logic, and a final output stage of exclusive-OR gates. The higher-radix structure has been previously suggested for both Kogge-Stone adders [10] and Skylansky adders [8]. Compared to a radix-two version, which would require six *Asynchrobatic* pipeline stages, this adder uses only four. This trade-off uses a more complex logical implementation that requires more inter-stage wiring, but should be both faster and more power efficient because there are less controller stages which consume most of the power used in this design style. To fully exploit the potential gains of this approach, a very wide data-path widths with

---

This research was partially funded by a Quintin Hogg Research Scholarship from the University of Westminster.

978-1-4244-1983-8/08/$25.00 ©2008 IEEE

complex pipeline stages will need to be deployed in designs undertaken using this design style. Whilst even fewer pipeline stages could be used by increasing the radix further. This was not done in order maintain circuit reliability by keeping the number of series nFETs to four or less.

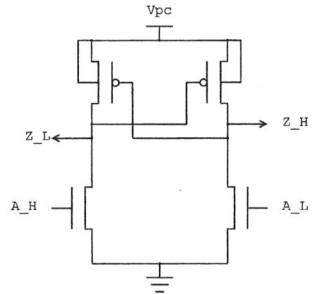

Figure 1. An ECRL buffer [2].

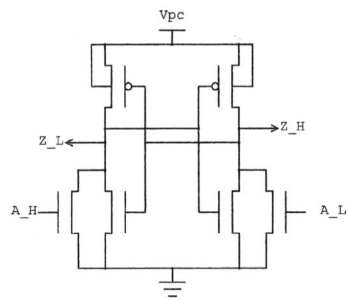

Figure 2. A 2n-2n2p buffer [3].

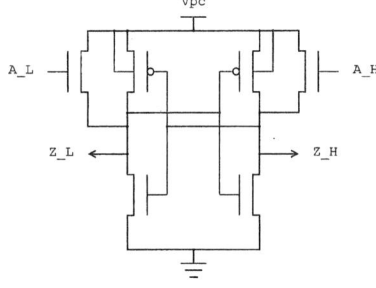

Figure 3. A PFAL buffer [4].

A. Adder cells

This adder structure uses the following data-path cells: buffer, two-input XOR, two-, three-, and four-input AND, and two-, four-, and six-level AND-OR type structures. The construction of the evaluation structures of the three most complex gates {two-input XOR, four-input AND and six-level AND-OR) are shown in Fig. 6, Fig. 7 & Fig. 8. From these, the design of the other gates can be easily derived. These cells were implemented using the PFAL design style.

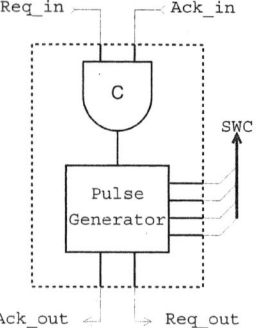

Figure 4. An asynchronous controller & pulse generator [1].

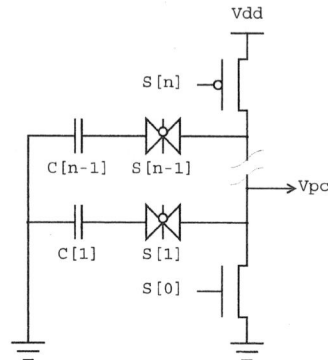

Figure 5. A SWC circuit [14].

They are combined to form the half-adder, constructed from a two-input AND and a two-input XOR, and the parallel-prefix Propagate/Generate Logic circuits. A fourth-order Propagate/Generate circuit (PG4) is constructed from an AND4 and a six-level AND-OR, whilst a first-order version (PG1) is simply a pair of buffers. Due to the dual-rail nature of these cells, this relatively small demonstration circuit shows that the majority of common combinational data-path functions are viable. However, based upon the previous caveat of no more than four series nFETs, it can be seen that not only is every possible logic function of four or less inputs viable, but that other potentially useful logic functions like multi-stage AND-OR and eight-way MUX can be implemented. Furthermore, due to the dual-rail nature of this logic style, a complete four-input library can be implemented with relatively small number of cells. With only 222 different cells required to implement every one of the 65,536 functions (including degenerate functions with one or more static inputs) of four inputs.

With the exception of the Exclusive-OR gate, which was designed using a Reduced Ordered Binary Decision Diagram (ROBDD) [12] method, these cells were designed using the Quine-McClusky [13] method. This allowed the six-level AND-OR structures which have seven inputs to be implemented with no more than four series nFETs.

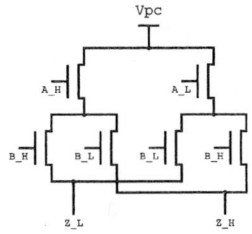

Figure 6.   nFET tree for two-input XOR.

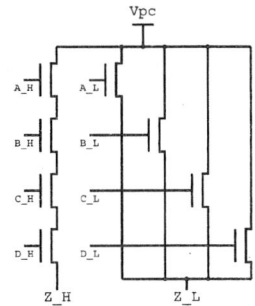

Figure 7.   nFET tree for four-input AND.

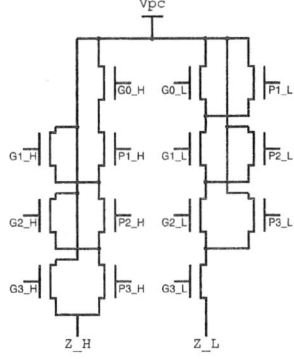

Figure 8.   nFET tree for four-level AND-OR.

The control logic is constructed using the asynchronous SWC circuit detailed above, and implemented using three tank capacitors each having a capacitance of 10pF. The choice of this value is a trade-off between the stability of the tank-capacitor voltage verses the time taken to supply the initial charge, and was arrived at by simulation studies. Furthermore, this can be achieved with on-chip capacitors in today's CMOS processes. For simplicity in this example the Carry input has been tied to zero and it has been assumed that validity of both the main adder inputs is represented by a single handshake signal, but in a more complex system, an asynchronous *join*-function could be implemented if each input had its own handshake signal. This could easily be done by using the appropriate multi-input C-element within the control logic. The high-level structure of the complete adder is shown in Fig. 9, the boxes labeled "HA" represent half adder circuits, the boxes labeled "X" represent XOR

gates, and the boxes labeled with numbers represent the Generate/Propagate logic of that order.

B.   Modeling the adder

The adder was initially described using Verilog to check that it was functionally correct, and then modeled using SPICE to allow functional circuit-level simulation. The use of Verilog models allow both a high level model and a cell accurate model to be created; the model could also be extended to switch-level modeling which would allow fully accurate, dual-rail models to be created. The cell accurate model implements the individual quasi-adiabatic cells as a rising-edge triggered flop with logic-processing inputs. This can be extended to incorporate a reset action on the outputs triggered by the negative edge of the local-power clock. The incorporation of the reset action adds an extra beneficial cross-check.

The SPICE implementation used Alcatel (AMIS) 0.35μm models. The current simulations were performed using pre-layout netlists, and do not include any parasitic elements.

C.   Testing the adder

The adder was tested by driving it with vectors generated using two differently-seeded Linear Feedback Shift Registers (LFSR), one to drive each of the adder's inputs. This ensured that identical data-streams were presented to each adder input in all simulation runs, irrespective of the operating conditions of the circuit under test. The control logic was connected so that the adder would run freely at a speed determined by the Process, Voltage and Temperature (PVT) conditions. The adder was tested at nominal voltage (3.3V) in the fast (ff, -40°C), typical (tt, 25°C) and slow (ss, 125°C) corners, in four skew corners (sf or fs, -40°C or 125°C) and at typical with different levels of bias applied to the delay circuits in the controller's pulse generator.

IV.   RESULTS

The power and performance figures were obtained from the netlist-only fast-SPICE (Mentor Graphic's Eldo Mach) simulations, and were calculated according to (1).

$$ P \; = \; \frac{V}{(T_1 - T_0)} \int_{T_0}^{T_1} I \, dt \qquad (1) $$

It can be clearly seen that the effective operational frequency is dependent upon both the PVT conditions and the control voltage applied to the delay elements. This confirms that the design is operating asynchronously. It can also be seen that the tank capacitors converge to an operating voltage, which again is dependent upon the PVT conditions and the control voltage, but also shows minor data-dependency. Under typical PVT conditions (tt, 3.3V, 25°C), the power consumption of a single cycle of a single SWC is 12.1pW and the power used in the adder circuit is 20.7pW. Although a full range of process conditions were analysed,

978-1-4244-1983-8/08/$25.00 ©2008 IEEE

the results presented in Table I keeps the voltages fixed at nominal value of 3.3V. This is to keep the bias voltages identical in all cases. Results are presented for fast, slow, typical and skew corners. Table II shows the effect of varying the bias voltage.

TABLE I. PERFORMANCE OVER PVT CONDITIONS.

Corner †	Effective Frequency (Hz)	Controller Power (W)	SWC Circuit Power (W)
{tt, 25°C}	2.20×10^7	4.83×10^{-11}	2.07×10^{-11}
{ff, -40°C}	4.99×10^7	4.55×10^{-11}	2.16×10^{-11}
{fs, -40°C}	2.11×10^7	5.37×10^{-11}	2.61×10^{-11}
{fs, 125°C}	2.38×10^7	4.77×10^{-11}	2.34×10^{-11}
{sf, -40°C}	1.96×10^7	4.46×10^{-11}	2.65×10^{-11}
{sf, 125°C}	2.09×10^7	4.22×10^{-11}	2.46×10^{-11}
{ss, 125°C}	9.96×10^6	4.61×10^{-11}	2.14×10^{-11}

† V_{dd}=3.3V V_{bias}=900mV

TABLE II. PERFORMANCE WHEN VARYING V_{BIAS}.

V_{bias} (V) ‡	Effective Frequency (Hz)	Controller Power (W)	SWC Circuit Power (W)
0.850	1.51×10^7	5.19×10^{-11}	1.96×10^{-11}
0.900	2.11×10^7	4.88×10^{-11}	2.02×10^{-11}
0.950	2.74×10^7	4.63×10^{-11}	2.13×10^{-11}

‡ PVT {tt, 3.3V, 25°C}

V. CONCLUSIONS

It has previously been shown that the *Asynchrobatic* logic style can be used to implement simple data-path structures like inverter and buffer chains. This opus extends the work described in that paper and demonstrates that within necessary process-related design constraints, arbitrarily complex logic functions can be implemented using *Asynchrobatic* logic. It also suggests a method for creating wider higher-radix adders by extending Knowles Adders to higher radices, allowing the designer to find an appropriate trade-off between wiring flux and fan-out.

REFERENCES

[1] D.J. Willingham and I. Kale, "Asynchronous, quasi-adiabatic (*Asynchrobatic*) logic for low-power very wide data width applications", Proc. ISCAS 2004, vol. 2, pp. 257-260.

[2] Y. Moon and D. Jeong, "An efficient charge recovery logic", IEEE J-SSC, vol. 31(4), pp. 514-522, 1996.

[3] J.S. Denker, "A review of adiabatic computing", Proc. ISLPE 1994, pp. 94-97.

[4] A. Vetuli, S.D. Pascoli and L.M. Reyneri, "Positive feedback in adiabatic logic", Elec. Lett., vol. 32(20), pp. 1867-1869, 1996.

[5] D.E. Muller and W.S. Bartky, "A theory of asynchronous circuits", Proc. Int. Symp. Theory of Switching, Havard University Press, 1959.

[6] L.J. Svensson and J.G. Koller, "Driving a capacitive load without dissipating fCV²", Proc. ISPLE 1994, pp. 100-101.

[7] R.E. Ladner and M.J. Fischer, "Parallel prefix computation", J-ACM, vol. 27, no.4, pp. 831-833, 1980.

[8] J. Skylansky, "Conditional sum addition logic", IRE T. Elec. Comp., vol. 9(6), pp. 226-231, June 1960.

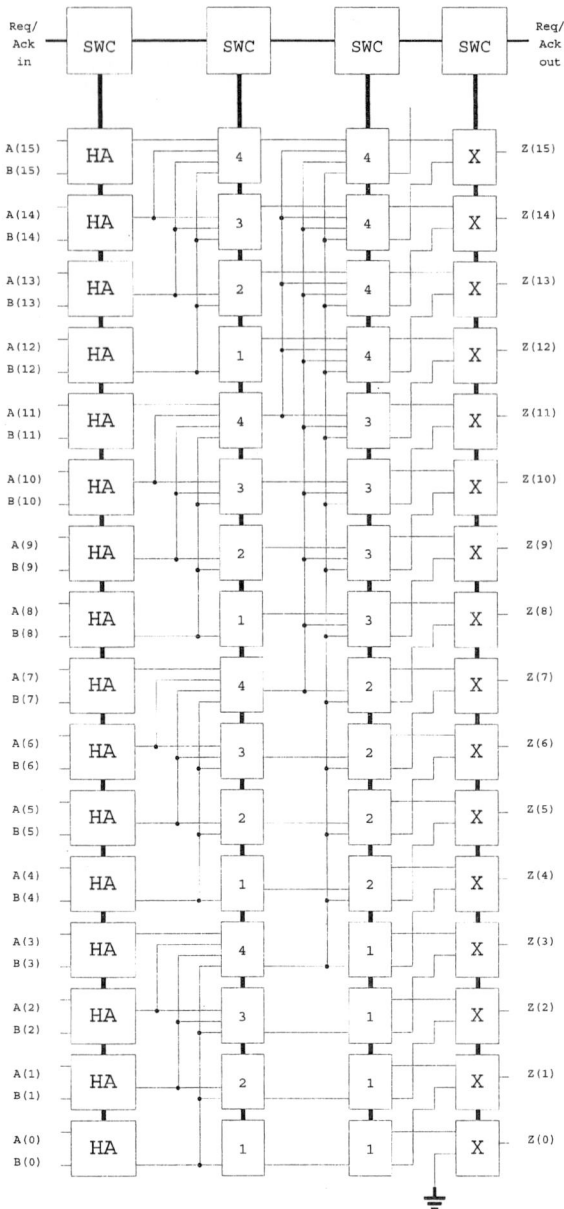

Figure 9. Top-level structure of the adder.

[9] S. Knowles, "A family of sdders", Proc. Symp. Comp. Arith. 1999, pp. 30-34.

[10] F.K. Gurkaynak, Y. Leblebicit, L. Chaouati and P.J. McGuinness, "Higher radix Kogge-Stone parallel prefix adder architectures", Proc. ISCAS 2000, vol. 5, pp. 609-612.

[11] D.J. Willingham, "Adiabatic CMOS 8x8 multiplier", MSc report, University of Westminster, 1999.

[12] R.E. Bryant, "Graph-based algorithms for boolean function manipulation", IEEE Trans. Comp., vol. 35(8), pp. 677-691, 1986.

[13] K.M. Chu and D.L. Pulfrey, "Design procedures for differential cascode voltage switch logic circuits", IEEE J-SSC, vol. 21(6), pp. 1082-1087, 1986.

A YAPI-KPN Parallel Model of a H264/AVC Video Encoder

Hajer Krichene Zrida, Mohamed Abid
Electrical Engineering Deptartment
CES Laboratory, ENIS Institute
Sfax University, Tunisia
hajer_kri@yahoo.co.nz, mohamed.abid@enis.rnu.tn

Ahmed Chiheb Ammri, Abderrazek Jemai
National Institute of Applied Sciences and Technology (INSAT)
7 November- Carthage University, Tunisia
chiheb.ammari@insat.rnu.tn,abderrazek.jemai@insat.rnu.tn

Abstract— **H264/AVC (Advanced Video Codec) is a new video coding standard developed by a joint effort of the ITU-TVCEG and ISO/IEC MPEG. This standard provides higher coding efficiency relative to former standards at the expense of higher computational requirements. This paper presents first a high-level complexity analysis of a H264 video encoder allowing for complexity reduction at the high system level. The complexity of the obtained cost-efficient configuration outlines the potential of using multi processor platforms for the execution of a parallel model of the encoder. For this, a YAPI-level parallel Kahn process network (KPN) model is proposed, implemented and validated at high-level using the YAPI library Programming Interface.**

I. INTRODUCTION

The H264/AVC has been designed with the goal of enabling significantly improved compression performance relative to all existing video coding standards [1]. Such a standard uses advanced compression techniques that in turn, require high computational power [1]. For a H264 encoder using all the new coding features, more than 50% average bit saving with 1–2 dB PSNR video quality gain are achieved compared to previous video encoding standards [2]. However, this comes with a complexity increase of a factor 2 for the decoder and larger than one order of magnitude for the encoder [2].

Implementing a H264/AVC video encoder represents a big challenge for resource-constrained multimedia systems since this requires very high computational power to achieve real-time encoding. While the basic framework is similar to the motion compensated hybrid scheme of previous video coding standards, additional tools improve the compression efficiency at the expense of an increased implementation cost. For this, the exploration of the compression efficiency versus implementation cost design space is needed to provide early feedbacks on the standard bottlenecks and select the optimal use of its coding features.

In a previous study [3], a high-level performance analysis of a H264 video encoder was performed to evaluate its compression efficiency versus its implementation complexity and to highlight important properties of the H264 framework allowing for complexity reduction at the high system level. In this study, the complexity analysis covered major H264 encoding tools. Each new tool has been tested independently comparing the encoding performance and computational complexity

of a complex configuration to the same configuration minus the tool under evaluation.

Absolute complexity values of the obtained cost-efficient configuration of the H264 encoder confirmed the big challenge of its cost-effective implementation. Given this, we will motivate the use of a multiprocessor approach to share the encoding time between several embedded processors. For this purpose, the sequential encoder reference code has to be distributed using a parallel programming model over a multiprocessor architecture. Prior to task-level decomposition and parallelization of the sequential H264 reference code, a computational profiling shall be considered to identify the most computationally-expensive tasks and to give a clear picture of the critical code parts candidate for parallelization. Based on the obtained profiling results and using the two predominant forms of parallelism (task and data levels), a first parallel model of the encoder will be proposed and validated using Kahn process networks (KPN) models [4] implemented by the Y-chart Applications Programmers Interface (YAPI) library of the multi-threading environment [8].

The paper is organized as follows. The next section presents the performance and complexity of the H264 major encoding tools. Section 3 discusses main aspects and issues for getting a high level parallel model of the H264 video encoder. A parallel YAPI-KPN model is proposed and the associated functional simulation results are presented in section 4. Concluding remarks are given in the final section.

II. PERFORMANCE AND COMPLEXITY ANALYSIS

In this section, the performance and complexity of the H264 major encoding tools are presented. The coding performance is reported in terms of PSNR (Peak Signal-to-Noise Ratio) and bit rate output, while the complexity metrics focus mainly on the amount of computing time required to encode a given test sequence on GPP (General-Purpose Processor) platform based on an INTEL Centrino 1.6 GHZ running a Linux operating system.

When combining the standard new coding features, the implementation complexity accumulates, while the global compression efficiency becomes saturated [2]. To find an optimal balance between the coding efficiency and the implementation cost, a proper use of the H264/AVC tools is needed to maintain the same coding performance as the most complex reference configuration while considerably reducing complexity. A parametric influence of major

---

Identify applicable sponsor/s here. *(sponsors)*

encoding tools of this standard on performance and computing time complexity has been evaluated in [3]. In this analysis, 7 test sequences in a 4:2:0 YUV format with different grades of motion characteristics and frame rate are used as given in table1. "Bridge-far", "container" and "Mother & Daughter" offer a wide variety of video QCIF content occurring in low-bit-rate applications of tens of Kbps. "Foreman" is a good medium complexity QCIF test sequence for medium bit rate applications of hundreds of Kbps. The CIF version of "Paris" and "Bridge close" are useful test cases for middle-rate applications. Finally, "Mobile" is a high-complexity CIF sequence with lot of movements including rotation and is a good test for high-rate applications of thousands of Kbps.

TABLE I.   USED TEST VIDEO SEQUENCES

Sequence	Format (Pixel)	Frame Rate (Hz)	Frame Coded
Bridge close (Bc)	CIF (352x288)	15	2000
Mobile (M)	CIF	25	300
Paris (P)	CIF	15	1065
Bridge far (Bf)	QCIF (176x144)	15	2101
Container (C)	QCIF	25	300
Foreman (F)	QCIF	25	400
Mother and Daughter (MD)	QCIF	25	961

For an optimal trade-off between coding efficiency and implementation complexity, the obtained cost-efficient configuration is given as follows. A 3Level-5B frames pyramid coding structure, an UMHexagonS fast motion estimation scheme, a search range fixed to 8, 4 variable block sizes, 3 reference frames, R-D Lagrangian optimization activated, Hadamard transform disabled, motion vector fractional pixel accuracy enabled, P and B frames weighted prediction with bi-prediction motion estimation disabled, a QP value fixed to 28, and CAVLC entropy coding technique used. Results of this parametric influence analysis are given in table 2.

TABLE II.   PERFORMANCE/COMPLEXITY OF THE COST-EFFICIENT CONFIGURATION VS THE REFERENCE CONFIGURATION

Res	Seq	ΔTEC (%)	ΔMEC (%)	fps (opt conf)	ΔBit rate (Kbps)	ΔPSNR-Y (dB)
CIF	Bc	-87,21	-72,7	0,81	15,82	-0,1
	M	-83,87	-68,11	0,6	89,04	-0,29
	P	-84,69	-69,68	0,73	20,39	-0,22
QCIF	Bf	-88,36	-74,16	3,51	-0,16	0
	C	-87,17	-72,88	3,22	2,66	-0,27
	F	-87,32	-70,01	3,05	8,69	-0,3
	MD	-86,49	-71	3,32	2,87	-0,25

ΔTotal Encoding Complexity (TEC) = Encoding Complexity (with cost-efficient configuration) – Encoding Complexity (with reference configuration), idem for ΔME Complexity (MEC)
ΔBit rate = Bit rate (with cost-efficient configuration) – bit rate (with reference configuration), idem for ΔPSNR-Y

In comparison with the most complex configuration, a complexity reduction of more than 80% has been achieved with less than 10% average bit rate increase for all the CIF and QCIF used test sequences. However, for this optimal configuration and even for the very low bit rate QCIF "bridge far" sequence, the associated encoding performance in frames per second is of 3.51 fps. Even with this configuration offering an optimal trade-off between coding efficiency and implementation complexity, we are still very far from a real time performance of 25 frames per second. Implementing this configuration of the encoder represents thus a big challenge for resource-constrained multimedia systems.

III.   PARALLEL KPN-BASED H264 MODELING ISSUES FOR MULTIPROCESSOR IMPLEMENTATION

In the previous section, we motivated the implementation of H264/AVC encoder application on a multiprocessor platform. The availability of many processing cores in such a multiprocessor system speeds-up the processing by introducing the predominant "task and data levels" forms of parallelism These two types should be used to get an optimal parallel model of the H264 video encoder. Any formal parallel model will be represented as a Kahn process network.

Typically, an optimal parallel khan process network model is obtained while extracting the concurrence available in the sequential reference code. For this, a computational profiling is considered to identify the most computationally-expensive tasks of the cost-efficient encoder configuration. Based on the profiling results, the task and data parallelism are exploited to maximize the parallelism between processes. Finally, the communication granularity must be determined for an optimal trade-off between the execution and the inter tasks communication time.

A. Computational profiling the execution of the cost-efficient encoder configuration

To identify the most computational-expensive tasks and to give a clear picture of the critical code parts candidate for parallelization, a computational "Gprof" GNU profiling is performed for the 300 frames QCIF "Container" sequence using the cost-efficient H.264 encoder configuration. The obtained results are reported in figure 1 in terms of the CPU time percentage spent in the execution of each module.

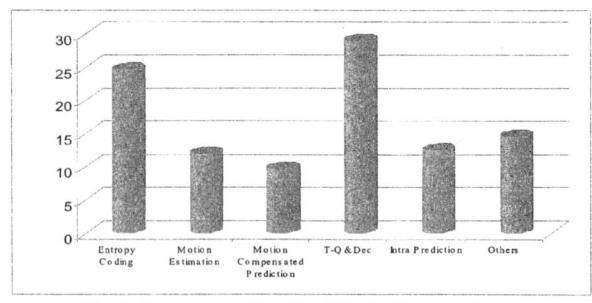

Fig. 1. Computational Profile of H264 video encoding

The obtained profile shows that the motion estimation and compensation (MEC), DCT transform module, the entropy coding, and the intra-prediction modules are the most time-consuming modules. These tasks constitute the major bottlenecks of the encoder. It is shown also from figure 1 that the most computationally intensive task is the "T-Q & Dec" module. For our case, this module represents

978-1-4244-1983-8/08/$25.00 ©2008 IEEE   110

the DCT transform T, the quantization Q, and the decoder rescaling and inverse transform blocks.

B. Task and Data parallelism

The application block diagram always serves as a starting point for extracting the task-level parallelism [5]. This type of parallelism may be achieved by decomposing the whole application into separate blocks. Each block defines one single task that runs a separate stage of an algorithm. For this case, the sequential H264 algorithm is first split into concurrent tasks that may be executed at the same time, and then the necessary inter task communication is established using KPN primitives [4].

In addition, using the obtained profiling results, as the "T-Q & Dec" module represents the most time-consuming module with about 30% of the total encoding time, data-level parallelism is also considered to increase the throughput and decrease the latency of the H264 optimal encoder application. The idea behind data parallelism is to perform the same transform on different data elements in parallel. For instance, it is possible to perform the "T-Q & Dec" Transform, quantization and decoding in parallel for the luminance plane (Y), the chrominance plane (U) and the chrominance plane (V). However as the combined size of the two chrominance planes is smaller than the size of the luminance plane for the used 4:2:0 coding format, it is possible to process the two chrominance planes consecutively in one chain.

C. Communication granularity

The third aspect to consider when extracting parallelism is the granularity at which data is communicated. The optimal communication granularity between tasks must be correct determined to prevent tasks from waiting and avoid spending too much time on synchronization.

Many previous works elaborated on the task-level parallelization of different coding applications like those presented by Shen in [6] and Bozoki in [7] showed that the GoP-level parallelization for such a distributed system provides encoding performances better compared to those obtained with the others communication levels. However, this is not suitable for embedded System-on-Chip (SoC) implementations seeing that a substantial on-chip and shared memory capacity are required for the parallel compression of a group of pictures. The better granularity for such a SoC is thus threading at the fine grain level, i.e. at Marco-Block (MB) level, since only the current and reference frames needs to be stored. Each frame is considered as the current workload, and the encoding process of each frame is divided between the processors.

D. The proposed parallel KPN model

In our approach, Kahn Process Networks (KPNs) are used as a parallel programming model of computation. The KPN model assumes a network of concurrent autonomous processes that communicate in a point-to-point fashion over unbounded FIFO channels, using a blocking-read and write synchronization primitives. Read actions from these FIFOs block until at least one data item becomes available. The execution of a Kahn Process Network is deterministic

and independent of process interleaving, meaning that for a given input always the same output is produced and the same workload is generated, irrespective of the execution schedule [4]. The key characteristic of the KPN model is that it specifies an application in terms of distributed control and distributed memory which allows us in a future work to map the application onto a multiprocessor platform in a systematic and efficient way.

Given the functional blocks diagram of the H264 video encoder, the sequential C-code specification and the results of the computational profiling, we propose the first parallel Kahn process network model of figure 2.

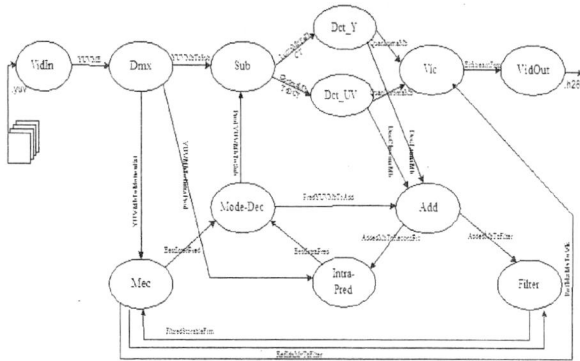

Fig. 2. Proposed Parallel KPN-H264 model

The "video_in" process shown in figure 2 represents the input of the encoder. It is responsible for collecting the video data (YUV frames) from the input file (video sequence with YUV format), the frame width and height dimensions, the total frames number, and the frame rate information. Each frame is divided into macro-blocks of 16x16 pixels. The "dmx" process forwards the YUV-MBs to the "substract", "mec", and "intra prediction" processes. The "substract" process reads the predicted MB, subtracts it from the current MB, and sends the luminance residual data into the "dct_y" task and the chrominance data into the "dct_uv" task. The "dct_y" and the "dct_uv" processes perform associated transforms, respectively on the Y luminance and the UV chrominance MBs. These MBs are received via dedicated data channels. Each 4x4 block is first transformed into block of DCT coefficients using an integer transform, then Q quantized, and finally decoded via a rescaling and an inverse transform. The "vlc" process receives the quantized DCT coefficients via two channels, applies on the CAVLC entropy coding method and transmits the resulting compressed bit stream into the "video_out" process. Finally, the "video_out" process sends the H264 compressed data bit stream to the output file (.h264).

The "adder" process uses the decoded chrominance and luminance MBs to reconstruct the previously encoded MB. Using the current MB "mb_ipred" output of the "dmx" process and the reconstructed (but un-filtered) previously encoded MB "rec_mb" from the "adder" process, the "intra_prediction" process performs an intra-prediction on each macro-block using 9 prediction modes for 4x4 luma blocks, 4 prediction modes for 16x16 luma blocks, and 4 modes for 8x8 chroma blocks. The best intra-prediction mode cost obtained and the associated predicted MB are sent to the "mode_decision" process. Parallel to the intra-prediction, the best inter-prediction mode cost obtained and the corresponding predicted MB are sent to the "mode_decision" process. Using the best intra- and inter-prediction modes, the "mode_decision" process selects the best optimal predicted MB between them.

IV. YAPI-LEVEL FUNCTIONAL SIMULATION OF THE PARALLEL KPN-BASED IMPLEMENTATION

The parallel H264 KPN model of figure 2 is implemented using the C++ YAPI multi-threading environment library. The YAPI implemented parallel KPN model is then validated by high level functional simulation. Typically, such a parallel thread-level implementation could not be performed in an automatic way only for small range of applications. For this case, the sequential code is modified and structured by hand to describe the KPN in C++. Each Kahn process is described by a set of associated functions got from the original code. The inter process communication is performed using solely the YAPI "read" and "write" primitives. Using global variables for this purpose is not allowed. Thus, to ensure inter process communication, all of the shared variables used in the sequential code are grouped into associated data structures for communication over FIFO channels.

The proposed parallel model of figure 2 has been validated at YAPI system level. At this level, when this model is executed, the YAPI "read", "write", and ""execute" functions generate information on computation and communication workload of the application. For a QCIF "Bridge close" of 13 YUV frame sequence, the communication workload analysis is obtained, and is shown in the figure 3.

```
                        size Tsize Wtokens Wcalls T/W Rtokens Rcalls T/R
 h264.YUVMB               1    768   1287    1287   1   1287    1287   1
 h264.YUVMbToSub          1    768   1287    1287   1   1287    1287   1
 h264.PredYUVMbToSub      1    640   1287    1287   1   1287    1287   1
 h264.PredYUVMbToAdd      1    640   1287    1287   1   1287    1287   1
 h264.LumaMbToDCT         1  38604   1287    1287   1   1287    1287   1
 h264.ChromaMbToDCT      14     36   1287    1287   1   1287    1287   1
 h264.QuanLumaMb          1  38604   1287    1287   1   1287    1287   1
 h264.QuanChromaMb       14     36   1287    1287   1   1287    1287   1
 h264.DecLumaMb           1  38604   1287    1287   1   1287    1287   1
 h264.DecChromaMb        14     36   1287    1287   1   1287    1287   1
 h264.AddedMbToReconsPic  1    768   1287    1287   1   1286    1286   1
 h264.YUVMbToIntraPred    1    768   1287    1287   1   1287    1287   1
 h264.BestIntraPred       1    648   1287    1287   1   1287    1287   1
 h264.AddedMbToFilter     1    768    693     693   1    693     693   1
 h264.RefIdxMvToFilter   42     12    594     594   1    594     594   1
 h264.RefIdxMvToVlc       1    304   1188    1188   1   1188    1188   1
 h264.FiltredStorableFrm  1   3284      7       7   1      7       7   1
 h264.YUVMbToMotionEst    1    768   1188    1188   1   1188    1188   1
 h264.BestInterPred       1    648   1188    1188   1   1188    1188   1
 h264.BitStreamMb        12     40     13      13   1     13      13   1
```

Fig.3. Communication workload of the implemented parallel model

This figure describes the total number of tokens communicated between tasks via dedicated data channels. In the validated parallel model, the number of tokens per call is equal to 1 for all the "reading" and "writing" operations. Each QCIF frame consists of 99 MBs of 16x16 pixels. Every MB token contains information about the Luminance (Y) and the Chrominance (UV) in a 4:2:0 format. That means that one MB contains two 8x8 blocks of UV data, and one 16x16 pixels block of Y data. Given this, it is shown that **1287** (99*13frames) intra and inter MBs received by the "dmx" process via the "YUVMB" FIFO from the "video_in" process. However, there are only **1188** inter predicted and bidirectional MBs sent from the "dmx" process into the "mec" process using the "YUVMbToMotionEst" channel. Among **13** encoded YUV frames, there are **7** references frames (one I-frame, 2 P-frames, and 4 B-frames) received by the "mec" process from the "loop_filter" process via the "FiltredStorableFrm" FIFO. These references frames are reconstructed from **693**

previously encoded MBs communicated between the "adder" and the "loop_filter" processes.

V. CONCLUSION

In this paper, a high-level performance analysis of a H264 video encoder is first performed to find an optimal balance between the coding efficiency and the implementation cost allowing for a complexity reduction at the high system level. The obtained results have provided the best parameter configuration for an optimal use of the AVC tools. For this cost-efficient configuration, the obtained absolute complexity values confirmed the big challenge needed for its effective implementation. Given this, we proposed the use of a multiprocessor approach to share the encoding time between several processors. For efficient parallel code decomposition, a "Gprof" execution profiling and a study of the predominant forms of parallelism have been explored. Based on this, a first parallel model of the H264 encoder was proposed and validated using khan process networks (KPN) models implemented by the YAPI multithreading library.

For an optimal parallel H264 encoder specification, we are actually working on the optimization of the proposed first parallel model to get one with more balanced workload, better communication behavior, and maximum parallelism between tasks. For this purpose, the system-level software CAST [9] tool will be used.

REFERENCES

[1] M. Alvarez, A. Salami, A. Ramirez, M. Valero, "A Performance Characterization of high Definition Digital Video Decoding using H264/AVC". Proceeding of the IEEE International, Symposium on Workload Characterization. pp. 24–33, 6-8 Oct 2005.

[2] S. Saponara, K. Denolf, G. Lafruit, C. Blanch, J. Bormans, Performance and Complexity Co-evaluation of the Advanced Video Coding Standard for Cost-Effective multimedia communication, EURASIP Journal on Applied Signal Processing, pp. 220-235, 2004:2.

[3] H. Krichene Zrida, A.C. Ammari, A. Jemai, M. Abid, Performance/Complexity Analysis of a H264 Video Encoder, International Review on Computers and Software (IRECOS), Vol 2 n°4, pp n°401-414, July 2007.

[4] G. Kahn, The semantics of a simple language for parallel programming, proceeding of the IFIP Congress 74. 1974, North-Holland Publishing Co.

[5] M. Pastrnak, P.H.N. de With, S. Stuijk, and J. van Meerbergen. "Parallel Implementation of Arbitrary-Shaped MPEG-4 Decoder for Multiprocessor Systems", Visual Communications and Image Processing (VCIP'06). pp 60771I-1 - 60771I-10, 2006.

[6] K. Shen, L.A. Rowe, and E.J. Delp. 1995. "A Parallel Implementation of an MPEG1 Encoder: Fater than Real-Time", Proceedings of SPIE Conference on Digital Video Compression: Algorithms and Technologies. San Jose (Feb).

[7] S. Bozoki, S.J.P. Westen; R.L. Lagendijk; and J. Biemond. 1996. "Parallel algorithms for MPEG video compression with PVM". In EUROSIM: Delft. The Netherlands 315-326.

[8] E.A. Kock, G. Essink, W.J.M. Smits, P. van der Wolf, J.-Y. Brunel, W.M. Kruijtzer, P. Lieverse, and K.A. Vissers, "YAPI: Application modeling for signal processing system," in Proc. 37th Design Automation Conference (DAC'2000), Los Angeles, CA, June 5-9 2000, pp. 402–405.

[9] S. Stuijk, and T. Basten, "Analyzing Concurrency in Computational Networks (Extended Abstract), Formal Methods and Models for Codesign". 1st ACM & IEEE International Conference, MEMOCODE'2003, Proceedings. Mont Saint-Michel, France, 24-26 June 2003. IEEE Computer Society Press. Los Alamitos, CA, USA.

978-1-4244-1983-8/08/$25.00 ©2008 IEEE

An evaluation methodology for the security of cryptosystems

Selma Laabidi *, Bruno Robisson, Michel Agoyan
Ecole des Mines de St Etienne - Site Georges Charpak, Laboratoire SESAM,
820 route de Mimet, 13120 GARDANNE, FRANCE
Email : laabidi@emse.fr

Abstract—This paper describes an integrated circuit design methodology which evaluates at the gate level the resistance of the circuit to a class of attacks, among them is the well-known Differential Power Analysis (DPA). This is of particular concern since it enables the designer to detect design strengths and weaknesses and to compare and choose different circuit's architectures at an early stage of the design flow.
The proposed methodology has been applied to different hardware models of the Advanced Encryption Standard (AES) and highlights substantial differences between the models in terms of area and security.

I. INTRODUCTION

Secure integrated circuits such as smart cards or SIMs cards have widely been used during this last decade. Cryptographic algorithms like triple DES, AES or RSA are embedded on these devices to ensure confidentiality, authenticity and integrity. The physical implementation of these algorithms is subject to attacks which may allow to recover the key. Many attacks have been described in the literature. Side channel analysis are physical attacks which exploit information leakage such as power consumption [6], timing computation [5] or electromagnetic radiation [11], in order to recover the key. Fault attacks, thanks to power or clock glitches or optical fault injection [1], create errors into the device to get exploitable abnormal results. Recently, some hybrid attacks have been published: they exploit the property of fault injection attacks and the statistical treatment of side channel attacks [12]. Many efforts were deployed to keep cryptographic devices secure against these attacks. To evaluate these devices, security test is done after the chip is realized with respect to normalized methods and criteria [10]. This approach is time consuming and very expensive. Moreover, if the device fails the test, the designer has to design again his circuit. To reduce the probability of such post-production security failure, we propose to test the security of an integrated circuit at an early stage of the design flow, which enables to detect design weaknesses as soon as possible. Nevertheless, the obtained results are to be confirmed by experimental measurements.
We apply our methodology to different gate-level models of a secret key cryptographic algorithm, the Advanced Encryption Standard (AES) [14] in order to test their robustness against some of the most powerful attacks. These attacks use the

[1]This work is supported by the FSE (Fonds Social Européen) through the BTRS project

correlation between a physical information and a model parameterized by the value of a small number of bits of the key, further called "correlation attacks".
This paper is organized as follows. In section 2, we describe our methodology. Section 3 presents a brief description of different implementations of the AES. Section 4 presents the obtained results. We conclude in section 5.

II. EVALUATION METHODOLOGY

A. Related work

Different design-time security evaluation methodologies have been proposed: to examine electromagnetic leakage of secure processors [7], to evaluate the security of secure processors against optical fault injection attacks [8] and to test the security of asynchronous processors against side channel attacks [4]. Another approach uses an information theoretic metric to evaluate side channel resistant logic styles [9]. This metric evaluates the amount of information provided by a circuit and the possibility for the attacker to turn this information into a successful attack. The proposed methodology, which was applied on the gate level, has allowed to compare several countermeasures against side channel attacks. The importance of simulating power analysis attacks at a early stage of the design flow has been demonstrated in [2].
Contrary to these approaches in which the information leakage is detected through the whole circuit's model, our methodology considers each binary value of a bit as a local signal which can be used by the attacker. The objective is to find all the signals in a circuit which make the attack possible. In fact, for correlation attacks, the attacker predicts a value of an intermediate variable, called attack bit, which is expressed as a function of a small number of bits of the key. Then, he correlates this value to a physical measurement in order to deduce information about a few bits of the key. Our methodology detects all the correlated bits to the attack bit. We call these bits "vulnerable bits" and we consider their number as a security criterion. For example, in DPA attack [6], these correlated bits increase potentially the side channel signal, and therefore may help the attacker to mount his attack. In DBA attack [12], the attacker has to inject the same fault on the same bit for a set of plaintexts; he increases his attack success rate by injecting his fault on any bit of the set of correlated bits. In the two cases, these bits are a threat for the security of the device.

978-1-4244-1983-8/08/$25.00 ©2008 IEEE

B. Description of the methodology

Our proposal methodology aims to retrieve all the correlated bits to a given attack bit. So, we have before to define a function which allows to retrieve these bits. Let:

- T, the set of plaintexts t_i (chosen or not but known);
- K_0, the secret key used by the algorithm;
- R, the set of bits r manipulated by the cryptographic algorithm;
- B, the set of attack bits b. These bits are known in the literature and give successful correlation attacks;
- $f(K_0, t_i, r)$, the function that computes the value of $r \in R$;
- $d(K_0, t_i, b)$, the function that computes the value of $b \in B$.

We define the following correlation function :

$$\Delta(T, r, b) = \frac{\sum\limits_{ti \in T} f(K_0, t_i, r) * d(K_0, t_i, b)}{|T|/2}$$

$$- \frac{\sum\limits_{ti \in T} f(K_0, t_i, r) * (1 - d(K_0, t_i, b))}{|T|/2}$$

We assume that the bits r for which $\Delta(T, r, b) = \pm 1$ are perfectly correlated to the attack bit. Then, they form the set of all the sensitive bits.

The security test for correlation attacks is shown Fig 1.

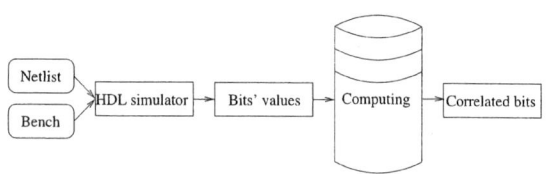

Fig. 1.   Correlation simulation analysis

The test bench gives test vectors for the set of plaintexts and extracts the bits b and r. The HDL simulator runs the bench with the netlist and computes $f(K_0, t_i, r)$ and $d(K_0, t_i, b)$ at a precise instant t. The processing step computes $\Delta(T, r, b)$.

III. DESCRIPTION OF THE STUDIED ARCHITECTURES

In this section, we briefly present the different architectures of AES that are considered for our study.

The Rijndael AES algorithm is a secret key block cipher which operates on a 128-bit data with a cipher key of length 128, 192 or 256 bits. In this paper, we will only consider a 128-bit key AES. The AES algorithm is divided into two processes : a datapath function and a KeyExpansion routine. The datapath function is a sequence of ten sub-processes called round. Each round is composed of four different transformations : AddRoundKey, SubBytes, ShiftRows and MixColumns. The KeyExpansion routine generates the different round keys from the cipher key. The data processed by these two processes is organized into two-dimensional array of bytes called state. The studied architectures depend on the implementation of

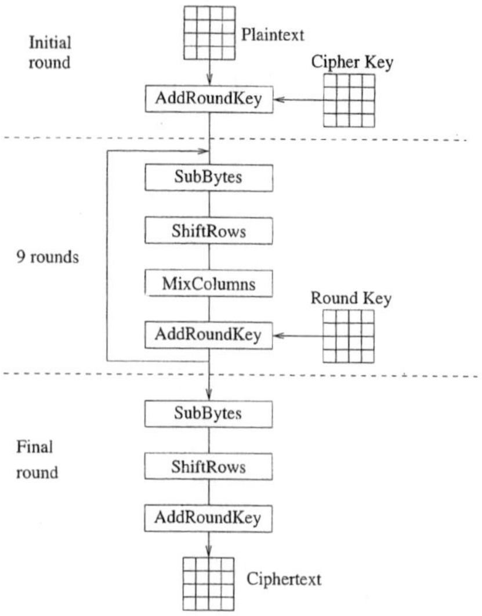

Fig. 2.   128 bit-AES datapath

the MixColumns transformation.

The MixColumns transformation is a matrix multiplication of one column of the current state by a constant matrix over $GF(2^8)$. The four bytes of one column are replaced by the following:

$$S'_{0,c} = (02 \bullet S_{0,c}) \oplus (03 \bullet S_{1,c}) \oplus S_{2,c} \oplus S_{3,c} \quad (1)$$

$$S'_{1,c} = S_{0,c} \oplus (02 \bullet S_{1,c}) \oplus (03 \bullet S_{2,c}) \oplus S_{3,c} \quad (2)$$

$$S'_{2,c} = S_{0,c} \oplus S_{1,c} \oplus (02 \bullet S_{2,c}) \oplus (03 \bullet S_{3,c}) \quad (3)$$

$$S'_{3,c} = (03 \bullet S_{0,c}) \oplus S_{1,c} \oplus S_{2,c} \oplus (02 \bullet S_{3,c}) \quad (4)$$

We propose three different implementations for the Mix-Columns : two of them use the multiplication in $GF(2^8)$ and one uses the multiplication in $GF(2^4)$.

1) Operations in $GF(2^8)$: in the MixColumns, multiplications in $GF(2^8)$ are computed with three values 01, 02 and 03. Knowing that $03 = 02 \oplus 01$, we only implement the multiplication by 02. This later can be described in two different ways.

- Conditional addition: the multiplication by 02 is realized with the function $xtime$. This later is implemented as a left shift and a subsequent conditional xor with $1b$. Let a byte $a = \{a_7 a_6 a_5 a_4 a_3 a_2 a_1 a_0\}$.

$$02 \bullet a = \begin{cases} a_6 a_5 a_4 \bar{a}_3 \bar{a}_2 a_1 \bar{a}_0 a_7 & \text{if } a_7 = 1 \\ a_6 a_5 a_4 a_3 a_2 a_1 a_0 a_7 & \text{otherwise} \end{cases} \quad (5)$$

- Non-conditional addition: in this implementation, we avoid the conditional test on a_7. So $02 \bullet a = a_6 a_5 a_4 a_3 a_2 a_1 a_0 0 \oplus 000 a_7 a_7 0 a_7 a_7$.

2) Operations in $GF(2^4)$: in this structure, two elements a and b in $GF(2^8)$ are represented as a linear polynomial in $GF(2^4)$, $a \cong a_h + a_l$ and $b \cong b_h + b_l$. A multiplication

over $GF(2^8)$ is realized with three general multiplications, four additions and one constant multiplication over $GF(2^4)$ [13].

IV. RESULTS AND INTERPRETATIONS

We have applied our methodology to different gate-level implementations of the AES. Logic synthesis is realized thanks to Synopsys Design Vision® with standard cells library from AMS in 0.35 μm.

T is chosen as the whole set of the $2^8 = 256$ distinct values which exhausts all the possible values of the entry of one S-box. R is the set of bits of the datapath process. B is the set of the eight bits at the output of the first S-box. The different computations are done at the end of the first round of the algorithm with Modelsim® simulator. The results are summarized in the Table I. We can see that the number of the

TABLE I

NUMBER OF SENSITIVE BITS FOR DIFFERENT AES IMPLEMENTATIONS

AES Implementation	Bit number							
	b_0	b_1	b_2	b_3	b_4	b_5	b_6	b_7
Conditional Addition	16	24	16	16	24	24	24	**25**
Non-conditional Addition	16	22	16	16	22	22	22	**23**
Op. in $GF(2^4)$	23	**25**	18	16	24	24	24	24

vulnerable bits depends on the attack bit and the architecture of the MixColumns.

In order to understand the dependency between the attack bit and the number of correlated bits, let consider the equations defined in Section 3. In our case, b belongs to $S_{0,0}$ and the bytes $S_{1,0}$, $S_{2,0}$ and $S_{3,0}$ are constant for the set of executions. Thus, according to equations 2 and 3, $S'_{1,0}$ and $S'_{2,0}$ are always correlated to $S_{0,0}$. The equation number 1 shows that $S'_{0,0}$ depends on the result of $02 \bullet S_{0,0}$. According to equation 5, $S'_{0,0}$ is correlated to $S_{0,0}$ for b_7, b_6, b_5, b_4 and b_1.

The number of vulnerable bits increases considerably when the area of the circuit grows as shown in Table II. This may explain the relationship between the number of vulnerable bits and the chosen architecture for the MixColumns transformation.

TABLE II

SENSITIVITY OF DIFFERENT AES IMPLEMENTATIONS

AES implementation	Area (mm^2)	Nb. of all vulnerable bits
Non-conditional addition	1,823	159
Conditional addition	1,828	169
Operations in $GF(2^4)$	2,348	178

V. CONCLUSION

This paper applies to several architectures of the AES a methodology which evaluates, as soon as possible during the design process, the resistance of circuits against correlation attacks. The obtained results show that the proposed methodology helps the designer to trade-off security and cost implementation. In particular, we have underlined that the

"Conditional addition" and "Non-conditional addition" implementation of the MixColumns block of the AES are better in terms of security and cost compared to implementation using multiplications in $GF(2^4)$.

A future work will consist in evaluating the impact of the place and route design step on the security. We also intend to test the security of widely used countermeasures such as masking methods [3].

REFERENCES

[1] H. Bar-El, H. Choukri, D. Naccache, M. Tunstall, C. Whelan, D.T. Ltd, and I. Rehovot. The sorcerer's apprentice guide to fault attacks. *Proceedings of the IEEE*, 94(2):370–382, 2006.

[2] M. Bucci, R. Luzzi, F. Menichelli, R. Menicocci, M. Olivieri, and A. Trifiletti. Testing power-analysis attack susceptibility in register-transfer level designs. *Information Security, IET*, 1(3):128–133, 2007.

[3] S. Chari, C. S. Jutla, J. R. Rao, and P. Rohatgi. Towards sound approaches to counteract power-analysis attacks. In *CRYPTO '99: Proceedings of the 19th Annual International Cryptology Conference on Advances in Cryptology*, pages 398–412, London, UK, 1999. Springer-Verlag.

[4] J. J. A. Fournier, S. Moore, H. Li, R. Mullins, and G. Taylor. Security Evaluation of Asynchronous Circuits. *Proceedings of Cryptographic Hardware and Embedded Systems-CHES2003*, pages 137–151, 2003.

[5] P. Kocher. Timing Attacks on Implementations of Diffie-Hellman, RSA, DSS, and Other Systems. *Advances in Cryptology–CRYPTO*, 96:104–113, 1996.

[6] P. Kocher, J. Jaffe, and B. Jun. Differential Power Analysis. *Proceedings of the 19th Annual International Cryptology Conference on Advances in Cryptology*, pages 388–397, 1999.

[7] H. Li, A. T. Markettos, and S. Moore. Security evaluation against electromagnetic analysis at design time. In *HLDVT '05: Proceedings of the High-Level Design Validation and Test Workshop, 2005. on Tenth IEEE International*, pages 211–218, Washington, DC, USA, 2005. IEEE Computer Society.

[8] H. Li and S. Moore. Security evaluation at design time against optical fault injection attacks. *IEE Proceedings - Information Security*, 153(1):3–11, 2006.

[9] F. Mace, F. Standaert, and J. J. Quisquater. Information Theoretic Evaluation of Side-Channel Resistant Logic Styles. *Proceedings of Cryptographic Hardware and Embedded Systems: CHES2007*, 4727:427, 2007.

[10] A. Merle and J. Clédière. Security testing for hardware products: The security evaluations practice. In *IOLTS*, pages 122–125, 2005.

[11] J. J. Quisquater and D. Samyde. ElectroMagnetic Analysis (EMA): Measures and Counter-Measures for Smart Cards. *Proceedings of the International Conference on Research in Smart Cards: Smart Card Programming and Security*, pages 200–210, 2001.

[12] Bruno Robisson and Pascal Manet. Differential behavioral analysis. In *Proceedings of Cryptographic Hardware and Embedded Systems: CHES2007*, pages 413–426. Springer, 2007.

[13] E. Soljanin and R. Urbanke. An Efficient Architecture for Implementation of a Multiplier and Inverter in $GF(2^8)$. *Bell-Labs Technical Memo, BLO11217-960308-O8TM*, 1996.

[14] N.F. Standard. Announcing the ADVANCED ENCRYPTION STANDARD (AES). *Federal Information Processing Standards Publication*, 197, 2001. http://csrc.nist.gov/publications/fips/fips197/fips-197.pdf.

BLANK PAGE

A Multi-Precision Floating-Point Adder

Metin Mete Özbilen
Department of Computer Engineering
Mersin University, Çiftlikköy, 33342, Mersin, Turkey
Email: mmozbilen@mersin.edu.tr

Mustafa Gök
Department of Electrical and Electronics Engineering
Cukurova University, Balcalı, 01330, Adana, Turkey
Email: musgok@cu.edu.tr

Abstract— **This paper presents a multi-precision floating-point adder that can perform a high-precision floating-point addition, or multiple low-precision floating-point additions in parallel. The proposed design eliminates time consuming format conversion operations when it is operating in low-precision modes. The proposed multi-precision floating-point adder has delay approximately equal to a standard double-precision floating-point adder.**

I. INTRODUCTION

The floating-point hardware in general-purpose processors and DSPs complies with *IEEE-754* standard which describes double and single precision formats [1]. Recently, there has been growing interest in 3D applications that execute SIMD type floating-point operations. Many general-purpose processors have multimedia extensions that increase the performance for 3D applications [2], [3], [4], [5], [6]. AMD's 3DNow! [3] and Intel's SSE3 and SSE4 [5], [6] extensions support execution of two or more single-precision floating-point operations. Major GPU manufacturers also respond to this trend by offering direct hardware support for 3D applications in their new architectures. NVIDIA's GeForce eight processor family consists of a specific hardware that increases the performance for Microsoft's DirectX 10 technology commonly used in 3D gaming programs [7]. ATI's FireGL 3D processor family is also certified by a variety of 3D application vendors [8]. A general characteristic of graphics processing is that it performs large numbers of low-precision floating-point operations. Thus, the performance for graphics processing can be significantly increased by using floating-point hardware tailored to perform fast low-precision operations. Floating-point addition is one of the most used operations in these type of applications. A low-precision floating-point addition can be performed by using a high-precision floating-point adder; however, this requires conversion of single-precision operands to double-precision and then conversion of the result back to single-precision format. On the other hand, it is practically not feasible to perform a double precision floating-point addition using a single-precision floating-point hardware. A research that focus on dual-precision floating-point addition is presented in [9]. A multi-mode floating-point adder that can perform add, subtract and several comparison operations is presented in [10]. This paper presents a multi-precision floating-point adder design that overcomes the performance degradation caused by format conversion operations. The proposed multi-precision floating-point adder design can

perform four half-precision (in NVIDIA format) floating-point additions or two-single precision floating-point additions or a single double-precision floating-point addition. In low-precision operation modes, the results are generated in parallel. To the best of our knowledge, a floating-point adder with the proposed functionality is not reported in the literature. The remainder of the paper is organized as follows: Section 2 and 3 demonstrate the multi-precision floating-point addition and the proposed implementation, respectively. Section 4 presents the syntheses results and compares the proposed design with a conventional double-precision floating-point adder. Section 5 presents conclusions.

II. FLOATING-POINT ADDITION

The addition of two floating-point numbers $F1$ and $F2$, $FT = F1 + F2$, is defined as

$$M_{FT}^* = \begin{cases} (M_{F1}^* \pm (M_{F2}^* \times b^{(E_{F2}-E_{F1})}) & \text{if, } E_{F1} \geq E_{F2} \\ (M_{F1}^* \pm (M_{F2}^* \times b^{(E_{F1}-E_{F2})}) & \text{if, } E_{F1} < E_{F2} \end{cases}$$
(1)

and

$$E_{FT} = max(E_{F1}, E_{F2})$$
(2)

where $M^* = (-1)^S \cdot M$ and S, E, and M represent the sign, exponent and the mantissa of a floating number, respectively. The floating-point addition operation is more complex than the floating-point multiplication since the alignment of mantissas is required before mantissa addition. The complexity increases the area and the delay of the hardware implementations. The basic implementation of the floating-point addition consists of the following units:

- The difference between the exponents is computed by an adder. The biggest exponent is selected as the exponent of the result.
- The mantissas are aligned by using a right-shifter. A swap unit selects the mantissa with the smallest exponent based on the sign of the difference between exponents and a right shifter shifts this mantissa.
- A sign-magnitude adder adds or subtracts the aligned mantissas based on the the effective operation (EOP).
- A leading-one detector LOD counts the number of leading-zeroes and sends this count to a left/right shifter. The left shift is required when the EOP is a subtraction. A one digit right shift is required when the EOP is an addition; in this cases the exponent is also incremented by an exponent update unit.

978-1-4244-1983-8/08/$25.00 ©2008 IEEE

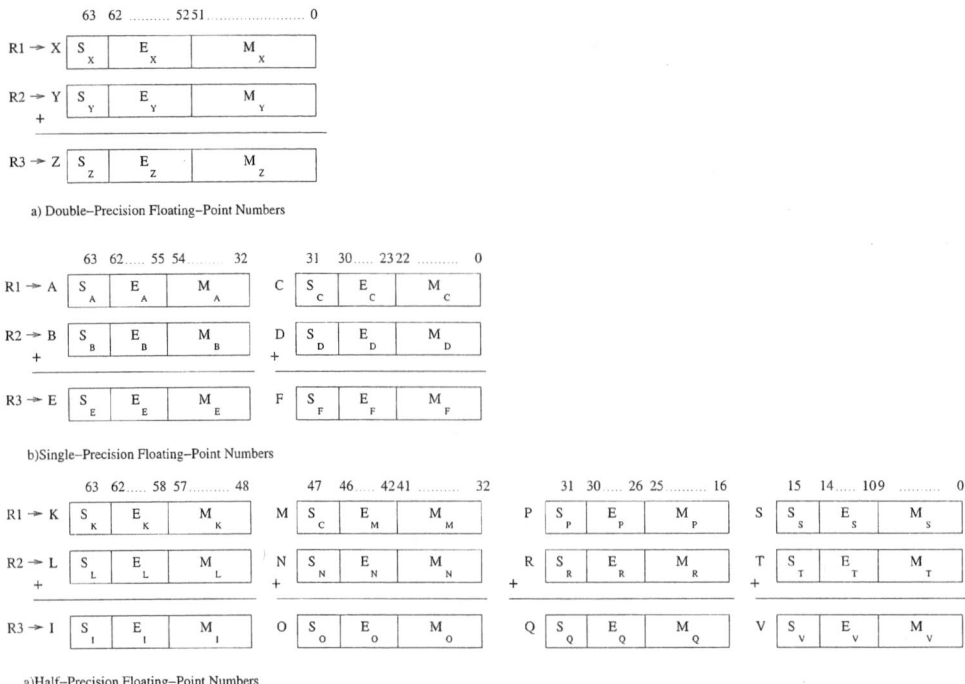

a) Double–Precision Floating–Point Numbers

b)Single–Precision Floating–Point Numbers

a)Half–Precision Floating–Point Numbers

Fig. 1. The Alignments of Double, Single, and Half Precision Floating-Point Numbers

- A rounding unit performs *IEEE-754* rounding, in some cases a 1 *ulp* is added to the mantissa by using an incrementer, if overflow occurs because of the increment operation, the mantissa is normalized again, and the exponent is incremented by the exponent update unit.

III. THE MULTI-PRECISION FLOATING-POINT ADDER

The input operands for the multi-precision adder are packed based on the operation mode. Figure 1 presents the alignments of double, single, and and half precision floating-point numbers and their sums in three 64-bit registers $R1$, $R2$, $R3$. The registers are used for demonstration purpose they are not a part of the actual implementation. In Figure 1.a three double precision floating-point numbers X, Y and their sum are shown. In Figure 1.b four single-precision floating-point numbers A, B, C, D and their sums, E, and F are shown. In Figure 1.c, eight half-precision floating-point numbers K, L, M, N, P, R, S, T, and their sums I, O, Q, and V are shown in NVIDIA half-precision format [7]. Though it is popular, NVIDIA's half-precision format is not included in the *IEEE-754* standard.

Figure 2 presents the block diagram for the proposed multi-precision floating point adder. The design of this adder is based on a modified a version of the single-path floating point adder presented in [11]. The mode of operation is selected by using a control signal, M. When $M = 01$ (Mode 1), a double-precision floating-point addition is performed. When $M = 10$ (Mode 2), two parallel single-precision floating-point additions are performed. When $M = 11$ (Mode 3), four parallel half-precision floating-point additions are performed. *EOP*

represents the effective operation. To reduce the complexity of the figure, the inputs of the units in Figure 2 are plainly designated as $R1$ and $R2$. In the actual implementation only the parts of the vectors that are used in the unit are connected. The location of these parts can be observed from Figure 1. The functionality of the main units and data flow are explained as follows:

- The *exponent subtracter* unit computes the differences of the operands' exponents in all modes. These differences are represented as $d_3 - d_0$. The signs of the differences $sd_3 - sd_0$ are send to the *swap unit*.
- The *swap unit* changes the places of the mantissa if the sign of the difference is negative. By this way only the mantissa with the smaller exponent is right-shifted. Based on the operation mode, the *swap unit* operates on different operands.
- The *compare* unit compares the magnitudes of the operands when the difference or differences between the exponents are zeroes.
- The *bit invert* unit inverts the mantissa (or mantissas) with the smallest exponent so that the result (or results) is always positive. The addition of 1 ulp required for two's complement conversion is performed in the *mantissa adder*.
- The *mantissa generator* unit prepares the mantissa bits for operation in all modes. The mantissas are converted into two's complement format and they are also shifted for alignment.
- The *mantissa adder* is a two's complement adder that can

perform an addition on 53-bit operands or two parallel additions on 24-bit operands, or four parallel additions on 10-bit operands. The signs of the results are generated in the mantissa adder.

- The *leading one detector* (LOD) units compute the number of right-shifts to normalize the result when the *EOP* is a subtraction. *LOD1* operates in all modes, *LOD2* operates in Modes 2 and 3, and *LOD3* operates only in Mode 3. *LOD3* operates on two half-precision operands in Mode 3.
- The *normalize* units are normalizing shifters. The mantissas are either left shifted with the amount determined in *LOD* units or right shifted by one digit when addition overflow occurs.
- The *flag* units determine the rounding flags with respect to the rounding mode that is selected. Since all *IEEE-754* rounding modes are supported a flag for each rounding mode is generated.
- The *rounding* units perform the addition of 1 ulp when it is necessary to perform rounding. These cases are indicated by the flags generated by the flag unit. The overflow due to the addition in rounding units is also checked here and adjustment shift is performed when necessary.
- The *exponent update* units update the exponent strings which are prepared in exponent generator unit.
- The *sign* unit generates the sign of the result or results based on the signs of the operands with greater magnitude. To do that the sign of the operands (indicated as Rs in the diagram), the sign of the differences (sd), and signals that are set when the differences are zeroes (zd) are input to the unit.
- The sign, exponent and mantissa of the result (or results) are represented as S^*, E^*, M^*.

IV. RESULTS

This section presents the syntheses results obtained for the proposed multi-precision floating-point adder and single-path floating-point adders. In addition to the double-precision floating-point adders, single-precision floating adders are also designed. The second multi-precision design performs a single-precision floating-point addition or a two half-precision floating-point adders in parallel. All designs are modeled with VHDL. Syntheses are done using TSMC 0.18 micron standard ASIC library and Leonardo Spectrum program. Both syntheses are tuned for delay optimizations with maximum effort. The area and delay estimates are presented in Table-1. In this table, the unit for area is the number of gates and the unit for delay is nanoseconds (ns). According to the given estimates the first multi-precision design has approximately 68 % more area and has less than 3 nanoseconds more delay than the reference double precision design and the second multi-precision design has approximately 38 % times more gates and has less than half nanoseconds more delay than the reference single-precision floating-point adder. The delay differences between the proposed designs and the reference designs are expected

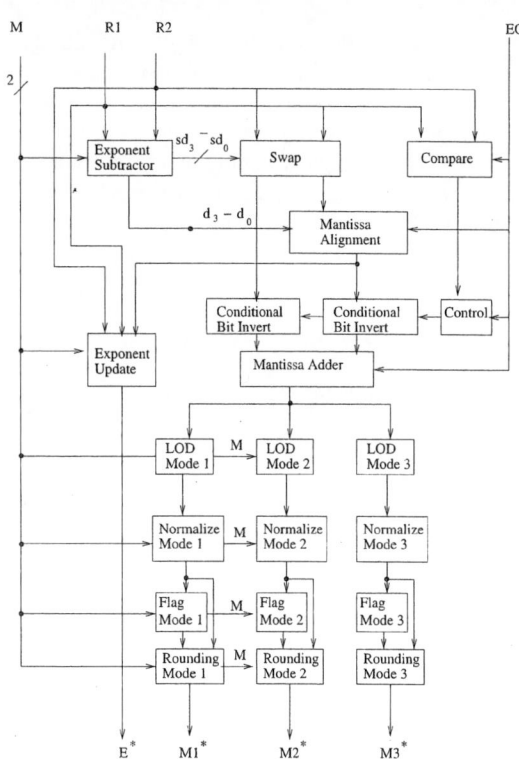

Fig. 2.   The Block Diagram of Multi-Precision Floating-Point Adder

TABLE I

AREA AND DELAY ESTIMATES

Adder Designs	Area	Delay
Double-Precision	4868	14.65
Multi-Precision 1	8195	17.33
Single-Precision	2056	9.33
Multi-Precision 2	2854	9.51

to decrease if the designs are pipelined. A question that can be raised is why not use one double-precision, two single-precision and four half-precision floating-point adders instead of the multi-precision 1 floating-point adder. The proposed unit is expected to use approximately 20 % less gates than the total gates required to design all separate units (Assuming a half-precision floating-point adder can be designed by using approximately 500 gates). Also, the dedicated bus requirement for all the units can be a serious design problem since the wire delay gets significant as the transistor sizes decreases.

V. CONCLUSION

This paper presents a multi-precision floating-point adder that can perform low-precision floating-point additions in parallel. The area and delay for the presented design is slightly bigger than a standard double-precision floating-point adder. The proposed design eliminates the type conversion requirement and generates multiple results in parallel. The presented design is especially expected to increase the performance for 2D and 3D applications since these applications performs intensive floating-point additions on low-precision floating-point operands. On the other hand, the performance of the proposed design is slightly less than a standard double-precision floating-point adder when the double-precision operation mode is used. Our future work will focus on design methodologies for low-precision multi-operand floating-point adders and effective pipelining for the presented designs.

REFERENCES

[1] "ANSI/IEEE 754-1985 standard for floating-point arithmetic," 1985.

[2] X. Yang and R. B. Lee, "PLX:FP: an efficient floating-point instruction set for 3d graphics," in *IEEE International Conference on Multimedia and Expo, ICME'04*, Taipei, Taiwan, 2004, pp. 137–140.

[3] "AMD-3DNow! technology manual," 2000. [Online]. Available: http://www.amd.com

[4] "Intel 64 and IA-32 architectures software developer's manual," 2007. [Online]. Available: http://www.intel.com/design/processor/manuals/253667.pdf

[5] "Intel 64 and IA-32 architectures software developer's manual," Online, 2007. [Online]. Available: http://www.intel.com/design/processor/manuals/253667.pdf

[6] "Intel SSE4 programming reference," 2007. [Online]. Available: http://www.intel.com/design/processor/manuals/253667.pdf

[7] "NVIDIA GeForce family," 2007. [Online]. Available: http://www.nvidia.com/object/geforce_family.html

[8] "FireGL technical specifications," 2007. [Online]. Available: http://www.ati.amd.com/products/workstation/techspecs2.html

[9] P.-M. Seidel and G. Even, "On the design of fast ieee floating-point adders," 2001, p. 0184.

[10] "An ieee floating-point adder for multi-operation."

[11] M. D. Ercegovac and T. Lang, *Digital Arithmetic*. San Fransisco, USA: Morgan Kaufmann, 2004.

Integration of a microfluidic flow cell on a CMOS biosensor for DNA detection

Alessandra Caboni
Massimo Barbaro
Department of Electrical and Electronic Engineering
University of Cagliari
Piazza d'Armi, 09123 Cagliari (Italy)
Email: (alessandra.caboni, barbaro)@diee.unica.it

Alexandra Homsy
Peter van der Wal
Vincent Linder
Nico de Rooij
Institute of Microtechnology
University of Neuchâtel
Jaquet-Droz 1, 2007 Neuchtel (Switzerland)

Abstract—**This paper describes the fabrication technique for the realization a microfluidic flow cell to be integrated on a CMOS biosensor for DNA hybridization detection. The main element of the microfluidic system is made in polydimethylsiloxane (PDMS) elastomer and takes up an area of 5 mm^2. PDMS is cast against a silicon master patterned by Deep Reactive Ion Etching (DRIE) and then bonded on the chip by means of oxygen plasma activation. The micro channels patterned in the flow cell are connected with capillary tubes that can be easily interconnected to a common syringe.**

I. INTRODUCTION

Over the last decade, considerable effort has been directed to miniaturize and integrate analytical devices for biological and biomedical applications. This trend has been particularly evident in the field of DNA analysis, where a number of new approaches for direct, label-free detection of DNA hybridization have been proposed. The most promising way for the realization of simple, portable, inexpensive detection platforms seems to be the exploitation of the advanced technology of consumer electronics [1]-[3]. The target is to replace the cumbersome equipment commonly used in the laboratory, with integrated devices, which handle small amounts of fluids, able to provide ultrasensitive detection at significantly lower cost per assay. In order to achieve high throughput while reducing the consumption of reagents and biological samples, microfluidic systems for dispensing small volumes of fluids must be developed. Such systems have to allow the exposure of the sensitive area of the biochips to various reagents and the sample being analyzed. The rapid advances in microfluidics and microfabrication techniques play a key role in this context.

In [4] we presented a DNA-chip hosting 80 sensing sites, entirely realized in a standard, commercially available, CMOS process, which allows direct electronic read-out of the outputs. In this paper we present the development and integration on the chip of a microfluidic system for the delivery of the reagents and the samples on the active surface of the sensor. A description of the CMOS biosensor, chip architecture and layout, is presented in Section II, whereas the work performed for the development of the microfluidic flow cell will be the topic of Section III.

II. SENSOR SYSTEM

The chip hosts 80 sensors for DNA hybridization detection. Each biosensor is made up of a FET device whose current is modulated by DNA electric charge. The chip incorporates integrated temperature detection for precise assay control and features programmable signal conditioning, amplification and A/D conversion.

A. Chip Architecture

A block diagram of the biochip is shown in Fig. 1. The 80 DNA sensors represent the core of the chip. They were subdivided in two arrays of 40 sensors each in order to make differential measurements and test specificity of the detection. The first array will be used to detect the target DNA and the second array will be used as a reference to verify robustness to non-specific binding. The sensor selected with the decoder circuitry is forwarded to 2 different read-out circuits with different characteristics. The outputs of such buffers are multiplexed with the outputs from 4 different temperature sensors (PTAT). The resulting signal is amplified with a programmable gain and converted into a digital word. Finally, an output multiplexer is used in order to allow test access to internal signals directly from output pads.

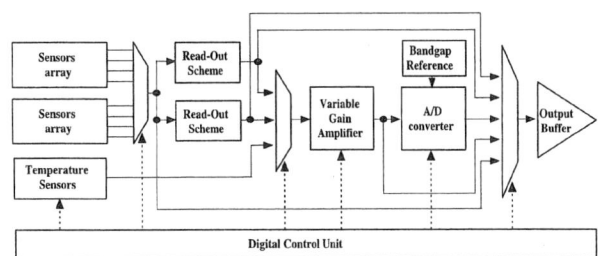

Fig. 1. Chip block diagram.

A digital control unit is needed for addressing a specific sensing site (either DNA or temperature sensor), driving the multiplexers, configuring read-out schemes and biases, programming the gain and sequencing operation of the ADC. Design issues related to the mentioned functional blocks are explained more in detail in the following.

978-1-4244-1983-8/08/$25.00 ©2008 IEEE

B. Chip Layout

The circuit was implemented in a standard $0.35\mu m$ CMOS process with two poly and five metal layers from AMI Semiconductors. A microphotograph of the chip is shown in Fig. 2: most important contribution to the chip area is due to the two sensor arrays. The distance between the blocks was selected in order to complete the chip with a micro-fluidic system. The chip measures $5mm^2$ including pads. It contains a total number of 13569 transistors, 10 BJTs, 128 poly1-poly2 capacitors and 136 resistors.

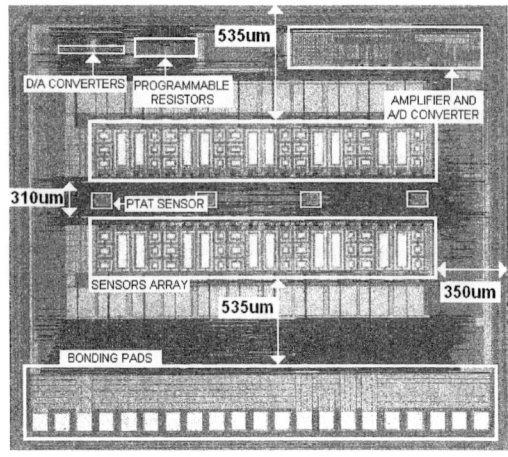

Fig. 2. Microphotograph of the chip.

III. MICROFLUIDIC SYSTEM DEVELOPMENT

The role of the microfluidic system is to allow the delivery of different reagents and samples on the two active clusters of the sensor. The need of such a system was clear from the beginning. For this reason the masks layout of the chip was designed in order to make feasible the binding of a microfluidic system on it: the contact pads for the wire bonding were all located along one side and the sensitive areas were placed in the center of the remaining area, maximizing the most critical distances. Despite all these expedients, the integration of a flow cell on the chip is a tricky task because of the small dimensions of the chip ($2450\mu m \times 2110\mu m$ including the pads) and the sensitive areas ($1743\mu m \times 214\mu m$). The chip is directly mounted on a test board (glass slide dimensions), without packaging.

A. Fabrication Process

The microfluidic flow cell has been developed as a system made up of three elements: the main body featuring two channels, 4 capillary tubes, and glue to fix the tubes. In Fig. 3 the main steps of the technology process are shown.

The first step consists in the realization of the main body. This has been thought as an element that takes up an area equal to that available on the chip for the bonding (the chip area without the region of the pads), that features two microchannels, as long as the chip width ($1743\mu m$), with a

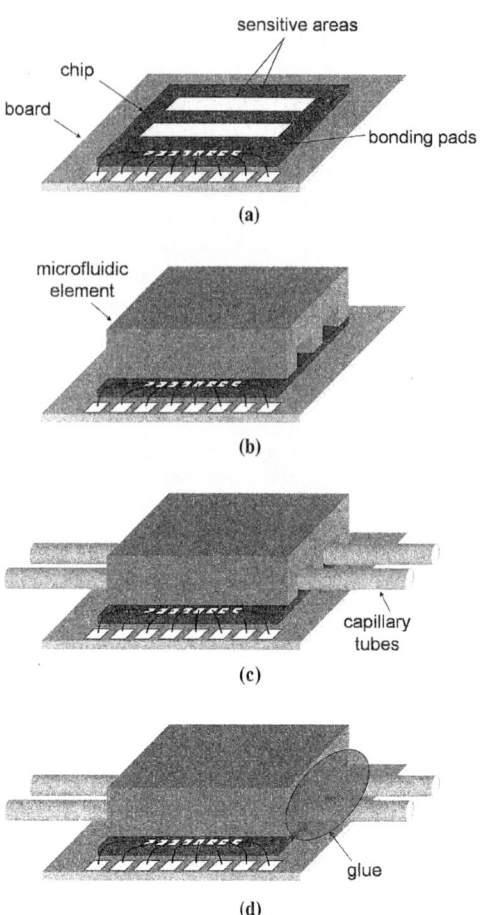

Fig. 3. Steps for the realization of the microfluidic system: a) Starting system: sensor mounted on board, b) Binding of the fluidic element, c) Placement of the capillary tubes at the channels' entrances, d) Glue placement.

square section of side of $214\mu m$, that is the height of the sensors' arrays. The reason behind this choice is due to the configuration of the system, shown in Fig. 4. The chip is mounted on a board by COB (Chip on Board Mounting), therefore between the surface of the chip and the surface of the board there is a gap equal to die thickness ($625\mu m$). This makes unfeasible the realization of a typical microfluidic structure, since the channels patterned in the flow cell would not be enclosed by a bottom surface because of the gap. We overcame this problem by making a simple flow cell, of the exactly the same dimensions of the chip, and gaining the access to the flow cell by means of capillary tubes.

The flow cell is obtained casting a mixture of 10:1 PDMS prepolymer and curing agent (Dow Corning Sylgard 184), on a silicon master patterned by DRIE. The mask for the etching was drawn using a computer-aided design program, Expert Layout Editor by SIMUCAD Design Automation, and then printed onto a transparency film using a high resolution printer. The prepolymer mixture is poured onto the mold, degassed,

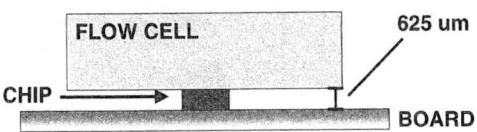

Fig. 4. Schematic illustration of the placement of a microfluidic flow cell on the chip mounted on board.

and cured for 15 min at $150°C$ on a hot plate. In order to facilitate the task of peeling off the PDMS replica from the mold, this last was previously silanized in a dessicator under vacuum with a vial containing a few drops of tridecafluoro-1,1,2,2,-tetrahydrooctyl1-1-trichlorosilane [5].

Then the PDMS element is directly immobilized on the chip, without any interface layer, so that the two microchannels are aligned with the two active areas. To promote the adhesion, the PDMS surface and the topmost layer of the chip, that is Si_3N_4, are activated by means of a treatment with oxygen plasma, before bonding. Bringing the two oxidized surfaces into conformal contact forms a tight, irreversible seal. To get the right alignment between the channels in the flow cell and the active areas on the chip, a variant of the technique experimented in [6] is used. We use water instead of methanol as surfactant: a small drop is placed on the chip, after activation with plasma, before the two surfaces are brought into contact. The alignment is performed by hand, by making one surface slide on the other, under a microscope. Once alignment is finished the system is left for one day at room temperature to evaporate the water. After the bonding the two channels are enclosed by the contact between the flow cell and the chip, and the liquid can flow without leakage.

Thus at the inlets and the outlets of the two channels 4 capillary tubes are inserted. We used fused silica capillary tubes, purchased from Postnova Analytics GmbH, with an inner diameter of $100 \mu m$ and an outer diameter of $193 \mu m$. The placement of the tubes is particularly difficult because they can be positioned only a length equal to $350 \mu m$, that is the distance between the active area and the edge of the chip.

The capillary tubes must be hermetically sealed to the fluidic element. The technique consists in putting a small drop of glue at the entrance of the channels. The dimensions of the gap between the tube and the walls of the channel patterned in the PDMS are such that the glue flows into the channel by capillarity, filling the gaps. The difficulty of this technique is that, if the glue does not stop flowing before it gets the end of the tube, and goes further on, the channels casted in the PDMS flow cell are closed. In order to find the right material, three different glues were investigated:

· Dow Corning Sylgard 184
· Dow Corning Sylgard 186
· Dow Corning 3140 RTV Coating.

The first glue investigated is the common PDMS, the same material used for the fabrication of the main body. It is characterized by a dynamic viscosity equal to 3900 Centipoise.

The second material, Sylgard 186, has been chosen because it has almost the same properties of Sylgard 184 (it is still made up of two parts that have to be mixed in 10:1 ratio), but it is characterized by a higher dynamic viscosity (65000 Centipoise). Finally, the 3140 RTV Coating is a flowable silicon coating, one part made, whose dynamic viscosity is 31000 Centipoise.

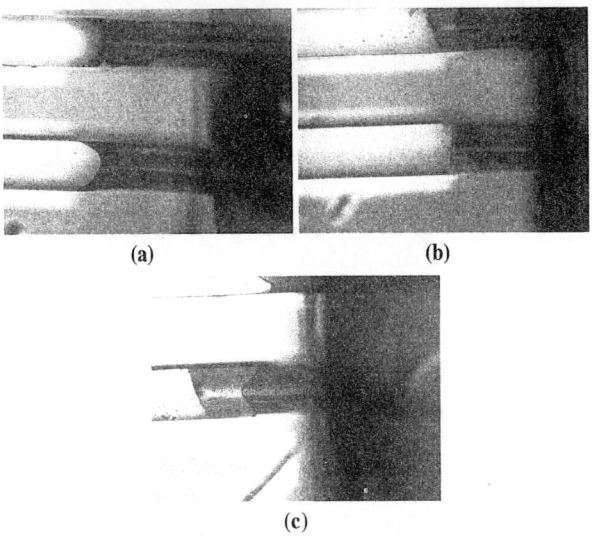

Fig. 5. Results of the experiments carried out with the three glues for the immobilization of the capillary tubes on the PDMS element: zoom on the channels' entrances after the glue placement: a) Sylgard 184, b) Sylgard 186, c) 3140 RTV Coating.

The pictures in Fig. 5 show the results of the experiments carried out with the three materials. It is possible to see how the Sylgard 186 and the 3140 RTV Coating almost gave the same good results (they both can be used for our purpose), whereas the Sylgard 184 turned out to be unsuitable for the task because it tends to go over the end of the tube, closing the channel.

The system so described was realized and then integrated on the chip. The zoom on one of the two sensors arrays, shown in Fig. 7, allows to see how the channels in the microfluidic element are thinner than the active areas and thereby how part of the sensors turn out to be covered. This phenomenon is due to the fact that the channels' walls are not perfectly vertical, rather they are characterized by a sizeable sidewall slope as shown in Fig. 8. Such a profile can be due to two different phenomena: a real sidewall slope in the master or PDMS shrinkage phenomena. The problem was solved by realizing a new silicon master with larger channels.

B. Results and Discussion

The chemical resistance of the microfluidic flow cell was tested with the chemicals involved in the testing with DNA. Using the fluidic system the hybridization reaction was induced on chip, following the common procedure provided for the detection with the CMOS sensor. Since our purpose

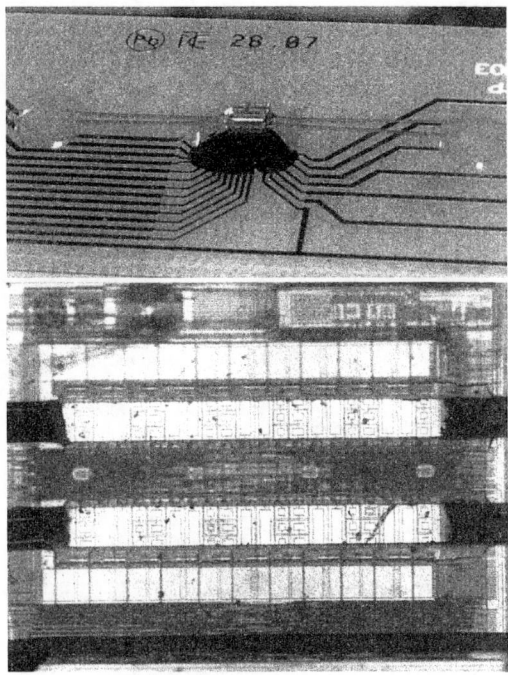

Fig. 6. Microphotographs of the CMOS sensor with the microfluidic system.

Fig. 7. Microphotograph of the CMOS sensor with the microfluidic system: zoom on one sensors array.

was the testing of the fluidic system rather than the sensor, we used a standard technique for the DNA detection, the optical one. We injected a solution $0.1pM/\mu L$ of DNA strands labeled with fluorescent dyes (Cy3) into the capillary tubes and then the system was observed under a fluorescent microscope. As shown in Fig. 9 the capillary tubes are characterized by background autofluorescence that, however, does not prevent the detection of the signals coming from the labels attached to DNA strands. The solution was then pushed onto the chip. The image acquired through the PDMS flow cell shows how the solution keeps confined on the sensor array, without any leakage.

IV. CONCLUSION

In conclusion, the microfluidic system was developed, realized and integrated on different CMOS sensors. Such a system satisfy the initial requirements since it allows the delivery of different solutions on the active clusters. The complete system, made up of the CMOS sensor and the microfluidic system, is now ready for the testing with the DNA.

Fig. 8. Cross section of the PDMS flow cell.

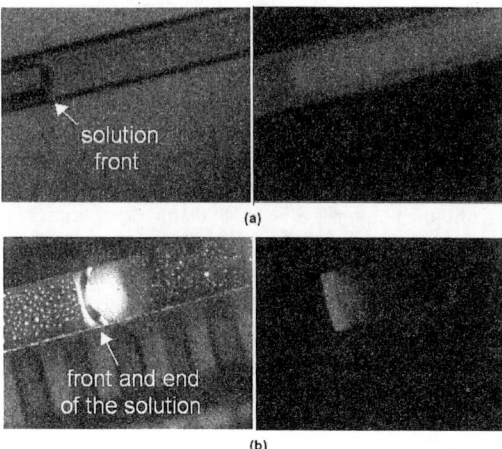

Fig. 9. Images of the experiment carried out injecting into the microfluidic system DNA strands labeled with fluorescent dyes (on the right the images acquired with the fluorescent microscope): a)Fluorescent solution in the capillary tube, b) Fluorescent solution in the channel in the PDMS flow cell, over the sensors' array.

ACKNOWLEDGMENT

The authors would like to thank Prof. Imrich Bark, of the Slovak Academy of Sciences, for his contribution in the experiment with DNA. This work is partially supported by SNSF and CNR, within the framework of an exchange program (PIIT1-117467), and by the project SHAPES (IST-FET-26285).

REFERENCES

[1] R. Moeller and F. W., "Chip-based electrical detection of DNA," *IEE Proc.-Nanobiotechnol.*, vol. 152, pp. 47–51, 2005.
[2] D.-S. Kim, Y.-T. Jeong, H.-Y. Park, J.-K. Shin, P. Choi, J.-H. Lee, and G. Lim, "An fet-type charge sensor for highly sensitive detection of DNA sequence," *Biosensors and Bioelectronics*, vol. 20, pp. 69–74, 2004.
[3] M. Schienle, C. Paulus, A. Frey, F. Hofmann, B. Holzapfl, P. Schindler-Bauer, and R. Thewes, "A fully electronic dna sensor with 128 positions and in-pixel a/d conversion," *IEEE Journal of Solid-state circuits*, vol. 39, no. 12, pp. 2306–2317, 2003.
[4] M. Barbaro, A. Caboni, and D. Loi, "A cmos integrated circuit for DNA hybridization detection with digital output and temperature control," *Proc. of 3rd IEEE Conference on Ph.D. research in Microelectronics and Electronics*, pp. 137–140, July 2007.
[5] Y.-C. Su and L. Lin, "A water-powered micro drug delivery system," *IEEE J. Microelectromech. Syst.*, vol. 13, no. 1, pp. 75–82, February 2004.
[6] B.-H. Jo, L. Van Lerberghe, K. Motsegood, and J. Beebe, D, "Three-dimensional micro-channel fabrication in polydimethylsiloxane (pdms) elastomer," *IEEE J. Microelectromech. Syst.*, vol. 9, no. 1, pp. 76–81, march 2000.

Design and Integration of a Bimorph Thermal Microactuator with Electrostatically Actuated MicroTweezers

Mehmet Yilmaz
Department of Mechanical Engineering
Columbia University
New York, NY 10027, USA
my2232@columbia.edu

Michalis Zervas
Microelectronic Systems Laboratory
EPFL
CH-1015 Lausanne, Switzerland

B. Erdem Alaca
Department of Mechanical Engineering
Koc University
34450 Sariyer, Istanbul, Turkey

Arda D. Yalcinkaya
Department of Electrical and Electronics Engineering
Bogazici University
34342 Bebek, Istanbul, Turkey

Yusuf Leblebici
Microelectronic Systems Laboratory
EPFL
CH-1015 Lausanne, Switzerland

Abstract—**A multi-digit gripper is proposed that consists of two electrostatically actuated end-effectors operating in the plane of the device and three thermal end-effectors operating out of plane. The integration of thermal and electrostatic actuation mechanisms is realized by using a three-mask monolithic process. First mask is used to define the silicon electrostatic actuator on SOI wafer. Second mask is used to obtain the bimorph thermal microactuator made of polyimide and aluminum layers on top of the electrostatic actuator. Third and the final mask is used to release the integrated electrostatic and thermal microactuators.**

I. INTRODUCTION

Microgrippers or tweezers are devices that find many applications in the areas of manipulation for assembly, testing and characterization purposes [1], [2], [3], [4]. Enhancing the gripping capability of such devices requires a careful design of end-effectors. In this study we propose a design that consists of five end-effectors where each can be individually actuated. Two of the end-effectors are connected to a pair of electrostatically actuated combdrives and operate in the plane of the device. A separate layer is processed on top of these combdrives where three additional end-effectors are placed. These fingers are actuated by thermal means and consist of a bimorph structure. The associated technology requires the combination of silicon processing with polymer fabrication. In this work, the design and fabrication of the tweezers will be explained with some of the characterization work highlighted.

II. MEMS DESIGN

The integrated bimorph thermal and electrostatic actuators can generate an out-of-plane displacement at the tip of the thermal actuator, along with an in-plane excursion due to the electrostatic actuator. It is vital to have the bottom layer of the thermal actuator as an insulator to prevent the flow of current from the upper Al metallization layer of the bimorph device to the electrostatically actuated comb-fingers. Specifications of the insulating material include low coefficient of thermal expansion ($\sim$ 3 ppm) and high dielectric constant [5] to prevent the dielectric breakdown during the simultaneous actuation of the bimorph thermal finger and electrostatic comb-finger actuator.

A. Thermal Actuator

Equation (1) is written below [6], and is used to calculate the out-of-plane displacement of a thermal bimorph actuator.

$$d = \frac{L^2}{(h_1 + h_2)} \cdot \frac{3(1+mn)^2}{[3(1+m)^2 + (1+mn)(m^2 + 1/(mn))]} \cdot (\alpha_1 - \alpha_2)\Delta T \quad (1)$$

where, d is the out-of-plane tip deflection L is the length of a thermal actuator, $m = h_1/h_2$ is the ratio of the lower layer thickness (h_1) to the upper layer thickness (h_2), $n = E_1/E_2$ is the ratio of the lower layer Young's modulus (E_1) to the upper layer Young's modulus (E_2), α_1 and α_2 are the thermal expansion coefficients (CTE) of the lower and the upper layers and ΔT is the temperature difference between initial and final states. The equation given in (1), models an out-of-plane tip deflection directionally proportional to the difference between CTEs of the materials. Thus, the materials of the bimorph actuator are chosen to maximize the $(\alpha_1 - \alpha_2)$ factor thereby increasing the deflection.

B. Electrostatic Actuator

The required electrostatic force (F_{es}) is given as [7]

$$F_{es} = -N\varepsilon_0 h(\frac{w}{(g_x - x)^2} + \frac{1}{g_y})V^2 \qquad (2)$$

where, N is the number of comb fingers in each comb, ε_0 is the permittivity of free space, h is the height of the comb fingers, w is the width of the comb fingers, g_x and g_y are air gaps in the longitudinal (translational) and lateral directions, respectively, V is the applied voltage difference between the fingers of the comb pairs and x is the displacement in the translational direction, which is given as

$$x = \frac{F_{es}}{k_{mech,x}} \qquad (3)$$

where, $k_{mech,x}$ is the spring constant of the double-folded beams that resist the electrostatic force attraction.

III. LAYOUT (MASK) DESIGN

The fabrication process uses 3 photomasks for lithographic definitions [8]. The first mask defines the contours of the devices (Fig. 1 a). Next, the mask for the definition of the thermal actuators is used (Fig. 1 b). The third mask (Fig. 1 c) is used to release both the thermal and electrostatic actuators of the device. Fig. 2 shows the overall view of the layout used for the micro/nanofabrication.

IV. PROCESS FLOW

During this study, we attempted to fabricate the devices with 3 main process steps [9].

A. Contour Definition

Conventional photolithography (PL) steps for the definition of the contours on photoresist (PR) are performed (Fig.3 a, b, and c). After the development of the PR (Fig. 3 d), the device layer of the SOI wafer (Device layer thickness: 50 microns, buried oxide layer thickness: 2 microns) is directionally etched until the Buried Oxide (BOX) surface is reached (Fig. 3 e). Following this process flow (in other words, defining the contours at first) introduces a low topography on the surface of the substrate and helps reduce problems that can be encountered in subsequent PL steps.

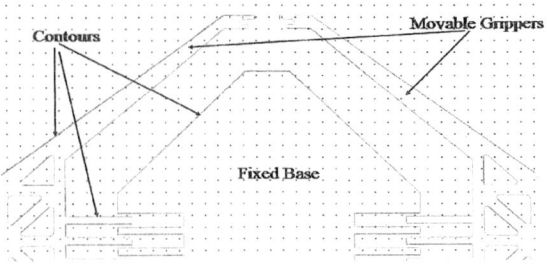

(a)

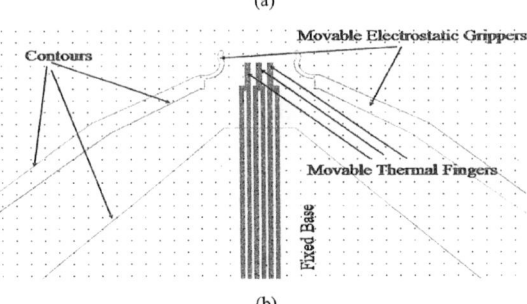

(b)

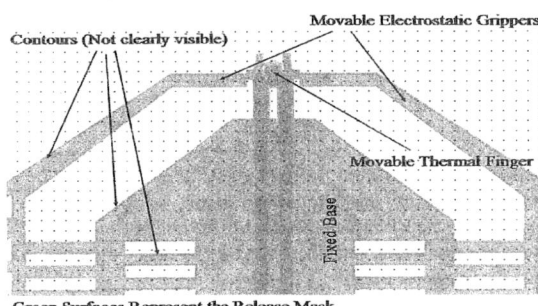

(c)

Figure 1. (a) First mask (Contours) (b) Second mask defining the thermal actuator (c) Third mask used for release process.

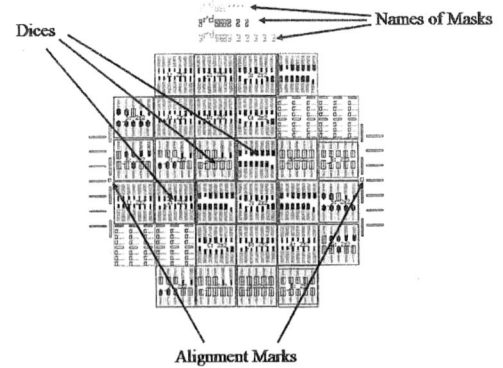

Figure 2. Overall view of the layout.

Figure 3. Definition of the contours of the devices.

B. Preparation for PI Spinning and Thermal Actuator Definition

After the definition of the contours for the electrostatic actuators, all of the exposed surfaces on the wafer are covered with thermal oxide, sputtered LTO, and BPSG with reflow at 1050°C. We have two major reasons to use this kind of oxide formation sequence with a final reflow step: (1) Instead of forming a mechanical barrier for the isotropic Si etching (Such as PR spinning), we form a thin chemical barrier which will serve the same purpose as a mechanical stop with the additional capability of reducing the surface topology for the coming steps of the process, (2) Without the reflow, PI will penetrate into the small gaps. However, this process is not controllable and affects the quality of subsequent steps. As a solution, we decided to do a final reflow step at 1050 °C and close these small gaps with reflow.

After the oxide layer formation on the surface of SOI wafer, we spin-coat PI and perform the curing process for PI as it is stated in [5] (This is the base layer of the thermal actuator, and final thickness of PI layer is ~ 2 microns) (Fig.4 a). Next, aluminum evaporation is performed (This is the top layer of the thermal actuator, and Al layer is ~ 1 micron thick) (Fig.4 b). Then conventional lithography steps are utilized to define the mask for the formation of the thermal actuator (Fig.4 c, d, and e). Following that, aluminum, PI, and oxide layers are directionally etched (Fig.4 f, g, h, and i).

At this point, the definition of the thermal actuator is completed and the only remaining step for the actuation of the thermal actuator is the release of the end-effectors.

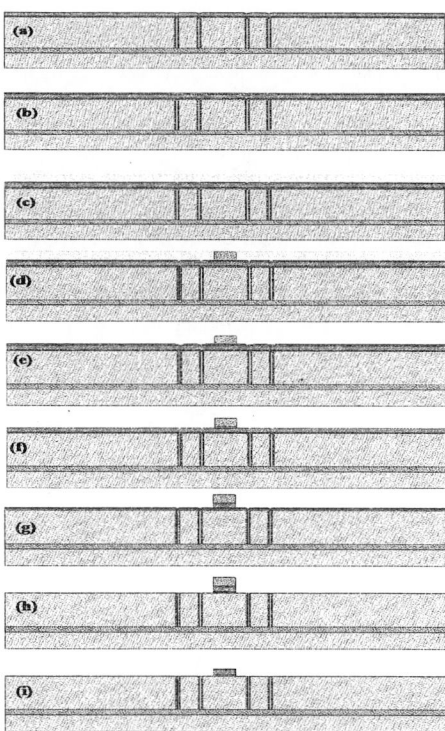

Figure 4. Process steps for the thermal actuator fabrication.

C. Release of the actuators

First, the conventional PL steps are performed to define the mask (Fig.5 a, b, and c). Then isotropic silicon etch is performed (Fig. 5 d). After this step, the electrostatic actuator is completely defined but it is still not released from the BOX layer of the SOI wafer. Directional oxide etch and isotropic oxide etch are the following steps to release the electrostatic device (Fig. 5 e). The final step is the stripping of the PR (Fig. 5 f).

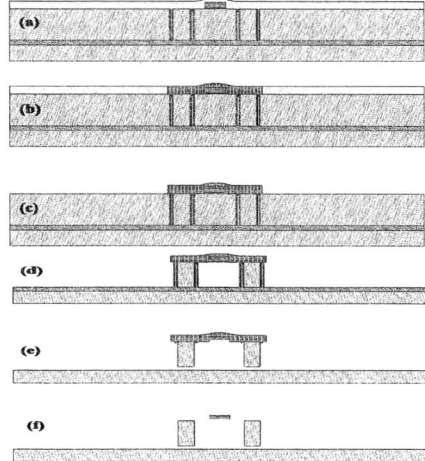

Figure 5. Preperation for the release, and release.

978-1-4244-1983-8/08/$25.00 ©2008 IEEE   127

V. MICRO DEVICE AND EXPERIMENTAL RESULTS

The tip region of the micro tweezers with nanoscale end-effectors is as given in the scanning electron micrograph in Fig. 6. Two electrostatically actuated fingers have the capability to move in-plane, while the three electro-thermally actuated, bimorph fingers have the ability to move in the out-of-plane direction. Furthermore, all of these fingers (both electrostatic and bimorphs) can be individually actuated turning this device into a microhand for different purposes. To state the idea more clearly, it is possible to actuate simultaneously only the right electrostatic and right bimorph fingers if the intention is to grab a microscale object which is close to the right-hand side of the device, or it is similarly possible to actuate the right and left electrostatic fingers and the middle bimorph finger simultaneously. One can also actuate all the bimorph fingers simultaneously to grab an object which is long in one direction compared to the other directions. In the final actuation mode stated above, the bimorph fingers help to touch the object at more than one point and help for a better control on the grasping of the objects.

Fig. 7 shows the actuated thermal actuator under an optical microscope with an interferometric lens to see the successful actuation of the device. Interference pattern at the tip of the actuator shows the deformation of the tip towards the substrate.

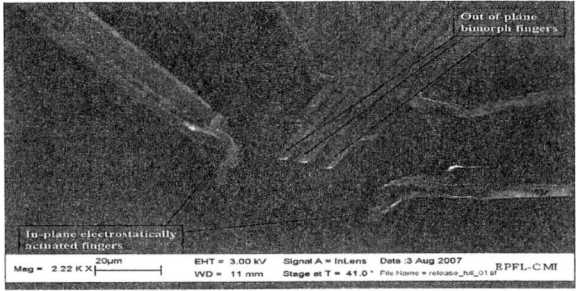

Figure 6.   SEM micrograph of the actual micro/nanofabricated integrated actuators.

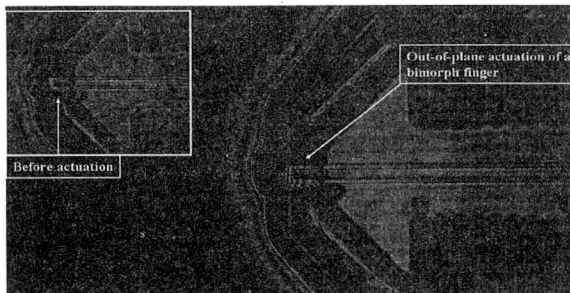

Figure 7.   Homemade optical microscope image of the thermal actutor in action (Interferometric lens is used to see the out-of-plane deflection of the thermal actuator). Inset picture at the upper-left corner shows the unexcited state of the bimorph actuator.

VI. CONCLUSION

This study presents several novelties in the field related to manipulation of objects in microscale. We proposed two different actuation mechanisms integrated together to obtain both in-plane and out-of-plane displacements simultaneously. Furthermore, using the microfabrication method presented in the study, it became possible to construct the bimorph thermal actuator on top of the electrostatic actuator without encountering any serious integration problems related with high topology, since defining the contours of the electrostatic actuator and then closing the contour trenches with successful thin layers of thermal oxidation, sputtered LTO, and BPSG with reflow at 1050°C enhanced a uniform surface for the microfabrication of the thermal actuator.

ACKNOWLEDGMENT

This study is supported in part by TUBITAK under grant No.104 M216. M. Yilmaz thanks to TUBITAK because of their pecuniary support and J. Craig from HD Microsystems. C. Ataman is acknowledged for his help in the test of the devices. We thank Prof. P. Boggild of DTU for valuable discussions in the initial phase of the project.

REFERENCES

[1]   P Bøggild, T. M. Hansen, C. Tanasa and F. Grey, "Fabrication and actuation of customized nanotweezers with a 25 nm gap", Nanotechnology, 12, 331-335, 2001.

[2]   B.E. Volland, H. Heerlein, I.W. Rangelow, "Electrostatically driven microgripper",Microelectronic Engineering, 61–62, 1015–1023, 2002.

[3]   N. Chronis and L. P. Lee, "Electrothermally Activated SU-8 Microgripper for Single Cell Manipulation in Solution", J. Microelectromech. Syst., 14, 857–63,2005.

[4]   K. Mølhave and O. Hansen, "Electro-thermally actuated microgrippers with integrated force feedback", J. Micromech. Microeng. 15, 1-6, 2005.

[5]   HD Microsystems PI2611/10 material bulletin.

[6]   S. Baglio, S. Castorina, L. Fortuna, and N. Savalli, "Modeling and Design of novel photo-thermo-mechanical microactuators", Sensors and Actuators A, 101, 185-193, 2002.

[7]   Johnson W A and Warne L K "Electrophysics of micromechanical comb actuators" Journal of Microelectromechanical Systems, 4, 49–59, 1995.

[8]   M. Zervas, "Clean-room notebook," unpublished.

[9]   M. Yilmaz, MSc.Thesis, Koc University, August 2007.

Elimination of notching phenomenon which occurs while performing deep silicon etching and stopping on an insulating layer

A. Summanwar[abc], F. Neuilly[b], T. Bourouina[c]

[a] Université Paris-Est, Cité Descartes, 5 Bd Descartes, Champs-sur-Marne, 77454 Marne la Vallée, France
[b] NXP Semiconductors, 2, rue de la Girafe, B.P. 5120, 14079 Caen, France
[c] ESIEE, Ecole Supérieure d'Ingénieurs en Electrotechnique et Electronique, ESYCOM-EA 2552, 2Bd Blaise Pascal, 93162 Noisy-le-Grand, France

Principal & corresponding author Tel.: +33 (0)2 31 45 23 89; Fax: +33 (0)2 31 45 21 12
E-mail address: anand.summanwar@nxp.com (A. Summanwar)

Abstract—**The notching phenomenon has been observed during high aspect ratio silicon etching while performing an etch stop on a dielectric layer. It is generally considered as a critical issue in the fabrication of MEMS structures on SOI substrates. This article reports a novel solution for the elimination of the notching while using conventional non-pulsed RF substrate biasing.**

I. INTRODUCTION

Anisotropic etching of high aspect ratio structures on a silicon substrate is a key step in the fabrication of MEMS devices. High aspect ratio structures in silicon are also finding a growing number of applications in semiconductor devices. To perform the deep anisotropic etching of silicon, a technique known as the Bosch process [1] is commonly employed. It is a dry etching technique making use of high density plasmas and consists of short alternating steps of etching and sidewall passivation to achieve an anisotropic profile.

A. RIE Lag, Microloading & other non-uniformities

While performing deep Si etching of high aspect ratio structures, it has been observed that as the aspect ratio of a structure increases the etch rate decreases. This is known as the aspect ratio dependence of etch rate or 'ARDE'. Due to ARDE, when patterns with different critical dimensions (CDs) are etched simultaneously, it is found that patterns with smaller CDs lag behind in etch rate compared to patterns with larger CDs. This leads to differences in their etch depths as the etching progresses. This phenomenon is known as 'RIE lag'.

Even while etching patterns having the same CD, it has been observed that the local pattern density has an influence on the etch rate. For example if a certain region on the wafer has a high density of patterns to be etched and at another region there are only a few isolated patterns, it is likely that the isolated patterns will be etched faster. This is known as 'microloading'. In addition to these phenomena, it is also possible that due to the characteristics of the etch chamber or the etch recipe being used there is a non uniformity in etch rate between patterns at different locations on the wafer.

Due to the existence of these non-uniformities during deep silicon etching, SOI substrates are often employed for fabrication of MEMS devices. An etch stop is performed on the buried oxide layer in order to ensure a uniform etch depth for all patterns. In other MEMS/semiconductor applications for example in the fabrication of via interconnections, it is necessary to perform an etch stop on a dielectric layer such as SiO_2 or Si_3N_4, deposited on the back surface of the wafer. Normally, if a conventional non-pulsed RF substrate biasing is used during the etching process, a phenomenon known as notching is observed.

B. The Notching phenomenon

Notching is the lateral etching of the sidewalls of the structures etched in silicon, in the region near the interface between the silicon and the dielectric etch stop layer as shown in Fig 1.

When a dielectric etch-stop layer such as SiO_2 or Si3N4 gets exposed to the plasma, it gets charged by the positive ions flux towards substrate. Due to this charging of the dielectric layer, the ion trajectories are distorted causing ions to get deflected towards the sidewalls of the etched structure [2]. These ions remove the passivation layer on the sidewalls of the etched structure in the region near the interface

978-1-4244-1983-8/08/$25.00 ©2008 IEEE 129

between the silicon and the dielectric layer and thus the structure starts to get etched laterally in this region.

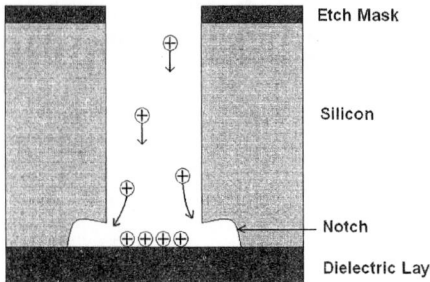

Figure 1: Notching mechanism

Although notching could be an interesting way to release MEMS structures [3], in general it is considered as a defect that causes problems in post-processing steps or degrades device performance or renders devices completely non-functional.

C. Solutions for eliminating Notching

For the elimination of the notching phenomenon, a possible solution is to use a pulsed substrate biasing [4],[5] instead of conventional non-pulsed biasing. This solution relies on suppressing the charge build-up on the insulating layer. However, in the present article, it has been shown that notching can be reliably eliminated even while using conventional non-pulsed RF biasing, by making use of a two step etch recipe.

II. ETCHING OF HIGH ASPECT RATIO VIAS

In the present article, we consider the fabrication of high aspect ratio via interconnects. In the fabrication process it was necessary to etch vias traversing the silicon substrate and perform an etch stop on a 1.5µm thick SiO_2 layer deposited on the back surface of the wafers. This etching was performed with the Bosch process using conventional non-pulsed RF substrate bias on an Aviza DSiE high density inductively coupled plasma (ICP) etch tool.

Due to the higher etch rate at the center of the wafer as compared to the edge, the vias at the center were inevitably subjected to a certain amount of over-etch, to ensure that all vias on the wafer were completely etched. Vias in which the SiO_2 layer was exposed during the over-etch period showed a significant amount of notching as shown in Fig 2 below. This notching poses a critical issue for post-processing of the wafers.

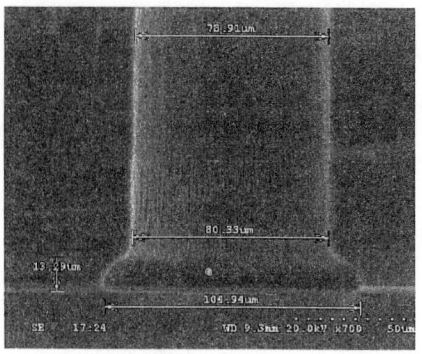

Figure 2: Notching observed at the interface between the silicon and the SiO_2 layer.

A. Development of a two step etch recipe to eliminate notching

A novel solution consisting of a two step etch recipe was developed which allowed the etching of the vias without notching. The first step of the recipe uses the Bosch process thus taking advantage of the high etch rate and selectivity of this process. During the first step, the goal is to attain the maximum etch depth possible, while making sure that the etching is voluntarily stopped before exposing the etch stop layer. The second step of the etch recipe makes use of a continuous etching process in which the etching and passivation gases are used simultaneously yielding an anisotropic etch profile with vertical sidewalls with complete elimination of notching.

B. Defining conditions for the continuous etch process

It was demonstrated that anisotropic silicon etching could be achieved by employing the etching gas (SF_6) and the passivating gas (C_4F_8) together in a continuous etch process. As the percentage of C_4F_8 with respect to SF_6 is increased the profile changes from isotropic (with 0% C_4F_8 addition) and becomes progressively anisotropic as can be seen in the Fig 3a (with 0% C_4F_8 added), Fig. 3b (with 10% C_4F_8 added), Fig. 3c (with 20% C_4F_8 added) & Fig 3d (with 30% C_4F_8 added). In Figs 3c & 3d it can be seen that the sidewalls of the structures have a well defined slope of 65° and 75° respectively. The use of this type of a continuous process could be an interesting technique to etch structures with sloping sidewalls and could find useful applications in MEMS or semiconductor fabrication.

978-1-4244-1983-8/08/$25.00 ©2008 IEEE

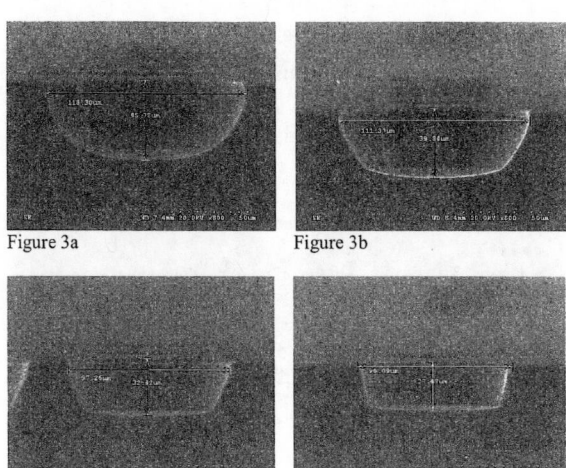

Figure 3a

Figure 3b

Figure 3c

Figure 3d

Figures 3a, b, c & d: Etch profiles obtained with increasing proportion of C_4F_8 gas added along with SF_6 gas in the plasma

With an even higher percentage (75%) of C_4F_8 added along with SF_6, it was demonstrated that an anisotropic etch profile with vertical sidewalls could be achieved as shown in Fig 4a & 4b below. In Fig 4b, it can also be seen that the level of mask undercut is extremely low.

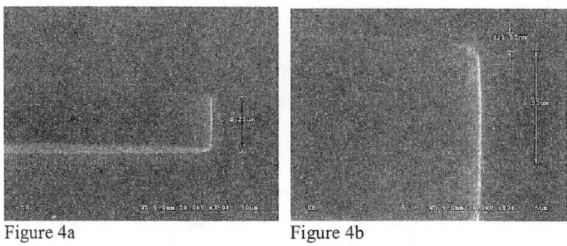

Figure 4a

Figure 4b

Figure 4a: Etch profile with 75% C_4F_8 added along with SF_6,

Figure 4b: Close-up of the top portion of the via showing the amount of mask undercut.

The drawback of this continuous process is that the etch rate is low. For the results shown below, the silicon etch rate is approximately 1.6μm/min.

C. Etch results obtained using the two step etch recipe

As the substrate bias used during the entire etch process is non-pulsed, the charging of the dielectric etch stop layer inevitably occurs once this layer gets exposed during the etching. Due to this the ions deflected towards the sidewalls of the structure start to remove the passivation layer from the sidewalls of the etched structure near the interface between the silicon and the dielectric layer. However, due to the presence of the large proportion of C_4F_8 gas in the plasma, passivation polymer precursors are readily available at all

times for replenishing the sidewall passivation layer. Due to this notching is completely eliminated even when the wafers are subjected to a long over-etch.

The two step recipe combining the Bosch process along with a continuous process was tested to etch via structures traversing the silicon substrate with an etch-stop on a 1.5μm thick SiO_2 layer. Figures 5a & 5b compare the results obtained with the two step recipe to previous results obtained using only the Bosch process.

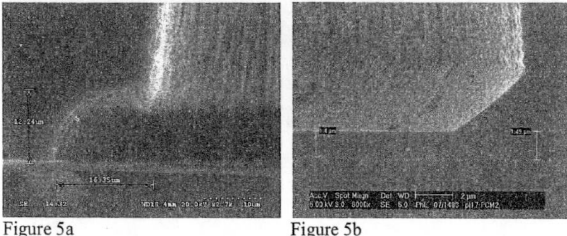

Figure 5a

Figure 5b

Figure 5a & 5b: Comparison of results obtained with (Fig 5b) and without (Fig 5a) using the two step etch recipe. Both images are close-ups of the silicon-SiO_2 interface

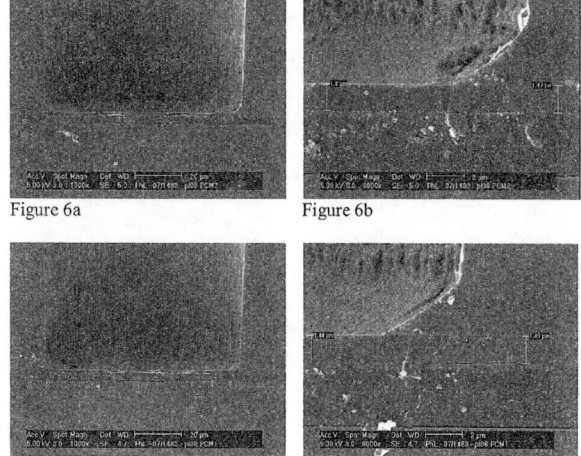

Figure 6a

Figure 6b

Figure 6c

Figure 6d

Figures 6a & 6c: Etch profile at the centre and edge of the wafer respectively.

Figures 6b & 6d: Close-up of the silicon-SiO_2 interface at the center and at the edge of the wafer respectively.

Figures 6a, 6b, 6c & 6d show the etch profiles at the centre and edge of the same wafer that was etched with the two step recipe.

From these results, it is evident that the notching has been completely eliminated at all locations on the wafer by using the two step etch recipe. Even at the centre of the wafer where the structures were subjected to the maximum over-etch, no notching was observed. It can also be seen that

978-1-4244-1983-8/08/$25.00 ©2008 IEEE

the consumption of the SiO_2 etch stop layer during the over-etch period is very minimal at both the centre and edge of the wafer which shows that the selectivity of the process to a dielectric etch-stop layer is very high.

III. CONCLUSION

A novel solution employing a two step etch recipe has been developed to eliminate the notching phenomenon which occurs during the silicon etching while performing an etch stop on a dielectric layer. This solution makes it possible to perform notch free etching of high aspect ratio MEMS structures using SOI wafers in an etch tool equipped with a conventional non-pulsed RF substrate bias. Using this solution, etching of vias traversing the silicon substrate, with an etch stop on a SiO_2 layer has been successfully demonstrated.

ACKNOWLEDGMENT

The authors would like to thank NXP Semiconductors for supporting this work and LAMIPS for their assistance in the characterization of the etch results.

REFERENCES

[1] Franz Lärmer, Andrea Schilp, Patents DE 4241045, US 5501893 and EP 625285

[2] G.S. Hwang, K.P. Giapis, "On the origin of the notching effect during etching in uniform high density plasmas", J. Vac. Sci. Technology B 15 (1997) pp.70-87

[3] J. Li, Q.X. Zhang, A.Q. Liu, "A novel DRIE fabrication process development for SOI-based MEMS devices", Symposium on Design, Test, Integration & Packaging of MEMS/MOEMS (2003) pp.234-238

[4] Franz Lärmer, A. Urban, "Challenges, developments and applications of silicon deep reactive ion etching", Microelectronic Engineering 67-68 (2003) pp.349-355

[5] Hopkins et al, Patent US 6187685

A Micro Power Generator with Planar Coils on Parylene Cantilevers

Ibrahim Sari[1], Tuna Balkan[1], and Haluk Kulah[2]

[1]Middle East Technical University, Dept. of Mechanical Eng., Ankara, TURKEY

[2]Middle East Technical University, Dept. of Electrical and Electronics Eng., Ankara, TURKEY

E-mail: isari@metu.edu.tr, balkan@metu.edu.tr, kulah@metu.edu.tr

Abstract—In this paper an electromagnetic vibration based micro power generator is presented. The proposed generator is composed of parylene cantilevers on which planar coils are fabricated. The system uses external vibrations to generate power by virtue of the relative motion between the cantilevers and a magnet. The parameters of the micro generator have been optimized for maximum output and it has been fabricated in micro scale. Initials tests show that 8.75 mV could be obtained from the proposed generator at a vibration frequency of 5.1 kHz.

I. INTRODUCTION

Following the recent improvements in MEMS technology, many of the electronic equipment that we use in our daily life can now be manufactured smaller. On the other hand, batteries being the major power source for these systems are not improved as fast. Thus, there is a mismatch between lately developed sensor and actuator systems and corresponding batteries in terms of size and weight. Besides, disposal and dirt has become an important problem. As a result, researchers are seeking for new alternatives that are clean, vast, small, light, and having enough energy to power up these systems. Various alternatives such as micro batteries, fuel cells, and energy scavengers (solar, thermal and vibration based) have already been proposed. Among these alternatives micro batteries and fuel cells are still problematic in terms of environmental issues and disposal, besides having a limited amount of energy. Energy harvesting from vibrations is a new and promising topic as it is clean, vast, and adequate power levels can be reached. In the literature, mainly three types of vibration-to-electrical energy conversion techniques have been proposed so far; electromagnetic, piezoelectric and capacitive. Among these techniques, electromagnetic conversion is particularly attractive because of easy interfacing and not requiring an external power supply.

In this study, an electromagnetic type vibration-based micro energy harvester is presented. The proposed device has been modeled, simulated, and optimized in Matlab to obtain maximum output from the generator. The micro power generator is implemented and tested for performance.

II. SYSTEM

Figure 1 illustrates the proposed micro power generator employing parylene cantilevers with planar coils fabricated on top. Voltage is generated across the terminals of the coils by virtue of the relative motion between the cantilevers and the magnet. These coils are connected in series to superpose the generated power from each cantilever. Similar generators have been proposed previously by using glass or silicon as the cantilever material [1]. In this study, Parylene C is used as the cantilever material as it is much flexible compared to silicon [2]. This will allow larger deflections before mechanic failure. Besides, parylene allows optimization of device parameters (resonance frequency, cantilever thickness, spring constant, etc.) in a much wider range with higher accuracy.

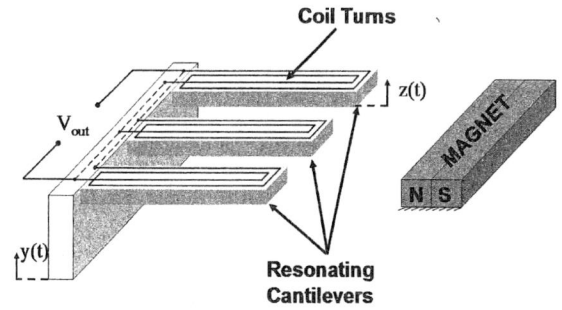

Figure 1. Illustration of the proposed generator.

Figure 2 shows the basic electrical model of a single cantilever. In this model it is assumed that, the coil's inductance is small compared to its resistance in the frequency range of interest and thus neglected. The induced voltage over a coil can be expressed by [3],

---

This work is funded by the Scientific and Technological Research Council of Turkey (TUBITAK) under grant number 104E119.

978-1-4244-1983-8/08/$25.00 ©2008 IEEE

$$\varepsilon = -\frac{d\Phi}{dt} = -\frac{d\left(\int \vec{B} \cdot d\vec{A}\right)}{dt} = -BL_P\dot{z} \qquad (1)$$

where Φ is the magnetic flux density, B is the magnetic field strength of the magnet, L_P is the practical coil length, and $\dot{z}$ is the relative velocity of the tip point of the cantilever with respect to the magnet. The electrical power can now be obtained from,

$$P = \frac{\varepsilon^2 R_L}{2(R_L + R_c)^2} \qquad (2)$$

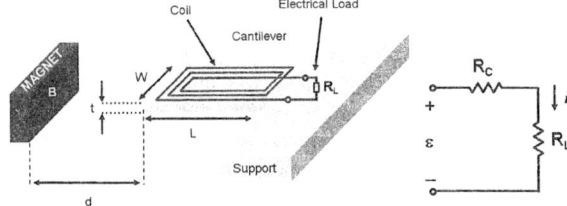

Figure 2. Electrical model for a single cantilever.

By substituting (1) in (2) the power term can be expressed in terms of the system parameters as,

$$P = \frac{1}{2}\frac{(BL_P)^2}{(R_L + R_c)^2} R_L \dot{z}^2 \qquad (3)$$

The relative velocity term in the last equation is obtained by constructing a suitable equivalent mechanical model of the system. For this purpose a second order mechanical equivalent of a single cantilever can be constructed as shown in Figure 3. In this model, the overall mass of the cantilever is lumped to the free end and the elastic behavior of the cantilever is represented by an equivalent spring.

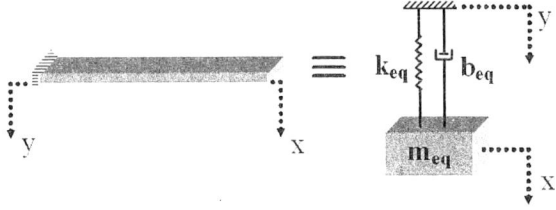

Figure 3. Equivalent mechanical model for a single cantilever.

The differential equation of motion for the model shown above can be determined by Newton's 2nd Law as,

$$m_{eq}\ddot{z} + b_{eq}\dot{z} + k_{eq}z = -m_{eq}\ddot{y} \qquad (4)$$

In this equation, m_{eq}, k_{eq}, and b_{eq} are the equivalent mass, stiffness, and damping, respectively. z is the relative displacement of the cantilever tip with respect to its fixed end and y is the base displacement of the support.

The equivalent mass and stiffness terms can be obtained from $m_{eq} = \frac{33}{140}m$ and $k_{eq} = \frac{3EI}{L^3}$, respectively [4]. In these equations E is the modulus of elasticity, I is the area moment of inertia, m is the mass, and L is the length of the cantilever. Using these two equations, the natural frequency of a single cantilever can be determined from,

$$\omega_n = \sqrt{\frac{k_{eq}}{m_{eq}}} = 3.57\sqrt{\frac{EI}{mL^3}} \qquad (5)$$

From the steady-state solution of (4), the relative velocity term can be obtained as,

$$\dot{z}(t) = \frac{\left(\dfrac{\omega}{\omega_n}\right)^2 \omega Y}{\sqrt{\left(1-\left(\dfrac{\omega}{\omega_n}\right)^2\right)^2 + \left(2\zeta_{eq}\dfrac{\omega}{\omega_n}\right)^2}}\cos(\omega t + \varphi) \qquad (6)$$

where ω is the vibration frequency, Y is the vibration amplitude, φ is the phase angle, and ζ_{eq} is the overall damping ratio. By substituting this equation in (3), the power term can be obtained in terms of the system parameters as,

$$P(t) = \frac{1}{2}\frac{(BL_P)^2 R_L}{(R_L + R_c)^2}\frac{\left(\dfrac{\omega}{\omega_n}\right)^4 \omega^2 Y^2}{\left(1-\left(\dfrac{\omega}{\omega_n}\right)^2\right)^2 + \left(2\zeta_m\dfrac{\omega}{\omega_n}\right)^2}\cos^2(\omega t + \varphi) \qquad (7)$$

Figure 4 shows the variation of power with respect to the frequency ratio for various values of damping. The plot is obtained by keeping the natural frequency constant and varying the input frequency only. As seen in the figure, for low damping ratios, as the input frequency is gradually increased, the power output makes a maximum at the resonance frequency and then makes a minimum and finally it keeps on increasing. It can be concluded that for maximum power generation there are two different operating points. The first one is the resonance point where the natural frequency of the generator is matched to the environmental vibration frequency, and the second one is far beyond the resonance point where the input frequency is much larger than the natural frequency of the generator. In the latter case, power levels greater than the resonance point can be obtained depending on the damping and the input frequency.

978-1-4244-1983-8/08/$25.00 ©2008 IEEE

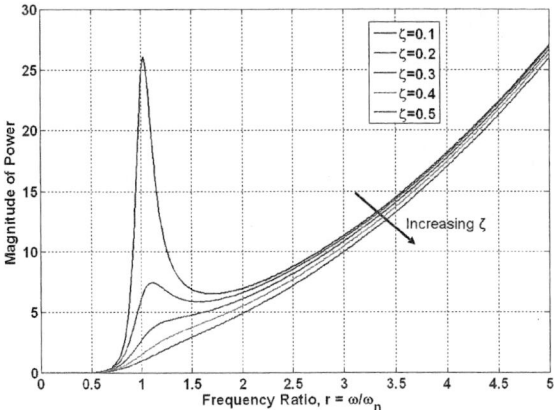

Figure 4. Plot of power as a function of frequency ratio for different damping ratios.

Equations (1), (5), (6) and (7) together with other system parameters are used to optimize the dimensions of the generator. For this purpose a Pattern Search Algorithm is utilized in Matlab. Using this algorithm all predetermined designs parameters are optimized to maximize the power or voltage output from the generator. During the optimization procedure all nonlinear and linear constraints together with upper and lower limits of the parameters are considered. Optimized parameters of the generator are given in Table 1.

TABLE I.   OPTIMIZED PARAMETERS OF THE GENERATOR.

Cantilever size (µm)	890x670x15
Natural freq. of cant.	4.53 kHz
Distance btw. cants. to magnet	50 µm
Coil width	20 µm
Coil resistance (single cantilever)	110 Ω
Coil length (single cantilever)	14 mm
Load resistance	5 kΩ
Overall voltage output	40 mV

III.   FABRICATION

The fabrication process for the proposed system is very simple and requires only 5 masks. Parylene is used as the structural material for the cantilevers due to its much lower modulus of elasticity compared to silicon [1], which allows much larger deflections and increased power generation. Also, it permits adjustment of cantilever parameters (e.g. stiffness and natural frequency) over a wide range. Figure 5 shows the fabrication outline for the generator [5]. First, an oxide layer is formed on the silicon substrate to provide isolation between metal pads and silicon. Next, a 1µm-thick parylene is patterned to form the cantilevers (a). Then, coils are formed by sputtering and patterning the first metal (b). Next, a second 1µm-thick parylene is deposited (c) to isolate the two metals. Then, the second metal is formed for electrical routing (d). Afterwards, 13µm-thick parylene layer is deposited to define the cantilever thickness (e). Next, the silicon is etched from backside by DRIE, and by etching the sacrificial oxide, devices are released (f, g). Figure 6 shows the fabricated prototype before the release.

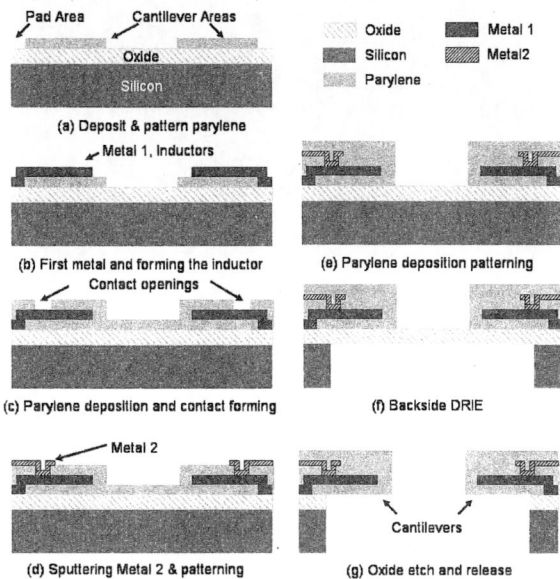

Figure 5. Process flow of the micro generator.

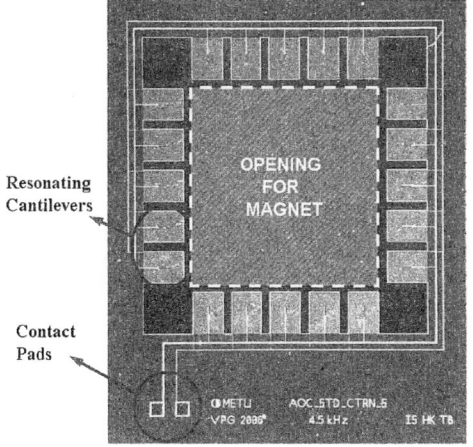

Figure 6. Fabricated device before the release step.

IV.   EXPERIMENTAL RESULTS

Figure 7 shows the fabricated prototype prepared for testing and Figure 8 shows the block diagram of the experimental setup. The tests are carried using a shaker table that is controlled in closed-loop to achieve desired vibration amplitude and frequency. The frequency is swept from 2.5 to 6 kHz at a constant displacement of 0.7 µm.

978-1-4244-1983-8/08/$25.00 ©2008 IEEE   135

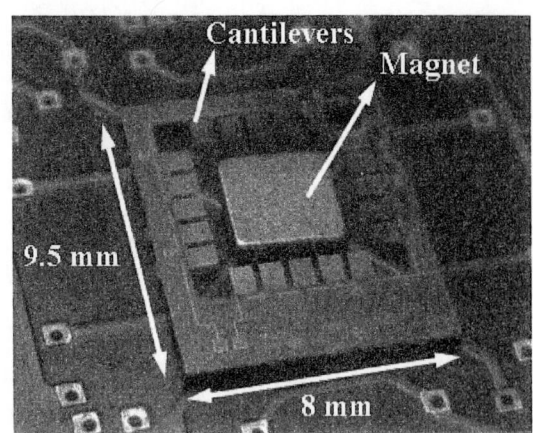

Figure 7.   Photograph of the fabricated prototype.

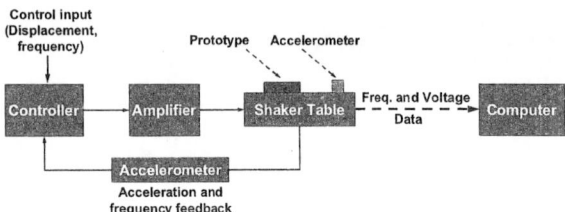

Figure 8.   Block diagram of the experimental setup.

Figure 9 shows the voltage output from the generator with respect to excitation frequency. The prototype generates a maximum voltage of 8.75 mV at a vibration frequency of 5.135 kHz. The bandwidth of the generator is measured to be 220 Hz with a damping ratio of 0.021.

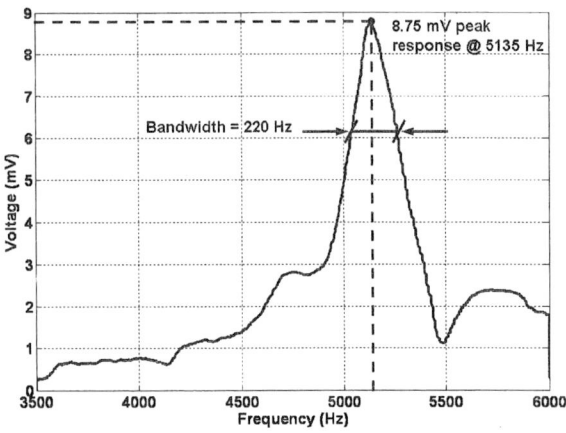

Figure 9.   Measured voltage output.

Table 2 compares the design and test parameters of the proposed generator. Notice that the resonance frequency and the voltage slightly differ from the expected values. This is mainly due to the deviation of the damping ratio and magnetic field values from calculations. These parameters

are quite sensitive to the experimental setup and may change the results significantly. First of all, the mechanical damping ratio is a complex quantity and as it depends on many variables, it is hard to estimate it exactly. Another parameter creating discrepancy is the magnetic flux density; it is estimated assuming that the distance between the coil and the magnet is 50 μm, but the actual distance is not controllable accurately.

TABLE II.   COMPARISON OF DESIGN AND TEST PARAMETERS OF THE GENERATOR.

	Design	Test
Size of the device (mm^3)	9.5x8x5	9.5x8x5
Natural freq. of cants. (kHz)	4.53	5.135
Cant. size (μm^3)	890x670x16	890x670x14
Number of cantilevers	20	20
Magnet size (mm)	4x4x4	4x4x4
Distance btw. cants. and magnet (μm)	50	900
Coil width (μm)	20	20
Coil resistance (single cantilever) (Ω)	110	350
Damping ratio	0.004	0.021
Max. voltage output (mV)	40	8.75

V.   CONCLUSION

In this work, the design, optimization, and implementation of an electromagnetic micro energy scavenger using an array of parylene cantilevers is presented. The coils located on the cantilevers are connected electrically in series to increase the voltage that is generated by virtue of the relative motion between the coils and the magnet. A detailed mathematical modeling and optimization of the design for various cases are carried out. It has been shown that the micro energy scavenger can generate a maximum voltage of 8.75 mV at a vibration frequency of 5.135 kHz. The deviation of test values from the design values is mainly due to the deviation of the damping ratio and magnetic field values from calculations.

REFERENCES

[1] M. Mizuno and D.G. Chetwynd, "Investigation of a resonance microgenerator", J. Micromech. Microeng., Vol. 13, 2003, pp. 209-216.

[2] Product specifications, Parylene knowledge, Specialty Coating Systems, Inc. IN, USA, 1-800-356-8260.

[3] I. Sari, T. Balkan and H. Kulah, "A wideband electromagnetic micro power generator for wireless microsystems," Transducers 2007, vol.1, pp. 275-278, June 2007.

[4] S.G. Kelly, Fundamentals of Mechanical Vibrations, McGraw-Hill, New York, International Edition, 1993.

[5] H. Kulah and K. Najafi, "An electromagnetic micro power generator for low-frequency environmental vibrations," MEMS 2004, pp. 237-240, 2004.

Self-Reconfiguration on Spartan-III FPGAs with Compressed Partial Bitstreams via a Parallel Configuration Access Port (cPCAP) Core

Salih Bayar
Computer Engineering
Boğaziçi University
P.K. 2 TR-34342 Bebek, Istanbul, TURKEY
Phone: +90 212 359 7780
Fax: +90 212 287 2461
Email: salih.bayar@boun.edu.tr

Arda Yurdakul
Computer Engineering
Boğaziçi University
P.K. 2 TR-34342 Bebek, Istanbul, TURKEY
Phone: +90 212 359 7224
Fax: +90 212 287 2461
Email: yurdakul@boun.edu.tr

Abstract—This paper presents an alternative approach for dynamic partial self-reconfiguration that enables a Field Programmable Gate Array (FPGA) to reconfigure itself at run-time partially through a parallel configuration access port (cPCAP) under the control of the stand alone cPCAP core within the FPGA instead of using an embedded processor. The cPCAP core with bitstream decompression module needs only 361 slices , which is approximately 18% of a Spartan-3S200 FPGA. The dynamic partial self-reconfiguration via cPCAP core works up to 50Mbyte/s. The compressed partial bitstream is stored in BlockRAM within the FPGA and decompressed via cPCAP core at the time of reconfiguration of the FPGA. This approach has been implemented on a pure Spartan-3 FPGA from Xilinx, but it can also be used for any other FPGA architectures, such as Virtex-II(Pro), Virtex-4, Virtex-5, etc.

I. INTRODUCTION

The dynamic partial self-reconfiguration (DPSR) concept is the ability to change the configuration of part of an FPGA device by itself while other processes continue in the rest of the device. A pure Spartan-3 FPGA, which doesn't have any multiboot capabilities and Internal Configuration Access Port (ICAP), cannot be reconfigured without any additional external hardware. However, some other FPGA series such as Spartan-3A(N), Virtex-II(Pro), Virtex-4, Virtex-5 FPGA series have this ICAP module on their predesigned hardware architecture[7][8][9][10][11]. In spite of the lack of an ICAP module on its architecture, dynamic reconfiguration is still supported in Spartan-3 via the external SelectMAP interface or JTAG. As a result, a component should be developed for pure Spartan-3 FPGAs, which acts as an ICAP and allows partial self-reconfiguration at run-time for Spartan-3 FPGA family.

In most cases a reconfigurable FPGA system consists of three main components: an external intelligent agent, some external (non-)volatile memory and a Complex Programmable Logic Device (CPLD). Such a reconfigurable FPGA system is described in detail in [5]. In some cases, systems may not require a CPLD if the used intelligent agent has a sufficient number of general purpose I/O (GPIO) pins. For these systems,

the FPGA can be (re)configured directly by the intelligent agent [5].

In this study, a custom soft-core is developed for self reconfiguration of Spartan-3 FPGAs. It is the extended version of our PCAP core [1] which controls the partial reconfiguration flow through SelectMAP port and supplies configuration clock for reconfiguration. cPCAP uses BlockRAMs to store partial configuration bitstreams. BlockRAMs provide on-chip fast memory in FPGAs. However, the number of BlockRAMs is limited. The new soft-core, cPCAP, maximizes the utilization of BlockRAMs by storing compressed partial bitstreams at initial configuration time and decompressing them during self reconfiguration. Due to compressed partial bitstream, more on-chip storage can be saved. Moreover, the reconfiguration clock speed of cPCAP is the same with that of PCAP in spite of the integrated decompression module. Because of being written entirely in VHDL, this cPCAP core is highly portable and can also be used for all other Xilinx FPGA architectures. Thus it is not necessary to use an external intelligent agent to control the partial reconfiguration flow.

This paper is organized as follows: In section 2, we briefly explain the main types of partial reconfiguration and give a few developed samples associated with DPSR concept. Section 3 presents the structure and functionality of our cPCAP core. Section 4 gives an example, where a run-time DCM reconfiguration via cPCAP is implemented. Finally, section 5 presents our conclusions.

II. DYNAMIC PARTIAL SELF-RECONFIGURATION (DPSR) CONCEPT IN XILINX FPGAS

Partial reconfiguration is only possible through either serial JTAG interface or parallel slave SelectMAP mode. Since parallel slave SelectMAP interface has higher performance than the serial JTAG interface, the SelectMAP port is used in this study. Parallel SelectMAP port is used for either complete configuration or partial reconfiguration for applications, where the performance is the most important consideration.

978-1-4244-1983-8/08/$25.00 ©2008 IEEE

To be able to perform dynamic partial self-reconfiguration on a FPGA-based system, there should be either an internal configuration access port or an equivalent port. Some FPGAs such as Spartan-3A(N), Virtex-II(Pro), Virtex-4, Virtex-5 from Xilinx, which have ICAP on their hardware, support self-reconfiguration without using an external intelligent agent. The fact that the pure Spartan-3 does not have such an internal configuration port, has rendered impossible the self-reconfiguration without using an external intelligent agent up to now apart from a few new studies, which are discussed below.

In [3], a soft ICAP, known as JCAP, has been developed in order to realize the self reconfiguration. As a reconfiguration interface they use serial JTAG interface which is very slow compared to parallel SelectMAP port. Though the ICAP on Virtex-II or Spartan3A devices have a reconfiguration speed 66MByte/s [7], JCAP only achieves a reconfiguration rate of 2Mbits/s. The reason of this huge performance difference between the ICAP and JCAP is the serial JTAG interface for JCAP.

In our study we have used parallel SelectMAP port instead of serial JTAG interface, thus we have developed a self-reconfigurable system on pure Spartan-3 series which should be at least 8 times faster than the developed system in [3].Since a serial configuration method is used in [3], they achieved to send one bit per configuration clock cycle. However, in our study we use a parallel configuration method, hence we send 8 bits at each configuration clock cycle.

In [2], a self-reconfiguration system on pure Spartan-3 has been developed. They have solved the lack of ICAP on Spartan-3 FPGAs by adding an external loopback, therefore they have used a GPIO core on MicroBlaze and 11 external wires to accomplish the interface through SelectMAP port. In order to store initial configuration bitstream and generate configuration clock signal they have also used a XCF configuration flash PROM. Under the control of GPIO core of MicroBlaze they reconfigure the target FPGA through SelectMAP port. Though they have achieved a speed as in ICAP, they have used an external PROM to store initial configuration bitstream, MicroBlaze soft core to control the configuration flow and a TFTP server and onboard SDRAM to store the partial bitstreams. Although we have also used 11 external wires and the SelectMAP port as a reconfiguration interface, we have used BlockRAM to store partial bitstream, cPCAP core to control the reconfiguration flow. Since they designed a MicroBlaze-based system, they used 4198 slices of an Spartan-3S2000 FPGA. Applying such a system to small FPGAs (e.g. to a Spartan-3S200) is impossible. However, our cPCAP core is very small, which is 361 slices and only 18% of a Spartan-3S200. Thus, we have accomplished a processor-independent run-time reconfigurable system and presented a very new approach for storing compressed partial bitstreams on BlockRAM within the FPGA.

Both [3] and [2] process uncompressed partial bitstreams that are stored in external memory. However, cPCAP can process compressed partial bitstreams that are stored in Block-

RAMs. This has two main advantages: 1)A BlockRAM can hold more compressed partial bitstreams than regular uncompressed partial bitstreams 2) Access time to a partial bitstream in a BlockRAM is much shorter than the access time to a partial bitstream in an external memory. Decompression is realized in cPCAP during reconfiguration time without sacrificing from reconfiguration speed.

III. CPCAP ARCHITECTURE

The Spartan-3 FPGA configures itself through its SelectMAP port under the control of cPCAP. Through the parallel SelectMAP interface, the partial reconfiguration information is accepted and the reconfiguration process is executed again by the same target FPGA, where this unique FPGA acts as not only a *slave* but also a *master* at the time of reconfiguration as shown in Figure 1.

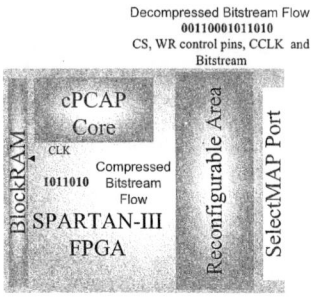

Fig. 1.   Hardware architecture of whole system

As shown in Figure 2, in order to generate configuration clock frequency a Digital Clock Manager (DCM) component is used. Since we generate CCLK signal within FPGA, it acts as master, but at the same time we accept the CCLK signal through the SelectMAP interface into FPGA as if the signal comes from other intelligent agent, where the FPGA acts as slave. The source of CCLK is not important for the FPGA. The cPCAP core reads a byte from the BlockRAM at each clock cycle. Under the control of CS, WRITE, CCLK signals, this byte is sent to SelectMAP interface. The bitstream information, which is accepted from BlockRAM, is in compressed manner. Since we achieve the decompression of bitstream information at the time of reconfiguration, we do not need any additional time for decompression. Therefore when the CCLK speed is set to 50 MHz, the reconfiguration speed is 50MByte/s. Note that the reconfiguration speed is independent from the size of partial bitstream. The BUSY signal is only used if the configuration clock frequency exceeds 50 MHz. In our study, the cPCAP core can be configured to operate up to 50 MHz. We have not exceeded 50 MHz yet, that's why we have not used BUSY signal. However, it will be examined in the future work. The configuration flow of an FPGA for run-time reconfiguration via cPCAP is shown in Figure 3. The uncompressed partial bitstreams are generated with the help of bitgen -r flow[6].

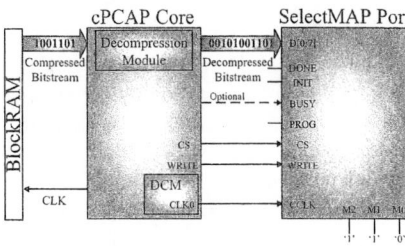

Fig. 2.   cPCAP Core and SelectMAP interface

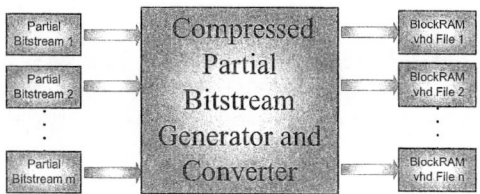

Fig. 4.   File conversion from partial bitstream file to BlockRAM coefficient file

The storage of partial reconfiguration bitstream requires also an additional external hardware for reconfigurable systems. In the most of reconfigurable systems the partial reconfiguration file is stored in an external non-volatile device. It can be read from there under the control of either an external intelligent agent or the FPGA itself, where FPGA acts as a *slave* and *master* respectively. Contrary to the standard methods based on storing partial reconfiguration bitstream on external non-volatile devices, the partial reconfiguration bitstream in this study is stored in BlockRAM within the target FPGA. As a result of this new approach, there is no need to use an additional external device for storing partial reconfiguration bitstream for any system, which works continuously after the initial configuration.

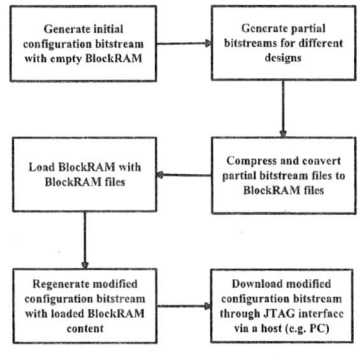

Fig. 3.   Configuration flow

A. File Compression and Conversion

There are a lot of different methods to store information in a BlockRAM. Using generic template from language templates in ISE is one of them. According to the bit-width and -depth of BlockRAM, there are various prepared pieces of code for BlockRAM available, which can be easily inserted into the BlockRAM HDL source file. Since we need to store partial bitstream bytes in BlockRAM, we have used one of these BlockRAM template.

After generating partial bitstream files, we have compressed these files and converted them to a suitable form with ".vhd" extensions using a file compression and converter module, which is written in Java language, in Figure 4. Note that

the number of partial bitstreams needs not to be equal to the number of BlockRAMs.

B. Dynamic Partial Self-Reconfiguration Flow

Since we have accomplished a dynamic partial self-reconfiguration through the SelectMAP port, the developed cPCAP core behaves in our study as if it is a mirror of SelectMAP port, as same in ICAP. The following flow in Figure 5 is very similar to "SelectMAP configuration Flow Diagram" in [5] except that PROG, INIT, DONE, BUSY pins are not taken into account in our study. The first three control signals PROG, INIT, DONE are only used during complete (re)configuration. BUSY signal is used if the configuration clock (CCLK) frequency is greater than 50Mhz. Due to the fact that there is only a 50Mhz oscillator available on Spartan-3 Starter Board we have chosen the CCLK for our system exactly 50Mhz, thus we do not need to use BUSY indicator signal. Under some circumstances, such as using BUSY indicator signal where it is needed, the developed design can be clocked with any other frequency values, which are supported by Spartan-3 FPGA and its SelectMAP interface. We have experimented, that our cPCAP core can run safely at all frequencies up to 50Mhz. However, the performance of the cPCAP core at higher frequencies will be tested after implementing the rest of the handshake signals in future.

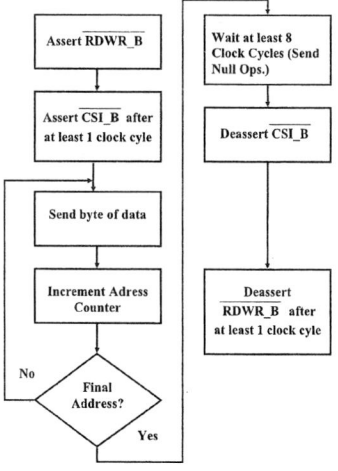

Fig. 5.   cPCAP core Configuration Control Flow Diagram

978-1-4244-1983-8/08/$25.00 ©2008 IEEE

IV. EXAMPLE: RUN-TIME DCM RECONFIGURATION

The soft cPCAP core, which is a pure VHDL code, has been synthesized on Spartan-3S200 Starter Kit Board. In this work we have reconfigured a clock output of Digital Clock Manager (DCM), which drives the complete system, at run-time. Such a DCM reconfiguration approach is described in detail in [4].

There are various applications, where clock frequency of a system should be change at run-time. Clock scaling method in [4] is one of them, which is mostly used to decrease FPGA power consumption by changing clock frequency of different components of system at run-time. Changing data transmission speed of a system, and other typical applications include speed drives, inverters, computers and computer controlled equipment, deep well pumps, industrial machinery, ships, aircraft.

In this work, a 4-bit up-down counter is taken as an example. This counter either works with 5 MHz or 50 MHz, which is accomplished by run-time DCM reconfiguration via cPCAP core. The size of each decompressed partial bitstream for DCM reconfiguration is 5 KByte. The reconfiguration speed is 50 MByte/s, which means that the DCM reconfiguration via cPCAP core takes approximately 0.1 ms. This counter is used solely to show that such a reconfiguration approach is possible for other reconfigurable systems where the frequency of a system or a component of system can be changed at run-time without affecting anything else in complete system.

To be able to do reconfiguration we have firstly generated complete bitstream files and then with the help of bitgen tool two fully routed NCD (Native Circuit Description) file for each different frequency values. Contrary to approach in [4] for generating partial bitstreams, the difference based approach is used in this work.

The implementation cost of our study is summarized in Table I. After adding the decompression module to the cPCAP core, we have needed to use only 1 BlockRAM to store two different compressed partial bitstreams within the BlockRAM. With this compression approach we have accomplished at least 76% space saving approximately, where the compression ratio is actually based on the structure and size of the partial bitstream.

TABLE I
OCCUPIED RESOURCES FOR PCAP AND CPCAP CORES

	PCAP (without compression)	cPCAP (with compression)
Occupied Resources		
BlockRAM	6 of 12, %50	1 of 12, %8
Slices	365 of 1920, %19	324 of 1920, %16
DCM	1 of 4, %25	1 of 4, %25

V. CONCLUSION

In this paper we have discussed dynamic partial self-reconfiguration of a pure Spartan-3 FPGA through the SelectMAP port and storing different partial bitstreams on Block-RAM within the target FPGA. The most important advantage of this study is to achieve a very fast partial reconfiguration compared to other serial JTAG interfaces and using a new approach such as storing compressed partial bitstreams on on-chip memory and reading them from there under the control of an cPCAP core instead of an external intelligent agent, which reduces hardware cost and power consumption simultaneously. Furthermore, with the capability of storing compressed partial bitstreams, many different partial bitstreams can be stored on-chip memory at a glance.

This kind of implementation, using no other additional external devices apart from FPGA, is a perfect solution for almost all systems based on cost and power consumption. What's also very impressive about this implementation is that the developed cPCAP core can be applied not only on a pure Spartan-3 and also on other FPGA series such as Virtex-II, Virtex-4, Virtex-5, Spartan-3A(N) and so on.

In addition to this big advantage related to size, this cPCAP core can be used anywhere in the FPGA, whereas the location of the ICAP module on Virtex-II devices and two ICAP modules on Virtex-4 are fixed [8][10].

In our future work, we plan to implement other handshaking signals on cPCAP for providing self-reconfiguration at frequencies higher than 50MBytes/sec. Decoupling of the memory architecture from SelectMAP interface in cPCAP will also be done in order to support different memory types and reconfiguration interfaces.

ACKNOWLEDGMENT

This work is fully supported by The Scientific and Technological Research Council of Turkey, TÜBİTAK (Project Nr.: 104E038) and Boğaziçi University Scientific Research Projects (Project Nr.: 06M105).

REFERENCES

[1] Salih Bayar and Arda Yurdakul. *Dynamic Partial Self-Reconfiguration on Spartan-III FPGAs via a Parallel Configuration Access Port (PCAP).* HIPEAC2008, Gothenburg, Sweden, January 2008.

[2] Ivan Gonzalez, Estanislao Aguayo, and Sergio Lopez-Buedo. Self-reconfigurable embedded systems on low-cost fpgas. *IEEE Micro*, pages 49 – 57, July-Aug. 2007.

[3] K. Paulsson, M. Hübner, G. Auer, M. Dreschmann, L. Chen, and J. Becker. *Implementation of a Virtual Internal Configuration Access Port (JCAP) for enabling Partial Self-Reconfiguration on Xilinx Spartan-III FPGAs.* FPL, Amsterdam, Netherland, August 2007.

[4] Katarina Paulsson, Michael Hübner, Salih Bayar, and Jürgen Becker. *Exploitation of Run-Time Partial Reconfiguration for Dynamic Power Management in Xilinx Spartan III-based Systems.* ReCoSoc2007, Montpellier, France, June 2007.

[5] Xilinx. *Using a Microprocessor to Configure Xilinx FPGAs via Slave Serial or SelectMAP Mode.* XAPP502, (v1.4) edition, November, 13 2002.

[6] Xilinx. *Two Flows for Partial Reconfiguration: Module Based or Difference Based.* XAPP290, (v1.2) edition, September, 9 2004.

[7] Xilinx. *Spartan-3 Generation Configuration User Guide.* UG332, (v1.2) edition, May, 23 2007.

[8] Xilinx. *Virtex-4 Configuration Guide.* UG071, (v1.9) edition, October, 1 2007.

[9] Xilinx. *Virtex-5 FPGA Configuration User Guide.* UG191, (v2.5) edition, October, 10 2007.

[10] Xilinx. *Virtex-II Platform FPGA User Guide.* UG002, (v2.1) edition, 28 March 2007.

[11] Xilinx. *Virtex-II Pro and Virtex-II Pro X FPGA User Guide.* UG012, (v4.1) edition, 28 March 2007.

978-1-4244-1983-8/08/$25.00 ©2008 IEEE

Generic Techniques and CAD tools for automated generation of FPGA Layout

Husain Parvez, Hayder Mrabet and Habib Mehrez

Laboratoire d'informatique de Paris 6; Université Pierre et Marie Curie, 4 Place Jussieu, 75005 Paris, France
E-mail : {parvez.husain, hayder.mrabet, habib.mehrez}@ lip6.fr

Abstract— **This paper presents an automated method of generating an FPGA layout. The main purpose of developing a generator is to reduce the overall FPGA design time with limited area penalty. This generator works in two phases. In the first phase, it generates a partial layout using generic parameterized algorithms. The partial layout is generated to obtain a fast bitstream configuration mechanism, an efficient power routing and a balanced clock distribution network. In the second phase, the generator completes the remaining layout using automatic placer and router. This two-phase technique allows better maneuvering of the layout according to initial constraints. The proposed method is validated by generating the layout of an island-style FPGA which includes hardware support for the mitigation of Single Event Upsets (SEU). The FPGA layout is generated using a symbolic standard cell library which allows easy migration to any layout technology. This layout is successfully migrated to 130nm technology.**

1. INTRODUCTION AND RELATED WORK

Developing a new FPGA is a time consuming and a challenging task. It is reported in [3] that a new FPGA creation involves approximately 50 to 200 person years, thus increasing the overall time to market of the final product. It is an interesting option to significantly reduce the time-to-market of the product at the expense of limited area penalty. One way to do this is by automating the complete FPGA design process. The work presented here discusses the automatic generation of FPGA layouts using open-source VLSI tools.

The generator presented here employs an elegant scheme to integrate manual intervention in the automated FPGA generation procedure. This is done with the help of generic parameterized algorithms which generate a partial layout. Later, the automated tools are used to complete the remaining layout. The partial layout is performed on those portions of the design that are either important in one aspect or other, or are too difficult to be handled properly by the automated tools. In this work the partial layout performs the power routing, clock distribution and the configuration memory placement.

A number of previous attempts have been made regarding automated generation of FPGAs. One of the major works in this domain is done in [6] [3]. They have demonstrated the complete automation of FPGA creation with significantly reduced manual labor. The GILES [3] tools are used to generate different tiles which are then abutted together to form a complete FPGA. The clock and power segments are later routed using SKILL [7]. Phillips and Hauck have focused on the automatic layout of domain specific reconfigurable systems [5]. They have reduced the amount of configurability

required by an application domain, and thus have generated smaller layouts.

These previous FPGA generators have used the commercial VLSI tools; whereas this work presents an FPGA generator based solely upon open-source VLSI tools. These tools can be adapted easily for specific demands. This work also defines a set of layout parameters to modify the layout according to the initial requirements.

2. FPGA GENERATION

This work focuses on the generation of Island style FPGAs. It comprises an array of configuration logic blocks (CLBs). Each CLB contains a 4-LUT followed by a by-pass flip-flop. Each CLB has 4 inputs (one on each side) and an output that derives adjacent channels on its top and right sides. The CLBs communicate with each other through a disjoint bi-directional routing network. All the inputs and outputs of a CLB connect with all the wires in a channel (i.e. Fc=1). The generated FPGA matrix can have '*Nx*' CLBs in X direction, '*Ny*' CLBs in Y direction, and a channel width '*Ch*'.

A. *Open-source VLSI tools:* An open-source VLSI tool kit ALLIANCE [1] and a python based language STRATUS [2] has been used for the development of this FPGA layout generator. Alliance is a complete set of free CAD tools and portable CMOS libraries for VLSI design. It includes a VHDL compiler and a simulator, logic synthesis tools, and automatic place and route tools. STRATUS generates parameterized VLSI modules. It extends the python language with a set of methods and functions for the procedural generation of netlist and layout views of structural cell based designs.

B. *Tile based approach:* A tile based approach is used to generate the desired FPGA architecture. In this approach a set of tiles are identified in the architecture which are repeatedly abutted to form the whole FPGA matrix. A set of 9 different tiles as shown in figure 1 are used for the generation of the target architecture. The principle tile is the 'basic' tile, whereas the other tiles are its derivations. The tiles on the leftmost column and the bottom row do not contain logic blocks. They only contain a channel which connects the adjacent IO pads and the adjacent logic block input. The rest of the tiles contain a top horizontal channel, a right vertical channel, a switch box, and a logic block. It can be seen in figure 1 that the horizontal repetition of 2nd column and the vertical repetition of 2nd row generate an FPGA of our desired size. An important aspect in the tile based design is that the adjacent sides of two abutted tiles must have same

978-1-4244-1983-8/08/$25.00 ©2008 IEEE   141

length. While deciding the sizes of the tiles, priority is given to the tile which is used the most; in this case it is the 'basic' tile. The sizes of the other tiles are adjusted accordingly.

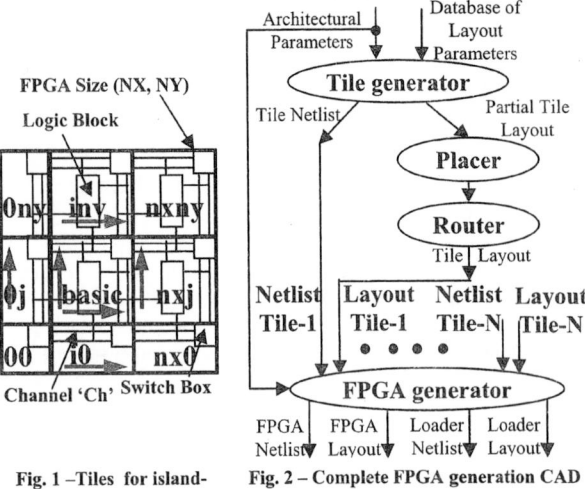

Fig. 1 –Tiles for island-style FPGA

Fig. 2 – Complete FPGA generation CAD Flow

C. Netlist generation: Each tile generator is written in the language STRATUS. The tile generator receives a set of architectural parameters as input. It then generates the netlist of the tile in accordance with the given parameters. Loops and conditional statements are used to generate a tile for different parameters. The netlist of each tile is generated directly using the standard cell library named SXLIB. It is a symbolic cell library which comes with the ALLIANCE tool chain. C++ routines are also merged in the tile generator for generating VHDL model of specific components. These components are synthesized by the Alliance synthesizer named BOOG. After synthesis these components are used by the tile generator. The generated netlists of all the tiles are passed to the FPGA generator which links them together to construct the netlist of a complete FPGA. This generated netlist may be integrated in any larger application.

D. Tile Layout: A tile generator generates both the netlist and the partial layout of a tile. The partial layout is generated with the help of parameterized algorithms which take a set of layout parameters as its input. Currently, the partial layout is performed for the generation of a fast bitstream configuration mechanism, proper buffering of few long wires, power routing and a balanced clock distribution network. Later-on the placer and the router are used to complete the remaining layout.

The partial layout generation algorithm places all the SRAM bits in rows and columns with a fixed distance between each row, as shown in figure 3. Each SRAM bit in a row receives a vertical data signal, and a horizontal strobe signal. The data bits are written in all the SRAMs of a row only when strobe is high for that row and the column is high for the complete tile. The column and strobe signals come from bitstream configurator (loader), which is discussed later in section 4. The column and the data signals from the top are buffered before they exit on the bottom side of the tile. Similarly a

strobe signals from the left is buffered before it exits on the right side of the tile.

The algorithm starts placing the bitstream configuration cells from a layout parameter named "Start Position". Similarly the height and width of a tile and the total SRAM bits are also variable parameters which change for each different channel width. These layout parameters change each time there is a change in the number of SRAM bits. For this purpose a small database is created which specify all these variables for different channel widths. The layout algorithm and the database specification are generic enough to handle other architectural parameters that are not yet generic.

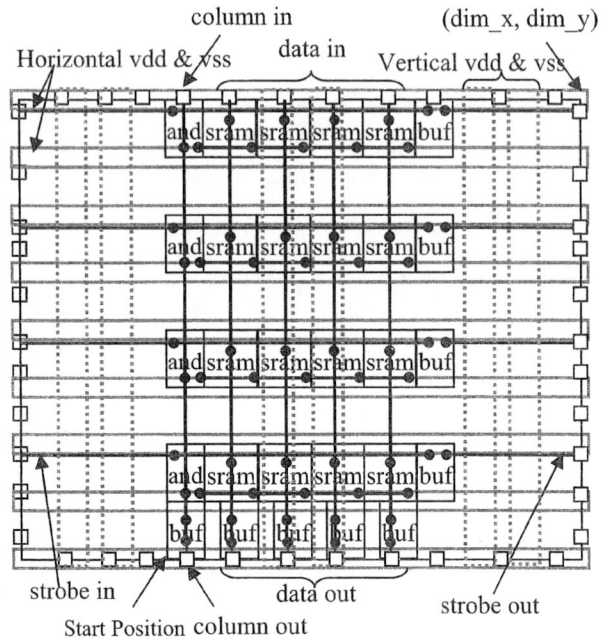

Fig. 3 –Partial layout of a sample FPGA Tile

E. Power routing: The layout generation algorithm generates horizontal and vertical power segments as shown in figure 3. The alternating VDD and GND segments in the horizontal direction are fixed whereas the placement of vertical power segments is supported by few layout parameters. The total number of vertical segments for power and ground in a tile, their positions and their widths are defined in the layout database. These values can be changed for tiles of different sizes. The horizontal power segments use the 1st and 2nd routing layer; whereas the vertical power segments use the 5th routing layer.

F. Clock generation: In this work, we have used a tile based approach for the routing of a symmetric H-tree clock distribution network. It is found that a group of 13 tiles can be used to generate a clock tree for a matrix of size $2^N \times 2^N$ where N>1. Each corresponding clock tile is automatically merged with the FPGA tiles during the partial layout phase. This results in the generation of multiple copies of the same FPGA tiles having different clock routings. After the merging of FPGA tiles and the clock tiles, 23 different tiles are produced.

These tiles can be abutted together to construct any FPGA of size $2^N \times 2^N$. All these clock tiles and the sample 8x8 clock distribution matrix is shown in figure 4 and 5. The main advantage of this mechanism is that we have a generic, tile based and a balanced clock distribution network. One of the disadvantage is that it limits the FPGA size in X and Y direction to be equal and power of 2. But since the clock generation algorithms and their merging with the FPGA tiles is totally automatic; we can always implement a generic algorithm for other clock distribution networks found in [4]. The only thing to consider for writing a new clock routing algorithm is that the placement of clock buffers must not overlap the partial layout. Currently the clock is routed in the 5th and 6th routing layer, whereas the partial layout is done on the first 4 routing layers.

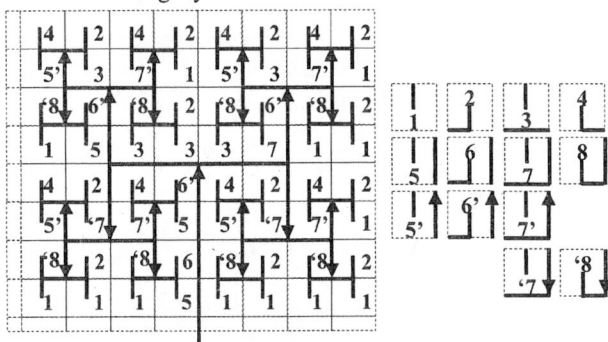

Fig. 4 – H-tree clock distribution network for 8x8 FPGA　　**Fig. 5 – Tiles for constructing clock H-tree**

G. *Pin generation:* In a tile based FPGA, the tiles connect together by abutment, and the pin locations on the boundaries of adjacent tiles must overlap. The positions of few of these pins are calculated on the basis of the layout parameters found in the database. Since the database is common for all the tiles, thus the pin abutment problem does not arise for these pins. There exist other pins which do not have fixed positions. Since the final automatic placement of all the tiles is done independently; it is difficult for the placer to correctly choose the pin locations of the tiles. So a generic algorithm is written to place all the remaining pins. This algorithm places the pins in all the four directions of the tile and ensures that the pins are not congested to a limited place. It utilizes all the available space and tries to distribute the pins with equal spacing.

H. *Automatic placement & routing:* After the partial layout generation of all the tiles; each tile is separately placed and routed with the help of ALLIANCE automatic placer and router named OCP and NERO respectively. The partial layout information is firstly given to the placer to place the remaining logic. If the placer is unable to place the design, the dimensions of the tile are manually increased in the database. The X and Y dimensions of the tile must be properly adjusted to make sure that a tile does not waste any extra space. The placer automatically adds the empty cells to fill up any extra space. After placement, NERO routes the whole design. All the tiles are successfully routed using 4 routing layers. Only the clock and the vertical power segments are routed on the 5

and 6 routing layer. The overall process of the netlist and layout generation is shown in fig 2.

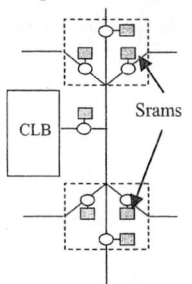

Fig. 6(a) – Standard sytem for deriving a single track

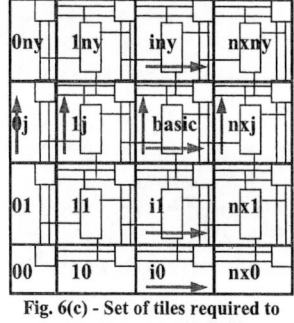

Fig. 6(c) - Set of tiles required to construct island-style FPGA

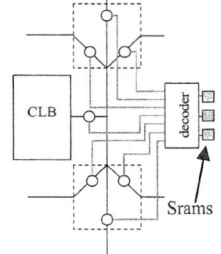

Fig. 6(b) – Decoder system for deriving a single track

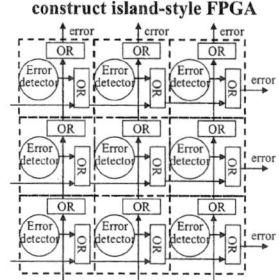

Fig. 6(d) – Scalable error detection method

3. ARCHITECTURE FEATURES

The above process of FPGA generation has been used to generate an island style FPGA with hardware support for the mitigation of Single Event Upsets (SEU) [8].

SEU are induced by energized particles hitting the silicon device. A particle hit with sufficient energy changes the logic state of the memory elements producing a transient error. An SEU on configuration bits may change the functionality of the look-up tables as well as the interconnect controlled by the SRAM cells, thus producing a hard error. These hard errors can be eliminated by using simple decoders, as shown in figure 6(a) and 6(b), to implement a system dependency between switches that derive the same track. An error detection system is integrated in each tile which enables an error signal whenever a change is detected in configuration bits. The error signal propagates through row and column, as shown in figure 6(d).

The addition of this architectural feature increases the total number of tiles to 16 as shown in 6(c). The merging of clock tiles with 16 different FPGA tiles produces a total of 34 different tiles. According to the final application requirements, these tiles are used to generate a 32x32 FPGA matrix with a channel width of 8.

4. VALIDATION

A. *Software flow:* A software flow is followed to test the functionality of the generated architecture. The sample application (in VHDL format) to be mapped onto the FPGA is the input to the software flow. Initially BOOG synthesizes the VHDL input into a netlist of gates VST. VST2BLIF and later

978-1-4244-1983-8/08/$25.00 ©2008 IEEE　　143

SIS is used to convert it into LUT form. T-VPACK and later VPR is used for the placement and routing of the netlist. A bitstream generator is written which generates a binary stream that contains all the required information for the configuration of the sample application onto the FPGA.

B. Bitstream configuration mechanism: An Nx by Ny FPGA contains (Nx+1) by (Ny+1) tiles; where Nx+1 is the total number of columns and Ny+1 is the total number of rows. Each FPGA tile comprises a set of SRAM bits arranged in multiple rows. The SRAM bits in a row are called a 'word'. For writing data to a word of a tile; a row number, a column number and a word number must be specified. The row and column numbers gives the location of the tile in a matrix, whereas the word number gives the location of word in a tile. All these three parameters are passed to the shift registers. The data to be written in a word is also specified in the same shift registers. With the help of the row, column and word decoders, the exact strobe and column signal is turned on. Thus when write enable turns high, the data is written onto the specific word of the requested tile. This process is repeated for all the words of all the tiles. The shift registers and decoder are implemented in a loader which is also generated by the FPGA generator.

C. Simulation: The generated FPGA netlist is tested on the ALLIANCE simulator called ASIMUT. Different test applications are mapped on the FPGA with the help of the sofware flow. Once the FPGA is programmed, the respective testbench of each test application is applied on the inputs of the FPGA and the outputs are compared. These simulations can also be easily performed on other commercial tools like SYNOPSYS.

D. Netlist layout comparison: The generated netlist and the generated layout must match with each other. For this purpose the ALLIANCE extraction tool COUGAR is used. It extracts a netlist from a layout. Later the ALLIANCE comparison tool LVX is used to compare the extracted netlist with the generated netlist. This confirms that the generated layout matches with its netlist. This method of layout verification is validated for a set of generated FPGAs. But the flattened 32x32 FPGA matrix is too large to be compared due to the limitations of COUGAR. So, instead of LVX, CALIBRE LVS is used to compare the 32x32 FPGA layout with its netlist.

E. Electric simulation: The ALLIANCE extraction tool COUGAR is used to extract the spice model of each tile. These models are later electrically simulated using ELDO. Our extraction tool is unable to support very large circuits. So it was impossible to electrically simulate the complete 32x32 FPGA. However for the proof of concept we successfully simulated the electric model of a smaller 4x4 FPGA matrix with channel width of 8.

5. TAPEOUT

The layout generation is done using symbolic standard cell library which works on unit λ (lambda). The ALLIANCE tool S2R (symbolic to real) is used to convert the symbolic design

to 130nm technology. The corresponding GDS and LEF files are also obtained. The 32x32 FPGA occupies an area of 3885.6 μm by 3882 μm. It is noticed that 19% of the FPGA area increases due to the hardware support for the mitigation of SEU. The generic symbolic design rules help easy migration to any technology but with some area penalty. Instead of symbolic library, if the netlist of the generated FPGA is laid out in ENCOUNTER using directly a 130nm technology library, 40% area reduction is noticed.

The generated FPGA layout can be used as a black box in any other larger system. For the proof-of-concept, it is used to lay out a complete chip. The pads are placed and routed using ENCOUNTER. The DRC and LVS verification is performed using CALIBRE. The final FPGA chip measures 23.86 mm^2.

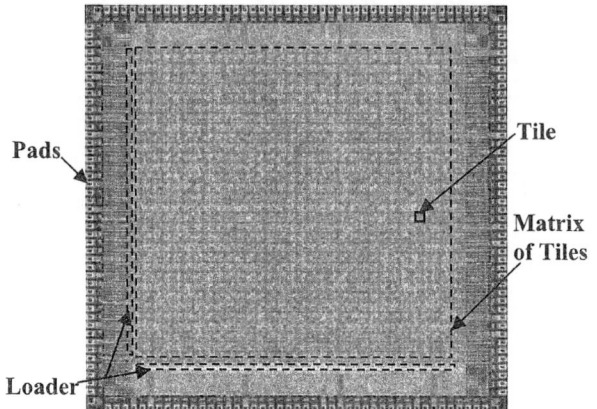

Fig. 7 – A Prototype FPGA chip layout

6. CONCLUSION AND FUTURE WORK

In this work we have presented a completely automatic method for the generation of an FPGA using an open-source VLSI tool-kit. We are able to generate FPGAs having different architectural parameters. In future, we intend to increase the number of variable architecture parameters. We also intend to add support for other clock distribution networks.

7. REFERENCES

[1] A. Greiner and F. Pecheux, "Alliance : A complete set of cad tools for teaching vlsi design", in Proceedings of 3rd Eurochip Workshop, 1992

[2] S. Belloeil, D. Dupuis, C. Masson, J.P. Chaput, H. Mehrez, "Stratus: A procedural description language based upon Python", in Proceedings of the 19th International Conference on Microelectronics, december 2007

[3] K. Padalia, R. Fung, M. Bourgeault, A. Egier, and J. Rose, "Automatic transistor and physical design of FPGA tiles from an architectural specification.", in Proceedings of 2003 ACM/SIGDA 11th international symposium on FPGAs, pp. 164-172. ACM press, 2003.

[4] E. G. Friedman, "Clock Distribution Networks in Synchronous Digital Integrated Circuits", in Proceedings of the IEEE, May 2001

[5] S. Phillips and S. Hauck. "Automatic layout of domain-specific reconfigurable subsystems for system-on-a-chip", In Proceedings of the 2002 ACM/SIGDA tenth International symposium on FPGAs, pages 165-173. ACM Press, 2002

[6] I. Kuon, A. Egier, J. Rose, "Design, layout and verification of an FPGA using automated tools", In Proceedings of the 2005 ACM/SIGDA 13th international symposium on FPGAs, 2005

[7] Cadence. SKILL Programming Language, http://www.cadence.com

[8] J.H. Elder, J. Osborn, W.A. Kolasinski, R. Koga, "A method for characterizing a microprocessor's vulnerability to SEU", IEEE Transaction on Nuclear Science, Dec 1988 v 35 n 6.

978-1-4244-1983-8/08/$25.00 ©2008 IEEE

Design and Optimization of PWL Circuits Used in Fuzzy Logic Hardware

Yankı Yalçın, Günhan Dündar
Boğaziçi University
Department of Electrical and Electronic Engineering
Bebek 34342, İstanbul, Turkey
dundar@boun.edu.tr

Bogdan M. Wilamowski
Auburn University
Department of Electrical and Computer Engineering
200 Broun Hall, Alabama 36849-5201 USA
wilam@ieee.org

Abstract—**This work is concerned with the design automation of analog circuits realizing piecewise linear functions (PWL) that may be used for fuzzy logic circuit design. There are several sources of systematic or random errors in the design of such functions. Various combinations of CMOS current mirror circuits are used to implement PWL functions. In order to simplify the optimization of implementation, PWL circuits are divided into smaller circuits which are assumed to be current mirrors in this work. This work presents a computer aided tool for calculation of optimized W and L values of current mirror transistors for various values of reference current within a specified error to find the best transistor parameters for possible minimum power dissipation. Results are tested on several applications to verify that the outputs of the computer aided system presented in this work match simulation results.**

I. INTRODUCTION

Over the last few years, analog implementation of neural networks and fuzzy logic with CMOS technology has enjoyed a lot of interest. This is partly due to the expected area and power advantages of analog circuits over their digital counterparts for such soft computing applications. However, the design of analog circuits is more complex than digital circuits because analog design requires creativity and expertise.

This work presents a computer aided design tool for designing a surface approximation circuit with the help of piecewise linear functions (PWL) that are used for fuzzy logic surface approximation. Usage of piecewise circuits is a known method but diode-resistor networks are difficult to be implemented in CMOS integrated circuits. Some methods are proposed for the solution of this problem. One of them is the use of current mirrors to obtain diode like curves where slopes are defined by the W/L ratios. In this work, the optimization of low power, current mode CMOS circuits for synthesis of arbitrary nonlinear functions [1], is performed. Various combinations of CMOS current mirror circuits are used to realize PWL functions. There are several sources of errors in design of such functions. In order to simplify the

---

This work was supported by NSF and TÜBİTAK under project number EEEAG-103E023.

---

optimization of these error calculations, PWL circuits are divided into smaller circuits which are assumed to be current mirrors. Implementation error which is caused by deviation from the real solution surface and mismatch errors between current mirror transistors due to difference between threshold voltages (V_{TO}), oxide capacitance (C_{OX}), width and length of transistor values are considered as the main sources of errors in this paper. In this work each of these errors is calculated independently of the other errors and in the end all of the calculated errors are combined. This final error is regarded as the total error and the W and L values are calculated according to this total value.

This work will make the overall design easier for the developer to pass by some stages faster with better assumptions, when several constraints are taken into consideration. Results of the system defined here are tested on several applications to verify that the results of the design tool presented in this work match with the simulation results. EKV analytical models are used in both calculations and simulations. This approach has been observed as a viable alternative to manual design of PWL circuits.

II. DESIGN METHODOLOGY

In this work, power dissipation and area are the cost functions as long as the circuit remains within certain error bounds. Power optimization and area can be adjusted by varying C_{OX}, V_T, W/L, and input current of the system. Other parameters are considered to be fixed. Here, the aim is to discover the error caused by these differences mentioned, which are going to be called design errors, and adjusting this error to a defined value while minimizing the power dissipation. Proper W and L values will be defined for the possible best optimized design for the output of this process. Surface approximation error will be considered as the implementation error. In approximation to solution surface some breakpoints are selected according to certain constraints defined by the user. This constraint controls the amount of breakpoints to be selected and implementation error. For example, defining more breakpoints will increase the accuracy of the approximation while increasing the transistor count which means an increase in power dissipation. This trade off must be defined and determined by the designer. In this work MATLAB libraries and compiler is used for software development and modeling. Input and output of this process are defined as text files so that these files can be used as a source for different development environments. In this work, EKV model for MOS transistors was chosen to define mathematical solutions of the circuits. This model is a compact analytical model for MOS transistors [3].

The system developed consists of a linear approximation algorithm to discover the breakpoints, a W/L calculation algorithm to find the optimum device sizes, and a circuit creation algorithm that uses the outputs of these two blocks to design the required circuit. These blocks will be discussed in the following sections.

III.   SURFACE APPROXIMATION

The approximation algorithm is designed for approximating the real solution surface with a minimum number of possible points. These points are regarded as the breakpoints and written to a file where they are going to be read to create a spice file. The algorithm is based on studying the variance of the output with respect to a pre specified threshold value. The second derivative is also used as a criterion. The output of this algorithm is the breakpoints and the total approximation error. Obviously, there is always going to be an error in approaching the desired function which can be decreased by increasing the number of breakpoints. This step is actually nothing but a simple PWL fit to an arbitrary nonlinear function.

IV.   TRANSISTOR PARAMETER CALCULATION

The EKV MOSFET model is based on the $V_p - V_{ch}$ which is defined to be the inversion charge Q. V_p is the pinch-off voltage and V_{ch} is the channel voltage.

Since the EKV model is continuous in all regions of the transistor, it allows the designer to use the same I_D equation in all regions. I_D in the EKV MOSFET model is expressed as the difference between the forward current I_F and the reverse current I_R. EKV MOSFET model is derived from the charge-sheet formulation [3].

$$I_D = 2 \cdot N \cdot \mu \cdot C_{ox} \cdot \frac{W}{L} \cdot (U_t)^2 \cdot \left[\ln^2 \left(1 + e^{\frac{\frac{V_G - V_T}{N} - V_S}{2U_t}} \right) - \ln^2 \left(1 + e^{\frac{\frac{V_G - V_T}{N} - V_D}{2U_t}} \right) \right] \quad (1)$$

where

$$V_P = \frac{V_G - V_{TO}}{N} \quad (2)$$

For a simple NMOS current mirror circuit, (1) can be evaluated for the reference transistor and the mirror transistor individually. The difference of the two currents yields the error due to variation in the drain potentials if both transistors are assumed to be equal. Another error source is the mismatch between the transistors. This can be defined as

having three components, one of them being the difference between C_{OX} values of each transistor, the second one being the difference between the V_T values, and the third one being the W and L value variations. It should be noted that these errors tend to work in opposite directions; that is, the V_T error effect is reduced with higher overdrive voltages (meaning narrower transistors or larger currents), whereas the W and L variations require larger transistors. Hence, an optimum design choice exists for a given error.

In calculation of transistor parameters, when short channel effects are taken into consideration, the simple models break down. Furthermore, minimum sized transistors cannot be used when matching is desired. Hence, 1μm was selected as the minimum size for both W and L values instead of the technology minimum. Therefore, in the rest of this work 1μm will be taken as the minimum size and calculations will be done for to this specification.

A sample input file for transistor parameter calculation contains information for the type of the MOSFET (n or p), the supply voltage, source potential of the transistors, the expected difference between C_{OX} values of current mirror transistors, the expected difference between V_{TO} values of current mirror transistors, step sizes and initial points for various design parameters, transistor parameters, and allowed total error. Based on these, the design tool will try to calculate the optimum transistor sizing. Please note that especially V_T mismatches are extremely sensitive to bias point and optimization for minimum power and minimum error is a must for best performance.

V.   CIRCUIT CREATION

This algorithm is used for the creation of the spice output file in order to establish a circuit whose output will be the approximated surface. Therefore, the created circuit will be a power optimized circuit which supplies the defined solution surface. In this work, a fuzzy solution surface is used as an input to the algorithm but any kind of surface may be approached.

Circuit creation is done in one and two dimensional modes. The two dimensional mode is included as an illustration of the generality of the presented system. Thus, the extension of the work to multidimensional inputs with dimension higher than two is feasible. In any design, the worst path, which is the longest path from input node to output node, is calculated. The remaining error (which is obtained by subtracting the approximation error from the total allowable error as described above) is divided by the current mirror count on the worst case path and the remaining error range for each current mirror is obtained.

VI.   APPLICATION EXAMPLES

A combination of circuits developed earlier [1] is used to approximate a surface. A sample curve to be approximated is shown in Figure 1. Breakpoint values are converted to actual current values by multiplying them with 10^{-5}.

978-1-4244-1983-8/08/$25.00 ©2008 IEEE

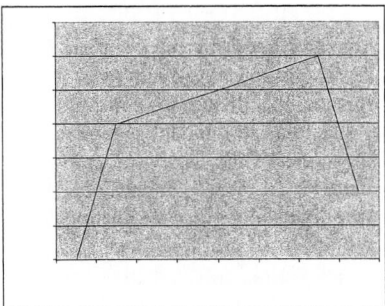

Figure 1. Expected solution surface

Circuit schematic of the automatically created and sized spice file is shown in Figure 2.

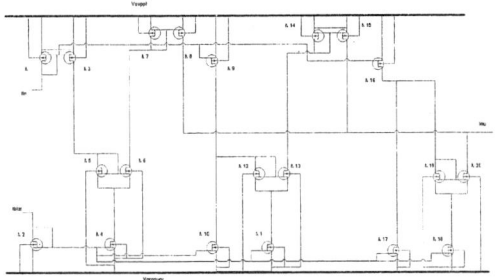

Figure 2. Circuit schematic of the sample circuit

The success of the approximation algorithm can be observed by inspecting Table 1.

Table 1. Breakpoints comparison of approximated and expected surface

Breakpoints	X_1	Z_1	X_2	Z_2	X_3	Z_3	X_4	Z_4	X_5	Z_5
Expected (10^{-5}A)	0.00	0.00	1.00	2.00	6.00	3.00	7.00	1.00	7.00	1.00
Approximated (10^{-5}A)	0.00	0.00	1.09	2.11	6.12	2.98	7.49	1.34	8.13	1.10

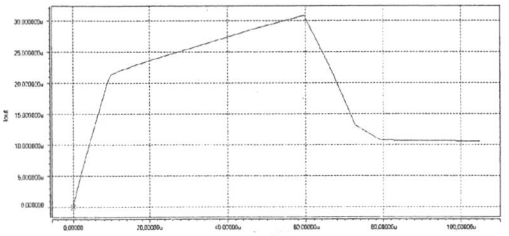

Figure 3. H-Spice simulation of the created circuit

Figure 3 shows the HSPICE simulation of the automatically designed circuit. This figure should be compared with Figure 1 which is the desired solution.

The purpose of the second example is to test the two dimensional approximation capabilities of the approach. Assuming the breakpoints given in Figure 4, where the first variable "xy" refers to the second input; "xx" refers to the

first input, and "xz" is the output, a circuit satisfying the desired criteria is automatically generated.

```
xy 1  xx 2   xz 0
xy 1  xx 3   xz 4
xy 1  xx 5   xz 3
xy 8  xx 4   xz 7
xy 8  xx 11  xz 9
```

Figure 4. Breakpoints of the two dimensional sample circuit

Two dimensional plots of the simulation outputs of the above design are shown in Figures 5 and 6. As it can easily be seen from these graphs, H-Spice simulations show perfect matching with the expected surface.

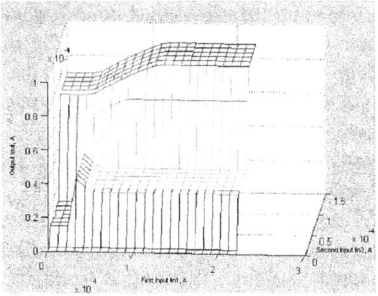

Figure 5. Sample two dimensional circuit two dimensional output graph-1

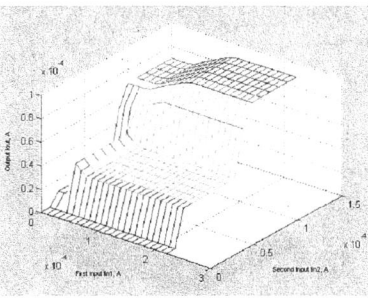

Figure 6. Sample two dimensional circuit two dimensional output graph-2

The third example presented is from the fuzzy logic demo represented in MATLAB R2006a library. This design will be constructed on the "Modeling Inverse Kinematics in a Robotic Arm" application. Kinematics is the science of motion. In a two-joint robotic arm, given the angles of the joints, the kinematics equations give the location of the tip of the arm. Inverse kinematics refers to the reverse process. Given a desired location for the tip of the robotic arm, what should the angles of the joints be so as to locate the tip of the arm at the desired location.

Since the scope of this work is not "Inverse Kinematics", it is not going to be explained in this work. For more information on this topic, MATLAB demo file will be a good start point.

The surface that is going to be approximated in Modeling Inverse Kinematics in a Robotic Arm is shown in Figure 7.

978-1-4244-1983-8/08/$25.00 ©2008 IEEE   147

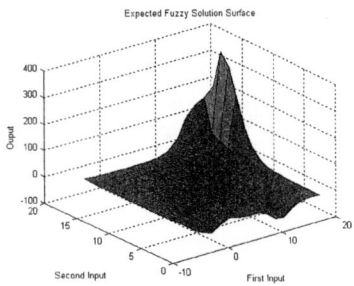

Figure 7. Fuzzy Solution Surface

The process of two dimensional surface approximation is the same as the process defined in the one dimensional mode. The only difference is that the second input is not fixed in this case. Solution surface defined in Figure 7 is the expected surface. Reconstructed surface created by the calculated breakpoints is normalized and the final output surface is shown in Figure 8.

The breakpoints are calculated and the remaining error for W and L calculations are the inputs of the Circuit Creation algorithm. In creation of two dimensional circuits, a kind of superposition is applied in that the output is created from two one dimensional designs. The output solution surface is shown in Figure 9. The actual generated circuit consists of more than 800 transistors. This circuit is optimum in terms of power and area as well as remaining within a predefined error bound. The creation time of the circuit is on the order of a few seconds.

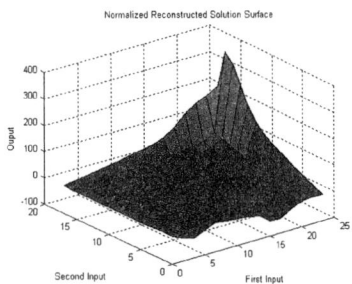

Figure 8. Normalized reconstructed two dimensional fuzzy solution surface

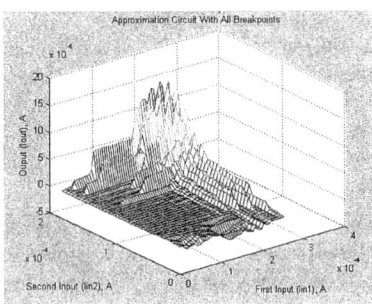

Figure 9. Fuzzy solution created by the Circuit Creation algorithm

VII. CONCLUSIONS

A design automation tool for the creation of one and two dimensional PWL response surfaces was described in this work. Performance specifications, constraints coming from the designer, and transistor model parameters are inputs of this tool. The success of the tool was demonstrated on three examples; namely, a one dimensional approximation problem, a two-dimensional approximation problem, and a fuzzy logic system. In all cases, the tool was able to generate circuits whose simulations fulfilled all the specifications.

Future work may progress in several directions. One direction will be the usage of this method in other blocks rather than current mirrors. Second direction is to take into consideration the gain bandwidth performance criteria in design of PWL circuits. Another possible direction is the integration of the tool defined in this work to other design tools to introduce a complete solution for fuzzy logic circuits. This work introduces a solution for two dimensional surfaces to show that n-dimensional solutions are possible. Extending this solution for n-dimensional surfaces may be another future research direction.

VIII. ACKNOWLEDGMENTS

This thesis has been supported by TUBITAK, NSF (National Science Foundation of USA), Bogaziçi University and Auburn University Research Project Fund.

IX. REFERENCES

[1] Wilamowski, B. M., E. S. Ferre-Pikal and O.Kaynak, "Low Power, Current Mode CMOS Circuits for Synthesis Of Arbitrary Nonlinear Functions", *9th NASA Symposium on VLSI Design,* 2000.

[2] Ahmadi, S., L. Sellami and R. W. Newcomp, "A CMOS PWL Fuzzy Membership Function", *IEEE International Symposium on Circuits and Systems*", Seattle WA, Vol. 3, pp. 2321-2324, April 30-May 3 1995.

[3] Christian C. E., F. Krummenarcher and E. A. Vittoz, "An Analytical MOS Transistor Model Valid In All Regions of Operation and Dedicated To Low Voltage and Low Current Applications", *Analog Integrated Circuits And Signal Processing,* 8, 83-114, 1995.

[4] Saski, M., T. Inoue, Y. Shirai and F. Ueno, "Fuzzy Multiple-Input Maximum and Minimum Circuits in Current Mode and Their Analyses Using Bounded- Difference Equations", *IEEE Transactions on Computers,* Vol. 39, No. 6, June 1990.

[5] Wilamowski, B. M., R. C. Jager and O.Kaynak, "Neuro-Fuzzy Architecture for CMOS Implementation", *IEEE Transactions on Industrial Electronics,* Vol. 46, No. 6, December 1999.

[6] Wilamowski, B. M., J. Binfet and O.Kaynak, "VLSI Implementation of Neural Networks", *Int. Journal of Neural Systems*, Vol. 10, No. 3, pp. 191-198, 2000.

[7] Yamakawa, T., "A Fuzzy Interface Engine in Nonlinear Analog Mode and Its Applications to a Fuzzy Logic Control", *IEEE Transactions on Neural Networks,* Vol. 4, No. 3, May 1993.

[8] Ota, Y. and B. M. Wilamowski, "Current-Mode CMOS Implementation of a Fuzzy Min-Max Network", *IEEE Transaction on Industrial Electronics,* Vol. 46, No. 6, pp. 1132-1136, Dec 1999.

[9] Ahmadi, S., L. Sellami and R. W. Newcomb, "A CMOS WL Fuzzy Membership Function", *International Symposium on Circuits and Systems*: pp. 2321-2324, 1995.

[10] Dharia, N., J. Gownipalli, B. M. Wilamowski and O. Kaynak, "Multi Dimensional Second Order Defuzzification Algorithm (M-SODA)", *The 28th Annual Conference Of The IEEE Industrial Electronics Society*: IECON '02, Nov. 5-8, 2002, Sevilla, Spain, pp.3215-3220, 2002.

On the Validation of Embedded Systems through Functional ATPG

Giuseppe Di Guglielmo
Department of Computer Science
University of Verona
Strada le Grazie 15, I-37134 Verona, Italy
Email: giuseppe.diguglielmo@univr.it

Abstract— **Increasing size and complexity of digital designs has made essential to address critical verification issues at the early stages of design cycle. Therefore, automated verification tools are necessary at higher levels of abstraction, but they are still in a prototyping phase. In this context, a valuable solution for the functional validation is represented by dynamic verification which exploits simulation-based techniques to stimulate the whole design under verification (DUV). To perform dynamic verification it is necessary to generate test sequences to be simulated on the DUV. This paper describes a functional test pattern generator which exploits two different paradigms: high-level decision diagrams (HLDDs) and extended finite state machines (EFSMs). HLDDs and EFSMs are deterministically explored by using propagation, justification, learning and backjumping. The integration of such strategies allows the ATPG to more efficiently analyze the state space of the design under verification and to generate very effective test sequences.**

I. Introduction

Gate-level automatic test pattern generation (ATPG) represents the state-of-the-art for digital system testing [1, 2]. However, many economical and practical reasons have induced the designers to apply automatic test pattern generation at higher abstraction levels [3] to implement hierarchical test strategies or to approach design validation, where design errors can be early identified and removed, saving time and money. Thus, many functional automatic test pattern generators (ATPGs) have been proposed in the literature to generate effective test sequences [4]. Generally, functional ATPGs can be divided in two main categories: random-based and deterministic. The first set adopts simulation-based strategies guided by genetic algorithms or other probabilistic-based techniques [5]. They rely on functional fault models or coverage metrics which require to accordingly instrument and simulate an HDL description (e.g., SystemC, VHDL, Verilog, etc.) of the design under verification (DUV). These ATPGs are fast, and they allow to quickly achieve an high coverage for easy-to-test designs. However, they tend to generate a large number of test sequences and they unlikely cover corner cases on complex DUVs. On the contrary, deterministic ATPGs exploit mathematical strategies tailored to allow a complete exploration of the DUV state space [4], thus covering corner cases, but they require a larger amount of timing and memory resources.

A possible way for limiting the resource consumption of deterministic ATPGs consists of implementing combined approaches that mix different state space exploration techniques and different computational models to address different DUVs and different areas of the same DUV. In this context, the paper presents a functional ATPG that relies on two computational models: high-level decision diagrams (HLDDs) [6] and extended finite state machines (EFSMs) [7, 8], and two ATPG engines which are based, respectively, on propagation and justification techniques across HLDDs, and learning and backjumping across EFSMs. Experimental results show that the joint use of such techniques allows us to improve the quality of generated test patterns and reduce the generation time.

The paper is organized as follows. Section II summarized related works about HLDD and EFSM-based ATPGs. Section III describes the main concepts related to the HLDD and EFSM computational

models. Section IV illustrates the ATPG framework and the integration of the HLDD and EFSM-based engines, which are then described in details, respectively, in Section IV-A and Section IV-B. Section V reports experimental results. Finally, Section VI is devoted to concluding remarks.

II. Related works

Recently, a number of works have been published on implementing assignment decision diagram (ADD) models combined with SAT methods to address register-transfer level test pattern generation [4]. In this paper, we take advantage of a different kind of representation, called high-level decision diagrams [6] that, unlike ADDs can be viewed as a generalization of binary decision diagrams (BDDs). HLDDs allow modeling of different abstraction levels from RTL to behavioral while ADDs are limited to RTL only. HLDDs have proven to be an efficient model for simulation [9] and fault modeling [3] as they provide for fast evaluation by graph traversal and for easy identification of cause-effect relationships.

As an alternative to decision diagrams, many works have been proposed also on finite state machine-based ATPGs, while few papers consider the use of extended FSMs. The reason that limits the use of EFSMs in the ATPG context depends on the fact that traversing an EFSM is more difficult than traversing an FSM. In fact, moving between EFSM states may require to satisfy conditions depending on primary inputs (PIs), but also on internal registers. Thus, the presence of conditions involving registers on the guard of transitions imposes that already existent approaches, developed for traversing FSMs, cannot be easily adapted to EFSMs.

In [10] different strategies are proposed to remove transitions whose guard involves conditions on registers (note as inconsistent transitions) for reusing FSM-targeted ATPGs. However, the removal of inconsistencies can lead to the state space explosion if the design under verification (DUV) description contains a large number of conditions. A different approach is proposed in [11], where the authors present a stabilization process to improve the traversing of the EFSM before applying a breadth first search to generate a set of test patterns which covers all the transitions on the stabilized EFSM. The main limitations of this approach are represented by the complexity of the stabilization process, which may lead to state explosion. Moreover, the breadth-first search-based approach is surely less efficient than strategies based for example on learning and backjumping [12]. In fact, these methods improve the performance of the ATPG by avoiding the starvation of DUV exploration in areas of the state space very far from the desired target.

III. Computational models

This section summarizes the basic concepts of HLDDs and EFSMs. These formalisms have been selected for the ATPG presented in this paper because they allow us to capture the main characteristics of state-oriented, activity-oriented and structure-oriented models [13].

978-1-4244-1983-8/08/$25.00 ©2008 IEEE

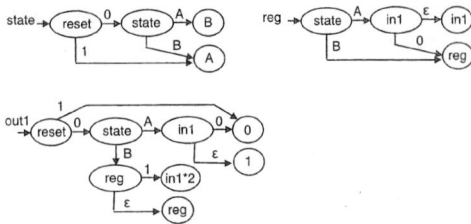

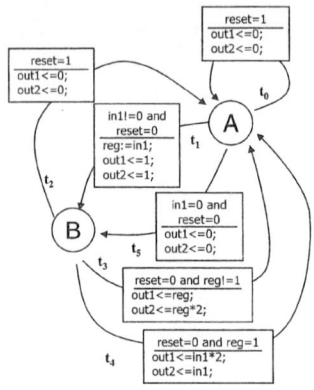

Fig. 1. HLDD models for variables *state*, *reg* and *out1*.

A. High-level decision diagrams

HLDDs are graph representations of discrete functions that can be considered as a generalization of BDDs. Unlike in BDDs, where nodes are labeled by Boolean variables and edges hold only Boolean values, in HLDDs, any scalar variable values (e.g. integer, floating point, enumeration type, etc.) are allowed and edges are labeled by partitions of the domains of respective variables. HLDDs have proven an efficient model for simulation and fault modeling as they provide for fast evaluation by graph traversal and for easy identification of cause-effect relationships. In the following we present the definition of HLDDs.

Definition 1: A HLDD representing a discrete function $y = f(x)$ is a directed non-cyclic labeled graph that can be defined as a quadruple $G = \langle M, E, X, D \rangle$, where M is a finite set of vertices (referred to as nodes), E is a finite set of edges, X is a function which defines the variables labeling the nodes and the variable domains, and D is a function on E. The function $X(m_i)$ returns a pair (x_i, X_i), where x_i is the variable letter, which is labeling node m_i and X_i is the domain of x_i. Each node of a HLDD is labeled by a variable. In special cases, nodes can be labelled by constants or algebraic expressions. An edge $e \in E$ of a HLDD is an ordered pair $e = (m_1, m_2) \in E^2$, where E^2 is the set of all the possible ordered pairs in set E. D is a function on E representing the activating conditions of the edges for the simulating procedures. The value of $D(e)$ is a subset of X_i, where $e = (m_i, m_j)$ and $X(m_i) = (x_i, X_i)$. It is required that $P_{m_i} = \{D(e) | e = (m_i, m_j) \in E\}$ is a partition of the set X_i. HLDD has only one starting node (root node), for which there are no preceding nodes. The nodes, for which successor nodes are missing are referred to as terminal nodes.

Figure 1 presents an example of an HLDD for three variables, `state`, `reg` and `out1` corresponding to the EFSM example shown in Figure 2.

B. Extended finite state machines

The EFSM model allows a more compact representation of the state space than traditional FSM, thus, the risk of state explosion that incurs to model a large design by using FSMs is sensibly reduced. A simple example of EFSM is reported in Figure 2.

Definition 2: An EFSM is defined as a 5-tuple $M = \langle S, I, O, D, T \rangle$ where: S is a set of states, I is a set of input symbols, O is a set of output symbols, D is a n-dimensional linear space $D_1 \times \ldots \times D_n$, T is a transition relation such that $T : S \times D \times I \rightarrow S \times D \times O$. A generic point in D is described by a n-upla $x = (x_1, ..., x_n)$; it models the values of the registers of the DUV. A pair $\langle s, x \rangle \in S \times D$ is called *configuration* of M.

An operation on M is defined in this way: if M is in a configuration $\langle s, x \rangle$ and it receives an input $i \in I$, it moves to the configuration $\langle t, y \rangle$ iff $((s, x, i), (t, y, o)) \in T$ for $o \in O$.

The EFSM differs from the classical FSM, since each transition does not present only a label in the classical form $(i)/(o)$, but it takes care of the register values too. Transitions are labeled with an *enabling* function e and an *update* function u defined as follows.

Fig. 2. A simple EFSM.

Definition 3: Given an EFSM $M = \langle S, I, O, D, T \rangle$, $s \in S, t \in T, i \in I, o \in O$ and the sets $X = \{x | ((s, x, i), (t, y, o)) \in T$ for $y \in D\}$ and $Y = \{y | ((s, x, i), (t, y, o)) \in T$ for $x \in X\}$, the *enabling* and *update* functions are defined respectively as:

$$e(x, i) = \begin{cases} 1 & \text{if } x \in X; \\ 0 & \text{otherwise.} \end{cases}$$

$$u(x, i) = \begin{cases} (y, o) & \text{if } e(x, i) = 1 \text{ and} \\ & ((s, x, i), (t, y, o)) \in T; \\ undef. & \text{otherwise.} \end{cases}$$

An update function $u(x, i)$ can be applied to a configuration $\langle s_1, x \rangle$ if there is a transaction $t : s_1 \rightarrow s_2$, labeled e/u, such that $e(x, i) = 1$. In this case we say that t can be *fired* by applying the input i.

Many EFSMs can be generated starting from the same HDL description of a DUV. However, despite from their functional equivalence, they can be more or less easy to be traversed. Stabilization improves the traversing easiness [14], but it can lead to the explosion of state space. Thus, in [15], a set of theoretically-based automatic transformations has been proposed to generate a particular kind of semi-stabilized EFSM (S²EFSM). It allows an ATPG to easily explore the state space of the corresponding DUV reducing the risk of state explosion.

The main problem of an S²EFSM with respect to a completely stabilized EFSM is related to the presence of transitions whose activation depends on conditions, involving registers, with a low probability of being satisfied without using backtracking or learning-based techniques. For example, let us consider transition t_4 of Fig. 2. To activate such a transition, an ATPG, after reset occurs, must move from A to B on t_1. If `in1` is a 32-bit integer, the ATPG fires t_1 by fixing `reset` at 0 and by choosing between $2^{32} - 1$ values, different from 0, for `in1`. Then, the ATPG can generate $2^{32} - 1$ different admissible configurations, when it moves on t_1. However, only the configuration where `reg=1` (obtainable by fixing `in1` at 1) is valid to fire t_4. Thus, the probability of firing t_4 is extremely low for ATPGs which implement an heuristic that exploits information local to the current configuration only. On the contrary, the EFSM-based engine presented in Section IV-B activates such a kind of transitions by exploiting learning and backjumping.

IV. ATPG FRAMEWORK

Figure 3 shows the proposed ATPG framework that joins an HLDD-based engine with an EFSM-based engine. The framework measures the quality of generated test sequences according to the well-known *bit coverage* fault model [16].

The HLDD-based engine is first applied in order to detect untestable areas within the DUV. The EFSM-based engine exploits information provided by the HLDD exploration to traverse, via

978-1-4244-1983-8/08/\$25.00 ©2008 IEEE

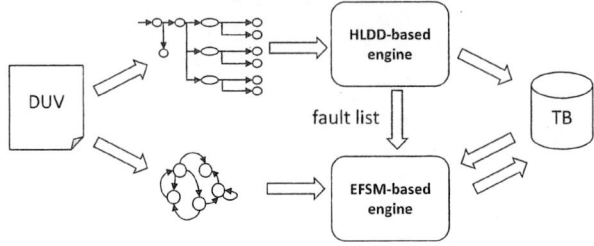

Fig. 3. The ATPG framework.

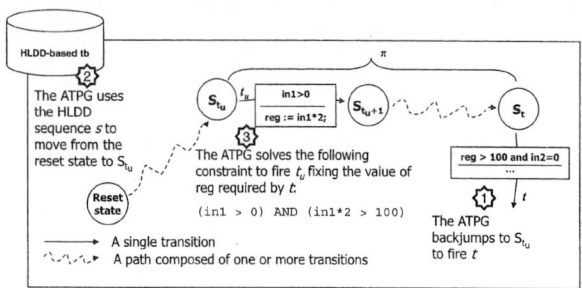

Fig. 5. The backjumping strategy.

backjumping, DUV areas that have not been explored. This generally provides higher fault coverage and lower execution time, as reported in the experimental results.

A. HLDD-based Engine

The test pattern generation algorithm implemented in the HLDD-based engine runs in two phases. During the first phase, a test path is activated to test a variable in the circuit and constraints required to activate it are extracted using HLDD models. At the second stage, the constraints are solved relying on the general purpose constraint solver ECLiPSe.

The test generation constraints can be divided into three categories: *path activation constraints*, *transformation constraints* and *propagation constraints*. Path activation constraints correspond to the logic conditions in the control flow graph that have to be satisfied in order to perform propagation and value justification through the circuit (in the Figure 4, $true = f(x_1, x_2)$ and $false = g(x_2, x_3)$). Transformation constraints, in turn, reflect the value changes along the paths from the variable under verification to the primary inputs of the whole circuit ($D = h(x_3, x_4)$). Finally, propagation constraints are necessary in order to calculate value transformation from the variable under verification until the primary output at the end of the currently activated test path ($y_3 = k(x_4, x_5, D)$). All three types of constraints can be represented by common data structures and manipulated by common procedures for creation, update, modeling and simulation.

As it was mentioned above, the HLDD-based engine is applied in order to detect untestable areas within the DUV. The tests are set up for variables and operations, which map to registers and functional units of the datapath of DUV, respectively. Relying on HLDD models the engine is capable of finding all the consistent high-level test paths represented as a set of constraints similar to the ones in the example of Figure 4. If the test path constraints of all possible paths are non-satisfiable then the variable is considered to be untestable.

B. EFSM-based Engine

The EFSM-based ATPG engine works in a three-step fashion traversing S^{2}EFSMs that model the DUV. *Phase 1: Learning* - The

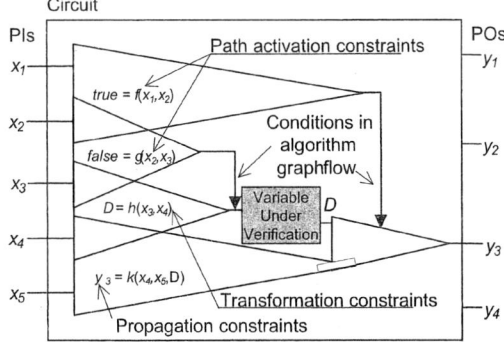

Fig. 4. HLDD-based path activation with constraint extraction.

set of S^{2}EFSMs representing the DUV are generated according to the approach described in [15], starting from a functional description of the DUV. An off-line learning phase is performed on the S^{2}EFSMs to collect information about location of registers within enabling and update functions.

Phase 2: Sequence simulation - Easy-to-traverse transitions (ETT) are traversed by using the test sequences generated by the HLDD-based engine. During this phase, information on state and transition reachability is also learned. In this way, the backjumping mode can exploits this information to move from the reset state to an already visited target state.

Phase 3: Backjumping - Finally, in the third phase, the information collected in the previous steps is exploited to fire transitions that have not been activated yet, by means of a backjumping-based approach. The EFSM-based engine changes to the backjumping mode when it exhausts the test set provided by the HLDD-based engine. The backjumping mode works as represented in Figure 5: let us assume t is a not fired transitions, out-going from a state S_t already visited during the HLDD sequence simulation phase. Let us also assume that the enabling function of t unsatisfiability depends on clauses involving a single register reg[1]. Then, extract an already visited transition t_u from the set of transitions T_{reg}^u whose update function updates reg. Load the test sequence, previously generated by HLDD-based engine, to move from the reset state to S_{t_u} (source state of t_u). Thus, the ATPG *backjumps* from S_t to S_{t_u}. Use the Dijkstra's shortest path search algorithm to provide a path π from S_{t_u} to S_t starting with t_u. Satisfy the enabling function of t_u according to the constraints derived from the enabling function of t as follows. Let us suppose that e_{t_u} is the enabling function of t_u and $e_t|_{reg}^{t_u}$ is the part of the enabling function of t which involves the clauses depending on reg, where each occurrence of reg has been substituted with the right-side expression of the assignment that updates reg in the update function of t_u. Invoking the constraint solver to satisfy the constraint $e_{t_u} \wedge e_t|_{reg}^{t_u}$ allows us to obtain a test vector which satisfies e_{t_u} and sets the value of reg in such a way that when simulation reaches transition t, following π, its enabling function will be correctly fired. The last observation may be false if there is a transition $t'_u \neq t_u \neq t$ in π, such that t'_u updates reg after t_u did. In this case, the ATPG moves the problem from t_u to t'_u requiring a solution for $e_{t'_u} \wedge e_t|_{reg}^{t'_u}$. Finally satisfy the enabling function of transitions included in π by iteratively applying the constraint solver to generate the corresponding test vectors. The test sequence obtained by joining s, to move from the reset state to S_{t_u}, and the test vectors generated to traverse π allows to fire t.

V. EXPERIMENTAL RESULTS

The efficiency of the proposed ATPG framework has been evaluated by using the benchmarks described in the first part of Table I,

---

[1]If the unsatisfiability of t depends on more than one register, the backjumping procedure is repeated for each of them.

DUT	PIs	POs	FFs	Gates	GT (sec.)	PD-ATPG			HLDD-EFSM-ATPG		
						SC%	FC%	T (sec.)	SC%	FC%	T (sec.)
ex1	66	32	130	10754	0.070	92.9	80.3	2.9	100.0	96.0	2.6
b00	66	64	99	1692	0.064	87.0	48.7	2.6	100.0	52.5	2.4
b04	13	8	66	650	0.168	95.0	99.0	8.7	100.0	99.0	7.3
b10	13	6	17	264	0.184	69.7	93.0	5.7	100.0	94.0	5.2
b11m	9	6	31	715	0.118	82.2	39.0	5.1	100.0	54.6	3.8
b00z	66	64	99	11874	0.084	75.9	44.3	5.0	100.0	51.8	4.0
fr	34	32	100	1475	0.182	86.7	70.4	4.9	100.0	84.0	4.1
dlx	29	31	25	232	0.212	63.9	46.7	3.2	100.0	59.5	2.7

TABLE I

COMPARISON BETWEEN A PSEUDO-DETERMINISTIC ATPG AND THE PROPOSED APPROACH.

where columns report the number of primary inputs (*PIs*), primary outputs (*POs*), flip-flops (*FFs*) and gates (*Gates*). Such benchmarks have been selected because they present different characteristics which allow us to analyze and confirm the effectiveness of the proposed approach. *b04*, *b09* have been selected from the well known ITC-99 benchmarks suite [17]. *b11m* is a modified version of *b11*, included in the same suite, created by introducing a delay on some paths to make it harder to be traversed. The original HDL descriptions of *b04*, *b09* and *b11m* contain a high number of nested conditions on signals and registers of different size. *ex1*, *b00*, *b00z* and *fr* contain conditional statements where one branch has probability $1 - \frac{1}{2^{32}}$ of being satisfied, while the other has probability $\frac{1}{2^{32}}$. Thus, they are very hard to be tested by a random ATPG. In particular, *ex1*, *b00* and *b00z* are internal benchmarks, while *fr* is a real industrial case, i.e., it is a module of a face recognition system. Finally, *dlx* is the controller of the well known RISC processor.

The effectiveness of the proposed ATPG framework has been evaluated by comparing it to a pseudo-deterministic ATPG [18], which outperforms pure random and genetic algorithm-based high-level ATPGs. It uses a constraint solver to traverse the DUV state space but it does not exploit neither propagation and justification strategies on HLDDs nor backjumping on EFSMs. The second part of Table I reports such a comparison. In particular, the Table shows the time required to automatically generate the HLDDs and the corresponding S^{2}EFSMs (*GT (sec.)*), the statement coverage (*SC%*), the fault coverage (*FC%*) and the test generation time (*T (sec.)*), by using respectively the pseudo-deterministic ATPG (*PD-APTG*), and the combination of the HLDD and EFSM-based ATPGs proposed in this work (*HLDD-EFSM-ATPG*).

It can be observed that the HLDD-EFSM-ATPG outperforms the PD-ATPG. The low statement coverage achieved by the PD-ATPG for some benchmarks is due to the presence of hard-to-traverse transition, whose enabling function has an infinitesimal probability of being traversed without backtracking or justification-based strategies. Such a problem is solved by the HLDD-EFSM-ATPG which exploits learning/backjumping technique. Indeed, the HLDD-EFSM-ATPG reaches 100% statement coverage for all benchmarks. Moreover, the fault coverage is sensibly increased for all benchmarks by adopting the HLDD-EFSM-ATPG. Moreover, the test generation time is reduced thanks to the capability of the HLDD-based engine to identify untestable areas, which are skipped during the subsequent test generation phase performed by the EFSM-based engine.

VI. CONCLUDING REMARKS

We presented an high-level ATPG framework which exploits two different computational models: HLDD and EFSM. The HLDD-based engine relies on fault propagation, fault activation and CLP, and it is particularly oriented to identify untestable areas in the DUV. On the contrary, the EFSM-based engine relies on learning, CLP and backjumping techniques, and it addresses DUV areas not covered by the HLDD exploration that have not been marked as untestable. The integration of the two engines allows us to improve

fault coverage, while testing time is reduced. The effectiveness of the proposed approach has been highlighted by comparing, at hight level, the results achieved by applying the HLDD-based and EFSM-based engines with a genetic algorithm-based ATPG.

REFERENCES

[1] C. Wang, S. Reddy, I. Pomeranz, X. Lin, and J. Rajski. *Conflict Driven Techniques for Improving Deterministic Test Pattern Generation*. In *Proc. of ACM/IEEE ICCAD*, pp. 87–93. 2002.

[2] B. Li, M. Hsiao, and S. Sheng. *A Novel SAT All-Solutions Solver for Efficient Preimage Computation*. In *Proc. of IEEE DATE*, pp. 272–277. 2004.

[3] R. Kapuer. *High Level ATPG is Important and Is on Its Way!*. In *Proc. of IEEE ITC*, pp. 1115–1116. 1999.

[4] I. Ghosh and M. Fujita. *Automatic Test Pattern Generation for Functional Register-Transfer Level Circuits Using Assignment Decision Diagrams*. IEEE Trans. on Computer-Aided Design of Integrated Circuits and Systems, vol. 20(3):pp. 402–415, 2001.

[5] A. Fin and F. Fummi. *Genetic Algorithms: the Philosopher's Stone or an Effective Solution for High-Level TPG?*. In *Proc. of IEEE HLDVT*, pp. 163–168. 2003.

[6] J. Raik and R. Ubar. *Fast Test Pattern Generation for Sequential Circuits Using Decision Diagram Representations*. In *JETTA, Kluwer, Vol. 16, No. 3*, pp. 213–226. 2000.

[7] G. Di Guglielmo, F. Fummi, C. Marconcini, and G. Pravadelli. *A Pseudo-Deterministic Functional ATPG based on EFSM Traversing*. In *Proc. of IEEE International Workshop on Microprocessor Test and Verification (MTV)*. Austin, TX, USA, 2005.

[8] G. Di Guglielmo, F. Fummi, C. Marconcini, and G. Pravadelli. *Improving Gate-Level ATPG by Traversing Concurrent EFSMs*. In *Proc. of IEEE VLSI Test Symposium*. Berkeley, 2006.

[9] R. Ubar, A. Morawiec, and J. Raik. *Back-Tracing and Event-Driven Techniques in High-Level Simulation with Decision Diagrams*. In *Proc. of the IEEE ISCAS2000 Conference, Vol. 1*, pp. 208–211.

[10] R. Hierons, T.-H. Kim, and H. Ural. *Expanding an Extended Finite State Machine to Aid Testability*. In *Proc. of IEEE COMPSAC*, pp. 334–339. 2002.

[11] K. Cheng and A. Krishnakumar. *Automatic Generation of Functional Vectors Using the Extended Finite State Machine Model*. ACM Trans. on Design Automation of Electronic Systems, vol. 1(1):pp. 57–79, 1996.

[12] G. Di Guglielmo, F. Fummi, C. Marconcini, and G. Pravadelli. *Improving High-Level and Gate-Level Testing with FATE: a Functional ATPG Traversing Unstabilized EFSMs*. Computers & Digital Techniques, IET, vol. 1:pp. 187–196, 2007.

[13] D. Gajski, J. Zhu, and R. Domer. *Essential Issue in Codesign*. Thecnical report ICS-97-26, University of California, Irvine, 1997.

[14] D. Lee and M. Yannakakis. *Online Minimization of Transition Systems*. In *Proc. of ACM Symposium on the Theory of Computing*, pp. 264–274. 1992.

[15] G. Di Guglielmo, F. Fummi, C. Marconcini, and G. Pravadelli. *EFSM Manipulation to Increase High-Level ATPG Efficiency*. In *Proc. of IEEE ISQED*, pp. 57–62. 2006.

[16] F. Ferrandi, F. Fummi, L. Gerli, and D. Sciuto. *Symbolic functional vector generation for VHDL specifications*. In *Proc. of IEEE DATE*, pp. 442–446. 1999.

[17] *High Time for High-Level Test Generation*. Panel at IEEE ITC, 1999.

[18] G. Di Guglielmo, F. Fummi, C. Marconcini, and G. Pravadelli. *Improving Gate-Level ATPG by Traversing Concurrent EFSMs*. In *Proc. of IEEE VTS*, pp. 172–179. 2006.

978-1-4244-1983-8/08/$25.00 ©2008 IEEE

Optimal Implementation of Combinational Logic on Look-up Tables

Kubilay Atasu, Tim Todman, Oskar Mencer and Wayne Luk
Department of Computing, Imperial College London
{kubilay.atasu,tjt97,o.mencer,w.luk}@imperial.ac.uk

Abstract—**We present a methodology for optimally implementing combinational logic equations on networks of look-up tables. Our work effectively extends optimality to span logic minimization and technology mapping. We restrict ourselves to 4-input look-up tables (LUTs) and enumerate all possible circuits up to a certain area or latency. Since simple-minded enumeration would take a long time, we develop levels of abstractions (steps) and we formulate the key step of enumeration as an Integer Linear Programming (ILP) problem. We show results on a set of ISCAS benchmarks.**

I. INTRODUCTION

We address the problem of optimization of designs for reconfigurable hardware. We use enumeration to optimize logic. In principle, we enumerate every possible configuration of a device. In practice, we simplify the enumeration to only consider configurations and connections of look-up tables (LUTs) on Field-Programmable Gate Array devices (FPGAs), under area and latency constraints. For this paper, we only consider combinatorial designs, including the state-transition logic of finite state machines.

The traditional approach in logic synthesis for FPGAs is based on two phases: technology independent optimization, and technology mapping [5]. The first phase aims to generate an optimal abstract representation of the logic circuit, a Boolean network most of the time. The second phase tries to transform the abstract representation into a network of primitive logic functions implemented by the available library, in our case 4-input look-up tables.

Although two level logic minimization can be done very efficiently [6], finding the optimum factored form of a logic function is a very complex problem, and existing methods for the first phase are heuristic or approximate.

A large body of research efforts has concentrated on the technology mapping problem for LUT-based FPGAs in the last decade. An algorithm to find delay-optimal mappings was described in [10]. On the other hand, it has been proven that the problem of finding area-optimal mappings for LUTs of input size four and greater is an NP-hard problem [7].

The early work on area minimization relied on decomposition of the circuit into a set of trees, and applied technology mapping on tree structures [8], [9]. Although area minimization on trees is much easily solvable, real circuits are rarely trees and this approach misses optimal solutions across tree boundaries. Cong et al. concentrated on enumeration of single output, K-input connected subgraphs (fanout free cones)

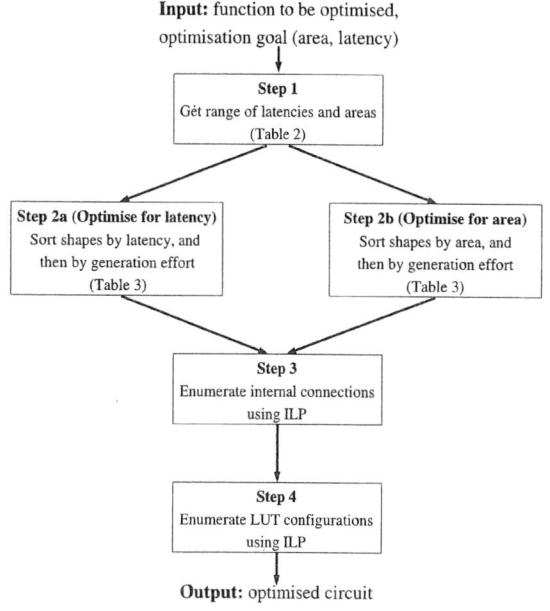

Input: function to be optimised, optimisation goal (area, latency)

Step 1
Get range of latencies and areas
(Table 2)

Step 2a (Optimise for latency)
Sort shapes by latency, and then by generation effort
(Table 3)

Step 2b (Optimise for area)
Sort shapes by area, and then by generation effort
(Table 3)

Step 3
Enumerate internal connections using ILP

Step 4
Enumerate LUT configurations using ILP

Output: optimised circuit

Fig. 1. Our approach to enumeration. Step 2 differs for area (step 2a) and latency (step 2b). In Steps 3 and 4, we use an ILP approach.

within the circuit, and proved that the problem can still be optimally solved by decomposing the circuit into maximal fanout free cones (MFFC), and enumerating separately on each MFFC in [11]. The proposed algorithm although very practical, had exponential worst case complexity, and restricted the solution to duplication free mappings where each circuit gate must be mapped to exactly one LUT. Later work by Cong et al. [12] introduced heuristics to reduce the runtime, and extended the approach to duplicable mappings.

More recent work, reformulating the technology mapping problem as a boolean satisfiability problem, has shown that state-of-the-art FPGA technology mapping algorithms miss optimal solutions [13]. Enumeration guarantees that all solutions are considered; one can obtain the absolute lower bound of logic resources needed to implement a particular problem.

Little related work has been reported on using enumeration for design optimization. A recent effort concerns an implicit technique for enumerating structural choices in circuit optimization based on re-wiring and re-substitution [14]. In [15],

978-1-4244-1983-8/08/$25.00 ©2008 IEEE

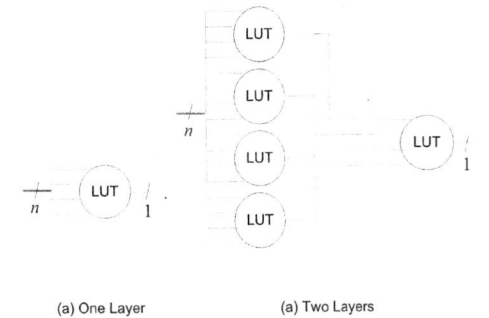

(a) One Layer   (a) Two Layers

Fig. 2. Examples of designs with one and two layers of LUTs.

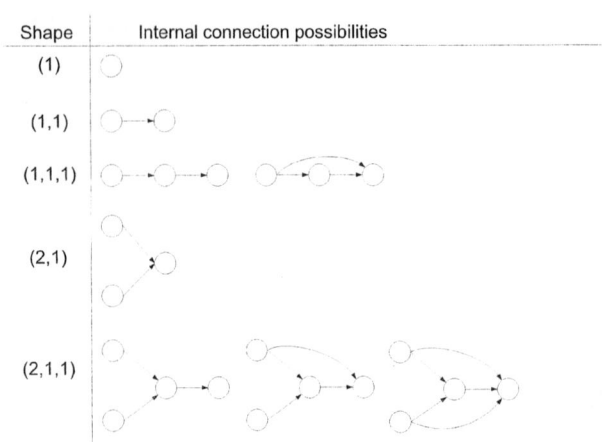

Fig. 3. Some internal connection possibilities for several shapes. Vertical lines separate the layers. (Step 3)

0	1	...	i	...	N-1	Y
0	0	0	0	0	0	y_0
:	:	:	:	:	:	:
$c_{0,t}$	$c_{1,t}$	...	$c_{i,t}$	...	$c_{N-1,t}$	y_t
:	:	:	:	:	:	:
1	1	1	1	1	1	y_{2^N-1}

TABLE I

TRUTH TABLE FOR AN N-BIT INPUT 1-BIT OUTPUT FUNCTION

a reconfigurable hardware implementation is proposed to accelerate circuit enumeration. Our research differs from [14], [15], since we use an ILP-based approach to implement the key step of enumeration.

Our three main contributions in this paper are:

- Extending optimality to span logic minimization and technology mapping.
- A 4-step process to enumerate all possible solutions.
- Implementation via software enumeration, and Integer Linear Programming.

Our overall approach to enumeration is illustrated in Figure 1, showing how we break the problem into several steps:

Step 1 Given a boolean input function and an optimization metric (area or latency), this step identifies observable inputs and limits the search space.

Step 2 Enumerate all circuit shapes within the search space from step 1, sort by (a) latency or (b) area,

Step 3 Enumerate all possible interconnections for each shape,

Step 4 Enumerate all possible LUT configurations for each *circuit*.

II. THEORY

In this section we develop expressions for the upper bound of the design space for enumeration, for each step of Figure 1.

We assume layers of LUTs (shapes) to realize a design, and we enumerate different LUT configurations and interconnections. Some possible shapes and internal connections for enumerating a design is shown in Figure 3.

We assume that the truth table for an N-bit input, 1-bit output function Y is given as in Table I. We assume that the function has already been reduced so all of the inputs are observable. Observability of an input can be computed using Boolean derivative as defined in [4]. We enumerate designs consisting of 4-bit LUTs (A 4-bit LUT has 4 inputs). We assume that the circuit is composed of H layers of LUTs, where layer h is composed of L_h LUTs. We define L_{tot} as the total number of LUTs contained in the design.

The design space for enumeration is large: a 4-bit LUT can be configured in 2^{2^4} ways. A logic function with 4 inputs can

be implemented with a single LUT. Logic functions with larger number of inputs require multiple LUTs. We further refine different steps of Figure 1 to handle N-input logic functions:

Step 1 We identify observable inputs and index into Table II to find the range of our design space with associated area and latency requirements.

The maximum latency and area requirements are calculated based on the following observations:

- a 4-input design can be implemented by a single LUT,
- an $n + 1$-input design can be implemented using two n-input designs, and an additional LUT multiplexing between the two using the $n + 1$th input.

The minimum area and latency requirements are calculated based on the following observations:

- each observable design input must be connected to at least one LUT input,
- at least one of the LUT inputs must be connected to a LUT output at a previous layer,
- there is a single LUT at the highest layer.

Step 2 We find all *shapes* for the range found in step 1 (See Table III). We sort the resulting list of shapes by latency (if optimizing for latency, step 2a) or area (if optimizing for area, step 2b). For example, to enumerate an 8-input design for minimum area, we first choose the smallest topology that will accept eight inputs: (2,1) in our terminology. If this fails, we

978-1-4244-1983-8/08/$25.00 ©2008 IEEE

function #inputs	optimize for latency		optimize for area	
	min	max	min	max
≤ 4	1	1	1	1
5	2	2	2	3
6	2	3	2	7
7	2	4	2	15
8	2	5	3	31
9	2	6	3	63
10	2	7	3	127
11	2	8	4	255
N	$log_4(N)$	$(N-3)$	$\lfloor (N+1)/3 \rfloor$	$2^{N-3}-1$
	$O(logN)$	$O(N)$	$O(N)$	$O(2^N)$

TABLE II

LATENCY (MAXIMUM NUMBER OF LUTs FROM INPUTS TO OUTPUT) AND AREA (NUMBER OF LUTs) FOR DIFFERING NUMBERS OF INPUTS. USER INPUT TO OPTIMIZATION IS # INPUTS AND OPTIMIZATION MODE (LATENCY OR AREA). (STEP 1)

	Latency				
	1	2	3	4	5
Area					
1	(1)				
2		(1,1)			
3		(2,1)	(1,1,1)		
4		(3,1)	(2,1,1)	(1,1,1,1)	
			(1,2,1)		
5		(4,1)	(3,1,1)	(2,1,1,1)	(1,1,1,1,1)
			(1,3,1)	(1,2,1,1)	
			(2,2,1)	(1,1,2,1)	
6			(4,1,1)	(3,1,1,1)	(2,1,1,1,1)
			(3,2,1)	(2,2,1,1)	(1,2,1,1,1)
			(2,3,1)	(1,3,1,1)	(1,1,2,1,1)
			(1,4,1)	(2,1,2,1)	(1,1,1,2,1)
				(1,2,2,1)	
				(1,1,3,1)	

TABLE III

ALL THE DIFFERENT SHAPES FOR ONE TO FOUR 4-LUTs, ARRANGED ACCORDING TO LATENCY AND AREA. (STEP 2)

choose one of the next smallest designs, and so on. Similarly, the minimum latency design can be found by iterating from the minimum latency topology to the maximum.

Step 3 We enumerate all interconnection possibilities. One of the LUT inputs must be connected to a LUT output at a previous layer. The remaining inputs may connect to the output of any LUT in a previous layer, or to a design input.

Step 4 For all graphs, we enumerate each configuration of each LUT. For L_{tot} LUTs, this is $2^{2^4 * L_{tot}}$.

The output of the final circuit must be identical to the N-bit function output specified by a truth table for each input over the input space of 2^N (See Table I).

To make enumeration more tractable, we only consider combinatorial designs (no registers or feedback) with single output. Enumeration can still be applied to the combinatorial parts of sequential designs. Multiple-output designs can still be considered by generating separate hardware for each output, followed by a common-subexpression elimination step to eliminate LUT configurations and connections common to several outputs.

This section has shown the size of the search space for enumeration. The next section introduces our ILP formulation to solve the LUT mapping problem exactly.

III. ENUMERATION USING ILP

We describe an Integer Linear Programming formulation to achieve Steps 3 and 4 together. The ILP formulation checks if there exists a feasible circuit given a shape from Step 2.

We define a binary decision variable $X_{h,l,k,i}$ which represents whether input i is connected to the kth input of the lth LUT at layer h. More formally:

$$X_{h,l,k,i} = \begin{cases} 1 & \textit{if input i is connected to the kth input} \\ & \textit{of the lth LUT at layer h} \\ 0 & \textit{otherwise} \end{cases}$$
$$h \in \{0..H-1\}, l \in \{0..L_h-1\}, k \in \{0..3\}, i \in \{0..N-1\} \tag{1}$$

We define a binary decision variable $XOUT_{h,l,k,hi,li}$ which represents whether output of the lith LUT at layer hi is connected to the kth input of the lth LUT at layer h:

$$XOUT_{h,l,k,hi,li} \in \{0,1\}, hi \in \{0..h-1\}, li \in \{0..L_{hi}-1\} \tag{2}$$

We associate a binary decision variable with each configuration bit of each LUT. For LUT l of layer h, we need to define 2^K new decision variables:

$$LUT_{h,l,j} \in \{0,1\}, j \in \{0..2^K-1\} \tag{3}$$

We associate a binary decision variable with each LUT output. The output of LUT l of layer h at time t is represented as $OUT_{h,l,t}$:

$$OUT_{h,l,t} \in \{0,1\}, t \in \{0..2^N-1\} \tag{4}$$

Each LUT input must be connected to exactly one function input or one LUT output at a lower layer:

$$\sum_{i \in \{0..N-1\}} (X_{h,l,k,i}) + \sum_{hi \in \{0..h-1\}} \sum_{li \in \{0..L_{hi}-1\}} (XOUT_{h,l,k,hi,li}) = 1 \tag{5}$$

We define a new binary decision variable $Z_{h,l,k,t}$ which stores the value assigned to the kth input of the lth LUT at layer h at time t:

$$Z_{h,l,k,t} = \sum_{i \in \{0..N-1\}} (X_{h,l,k,i} \wedge c_{i,t}) + \sum_{hi \in \{0..h-1\}} \sum_{li \in \{0..L_{hi}-1\}} (XOUT_{h,l,k,hi,li} \wedge OUT_{hi,li,t}) \tag{6}$$

We calculate the output of the lth LUT at layer h at time t as follows:

$$OUT_{h,l,t} = \begin{cases} (\overline{Z_{h,l,0,t}} \wedge \overline{Z_{h,l,1,t}} \wedge \overline{Z_{h,l,2,t}} \wedge \overline{Z_{h,l,3,t}} \wedge LUT_{h,l,0}) \vee \\ (\overline{Z_{h,l,0,t}} \wedge \overline{Z_{h,l,1,t}} \wedge \overline{Z_{h,l,2,t}} \wedge Z_{h,l,3,t} \wedge LUT_{h,l,1}) \vee \\ \vdots \\ (Z_{h,l,0,t} \wedge Z_{h,l,1,t} \wedge Z_{h,l,2,t} \wedge \overline{Z_{h,l,3,t}} \wedge LUT_{h,l,2^K-2}) \vee \\ (Z_{h,l,0,t} \wedge Z_{h,l,1,t} \wedge Z_{h,l,2,t} \wedge Z_{h,l,3,t} \wedge LUT_{h,l,2^K-1}) \end{cases} \tag{7}$$

The output of the highest layer LUT must be identical to the N-bit function output for the same set of inputs. The optimal solution returns an objective value equal to zero if and only if a circuit implementing the given functionality is found:

$$min \sum_{t \in \{0..2^K-1\}} |y_t - OUT_{H-1,0,t}| \qquad (8)$$

IV. RESULTS AND EVALUATION

Our tool chain starts with the truth table specification of a given logic function. We traverse Table III columnwise or rowwise depending on the optimization mode (latency or area respectively). For each shape we automatically generate the associated ILP problem, and solve using CPLEX Mixed Integer Optimizer [16]. The process is continued until a feasible circuit implementing the given function is found. Once a circuit is found, we automatically generate the hardware description in ASC [17], explicitly specifying the LUT configurations and interconnections to be mapped onto an actual FPGA. Additionally, for each circuit, we automatically generate a testbench, and simulate for the set of inputs specified in the truth table, comparing the result with the expected output. In this way we verify the correctness of the circuits we generate.

We have applied our algorithms on a set of ISCAS benchmarks shown in Table IV. The benchmarks describe circuits with multiple inputs and outputs. Most of the time an output is sensitive to changes in a subset of the inputs only (i.e., observable inputs). Therefore only a subset of the circuit inputs have to be considered during enumeration. The shapes automatically identified for the benchmarks, together with the execution time of the ILP solver are given in in the last two columns of Table IV. In all cases, where ILP completed successfully, only two LUTs (i.e., a shape of (1,1)) were sufficient to realize the output functions. We have observed that the number of constraints increased quickly with the number of inputs, and ILP did not complete within 24 hours in two of the cases.

V. CONCLUSION

We have described an enumeration approach for identifying optimal combinational circuit implementations on networks of look-up tables. We divide the enumeration into steps that enable efficient exploration of the search space. We explore different circuit topologies (shapes) in the order of latency or area, depending on the optimisation mode. We make use of the existing ILP technology [16] to carry out the key step of enumeration, where we identify whether a chosen shape is feasible for implementing a given logic function.

Our current and future work involves improving the speed of enumeration for the optimization of larger logic functions. In particular, we are exploring more efficient ILP formulations that can better exploit symmetries within a circuit. Additionally, we are planning to evaluate the performance of hybrid techniques that combine software enumeration, hardware enumeration [15], and ILP.

Name	#Inps	Output	#Obs.Inps	Shape	Run-time
c17	5	1	4	(1,1)	0 s
		2	4	(1,1)	0 s
s27	7	1	6	-	-
		2	5	(1,1)	42 s
		3	6	-	-
		4	3	(1,1)	0 s
b01	7	1	1	(1,1)	0 s
		2	1	(1,1)	0 s
		3	3	(1,1)	0 s
		4	5	(1,1)	17397 s
		5	5	(1,1)	972 s
		6	5	(1,1)	1120 s
		7	5	(1,1)	156 s
b02	5	1	2	(1,1)	0 s
		2	3	(1,1)	0 s
		3	4	(1,1)	0 s
		4	4	(1,1)	0 s
		5	4	(1,1)	0 s

TABLE IV

RESULTS FOR A SET OF ISCAS BENCHMARKS. RUN-TIMES ARE GIVEN IN SECONDS. ILP DID NOT COMPLETE IN TWO OF THE CASES.

Acknowledgements: The support of UK Engineering and Physical Sciences Research Council (Grant number EP/C509625/1 and EP/C549481/1), European FP6 Project hArtes, Agility, Celoxica and Xilinx is gratefully acknowledged.

REFERENCES

[1] D. E. Knuth, "A Draft of Section 7.2.1.3: Generating All Combinations", *The Art of Computer Programming: Pre-Fascicle 3A*, 2005
[2] R. L. Graham, D. E. Knuth, O. Patashnik, *Concrete Mathematics: A Foundation for Computer Science*, Addison-Wesley, 1989.
[3] S. J. Russell and P. Norvig, *Artificial Intelligence: A Modern Approach*, Prentice Hall, 1995.
[4] G. De Micheli, *Synthesis and Optimisation of Digital Circuits*, McGraw-Hill, 1994.
[5] A. Sangiovanni-Vincentelli, A. El Gamal, and J. Rose. "Synthesis methods for field programmable gate arrays". Proceedings of IEEE, pp. 1057–1083, July 1993.
[6] R. K. Brayton, C. McMullen, G. D. Hachtel, and A. Sangiovanni-Vincentelli. "Logic Minimization Algorithms for VLSI Synthesis". Kluwer Academic Publishers, 1984.
[7] A. Farrahi, and M. Sarrafzadeh. "Complexity of the Lookup-Table Minimization Problem for FPGA Technology Mapping". IEEE Transactions on Computer-Aided Design of Integrated Circuits and Systems, 13(11):1319–1332, 1994.
[8] K. Keutzer. "DAGON: Technology Binding and Local Optimization by DAG Matching". In DAC 1987, pp. 341–347.
[9] R. Francis, J. Rose, and Z. Vranesic. "Chortle-crf: Fast Technology Mapping for Lookup Table-Based FPGAs". In DAC 1991, pp. 227–233.
[10] J. Cong and Y. Ding. "An Optimal Technology Mapping Algorithm for Delay optimization in Lookup-Table Based FPGA Designs". In IEEE ICCAD, 1992.
[11] J. Cong, and Y. Ding. "On area/depth trade-off in LUT-based FPGA technology mapping". In DAC 1993, pp. 213–218.
[12] J. Cong, C. Wu, and Y. Ding. "Cut ranking and pruning: enabling a general and efficient FPGA mapping solution". In FPGA 1999, pp. 29–35.
[13] A. Ling, D. P. Singh and S. P. Brown, "FPGA Technology Mapping: A Study of Optimality", In DAC 2005.
[14] V. N. Kravets and P. Kudva, "Implicit Enumeration of Structural Changes in Circuit Optimization", In DAC 2004, pp. 438–441.
[15] T. Todman, H. Fu, O. Mencer, and W. Luk. "Improving Bounds for FPGA Logic Minimization". In FPT 2007, pp. 245–248.
[16] ILOG CPLEX. http://www.ilog.com/products/cplex/
[17] O. Mencer, "ASC, A Stream Compiler for Computing with FPGAs" *IEEE Transactions on CAD*, IEEE, 2006.

Phase Noise Behaviour of Fractional-N Synthesizers with $\Delta\Sigma$ Dithering for Multi-Radio Mobile Terminals

Václav VALENTA
Université Paris-Est[‡] and Brno University of Technology[*]
[‡]ESYCOM, ESIEE; [*]DREL
[‡]Noisy-le-Grand, France; [*]Brno, Czech Republic
valentav@esiee.fr

Geneviève BAUDOIN, Martine VILLEGAS
Université Paris-Est
ESYCOM, ESIEE
Noisy-le-Grand, France
baudoing@esiee.fr, villegam@esiee.fr

Abstract— **This paper presents phase noise behaviour and design aspects of PLL based frequency synthesizers with $\Delta\Sigma$ dithering for cognitive multi-radio mobile terminals. Principal features of PLL based frequency synthesizers and 1-bit $\Delta\Sigma$ dithering are presented and simulated. Moreover, frequency synthesizer requirements for main standards in the frequency band 800 MHz to 6 GHz are investigated as well.**

I. INTRODUCTION

During the recent past, there has been a significant progress in wireless communications in terms of integration of various communication standards into a single mobile terminal. Instead of using multiple transceivers for different standards, the goal is to employ a single reconfigurable radio transceiver that is able to achieve all requirements of different communication standards. The goal of this evolution is to reduce the number of external components and to increase the integration in the low-cost CMOS technology. Another stage which pursues the evolution towards the cognitive radio and implies flexibility of each stage of the communication chain is the cognitive multi-radio. Cognitive multi-radio has the capability of the multi-standard concept and moreover it is capable to perform an efficient environment spectrum scanning in order to switch accordingly to an appropriate communication standard.

Single chip radios that support WLAN 802.11 a/b/g standards have already been proposed in [1], [2]. Another example of integration of cellular standards GSM 900/1800 and UMTS in a single chip can be found in [3]. These multi-radio proposals are mainly based on the direct conversion technique which is the most suitable techniques for 2G, 3G and 4G multi-radio terminals [4]. This is due to the fact, that the direct conversion eliminates the sensitivity to the image frequency and hence, it is not necessary to build any additional filters for the image rejection. One of the most challenging tasks in the multi-radio transceiver is the design of the frequency synthesizer. Frequency synthesizer has to provide all necessary frequencies for the down and up conversion with proper channel spacing that corresponds to

the channel bandwidth and the raster of the communication standard. Frequency switching has to be performed agilely, with respect to the standard settling time requirements. Moreover, the local frequency synthesizer has to fulfil the tightest signal purity requirements which can be expressed in terms of the phase noise and the spurious output. These requirements given by the most diffused standards in the frequency band 800 MHz to 6 GHz, namely by GSM, UMTS, Bluetooth, WiFi and WiMAX, are summarized in Table 1 [5]-[9].

It can be seen, that the most critical local oscillator requirements in terms of the phase noise are imposed by the GSM standard. This is due to the fact that the powers of the in-band unmodulated interfering signals, so called blockers, are at very high level [10].

TABLE I.   RF SPECIFICATIONS FOR THE MULTI-RADIO FREQUENCY SYNTHESIZER

Standard	Frequency Band [MHz]	Channel / Raster [MHz]	Settling Time[µs]	Phase Noise [dBc/Hz]
GSM 900/1800	880-960 1710-1880	0.2/0.2	577 150 (GPRS)	-122@0.6 MHz -132@1.6 MHz -139@3 MHz
UMTS FDD/TDD	1920-2170 1900-2025	5/0.2	200	-132@3 MHz -132@10 MHz -144@15 MHz
Bluetooth	2402-2480	1/1	150	-84@1 MHz -114@2 MHz -129@3 MHz
Mobile WiMAX IEEE 802.16e	2300-2400 2305-2320 2469-2690 3300-3400 3400-3800	3.5-10 / 0.25	< 100 (HFDD)	Phase Jitter < 1° rms
WiFi IEEE 802.11a	5150-5350 5470-5825	20/20	500	-102@1 MHz -125@25 MHz
WiFi IEEE 802.11b	2412-2472	20/5	225	
WiFi IEEE 802.11g	2412-2472	20/5	225	

This work was supported by the French Government Ph.D. program Doctorat en Cotutelle and by the Grant Agency of the Czech Republic under the grant No. 102/08/H027.

978-1-4244-1983-8/08/$25.00 ©2008 IEEE

Various approaches for multi-radio frequency synthesizers have been proposed in [11]-[13], considering wideband tuning range VCO, use of multiple PLL loops or all digital PLL design. Although there are many techniques for frequency synthesis, the dominant technique used in wireless technology is based on the PLL principle. This technique is considered in this article. First, a linearized PLL model is described along with characteristic transfer functions. Furthermore, noise sources in the $\Delta\Sigma$ fractional-N PLL circuit are studied and simulated.

II. PLL ARCHITECTURE

A. PLL Linearized Model

Figure 1 displays a model of the $\Delta\Sigma$ fractional-N PLL. This model includes a charge pump phase-frequency detector CP/PFD with the gain K_d, a loop filter with the transfer function $F(s)$, a VCO with the gain K_0/s and a frequency divider dividing by N. The gain K_d can be written as follows:

$$K_d [A/rad] = I_{cp}/2\pi . \qquad (1)$$

The charge pump current I_{cp} is proportional to the phase error in the phase detector and after low pass filtering by $F(s)$, it is applied to the control input of the VCO. In a conventional PLL, the frequency at the PLL output equals $f_{ref} N$. Divide ratio N can be an integer or a fractional value. In a fractional-N divider, the fractional division is achieved by periodic altering the division value between two integer values N and $N+1$, hence the average division becomes a fraction. However, the periodic switching between two division values leads to a sawtooth phase error which creates several spurious fractional tones. This problem is solved with help of $\Delta\Sigma$ modulator, which randomize the division ratio in the PLL but on the other hand, it introduces quantization noise into the loop as described in Section III.

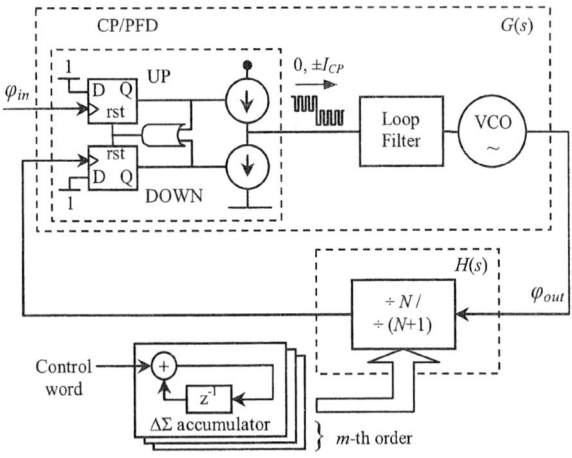

Figure 1. A Simplified model of fractional-N PLL frequency synthesizer with $\Delta\Sigma$ modulator that controlls the division ration of the divider.

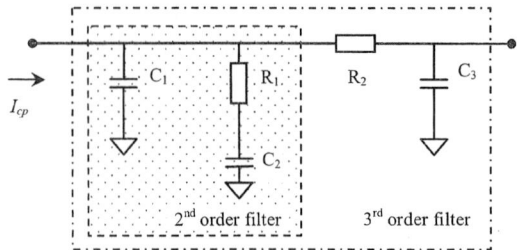

Figure 2. Second and third order passive filter.

According to the control theory, the effect of a closed feedback loop on the input signal φ_{in} can be described by the closed loop transfer function $T(s)$ as:

$$T(s) = \frac{\varphi_{out}(s)}{\varphi_{in}(s)} = \frac{G(s)}{1+G(s)H(s)} = \frac{\dfrac{K_d K_0}{s}F(s)}{1+\dfrac{K_d K_0}{s}F(s)\dfrac{1}{N}}, \qquad (2)$$

where the $G(s)$ represents the forward transfer function and the term $G(s)H(s)$ open loop transfer function. $H(s)$ corresponds to the division factor $1/N$. Since the input signal at the loop filter is a charge pump current I_{cp}, the filter transfer function represents the transimpedance. The transimpedance of the 2nd order filter shown in Fig. 2 reads:

$$F(s) = \frac{1+sC_2 R_1}{s(C_1+C_2)\left(1+s\dfrac{C_1 C_2 R_1}{C_1+C_2}\right)} . \qquad (3)$$

In order to boost the reference spur suppression at the output of the PLL, additional RC low pass stages can be added [14]. However, additional passive components will contribute to the total phase noise at the output of the PLL.

III. PLL PHASE NOISE SOURCES

Phase noise performance of the frequency synthesizer is very important issue for many applications, notably for OFDM systems. The phase noise phenomenon in the PLL is a result of the thermal noise, shot noise, $1/f$ noise in all the active or passive components. Hereafter, we will analytically describe the phase noise behavior in the PLL circuit for the following noise sources: reference oscillator noise, VCO noise, loop filter noise and $\Delta\Sigma$ quantization noise.

A. Reference Noise

Reference oscillator's phase noise performance has been predicted with help of the Leeson's model [15] as:

$$S_{ref}(f_m)[dBc/Hz] = 10\log\left[\frac{FkT}{2A}\left(1+\frac{1}{f_m^2}\left(\frac{f_O}{2Q}\right)^2\right)\left(1+\frac{f_f}{f_m}\right)\right], \qquad (4)$$

where f_m is the frequency offset, F represents the noise figure of the oscillator's amplifier, k is the Boltzmann's constant $1.38 \cdot 10^{-23}$ J/K, T is the absolute temperature, A is the power available to the amplifier from the resonator, Q is the loaded quality factor of the resonator, f_O is the carrier frequency and f_f is the flicker corner frequency. According to Fig. 1, the transfer function of the reference oscillator's noise S_{ref} injected at the input of the PLL corresponds to the low-pass filter function G_{LPF} with a loop bandwidth B_{PLL}. B_{PLL} is the frequency at which the open loop gain magnitude equals 1. Phase noise contribution $S_{ref,out}$ from the reference oscillator to the PLL output becomes:

$$S_{ref,out}(f) = S_{ref}(f)\left(\frac{1}{H(f)}\right)^2\left|\frac{G(f)H(f)}{1+G(f)H(f)}\right|^2 \quad (5)$$
$$= S_{ref}(f)N^2\left|G_{LPF}(f)\right|^2.$$

Fig. 3 demonstrates that the reference noise is low-pass filtered and hence the phase noise contribution at the PLL output becomes dominant at lower frequency offsets. Moreover, the reference phase noise is amplified by factor N, which is in this example equal to 2439 ($\approx$67.8 dB). In order to keep low in-band phase noise, the division factor should be set as low as possible. This problem is solved by fractional-N dividers that can provide arbitrary small frequency resolution without the need of low reference frequency. Hence, the in-band phase noise performance is improved significantly [5].

B. VCO Noise

VCO phase noise S_{VCO} was modeled according to the Leeson's model as in (4). The transfer function is calculated similarly as in the reference noise case. The noise injected this time at the output of the VCO is high-pass filtered and hence the contribution of this noise to the PLL output reads:

$$S_{VCO,out}(f) = S_{VCO}(f)\left|\frac{1}{1+G(f)H(f)}\right|^2 = S_{VCO}(f)\left|G_{HPF}(f)\right|^2. \quad (6)$$

Due to the high-pass filtering, the VCO's phase noise will be suppressed within the loop bandwidth B_{PLL} and the dominant noise contribution will appear at higher frequency offsets. This phenomenon is depicted in Fig. 4.

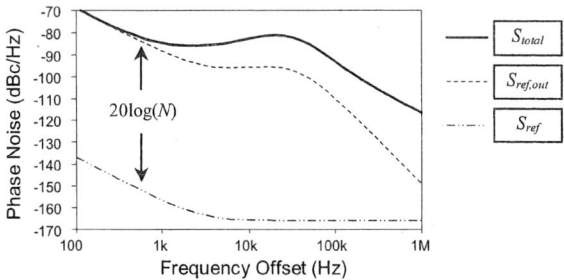

Figure 3. Reference oscillator's noise performance at the input and at the output of the PLL. Moreover, total phase noise at the PLL output is shown.

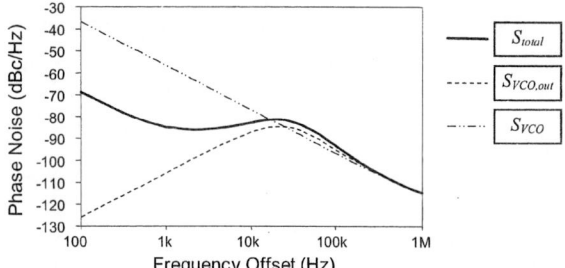

Figure 4. VCO's noise performance for open and closed loop case.

Furthermore, the out-band phase noise will drop to the VCO's noise floor which is defined by the oscillator's output power A and the system's noise figure F. Total integrated phase noise at the output of the PLL can be minimized by means of B_{PLL} reduction, however this will lead to settling time degradation. Hence, the tradeoff between the minimum loop bandwidth and the settling time has to be taken into account for design. A simplified tradeoff has been presented in [5] as $B_{PLL} = 4 / t_{lock}$, where t_{lock} is the settling time.

C. Loop Filter Noise

A passive loop filter consists of only resistive and capacitive components. Hence, the output voltage noise is the result of the thermal noise present in the real part of the complex admittance of the loop filter. Two sided PSD of current fluctuations is defined by the Nyquist equation as:

$$S_{fil}(f) \approx 2kT\,\text{Re}(Y(f)), \quad (7)$$

where Y is the loop filter admittance that corresponds to the inverse filter transimpedance $1/F(f)$. Furthermore, this thermal noise is band-pass filtered (see Fig. 5) by the PLL and the noise contribution to the output of the PLL reads:

$$S_{fil,out}(f) = S_{fil}(f)\left|F(f)\right|^2\left|K_0 / j2\pi f\right|^2\left|G_{HPF}(f)\right|^2. \quad (8)$$

Fig. 5 shows the loop filter noise contribution to the PLL output for two values of the CP current. If the CP current is increased while keeping the same PLL transfer function, the thermal noise will be reduced since the loop filter resistor becomes smaller in order to keep the same output voltage.

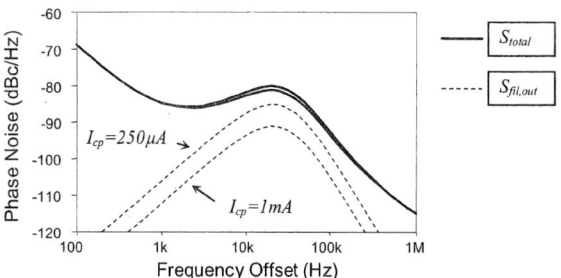

Figure 5. Noise contribution of the 2nd order filter for two values of I_{cp}.

978-1-4244-1983-8/08/$25.00 ©2008 IEEE

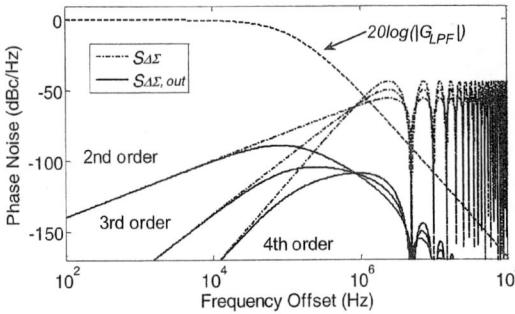

Figure 6. Noise spectrum induced by the 2nd, 3rd and 4th order $\Delta\Sigma$ modulator (dash dot) and the noise spectrum at the PLL's output (full line).

D. Quantization Noise from $\Delta\Sigma$ Modulator

The $\Delta\Sigma$ modulator used in a synthesizer is to randomize the instantaneous division ratio and hence push phase noise associated with the divider from low frequencies to high frequencies [16], see Fig. 6. The spectrum of the quantization noise $S_{\Delta\Sigma}$ is described for m-th order $\Delta\Sigma$ modulator as [17]:

$$S_{\Delta\Sigma}(f)[rad^2/Hz] = \frac{(2\pi)^2}{12F_{ref}} \left[2\sin\left(\pi f/F_{ref}\right)\right]^{2(m-1)}. \quad (9)$$

We consider the quantization noise $S_{\Delta\Sigma}$ to be injected before the frequency divider $N/N+1$. Hence, the noise transfer function of the phase corresponds to the low-pass transfer function G_{LPF} as derived in equation 5. Furthermore, the noise contribution to the output of the PLL reads:

$$S_{\Delta\Sigma,out}(f) = S_{\Delta\Sigma}(f)\left|\frac{G(f)H(f)}{1+G(f)H(f)}\right|^2 = S_{\Delta\Sigma}(f)\left|G_{LPF}(f)\right|^2. \quad (10)$$

Notice the 20, 40 and 60 dB/decade slopes for the 2nd, 3rd and 4th order of the $\Delta\Sigma$ modulator respectively. Higher $\Delta\Sigma$ order provides better spurious suppression. Moreover, it can be seen that the $\Delta\Sigma$ modulator of higher order contributes less quantization noise close to the carrier, but on the other hand, it adds more noise to the PLL's output at higher frequency offsets (around 12 dB difference between the 2nd and the 4th order at higher frequency offset).

IV. CONCLUSION

In order to meet demanding synthesizer requirements for considered standards, understanding of the phase noise behavior in PLL based synthesizers becomes critical. We have presented phase noise behavior of PLL components, including noise sources of the reference oscillator, VCO, loop filter and quantization noise of the $\Delta\Sigma$ modulator. It has also been shown that the noise contribution of the loop filter can be reduced by means of the charge pump current optimization and by optimizing values of loop filter

components. In this particular simulation, the noise contribution was reduced by 6 dB.

Moreover, frequency synthesizer requirements of the most diffused standards in the frequency band 800 MHz to 6 GHz have been presented as well. The most challenging synthesizer design issues are dictated by GSM and WiMAX standards. More precisely, GSM standard with very straighten phase noise requirements due to large allowed interferers and WiMAX standard due to very low settling time. Due to diverse requirements of considered standards, a multiple frequency synthesizers or reconfigurable loop filter might be considered. The most promising techniques include fractional-N $\Delta\Sigma$ approach which offers a significant improvement over integer-N synthesizers and it provides low phase noise performance and small step size which is indispensable for future cognitive multi-radio terminals.

REFERENCES

[1] Z. Xu et al., "A compact dual-band direct-conversion CMOS transceiver for 802.11a/b/g WLANs," in IEEE ISSCC Dig. Tech. Papers, Feb. 2005, pp. 98–99.

[2] P. Zhan et al., "A single-chip dual-band direct-conversion IEEE 802.11a/b/g WLAN transceiver in 0.18-μm CMOS," IEEE J. Solid-State Circuits,vol. 40, no. 9, pp. 1932–1939, Sep. 2005.

[3] J. Ryynanen et al., "A single-chip multi-mode receiver for GSM900 DCS1800, PCS1900, and WCDMA," IEEE J. Solid-State Circuits, vol. 38, no. 4, pp. 594–602, Apr. 2004.

[4] M. Brandolini et al., "Toward multistandard mobile terminals—fully integrated receivers requirements and architectures," IEEE Trans. on Microwave Theory and Techniques, vol. 53, no. 3, March 2005.

[5] Keliu Shu and Edgar Sánchez-Sinencio, "CMOS PLL Synthesizers: Analysis and Design," Springer 2005.

[6] Xiaopeng Li, Mohammed Ismail, "Multi-Standard CMOS Wirelles Receivers, Analysis & Design," Kluwer Academic Publishers, 2002.

[7] Y. Zhang, H. Chen, "Mobile WiMAX: Toward Broadband Wireless Metropolitan Area Networks," Auerbach Publications, 2008.

[8] G. Cantone, "0.25 μm 802.11a WLAN front end," MuMoR Workshop, Lausanne 2004.

[9] IEEE Standard for Local and metropolitan area networks, Part:16 Air Interface for Fixed Broadbanc Wireless Access Systems

[10] Digital Cellular Telecommunications System (Phase 2): Radio Transmission and Reception, GSM Standard 05 05, 1999.

[11] A. Koukab, Yu Lei, J. Declercq, "A GSM-GPRS/UMTS FDD-TDD/WLAN 802.11a-b-g multi-standard carrier generation system," IEEE Journal of Solid-State Circuits, vol. 41, no.7, July 2006.

[12] G. Itkin, A. Pestrayakov, "Multi-Band Frequency Synthesizer for Mobile Terminals," United States Patent no. US 6,828,836 B2, 2004.

[13] L. Syllaios, T. Balsara, R. Staszewski, "On the reconfigurability of all-digital phase-locked loops for Software Defined Radios," The 18th Annual IEEE International Symposium on Personal, Indoor and Mobile Radio Communications (PIMRC'07).

[14] D. Banerjee, PLL Performance, Simulation, and Design 4th Edition, 2006.

[15] D.B. Leeson, "A simplified model of feedback oscillator noise spectrum," IEEE, Volume 42, Feb. 1965, pp. 329-33.

[16] K. Shu et all., "A comparative study of digital $\Sigma\Delta$ modulators for fractional-N synthesis," ICECS 2001, vol.3, Sept. 2001.

[17] B. Miller, "A multiple modulator fractional divider," Transactions on Instrumentation and Measurements, 40, 578–583, 1991.

Statistical Performance of IIP2 in Active and Passive Mixers

Markus Voltti*, Tero Koivisto*†, Esa Tiiliharju*

*University of Turku, Department of Information Technology, Microelectronics Laboratory
FIN-20014, Turku, Finland, Email: mjvolt@utu.fi
†Turku Centre for Computer Science, TUCS

Abstract— In this work, we study two down conversion mixers, a passive ring mixer and an active Gilbert mixer from the point of view of even order distortion. First, the effect of mismatch on the IIP2 mean value and deviation of the mixer is studied by Monte Carlo simulations. The simulated results are compared to demands of the WCDMA standard, and the need for second order distortion compensation techniques is discussed. After this, we study how phase and amplitude errors of LO-signals affect the IM2 distortion of the mixer. The simulation results show that without an extra IIP2 compensation technique the mixers cannot meet the linearity requirements demanded by the system.

I. INTRODUCTION

With the increasing demand for wireless devices, there is a need for a highly integrated receiver solution. The direct-conversion receiver is appealing due to the fact that only one LO oscillator is needed and the use of an expensive off chip IF filter can be avoided [1]. These characteristics raise the integration level of the device, which makes the device cheaper to manufacture [2]. The direct conversion technique does have some drawbacks, e.g., DC offset, 1/f noise and a high IIP2 requirement [3], [4].

In this paper two well-known mixer structures, the passive ring mixer and the active Gilbert mixer are investigated. Since WCDMA IIP2 specification demands +60 dBm [1], this study gives special attention to the statistical simulations, which accurately predict reliable performance. Systems with a low IIP2 will suffer from a varying DC offset, which de-sensitizes the receiver [5]. Fully differential structures are used in mixers and therefore in an ideal situation there is no IM2 distortion. In practice, however, systematic and random errors in the manufacturing process cause differences in devices meant to be alike [6], and as a result, even order distortion arises, which deteriorates reception of the desired signal. This paper complements our previous paper on available mixer performance [7] with an estimation of practical available performance.

II. MIXERS

In this section, the passive and active mixers are introduced. The mixers are compared in voltage-mode since inputs and outputs are not matched. The simulations are performed in 0.13-μm nominal CMOS process. The main difference between mixer types is that the active mixer consumes power, whereas the passive one does not. To counterbalance the lack of gain we need more gain for the previous stage in order to keep noise low. From the RF metrics point of view, the main

difference is in the linearity performance and the required LO drive level. In order to switch properly, the passive mixer, in practice, demands a rail-to-rail LO signal. However, the linearity performance is superior as compared to the active mixer. At the end of this section the performance of active and passive mixers will be compared and IIP2 will be studied.

A. Passive mixer

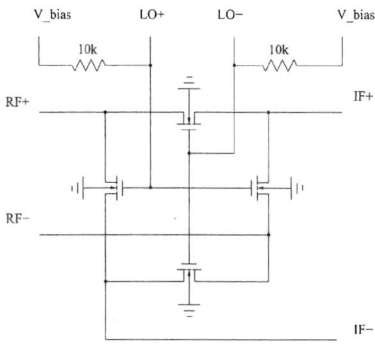

Fig. 1. The Passive Mixer

In Fig.1 the passive mixer is shown. It is easiest to understand the operation of this circuit if the FETs are viewed as switches. At any time, the two FETs on opposite sides of the ring are turned on, and the other two are turned off. In effect, the RF is multiplied by square wave. Since the LO signal must switch the switches on and off, a large LO power is required. The theoretical optimum conversion loss for an ideal double balanced passive mixer is equal to [8]

$$20 \cdot \log \frac{2}{\pi} = -3.9 \text{ dB.} \quad (1)$$

The result of the equation (-3.9 dB) is a theoretical upper limit, which in practise is only achived with ideal switches, which are connected by using square waves.

There is a trade-off between noise figure (NF) and required LO drive. A large W/L ratio will lead to a low NF, however, the required LO grows due to larger transistor capacitances. In this design, the dimensions (W/L) of the transistors are W = 20 μm and L = 0.13 μm. The LO transistors are biased in the vicinity of threshold voltage by providing bias voltage to the switching transistors gates via 10 kΩ resistors. The applied frequencies are 2.1375 GHz for F_{LO} and 2.1325 GHz for

978-1-4244-1983-8/08/$25.00 ©2008 IEEE

F_{RF} with the amplitudes of -9 dBV and -40 dBV respectively. The -9 dBV amplitude of the LO signal is a trade-off between noise figure and linearity. With these values, the passive mixer has -4.6 dB RF-to-IF conversion gain, 9 dB single-sideband NF (SSB@50MHz, only thermal noise will be considered), IIP3 = -4 dBV and P_{-1dB} = -10 dBV. The IIP2 value has not been simulated at this stage, since with nominal components no even order distortion is produced.

B. Active mixer

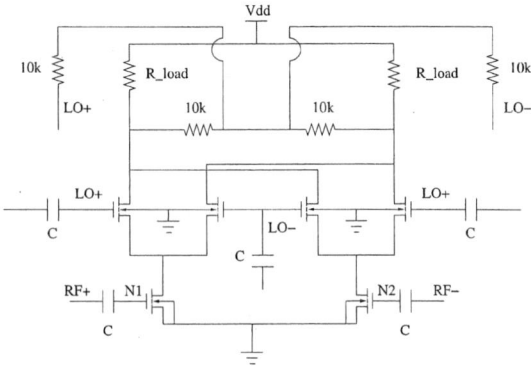

Fig. 2. The Active Mixer

In Fig. 2. the active (Gilbert cell) mixer is shown. The Gilbert double-balanced mixer configuration is widely used in RFIC applications because of its compact layout, good isolation properties and moderately high performance.

$$A_V = (2/\pi)g_m R_{load}, \qquad (2)$$

where R_{load} represents the load resistance and g_m the transconductance of the input transistor.

The switching quad transistor dimensions are W = 20 μm and L=0.13 μm, whereas the dimensions of the RF transistors are W = 100 μm and L = 0.13 μm and the load resistances are 250 Ω. Unlike a passive mixer, all the transistors operate now in the saturation region. With the same signals that were used with the passive mixer above the active mixer gives the following results: 8 dB RF-to-IF conversion gain, 9 dB SSB NF, IIP3 = -12dBV and P_{-1dB} = -21 dBV.

C. Comparison of mixer characteristics

The tabulated values of the mixer metrics are presented in Table I. The results show that with signals of the same frequency and strength the IIP3 value of the passive mixer is 8 dB higher and the P_{-1dB} is 11 dB higher than that of the active mixer. The NF in both mixers is around 9 dB, whereas the conversion gain of the active and passive mixer are 8 and -4.6 dB respectively. It would be possible to improve the passive mixer NF of 9 dB by approximately 3 dB by increasing the amplitude of the LO signal. This is not done, however, because the frequencies and amplitudes of the LO and RF signals are kept the same in order to make comparison possible.

TABLE I

RESULTS OF THE SIMULATED MIXERS.

	Passive Mixer	Active Mixer
Supply voltage (V)	1.2	1.2
DC Power consumption (mW)	0	4.8
RF-to-IF conversion gain (dB)	-4.6	8
SSB Noise Figure (dB)	9	9
P_{-1} (dBV)	-10	-21
IIP3 (dBV)	-4	-11.5
LO amplitude (dBV)	-9	-9
LO frequency (GHz)	2.1375	2.1375
RF amplitude (dBV)	-9	-9
RF frequency (GHz)	2.1325	2.1325

By comparing the conversion gain of active and passive mixer, it becomes clear why the active mixer is more popular than the more linear passive mixer. A direct conversion receiver consists of the RF front-end and the analog baseband part. The front-end consists of the LNA and the mixer. If the gain and noise figure of the all circuit blocks is known, the noise figure of the whole DCR can be calculated with the Friis formula [9]:

$$F_{DCR} = F_{LNA} + \frac{F_{MIX} - 1}{a_{V,LNA}^2} + \frac{F_{BB} - 1}{a_{V,LNA}^2 \cdot a_{MIX}^2}. \qquad (3)$$

In the formula $a_{V,LNA}$ stands for the LNA voltage gain, a_{MIX} stands for the mixer voltage gain, and F_{LNA}, F_{MIX} and F_{BB} for the noise figure of the LNA, mixer and baseband, respectively. It can be seen that to minimize the NF of the whole receiver, the gain of the LNA should be as high as possible.

D. IIP2

The input second order intercept point, IIP2 , is a measure of the second order nonlinearity of the receiver. The IIP2 is defined by the following equation

$$IIP2 = P_{in} + \Delta \qquad (4)$$

where P_{in} is the input power in dBm, Δ is the difference between the fundamental signal component and the second order harmonic distortion. The $IIP2$ value of the RF part of the whole receiver can be calculated using the following formula, provided that the gain and $IIP2$ values of each individual block are known:

$$\frac{1}{IIP2_{DCR}} = \frac{1}{IIP2_{LNA}} + \frac{a_{V,LNA}^2}{IIP2_{MIX}}. \qquad (5)$$

The formula shows that the last stage linearity is the most crucial one for the linearity of the whole receiver. In practise, we can block IM2 distortion of LNA by placing a capacitor between LNA and mixer [5]. This is why we can assume mixer is the dominant source of second order distortion in the receiver.

978-1-4244-1983-8/08/$25.00 ©2008 IEEE

III. STATISTICAL SIMULATIONS

Next we study how mismatch of devices affects the IM2 distortion of the active and passive mixers by using Monte Carlo simulations. The simulation has been performed for 2000 samples with ideal RF- and LO-signals. Based on this simulation, the mean value (μ) of passive mixer is 58 dBV, standard deviation (σ) is 9.6 dBV, minimum IIP2 is 42.3 dBV and the maximum IIP2 is 99 dBV. The corresponding values for the active mixer IIP2 are μ = 34.3 dBV, σ = 9.3 dBV, IIP2min = 14 dBV and IIP2max = 92 dBV, respectively. The simulated values of the mixers are presented in Table II.

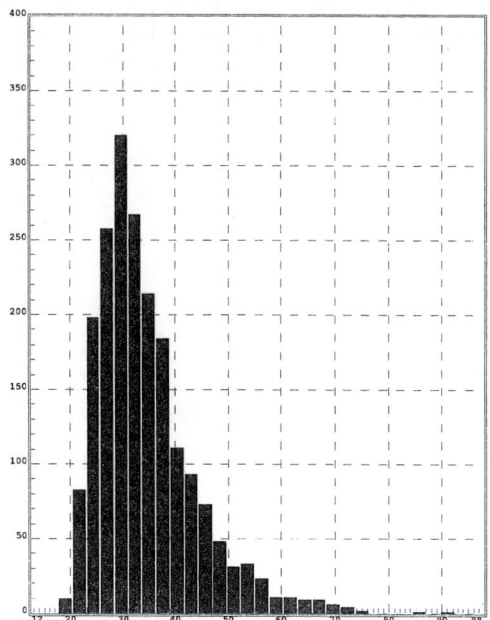

Fig. 4.   IIP2 histogram of the active mixer

TABLE II
SIMULATED IIP2 VALUES OF THE MIXERS

Mixer	(μ) IIP2/dBV	(σ) IIP2/dBV
Passive	57.7	9.2
Active	34.3	9.4

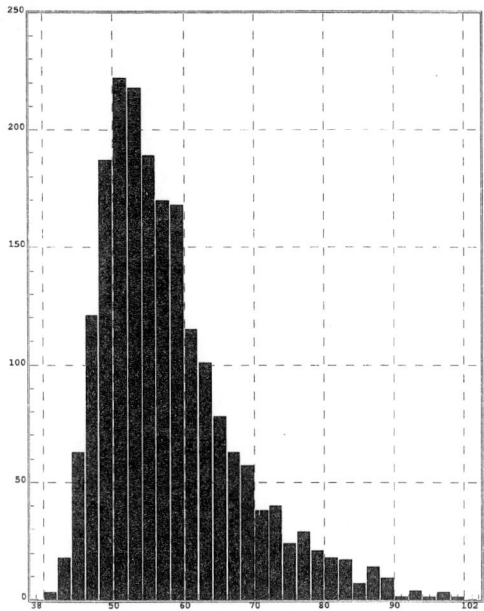

Fig. 3.   IIP2 histogram of the passive mixer

WCDMA standard demands +60 dBm IIP2 performance of the DCR mixer. Because our mixer is not matched, the simulation results are in dBV. 60 dBm converted to dBV is 50 dBV. Fig. 3 shows the histogram of IIP2 for the passive mixer and Fig. 4 for the active mixer. From the histograms it can be estimated that over 90 % of the passive mixers and only 20 % of the active mixers meet the IIP2 requirements of the WCDMA standard without extra IIP2 compensation techniques.

Several IM2 distortion compensation methods have been proposed [3], [9]. According to [7], the dominant source of IM2 distortion is the mismatch of LO quad (with ideal RF and LO signals). One possible way to increase IIP2 performance of the active mixer could be a digital compensation technique between LO transistors. The basic idea would be to remove DC-offset between the LO transistors.

A. Practical available performance

Double balanced mixers are measured using baluns which convert the signal from single-ended to differential. This conversion creates amplitude and phase errors to the signal.

The effect of phase and amplitude errors in the LO branch of the active mixer are studied via simulations. In Fig. 5 the phase error of the LO signal changes -3 degrees to +3 degrees on the horizontal axis. The IIP2 values in the Fig. 5 have amplitude errors of -0.5 dB, 0 dB and 0.5 dB. It can be seen that without the phase and amplitude errors of the LO signal, it is possible to reach a high IIP2 value with nominal components. Once a phase error is introduced to the LO signal, the IIP2 value drops quickly. In Fig. 5 the width of its peak on the left has a -0.5 dB LO amplitude error. A clear peak can be seen but instead of 0° it has moved to -0.8°. With a +0.5 dB amplitude error the peak has shifted to +0.8°. One can deduce from the Fig. 5 that the phase error of LO can be fixed to some extend with the amplitude error of LO signal. These simulations show that with phase and amplitude errors of 3° and 0.5 dB the IIP2 value drops to 66 dBV.

Next, we performed the same simulations than previously using ideal LO signal and non-ideal RF-signal. With ideal LO signal and non-ideal RF signal, we observed that the degradation of IIP2 was negligible when RF amplitude unbalance was 0.5 dB and RF phase unbalance was 3°. However, unbalance in RF signal is not completely negligible since in the case with non-ideal LO and RF signal the IIP2 performance of active mixer is substantially lower than in the case of ideal RF signal and non-ideal LO signal. According to [10], amplitude and phase errors of baluns are on the order of 1 dB and 6° at 2 GHz respectively, which makes it difficult to measure IIP2 accurately.

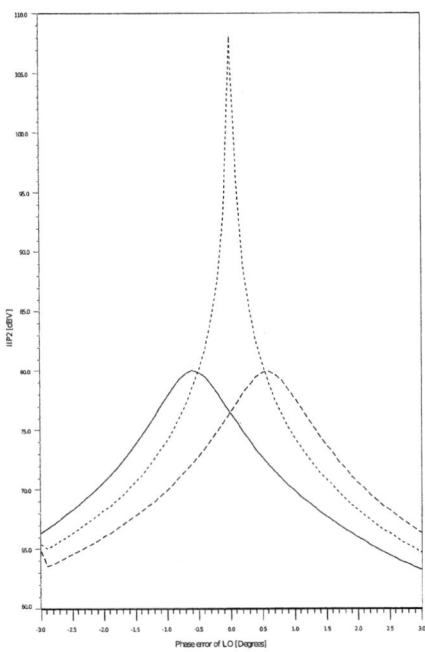

Fig. 5. The IIP2 values with amplitude errors of -0.5 dB, 0 dB and 0.5 dB.

IV. CONCLUSION

This work has studied the active and passive mixers. The IIP2 histograms of both mixers have been simulated. The conclusion is that only 20 % of the active mixers attain to demands of the WCDMA standard. The corresponding figure for the passive mixers is over 90 %. These results indicate that without the external IIP2 compensation technique it is very difficult to achieve a good yield. In addition, it has been discovered that the phase and amplitude error of LO and RF signals affect the IM2 distortion of the active mixer substantially, which makes it difficult to measure IIP2 accurately.

REFERENCES

[1] M. Brandolini, P. Rossi, D. Manstretta, and F. Svelto, "Toward multi-standard mobile terminals - fully integrated receivers requirements and architectures," *Microwave Theory and Techniques, IEEE Transactions on*, vol. 53, pp. 1026–1038, 2005.
[2] B. A. Floyd, S. K. Reynolds, T. Zwick, L. Khuon, T. Beukema, and U. R. Pfeiffer, "WCDMA direct-conversion receiver front-end comparison in RF-CMOS and SiGe BiCMOS," *Microwave Theory and Techniques, IEEE Transactions on*, vol. 53, pp. 1181–1188, 2005.
[3] E. E. Bautista, B. Bastani, and J. Heck, "A high IIP2 downconversion mixer using dynamic matching," *Solid-State Circuits, IEEE Journal of*, vol. 35, pp. 1934–1941, 2000.
[4] M. Hotti, J. Ryynänen, and K. Halonen, "RC-load analysis of the downconversion mixer IIP2," in *Circuit Theory and Design, 2005. Proceedings of the 2005 European Conference on*, vol. 1, 2005, pp. I/237–I/240 vol. 1.

[5] B. Razavi, "Design considerations for direct-conversion receivers," *analog and Digital Signal Processing, IEEE transactions on circuits and systems-II,*, vol. 44, pp. 428–435, 1997.
[6] P. R. Kinget, "Device mismatch and tradeoffs in the design of analog circuits," *Solid-State Circuits, IEEE Journal of*, vol. 40, pp. 1212–1224, 2005.
[7] M. Voltti, T. Koivisto, and E. Tiiliharju, "Comparison of active and passive mixers," *European Conference on Circuit Theory and Design, ECCTD'07*, 2007.
[8] V. Gefferoy, G. De Astis, and E. Bergeault, "RF mixers using standard digital CMOS 0.35um process," *analog and Digital Signal Processing, IEEE transactions on circuits and systems-II,*, vol. 44, pp. 428–435, 2001.
[9] K. Kivekäs, "Design and characterization of downconversion mixers and the on-chip calibration techniques for monolithic direct conversion radio receivers," Ph.D. dissertation, Helsinki University of Technology, P.O.Box 3000, HUT-02015, 2002.
[10] E. Tiiliharju and K. A. I. Halonen, "An active differential broad-band phase splitter for quodrature-modulator applications," *Microwave Theory and Techniques, IEEE Transactions on*, vol. 2, pp. 679–686, 2005.

Fully integrated coarse-fine wideband distributed Voltage Controlled Oscillator

F. Cannone, G. Avitabile, D. Cascella

Department of Electrical and Electronics Engineering
Politecnico di Bari
Bari, ITALY
f.cannone@poliba.it

Abstract—**A new fully integrated distributed voltage controlled oscillator is presented. The proposed topology allows wide tuning range and fine–coarse tuning. The design of an integrated DVCO prototype is discussed and results are reported showing good performance in terms of phase noise, tuning range and power consumption.**

INTRODUCTION

The modern TLC systems support multiple standards, as in the case of mobile telephony. This kind of situations implies that the frequency synthesis must cover wide bandwidths usually divided in sub-ranges, each of them being determined by the reference standards (GSM, PCN, EDGE, etc.). The wanted VCOs have to deal with a difficult trade-off between wide tuning bandwidth and phase noise, normally requiring low K_{vco}s and as linear as possible for PLL use.

Distributed VCOs (DVCO) have been proposed as an alternative approach to ring oscillators and LC tank architectures in order to obtain a wide tuning range and good phase noise performances[1][2]. The DVCO is arranged around a distributed amplifier in a phase shift configuration, in this way the DVCO could benefit of the extremely wide bandwidth of the amplifier. Moreover, the performances of passive devices in standard silicon technologies at frequencies higher than ten gigahertz do not improve at the same rate of the active devices with the advancement of technology. For these reasons, the D-VCOs allow to design fully integrated VCOs which operate at higher frequencies compared with the integrated VCO-LC for a specific technology[3]. The literature proposes several examples of such organization, while these solution take advantage of the possibility to avoid passive devices, due to both topologies and tuning techniques used, none of these examples exploits completely the potentiality of such class of oscillators in terms of wide tuning range. It is interesting to take advantage of the wide band of the distributed amplifier in order to design a multi-band VCO able to cover different band of different generations of wireless standards used into the modern telecommunication services. The topologies proposed in [4][5] are novel architectures of DVCO based on the distributed amplifier topology, which provide the capability to design a wide tuning range VCO.

The new topology introduced fully exploits the potentiality of the distributed amplifier design, obtaining a very wide bandwidth joint to extremely uniform performance all along that bandwidth as shown in [6]. The general condition of oscillation to design this novel topology is presented in [7].

This paper describes the first fully integrated wide tuning range DVCO based on the novel idea presented in [5] designed in S34D4M5 Austriamicrosystem process.

GENERAL CONDITION OF OSCILLATION

The topology of DVCO contains two blocks: the distributed amplifier followed by a synthetic line (fig.1). In order to determine the general condition of oscillation it is necessary to calculate the open loop gain[7]. The open loop gain is the product of the gains of each stage. The gain of the Distributed Amplifier(DA) is expressed by the following expression

$$A_{DA} = -\frac{G_m Z_d}{2}(1+\Gamma_{SL}) \cdot e^{-(\gamma_d l_d + \gamma_g l_g)} \frac{e^{-\gamma_d n l_d} + e^{-\gamma_g n l_g}}{e^{-\gamma_d l_d} + e^{-\gamma_g l_g}} \quad (1)$$

where G_m is the transconductance of the transistors; Z_d, γ_d, l_d are the characteristics impedance, the complex propagation constant and the single segment length of the drain line; γ_g and l_g are the complex propagation constant and the single segment length of the gate line, n is the number of transistors; Γ_{SL} is the reflection coefficient at the input of the synthetic line. By using the relation between the values of voltage along a transmission line, it is possible to express as follows the gain of the second stage

$$A_{SL} = \frac{(1+\Gamma_L) \cdot e^{-m\gamma_{sl}}}{(1+\Gamma_L e^{-2m\gamma_{sl}})} = \frac{(1+\Gamma_L) \cdot e^{-m\gamma_{sl}}}{(1+\Gamma_L^m)} \quad (2)$$

978-1-4244-1983-8/08/$25.00 ©2008 IEEE

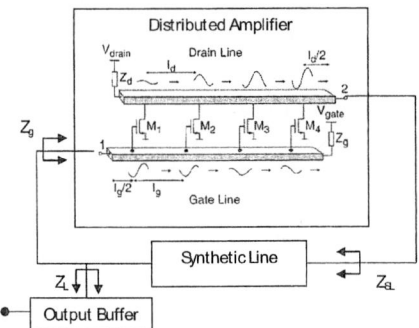

Figure1. DVCO topology

where Γ_L is the reflection coefficient at the output section of the synthetic line while γ_{sl} is the complex propagation constant of the synthetic line, the expression of γ_{sl} depends on the type of cell used into the synthetic line[8]; m is the number of cells used into the synthetic line.

Finally the general condition of oscillation is expressed by the following equation

$$\frac{(1+\Gamma_L)\cdot e^{-m\gamma_{sl}}}{(1+\Gamma_L^m)}\cdot\frac{G_m Z_d}{2}(1+\Gamma_{SL})\cdot e^{-(\gamma_d l_d + \gamma_g l_g)}\frac{e^{-\gamma_d n l_d}+e^{-\gamma_g n l_g}}{e^{-\gamma_d l_d}+e^{-\gamma_g l_g}}=-1\cdot \quad (3)$$

In the case of $\gamma_g l_g = \gamma_d l_d = \gamma l$ the previous equation becomes

$$\frac{(1+\Gamma_L)\cdot e^{-m\gamma_{sl}}}{(1+\Gamma_L^m)}\cdot n\frac{G_m Z_d}{2}(1+\Gamma_{SL})\cdot e^{-\gamma(n+1)l}=-1\cdot \quad (4)$$

This relation provides a good approximation of the oscillation condition. It has been used to design the DVCO presented in the next section.

DESIGN OF THE INTEGRATED DVCO

The DVCO has been designed in 0.35μm SiGe Austriamicrosystem technology. The single gain stage of the DA is constituted by a npn transistor biased through an active load. In the range of considered frequencies, it is possible to make the transmission lines joining the active stages as synthetic lines. Taking into account the input and output impedances of each transistor, it has been possible to use a single cell of synthetic line to connect the gain stages. In fig.2 is shown a generic elementary cell of a synthetic line.

A. Single cell topology

A design issue is the suitable choice of the topology of the elementary cell in order to achieve the better performances of the DA in terms of matching and cutoff frequency. The simpler topologies are the T-cell and Π-cell. Considering the electrical equivalent model of an integrated inductor, the Π-cell allows to obtain simultaneously two advantages: to reduce the number of inductors and to facilitate the matching. While the first advantage is an intrinsic feature of this topology, the second

advantage comes directly from the electrical equivalent model of an integrated inductor (fig.3). Indeed, it is intrinsically a Π-cell, then, the parasitic elements can be absorbed completely into

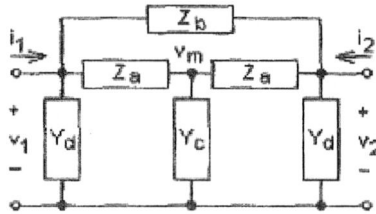

Figure 2. Generic cell of synthetic line

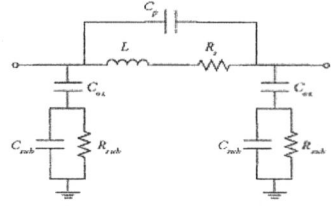

Figure 3. Model of shielded integrated inductor.

the basic cell components that constituted the synthetic line, in the same way used to absorb the input and output impedances of the transistors. Differently, in the case of T-cell based line, the parasitic capacitances at the edges of the synthetic line cannot be absorbed. For these reasons it is possible, by using Π-cells, to obtain more predictable transmission lines. The equivalent capacitance at the internal nodes of the synthetic lines can be expressed as follows

$$C_{int} = 2C_{p,ind} + C_{in/out} \quad (5)$$

Where $C_{p,ind}$ is the parasistic capacitance of the integrated inductor (the capacitance seen from each node to ground), and $C_{in/out}$ is the input/output capacitance of the transistors. The Π-cell topology line ends with a capacitance C_{ext} equal to C/2 provided by the $C_{in/out}$ and $C_{p,ind}$, then in order to have a perfect matching also at the edge of the synthetic line it is needful to add at each internal nodes a capacitance equal to $C_{in/out}$. In this way C_{int} is exactly two time the capacitance at the edge, C_{ext}. Therefore, for a wanted characteristic impedance it is necessary to increase the inductance of the cell consuming more area. Higher values of L and C reduce the value of cutoff frequency of the synthetic line. In general this reduction is bigger for the gate/base transmission line, in that $C_{\pi/gs}$ is bigger than the capacitance at the output node. This condition does not occur in any case. For example, in a design based on MOSFETs, if the size of device is big enough, then the parasitic capacitance source/drain-well/substrate make C_{out}

978-1-4244-1983-8/08/$25.00 ©2008 IEEE

comparable and sometimes greater than C_{in}. In general, the reduction of the cutoff frequency affects in the same way both the transmission lines. Indeed, in order to achieve the same phase velocity along the transmission lines, usually it is necessary to add a capacitance into the cell of the synthetic line which has a lower value of capacitance to compensate the value. The choice between the T-cell and Π-cell depends principally on the operating frequencies and on the input and output impedances of the transistors. At low frequencies and small values of capacitance, it is possible to choose the Π-cell topology in order to achieve maximum matching and cutoff frequency still high enough. In other situations it is possible to use the T-cell topology in order to maximize the cutoff frequency and characteristic impedance of the synthetic lines, admitting a little mismatch at the terminations of the DA. The transmission lines must be designed to obtain the same phase velocity along the lines into the DA, allowing to achieve the maximum gain. In the implementation of the presented DVCO, the following choices have been made: to use a Π-cell topology to reduce the number of inductors and to accepting a little mismatch at the edge of the synthetic lines, (a capacitance equal to $C_{in/out}$ has not been added at each internal node). In this way it has not been necessary to increase the inductances saving chip area and maintaining high the cutoff frequency.

B. Bias point

Another design issue is the better bias point. It is important to consider the choice of the bias point in order to reduce the harmonics of the generated signal inside the DA. Indeed, unlike in the case of classical DVCO, there is not an action of filtering inside the DA itself, the filtering which reduces the harmonics is made this time by the second block. Although the output of the DVCO is also in this case a clean waveform, it is important, in order to stabilize the oscillation, to use a suitable bias point.

As mentioned before the second block is a synthetic line, the phase shift introduced by this block is set by modifying the value of the phase velocity along this line. This change varies the cutoff frequency, f_{cSL}, of the line as well. Then, there is a direct relation between the filtering performed by this block and the phase shift needed for the wanted frequency of oscillation. The greater is the phase shift introduced by the second block, the smaller is the cutoff frequency and the higher is the filtering. For this reason the filtering is theoretical slightly more effective for frequencies of oscillation near the lower edge of the range of synthesis.

To separate the bias network to the signal path, MIM(Metal-Insulator-Metal) capacitances equal to 4 pF have been used. We decided to use MIM capacitance instead of poly-poly capacitances not only for the higher Q factor but primarily for the achievable big reduction of the bottom plate-well/substrate parasitic capacitance. Indeed, this has the same unwanted effect of the parasitic components of the inductors and it can lead to a reduction of the band of synthesis.

To reduce the phase noise, in agreement with the Hajimiri and Lee's time variant model[9], the DA has been designed to have an as symmetric as possible waveform at its outputs (lower conversion of flicker noise) with as low as possible harmonics level (lower conversion of thermal noise. Really, the direct linking between noise and phase noise is the Impulse Sensitivity Function (ISF), but this function depends on the spectrum of the periodic signal at the node). Taking into account the high level of the signal on the drain(collector) line into the DA and the low equivalent load impedance, it has been needed to set a bias current high enough to allow the complete swing on the output node of the DA in order to reduce the DC component into the produced signal. Moreover a degeneration resistance is been used to reduce the gain of each stage and in this way the harmonics. These choices improve the quality of the signal at the output of the DVCO. The drawback is the increase of the power consumption, however this is an intrinsic problem of the DAs.

C. Output buffer

Another design issue is how to take the output signal on a fifty ohms load. The output of the DVCO is the output of the second block that is the input of the DA. Then the output is taken on a node of a transmission line which has a given characteristics impedance. If the fifty ohms load is connected directly to the output node, it represents a loss for the line able to switch off the oscillation. For these reasons, in this case it is necessary to complicate the matching design of the transmission lines in order to consider this load and still fulfill the Barkhausen' criterion. Another possibility is to use an output buffer. Indeed, its high input impedance corresponds to a negligible loss along the transmission line and so this impedance do not interfere in the matching design and it is possible to match the structure by simply setting $Z_{g/b} = Z_{d/c} = Z_{SL}$.

D. Tuning range

After defined the better possible matching the value L and C of the second block are fixed. The tuning is made by varying the value of the capacitance C. This change modify besides the phase velocity also the characteristic impedance of the line decreasing slightly the matching. It is important to take into account this when the DA is designed. Since the gain falls at higher frequency, it is necessary to assure the maximum matching at the upper limit of the synthesis range and to accept a little mismatch at lower frequencies.

At frequencies where there is not a perfect matching as expressed in (1) the frequency of oscillation depend on the reflection coefficient at the input of the synthetic line. It is possible to exploit this relation in order to increase the tuning range for a given variation of the capacitance C of the cell into the synthetic line. This increase works until the Barkhausen criterion is satisfied.

In order to demonstrate the achievable wide tuning range, the variable capacitances $C(V_{tune})$ into the cells of the synthetic line have been implemented by using varactors A-MOS which allow a big ratio C_{max}/C_{min}. The drawback of this choice is a higher phase noise at the center band of synthesis due to the high value of the K_{VCO}. The DVCO has been designed and its performances have been simulated by using Cadence Spectre. As shown in fig. 4, the DVCO provides a tuning range of 400MHz centered at 1.5GHz. In the region from 1.34GHz to 1.60GHz (260MHz) the K_{VCO} presents a change similar to the variation of any commercial varactors. The output power has a variation less than a 1dB into the band of synthesis moreover, phase noise performances and harmonics suppression are good and uniform into the tuning range. In fig. 5 and fig. 6 are reported the harmonics suppression and phase noise at center of tuning range respectively which represents the worst case in the results provided by the DVCO. The prototype draws 6 mA from 3.3V . The size of the DVCO is (2,2x1.8) mm^2 .

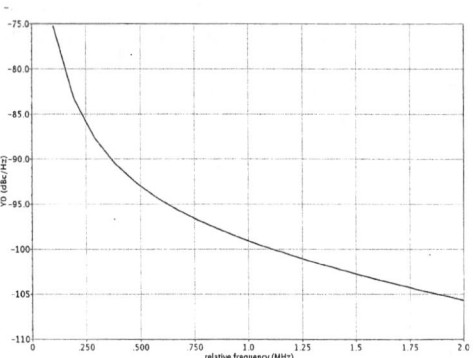

Figure 6. Phase noise at the center of tuning range.

CONCLUSION

A new approach to integrated wideband DVCO design has been described and validated through the post layout simulation results of a first fully integrated DVCO based on the AMS S35D4M5 process.

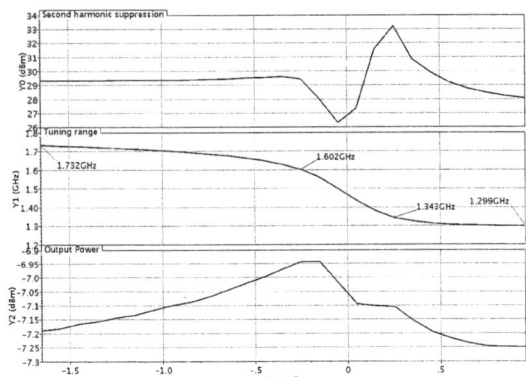

Figure 4.   Post layout simulation results. (up) Second harmonic suppression. (middle) Tuning range. (down) Output power.

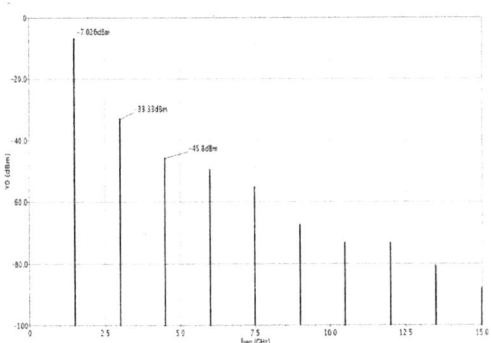

Figure. 5 Harmonics suppression at center of tuning range.

REFERENCES

[1]   Z. Skvor, S.R. Saunders and C. S. Aitchison, "Novel decade electronically tunable microwave oscillator based on the distributed amplifier , " Electronics Letters, vol. 28, no. 17, pp. 1647-1648, August 1992.

[2]   Hui Wu and A. Hajimiri, "Silicon-based distributed voltage-controlled oscillators, "IEEE J. of solid-state circuits, vol. 36, no. 3, pp. 493-502, March 2001.

[3]   G.P Bilionis, A.N. Birbas, M.K. Birbas," Fully integrated differential distributed VCO in 0.35-μm SiGe BiCMOS Technology, " IEEE Transaction on Microwave Theory Techniques, vol.. 55, no. 1, January 2007.

[4]   Z.A.Shaik, P.N.Shastry(S.N.Prasad),"A Novel Distributed Voltage-Controlled Oscillator for Wireless Systems,"IEEE Proc. of Radio and wireless symposium, pp. 423-426, January 2007.

[5]   G. Avitabile, F.Cannone, M.Capodiferro, L.Carella, N.Lofù, "A coarse-fine, wideband distributed voltage controlled oscillator for wireless applications, "Electronics Letters, vol. 42. no. 5, pp. 486-487, March 2006.

[6]   F. Cannone, G.Avitabile, N.Lofù, " New wideband distributed voltage controlled oscillator with a coarse-fine tuning, IEEE Proc. of the 9th European. conf. on wireless tech., pp.302-305, September 2006.

[7] G.Avitabile, F.Cannone, G. Coviello, D. Cascella," Analysis and design of a new architecture coarse-fine wideband distributed voltage controller oscillator," Proc. of international symposium on microwave and optical tech., pp. 225-228, December 2007.

[8]   Douglas R. Jachowski, Clifford M. Krowne, " Frequency Dependent Of Left-Handed And Right-Handed Periodic Trasmission", Proc of MTT-S., pp. 1831-1834, 2004

[9]   A.Hajimiri, T.Lee,"A General Theory of Phase Noise in Electrical Oscillators", IEEE J. of solid-state circuits, vol. 33, no. 2, pp.179-194, February 1998.

A Matching Circuit Tuned, Multi-Band (WLAN and WiMAX), Class – A Power Amplifier Using 0.25μm-SiGe HBT Technology

Mehmet Kaynak*, Ibrahim Tekin and Yasar Gurbuz

Sabanci University, Faculty of Engineering and Natural Sciences, Tuzla, 34956, Istanbul, Turkey
Tel: ++90(216) 483 9533, e-mail: yasar@sabanciuniv.edu

Abstract— In this work, a MOS based output matching network is designed and fabricated using IHP (Innovations for High Performance), 0.25μm-SiGe HBT process and measured which can give 4 different impedance values. Also, a multi-band, Class-A, power amplifier (PA) has been designed with same technology and the desired output impedances for matching network are taken from the load-pull simulation results of this PA. The behavior of the amplifier has been optimized for 2.4 GHz (WLAN), 3.6 GHz (UWB-WiMAX) and 5.4 GHz (WLAN) frequency bands for high output power. Multi-band characteristic of the amplifier was obtained by using MOS based switching network. Two MOS switches are used for changing the behavior of the matching network and 4 possible states are achieved. Post-Layout simulation results of the PA circuit provided the following performance parameters: output power of 28-dBm, gain value of 26-dB and efficiency value of %19 for the 2.4 GHz WLAN band, output power of 28-dBm, gain value of 22-dB and efficiency value of %20 for the 3.6 GHz UWB-WiMAX band, and output power of 27-dBm, gain value of 23-dB and efficiency value of %17 for the 5.4 GHz WLAN band.

Keywords— Power Amplifier, SiGe, Dual-Band, WLAN, WiMAX,

I. INTRODUCTION

Power Amplifiers (PAs) are key components in the wireless communications industry, which consume a great amount of power in overall transceiver. The PAs must achieve high operation efficiency in order to maximize the battery life and minimize the size and cost. The general PA design focuses on the whole system architecture and should achieve advantages in terms of performance, cost, and size. The integration of analog and digital blocks on the same PA chip is becoming increasingly important in all transceivers [1].

For the past several years AlGaAs/GaAs heterojunction bipolar transistor (HBT) technologies are currently the preferred bipolar technologies for the commercial development of linear power amplifier (PA) modules for wireless handsets due to their excellent linearity and power-added efficiency (PAE) [2]. However, these types of integrated circuits are relatively expensive and significant efforts are being exerted to displace these HBT technologies with alternative bipolar technologies, such as Silicon (Si) bipolar junction transistor (BJT), Silicon Germanium (SiGe) HBT [2]. SiGe/Si HBTs are more attractive primarily due to their high substrate thermal conductivity (150

W/m- C), comparable device performance (30 GHz and 50 GHz), lower emitter/base turn-on voltage (0.75 V), and significantly lower production cost [3]. Unfortunately, SiGe/Si HBTs have their own disadvantages: the substrate is very conductive, which adds significant parasitics to both active and passive components of the power amplifier. SiGe HBTs also have relatively low breakdown voltages (~ 4 - 7 V) [3].

Future mobile terminals are expected to operate in a wireless environment which combines a variety of wireless systems. This means that RF circuits would have to support various wireless environments, supporting multi-band/frequency operations. This brings additional challenges for RF circuits, supporting these platforms, especially for Power Amplifiers. One of these is the technology, used to realize the RF Circuits, compatible with integration of other wireless components/systems, able to provide both multi-frequency operations while also delivering the required power for each frequency at the desired quality, etc. Finding the technology that can provide solutions to some of this problem would be a great achievements and interest to multi-band wireless systems.

In data communication, WLAN and WiMAX are the two main communications standards. IEEE 802.11 b/g standards which operate 2.4 GHz are very close to the WiMAX operating frequency, 3.6 GHz. Multi-Band operation could be performed by sweeping the frequency interval using varactor which the capacitance of these varactors can be adjusted by applying DC voltage [4]. However, the upper band of the WLAN which the standard called IEEE 802.11a is moderately far away from these frequency ranges. Tuning components could not have large tuning ranges and not sufficient enough to cover large frequency ranges. Switches are the solution for getting rid of this problem by switching the different output matching networks [5]. However, realizing switches on Si-substrate is the main problem and some other techniques such as off-chip matching network topologies which include PIN diodes for switch mechanisms, MMICs, and RF MEM switches are commonly used in literature [6]. Moreover, MOS switches could also be used for changing the output impedance of the PA. In this work, MOS switch based second-order output matching network is designed, fabricated and measured as an switch-able output matching component. The desired impedance values are adjusted using two MOS switches for multi-band operation. Furthermore, Class-A, power amplifier

---

* M. Kaynak was with the RFIC group of Sabanci University, Istanbul Turkey. He is now with the Technology department of IHP Microelectronics, Frankfurt (Oder), 15236 Germany.
(e-mail: kaynak@ihp-microelectronics.com)

978-1-4244-1983-8/08/$25.00 ©2008 IEEE

(PA) has been designed using IHP's 0.25μm-SiGe HBT process technology which is the same technology with the output matching network, and waiting to fabricate in next run of IHP. The performance of this PA is simulated with the impedance levels which the fabricated output matching network could give. The measurement results show that two MOS switches which is able to give four different impedance values is adequate for multi-band PA operation and could be considered as a solution for multi-band RF front-end systems.

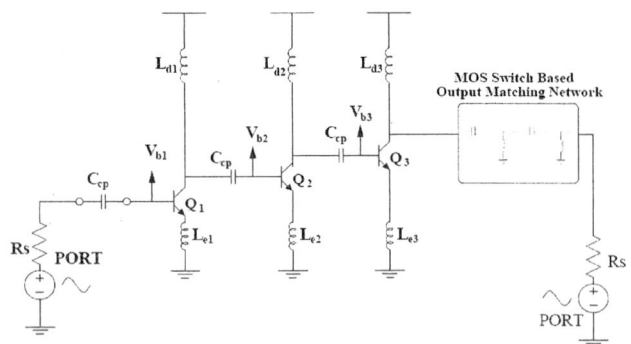

Figure 1 Three-Stage PA Schematic

II. DESIGN OF MATCHING CIRCUIT AND MULTI-BAND PA

Figure 1 shows the schematic view of the three-stage power amplifier. The design of the PA started with selecting the appropriate transistor in IHPs technology library. IHPs SiGe:C high voltage transistor has f_{max} of 70 GHz, β of 190 and collector-emitter breakdown voltage of 7V which is suitable for high-power, high-frequency circuit designs [7]. Transistor gives the maximum β while driven with 50-150 μA base current which also gives the maximum f_T values. DC operating point of the transistors also specifies the parasitic components and also changes the input and the output impedance of the transistor. The sizes of the transistors were adjusted to achieve the desired power level controlled with DC current level. For high output power levels, a higher-level DC current is required. This has been one of the major challenges of RFIC technologies due to the limitation on achievable metal-thickness.

A three stage topology for achieving the desired specifications of the power amplifier while performing a dual-band operation was chosen and presented in Fig. 1. Here in this topology, the first stage was used as a driving stage and also performs an impedance matching with the source. The use of a driver stage in high power amplifiers applications is suggested because the transistors generally don't give the maximum output power and maximum gain at the same output impedance [8]. The second stage is used as a gain stage while the third stage was for power delivery which could handle large amount of power.

To achieve a higher output power levels, larger sized transistors are necessary. However, this usage decreases the input impedance of the transistor and brings difficulties in input impedance matching to 50 Ω source. Also, increasing the maximum output power level may require a decrease in the gain of the amplifier. This can be explained by different required impedance values for load-pull simulations, specifying the maximum output power and gain-circles for the maximum gain conditions. Furthermore, multi-stage PAs includes a last stage with higher output power level and a high-gain driver stage. In this design, three-stage PA topology, as presented in Fig. 1, is used for achieving high-output power and high gain, simultaneously for the requirements of WLAN and WiMAX applications.

Load-Pull simulations are very important for finding the adequate impedance values which provides the maximum output power. However, since the PA is designed for multi frequency bands, load-pull simulations should be repeated for each band. This also specifies the required output matching network component values. Load-Pull simulation results for each band are shown in Fig. 3. The maximum output powers can be achieved with low impedance values, as expected. Specified impedance values are 5-10 Ω pure real for 2.4/5.2 GHz frequency bands and 5-10 Ω real, 0-10 capacitive imaginary part for 3.6 GHz band. These impedance values are not far away from each but the problem is the constant passive component value behaving differently with a change in frequency.

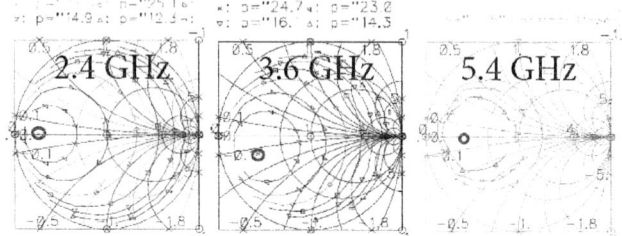

Figure 2 Load-Pull Simulations

For achieving a multi-band operation in the circuit, tuning or switching passive components is necessary for changing the impedance values at different frequencies. MOS-based varactors can be used for sweeping the bands and also MOS-based transistors can be used as switches for switching the matching networks. Varactors can be used for changing the capacitance values, directly affecting the matching circuit behavior. Although the tuning range of MOS varactors is limited, for close frequencies, it could be enough for achieving required output impedances [4]. As the applications with not close operating frequencies, some passive components are switched on and off for changing the effective impedance of the output matching network. In this study, MOS transistors are used as a switch for switching the matching network which provides multi-band operation of PA in the transmitter part. The topology of the output matching network is shown in Fig. 3. By switching on and off of the MOS transistors, the effective capacitance value could be changed. This topology allows avoiding the inductance losses which affects extremely when the inductors are used in series way.

In this work, a switching type output matching network is designed in order to maximize the output power of a single power amplifier in 2.4/3.6/5.4 GHz frequency bands which is illustrated in Fig.3. The main idea is to change the series capacitance values of the second order filter and achieve the desired output impedances in both frequencies. In order to specify the desired output impedance for maximize the performance of the power amplifier, load-pull simulations are performed to approximate the power contours of the amplifier in both frequencies.

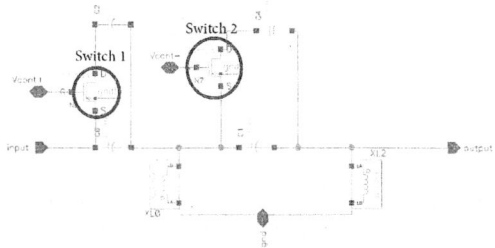

Figure 3 Schematic view of the matching circuit

After determining the optimum values of the output impedances, second-order filter is designed for both frequencies considering the change in series capacitances. Due to this, the fixed inductance values are specified and the switching capacitance values are determined for each frequency. In order to achieve different capacitance values for each frequency, MOS switches are designed. MOS transistor sizes are determined with the purpose of minimizing the insertion loss in on-state and maximizing the isolation in off-state of the switch. Also, the off and on state capacitance (parasitic capacitance of the MOS transistor) and on-state resistance are modeled using the high-frequency small-signal model of the nMOS for optimizing the switches. Power handling problem of the MOS switches are improved by using the MOS transistors in parallel way to the matching network. Post-Layout simulation results show that the MOS transistor can handle up to 24 dBm power with degrading linearity but the power levels when the transistors are used in parallel way could reach only 15-18 dBm.

The matching circuit has been fabricated in 0.25μ SiGe:C BiCMOS technology. Fig. 4 shows a photograph of the die, whose active area is approximately 500μm×540μm. All the measured S-Parameter results of matching network are shown in Fig. 5 for each frequency range. The measured data are taken from Agilent 8720ES Network Analyzer using Karl-Suss RF Probe Station. The simulated and measured data are also analyzed in Fig. 6.

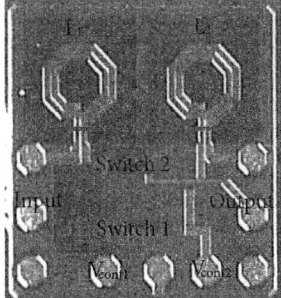

Figure 4 Die Photo of the Matching Circuit

The main idea of the switching is making the switch on and off to change the effective value of the capacitance. Switch1

switches 5 pF capacitor on/off and Switch2 switches 2.2 pF capacitor on/off. Detailed state of the switches for each frequency and the measured impedance values for these states are shown in Table1.

Table 1 Impedance values for Different States

	Switch 1	Switch 2	Impedance (Measured)
State 1 (2.4 GHz)	Short	Open	Z=9-j0.8 (@ 2.4 GHz)
State 2 (3.6 GHz)	Short	Short	Z=7.5-j6 (@ 3.6 GHz)
State 3 (5.4 GHz)	Open	Short	Z=12-j0.4 (@ 5.4 GHz)

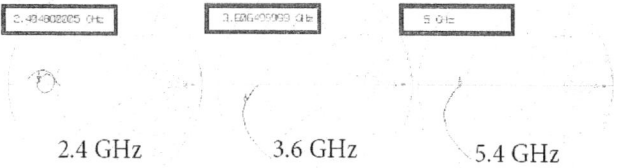

2.4 GHz   3.6 GHz   5.4 GHz

Figure 5 Measured Data for each frequency (S_{11})

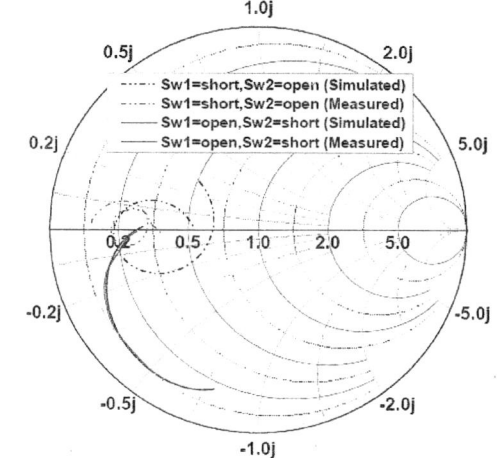

Figure 6 Simulated vs. Measured Data (S_{11})

As illustrated in Fig. 5, the required impedance values from the load-pull simulations can be approximately achieved. Switching the matching networks now gives the optimum impedance values for achieving the maximum output power level for the last stage of the PA. Second stage is designed as a typical gain stage with a gain value of about 10 dB. The important part while designing the second stage is the impedance matching between the second and last stage. Because the input impedance of the last stage is very low due to the large size transistors, achieving high gain in the second stage might be difficult without a using matching circuit between the second and last stage. Designing the first stage is similar with the second stage as it performs input matching of the overall PA and intermediate matching between the first stage and second stage. This stage also designed for high gain of about 10 dB. All three stages include a degeneration inductor for increasing the linearity. For high power applications, the handled current levels of inductors are very high and should be selected as low-resistance as possible to avoid higher voltage drop. Also, inductors should also be able to handle high DC and AC current levels.

III. LAYOUT AND POST-LAYOUT SIMULATION RESULTS OF MULTI-BAND PA

Design and simulation of the power amplifier is performed using Cadence® and Agilent Design System (ADS)® environments, supported by IHP technology library. Layout of the PA circuit is shown in Fig. 7, occupying an area of 1 x 2 mm² and waiting the next run of IHP to be fabricated. Some layout techniques, such as stacked metal layers, are used for increasing the current handling capability of these metal paths in IC technologies.

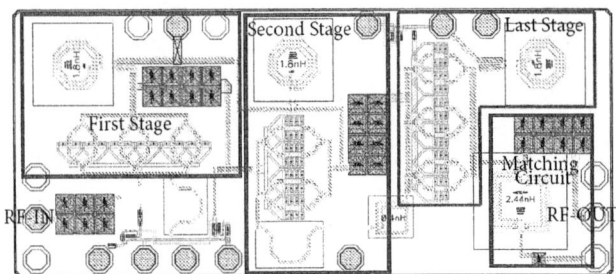

Figure 7 Layout of the three-stage PA

The power levels of the three-stage power amplifier are shown in Fig. 7 for 2.4 GHz frequency and all the specifications of the multi-band PA for each frequency are summarized in Table.2. All the results are taken from the post-layout simulations of the PA.

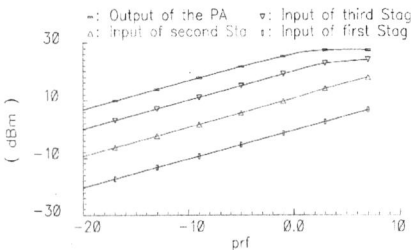

Figure 8 Power levels of overall PA at 2.4 GHz

Table 2 shows that, for the entire bands, the desired maximum output power of above 25 dBm is achieved. Switching the output matching networks is not only improves the maximum output power, but also improves the efficiency so much. Also, the S_{22} data is enough for transmitting the generated power. Because the lack of switching network in input of the amplifier, S_{11} of the PA is only lower than -9dB for each band but can be improved using a switch-able input matching network.

Table 2 Overall PA Specifications

Freq. (GHz)	Output Power, @ 1-dB comp. point (dBm)	Gain (dB)	PAE (%)	S_{11} (dB)	S_{22} (dB)
2.4	28	26	19	-12	-18
3.6	28	22	20	-9	-14
5.4	27	23	17	-10	-16

IV. CONCLUSION

In this study, a design methodology of a multi-band PA, using MOS switch based output matching network was described. Measurement results show that, using the MOS transistors as a switch is an adequate way for achieving the different desired impedance values for different frequencies. Using the measured impedance values, Post-Layout level simulation results of the PA circuit provided the following performance parameters: output power of 28-dBm, gain value of 26-dB and efficiency value of %19 for the 2.4 GHz WLAN band, output power of 28-dBm, gain value of 22-dB and efficiency value of %20 for the 3.6 GHz UWB-WiMAX band, and output power of 27-dBm, gain value of 23-dB and efficiency value of %17 for the 5.4 GHz WLAN band. These results proved that the more impedance values could be achieved using more switches and increasing the possible states of matching network.

ACKNOWLEDGMENT

This work was performed in the context of the network TARGET– "Top Amplifier Research Groups in a European Team" and supported by the Information Society Technologies Programme of the EU under contract IST-1-507893-NOE, http://www.target-org.net/

REFERENCES

[1] S. C. Cripps, *Advanced Techniques in RF Power Amplifier Design* Norwood: Artech House, INC, 2002.

[2] K. N. a. P. J. Zampardi, "Comparison of Linear Handset Power Amplifiers in Different Bipolar Technologies," *IEEE Journal of Solid-state Circuits*, vol. 39, pp. 1746-1754, OCTOBER 2004.

[3] A. J. J. Jeffrey B. Johnson, David C. Sheridan, Ramana M. Maladi, Per-Olof Brandt, Jonas Persson, Jesper Andersson, Are Bjorneklett, Ulrika Persson, Fariborz Abasi, and Lars Tilly, "Silicon-Germanium BiCMOS HBT Technology for Wireless Power Amplifier Applications," *IEEE Journal of Solid-state Circuits*, vol. 39, pp. 1605-1614, OCTOBER 2004.

[4] M. Kaynak, I. Tekin, Y. Gurbuz, Design of a Single-Chip, Dual-Band (2.4-2.5 GHz -WLAN and 3.3-3.9 GHz –WiMAX), Class – A Power Amplifier using 0.25μm-SiGe HBT Technology, TARGET DAYS 2006, http://www.target-net.org/target-days-2006

[5] Telegdy, A.; Molnar, B.; Sokal, N.O., "Class-E/sub M/ switching-mode tuned power amplifier-high efficiency with slow-switching transistor" Microwave Theory and Techniques, Volume 51, 6, pp:1662-1676 June 2003

[6] Fukuda, A.; Okazaki, H.; Narahashi, S.; Hirota, T.; Yamao, Y. "A 900/1500/2000-MHz triple-band reconfigurable power amplifier employing RF-MEMS switches", 2005 IEEE MTT-S International Stmposium

[7] IHP Microelectronics, "IHP SGB25 Process Specification Rev. 1.1 (51207) frontend modules V/ VD backend modules M4/thick M5," 2006.

[8] C. P. John Rogers, *Radio Frequency Integrated Circuit Design*: Artech House, 2003.

A 90nm-CMOS 1.8mW 87dB-SNR 3rd Order Analog Filter for GSM Receivers

M. De Matteis[1], S. D'Amico[1], P. Delizia[1], A.Baschirotto[1,3]

[1]Dep. of Innovation Engineering, University of Salento
Lecce – Italy
[3]Department of Physics, University of Milano Bicocca - Italy

C. Azeredo-Leme[2], A. Tavares[2]

Chipidea Microelectronica
Portugal

Abstract— **A 3rd order continuous time low pass filter for a GSM Receiver is presented. The filter uses the cascade of two active RC cells. The cut-off frequency deviation due to the technological spread, aging and temperature variation is adjusted by an on-chip tuning circuit. Fine definition (based on filter quality factor sensitivity functions) of the opamp frequency response and thermal/flicker noise optimization provides accurate filter transfer function. No miller cap tuning is needed, saving area and linearity. The device in 90nm CMOS technology consumes 1.8mW, features 87dB-SNR and 48μV$_{rms}$ in-band integrated noise. The supply voltage is 1.2V. The filter has been designed using an automatic Matlab design procedure which is validated by the agreement between the simulation results and the expected performance.**

Index Terms—**Analog filter, Cut-off frequency Tuning, GSM.**

I. INTRODUCTION

The increasing demand for a wireless telecommunications systems rapidly focused a significant effort on implementing a low-power, low-cost, and highly integrated (i.e., including both RF and baseband parts) mixed signal IC. The 0.18μm CMOS technology is the state-of-the-art for fully integrated System-on-Chip (SoC) for telecommunication systems [1]. Low voltage design is necessary for scaled technologies due to supply voltage reduction. In particular the technological scaling features a lot of advantages for the digital part of the mixed signal circuits. Higher data rate and power supply reduction are possible reducing the minimum length of the transistors MOS.

Otherwise the analog part of the mixed signal systems suffers the problems related with the technological scaling. These problems are well predicted by the International Technology Roadmap for Semiconductor. In fact the oxide thickness reduction implies critical variations of the transistors small signal parameters – g_m, g_{ds}, parasitic capacitances and resistances – with the temperature and the process variation. Furthermore the flicker noise increases due to the reduced minimum length and this is critical for the typical low frequency applications as the analog filters operating in the GSM Receivers. In these conditions the most important aspect for the analog designers in the scaled technologies is to guarantee the accuracy of the overall specifications under significant variations of the physical conditions – as temperature, technological spread, process and power supply. An accurate design enables the manufacture of SoC in scaled technologies for GSM transceiver.

The design of the GSM Receiver analog filter, presented in this paper, is developed in order to obtain high in-band linearity and strong out-band interferers rejection. The active RC topology appears the most feasible solution in order to achieve large linearity and cut-off frequency accuracy. The closed-loop gain makes the filter relatively independent on the transistors variation. In this way the cut-off frequency depends in the first approximation only on the RC time constant that is tuned by a proper master-slave circuits[2]. So that the precision of the frequency response depends on the accuracy of the tuning. In the 90nm CMOS circuits the accuracy of the tuning is critically dependent on the currents and voltage references used to evaluated the effective value of the RC time constant. When the references change for temperature, process or power supply variations, the algorithm of the tuning could not converge. Furthermore the precision of the transistors small signal parameters can affect the analog filters performance. Therefore some of the most important problems of this design are related with the technological scaling and are here reported.

- The R_{on} and R_{off} of the switches used for the cut-off frequency tuning can be comparable with the resistances used in the circuits.
- The current mirrors can feature significant errors. No cascode topology can be used due to the 1.2V supply voltage – the transistors V_{th} is around 350mV.
- The transistors V_{th} and small signal parameters – g_{ds}, g_m – are significantly dependent on the temperature and process variations.

In order to avoid performance degradation due to problems just commented, a careful design is needed. Then the accuracy of the cut-off frequency tuning is guaranteed by an efficient algorithm that forces the convergence also in the critical cases. The forcing is implemented in the digital part of the tuning machine. In this way a maximum 5% error on the cut-off frequency is obtained.

II. 3RD ORDER ANALOG FILTER

The most important specifications of the GSM Receiver baseband analog circuits are reported in the Tab. I. High f@-3dB accuracy is required for the GSM Receivers analog filters. The SNR is referred to the power of the maximum output signal with THD>50dBc.

978-1-4244-1983-8/08/$25.00 ©2008 IEEE

Parameter	Specification
V_{DD}	1.2V
Technology	90nm CMOS
f@-3dB	500kHz
f@-3dB Accuracy	+/-5%
G	0dB
SNR	>80dB

Tab. I – Analog Filter Specifications

A. Functional Scheme

The functional scheme of the GSM analog filter is reported in the Fig. 1.

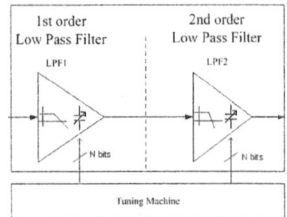

Fig. 1 – Functional Scheme

The circuit is composed by the 3rd order continuous time analog filter and by the Tuning Machine needed to compensate the cut-off frequency variation due to the technological spread. The tuning is obtained using variable capacitor set by the N bits word. The most important parameters of the filter transfer function and the plotting of the required LPF frequency response are reported in the Tab. II and Fig. 2 respectively.

	LPF1	LPF2
Order	1st	2nd
G	0dB	0dB
fo	567kHz	500kHz
Q	-	0.95

Tab. II – Frequency Response Parameters

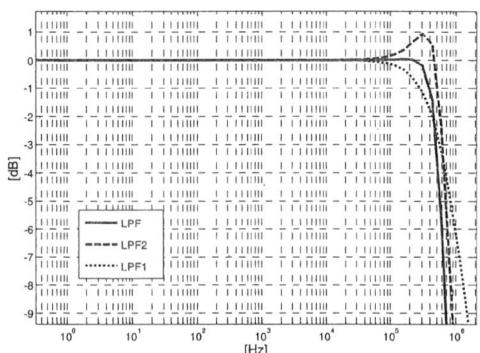

Fig. 2 – LPF Frequency Response

B. Circuital Topology

Starting from the SNR requirement reported in the Tab. I, the circuital topology selected for the LPF is the Active RC configuration. In fact the g_m-C filters does not appear suitable for this application due to the reduced in-band and out-band linearity performance. Otherwise these circuits feature good performance in terms of thermal noise because no resistors are used to synthesize the filter transfer function, but are affected by the flicker noise as the active RC filters. In the GSM analog circuits the thermal noise is important but the flicker noise is not negligible, due to the low in-band signal frequency – a few kHz. So it appears no reasonable to use the gm-C topology. The Active G_m-RC topology [1] assures good performance in terms of in-band linearity, but a linearity performance degradation can have when the input frequency is close to the filter cut-off frequency. Furthermore this cell needs an additional proper bias circuit which fix the g_m of the input stage of the opamp when the temperature changes. For the last reasons the better solution is to use the typical closed-loop configurations. In the Fig. 3 the schematic of the LPF filter is reported.

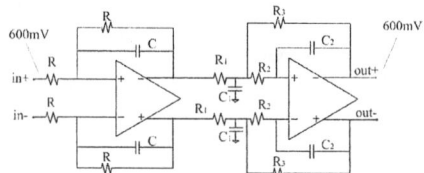

Fig. 3 – LPF Schematic

The filter is composed by the cascade of two active RC cells. The first cell synthesize the real pole at 567kHz and the second one is a Rauch cell where the complex poles are synthesized. The value of the passive components for both cells are reported in the Tab. III.

	LPF1	LPF2
R1	-	4.35 kΩ
R2	-	4.35 kΩ
R3	-	4.35 kΩ
R	13.5 kΩ	-
C	20.65pF	-
C1	-	208pF
C2	-	25.7pF

Tab. III – LPF1 and LPF2 Passive Components Values

The opamp used for both cells is a fully differential class A opamp. The MOS transistor output impedance of scaled-down technologies decreases and, as a consequence, also the achievable gain-per-stage decreases. In addition, at 1.2V with V_{th}=350mV, stacked configurations (for example cascode stages) are not possible. In this situation a sufficiently large opamp gain can be achieved by adopting two stage structures which, however, for stability reason, tend to have a relatively small bandwidth as compared with single stage structures. That is not a problem considering the 500kHz cut-off frequency requirement. A Class A Miller opamp has been designed. The requirements of the opamp in terms of minimum dc gain A_o, minimum UGBW and PM, IRN_{opamp} and current consumption, are obtained using an automatic Matlab procedure. In fact in the real life, the filter frequency response depends on the opamp finite bandwidth and dc gain. The opamp UGBW variation is due to the technological spread and opamp input stage g_m variation. Considering the opamp schematic, the g_m variation due to temperature, process and supply voltage can be compensated using proper constant g_m bias circuits[5], while technological spread is typically compensated using dedicated calibration/tuning circuits.

978-1-4244-1983-8/08/$25.00 ©2008 IEEE   174

Tuning circuits are based on variable capacitors, adequately set by a digital word. Supposing a maximum technological spread about +/-45%, that means an increasing of area for each capacitors. Rauch cell features three different capacitance: C1, C2, and miller Cc inside the opamp. It is needed to set C1 and C2 because they fix the f_{3dB} and Q. Calibration/Tuning circuit should set Cc also, which is responsible with the g_m of the opamp f_u variation. That increases area, as just explained, and complexity, degrading the dynamic range. In fact analog switches between the first and the second stage of the opamp are typically used to design variable capacitors. In this point the signal swing has been just amplified by the first stage, so that the linearity becomes critical. In this scenario a design approach, based on finding the minimum opamp UGBW f_u and dc gain A_o, in order to avoid significant variation on filter frequency response is interesting, because constant capacitance is used for miller compensation. For this reason the sensitivity function of the filter with respect the opamp UGBW and A_o has been calculated for both cells. Now a detailed explanation of the design procedure is reported.

The starting point of the Matlab algorithm is to fix the overall filter frequency response requirements. Then the sensitivity function of the quality factor Q and of the f_o is imposed to be ≤5%. Starting from this maximum value the minimum opamp dc gain A_o and UGBW f_u are calculated. The formulas (i) and (ii) report the sensitivity functions for the Rauch cell.

(i) $\quad S_{\omega_u}^{Q} = \frac{\partial Q/Q_o}{\partial \omega_u/\omega_u} \cong -\frac{(1+G)Q_o\omega_o}{\omega_u}$

(ii) $\quad S_{A_o}^{Q} = \frac{\partial Q/Q_o}{\partial A_o/A_o} \cong -\frac{G}{2\cdot A_o}$

G, f_o and Q are gain, cut-off frequency and quality factor of the cell respectively. The parameters obtained by the Matlab model are reported in the Tab. IV.

	LPF1 Opamp	LPF2 Opamp
Ao	≥45dB	≥45dB
UGBW	≥200MHz	≥300MHz

Tab. IV – Opamp Specifications

The time constant of the integrator used in the Rauch cell is $R_2 \cdot C_2$ and corresponds to 1.6MHz of unity gain frequency. At least 300MHz of f_u is needed in order to avoid cut-off frequency spread with the opamp frequency response variations.

III. TUNING CIRCUIT

Active-RC filters frequency response is determined by time constants $R \cdot C$ that are poorly controlled in an integrated technology. The uncertainty on the passive components nominal value due to the fabrication process, temperature variations and aging can reach the 45%, heavily affecting the filter frequency behavior. To adjust accurately the filter transfer function, additional tuning circuits are required. Since the low-pass filter cut-off frequency is determined by $R \cdot C$ products, any process variation and temperature dependencies can be compensated by tuning either the R's or the C's values. In this work, the tunability of the frequency response is

achieved by arranging capacitive elements in digitally programmable arrays, as shown in the Fig. 4.

The array value is, then, set using a digital code produced by an on-chip calibration circuit. The tuning scheme is based on the comparison between an isolated time constant $R \cdot C$ and a precise external clock period.

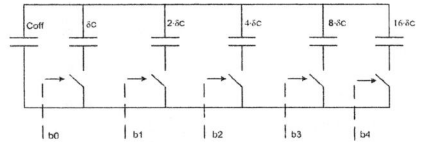

Fig. 4 – Capacitor Array

The resolution of the array in terms of number of the bits is fixed by the formulas:

(iii) $\quad \delta C = \frac{C_{max} - C_{min}}{2^N - 1}$, where C_{max} and C_{min} are the minimum and maximum capacitor array value and are given by:

(iv) $\quad C_{max} = \frac{C_{nom}}{1-\xi}$, $C_{min} = \frac{C_{nom}}{1+\xi}$.

ξ is the technological spread percent variation, that in this work it is supposed to be around +/-45%. Using N=5 bits the maximum percent error of the cut-off frequency depending on the passive components values is below the 3%. The basic schematic of the tuning circuit is presented in the Fig. 5.

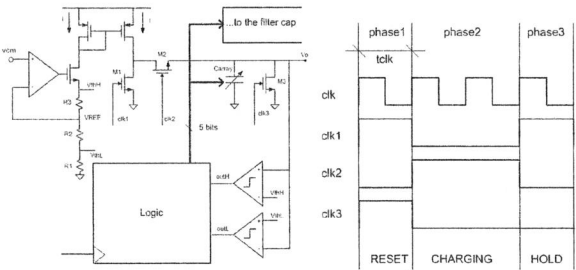

Fig. 5 – Tuning Machine Schematic

The behavior of the tuning algorithm is related with the clock phases of the switches M1, M2, M3 – clk1, clk2, clk3. The clock phases are also plotted in the Fig. 5.

The C_{nom} is related with the nominal reference current I and the nominal reference voltage V_{ref} by the formulas:

(iv) $\quad I = \frac{V_{ref}}{R_{nom}}$, $R_{nom} = R_1 + R_2$ (v) $\quad C_{nom} = \frac{I}{2\cdot t_{clk}}$

During the RESET the C_{array} is discharged. In the CHARGING phase the C_{array} is charged using the I reference current. During the HOLD phase the V_o is compared with two reference voltages V_{thH} and V_{thL}. If $V_o > V_{thH}$ the logic circuit decreases C_{array}. Otherwise the logic circuit increases C_{array}. So that the algorithm converges when $V_{thL} < V_o < V_{thH}$.
Simulation Results

The filter and the tuning machine circuit have been designed using the 90nm CMOS technology. The VDD is 1.2V. In order to validate the design and to verify the filter works under each physical condition variation, a lot of simulations have been run using the Ocean tool. Each simulation corner is obtained maintaining constant every parameter of the Tab. V and changing the remaining ones. In this way it is possible to mix every variation of each physical parameter. About 60 simulations for 60 different corners have been run to validate the design. The parameters and the correspondent variation are shown in the Tab. V.

Parameter	Nominal Case	Values
Supply Voltage	1.2 V	+/-10%
Temperature	27 °C	-45°C, 125°C
Reference Current	10 μA	+/-10%
Reference Voltage	600 mV	+/-10%
MOS Process	Nominal	Slow/Fast
Passive Components Technological Spread	-	+/-45%

Tab. V – Corners Simulation

In the Tab. VI the simulation results are reported, in the nominal case and in the worst case considering the entire set of simulations. In order to evaluate the efficiency of the tuning algorithm Fig. 6 shows the behavior of the cut-off frequency error for each case/corner. The maximum cut-off frequency error is 5%, and it depends on the quantization error of the capacitor array and on the variation of the opamp UGBW and dc gain.

This work can be compared to other continuous filters implementations by evaluating the figure-of-merit (FOM) defined as:

$$(vi) \quad FOM = \frac{P}{8 \cdot k \cdot T \cdot f_{-3dB} \cdot N \cdot DR}$$

where P is the total power consumption, f_{-3dB} is the cut-off frequency, N is the number of poles and DR is the dynamic range. Fig. 7 shows the reasonable result obtained with respect to the state-of-the-art. The favorable position in terms of FOM depends on two basic reasons:

• Design approach based on sensitivity function. Using miller compensated opamp, avoiding variable cap – typically needed to compensate UGBW variations due to the tech spread in active RC filters – allows saving area improving the SNR, because no switches are placed inside the opamp.

• Low Power Filter. Current consumption reduction is performed optimizing the thermal/flicker noise contribution. Since at 500kHz bandwidth flicker noise is not negligible, the optimization is based on increasing opamp thermal noise contribution while limiting the flicker noise. Thermal noise is inversely dependent on the input stage g_m and consequently on its current consumption, while flicker noise decreases with the MOS area increasing. Such area increasing is not important because at 500kHz the dominant area is due to the cap.

IV. CONCLUSION

In this paper a complete design of the 3rd order active RC analog filter operating in the GSM Receiver has been presented. The filter uses a proper tuning circuit that with a reference clock signal allows aligning the cut-off frequency compensating its variation due to the technological spread. The circuit has been tested under a lot of cases, imposing temperature, power supply and MOS process variation. The

filter performs +/-5% accuracy of the cut-off frequency, 85/87dB of SNR. The nominal current consumption is 2.27mA at 1.2V, considering the Tuning Machine and the Bias circuit current consumption.

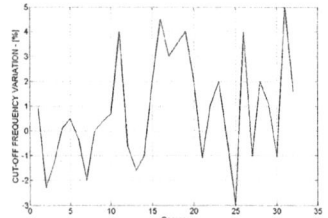

Fig. 6 – Cut-Off Frequency Error

Parameter	Worst Case	Nominal Case
Supply Voltage		1.2 V
Temperature		27 °C
Current Consumption	3.02 mA	2.27 mA
Cut-Off Frequency	476 kHz (-4.8%)	500kHz
Attenuation		-75 dB
Reference Current		10 μA
Reference Voltage		600 mV
Input Impedance	11 kΩ	13.3 kΩ
Input differential Amplitude		$1.5V_{pp,diff}$
SNR	85dB	87dB
THD v_{in}=$1.5V_{pp,diff}$@100kHz	-50.43 dBc	-56 dBc
In band Integrated Noise (1Hz÷500kHz)	$54\mu V_{rms}$	$48\mu V_{rms}$
Output Load	-	10 kΩ//5 pF
Reference clk		13 MHz
Maximum Tuning Time	-	100 μs

Tab. VI – Performance Resume

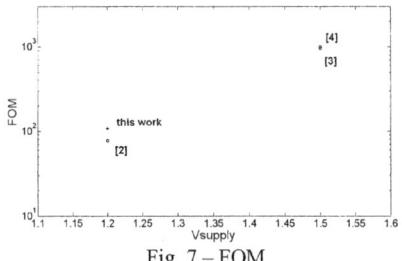

Fig. 7 – FOM

REFERENCES

[1] "A Fully Integrated SoC for GSM/GPRS in 0.13/spl mu/m CMOS" Bonnaud, P.; et. All. Solid-State Circuits, 2006

[2] S. D'Amico, et al. "A 4th-Order Active-G_m-RC Reconfigurable (UMTS/WLAN) Filter", *IEEE J. Solid-State Circuits*,July 2006.

[3] R. H. Zele and D. J. Allstot. "Low-power CMOS continuous-time filters," *IEEE J. Solid State Circuits*, vol. 31, no.2, pp. 157-168, Feb. 1996.

[4] U. Yodprasit and C. Enz, "A 1.5V 75 dB-dynamic range 3rd order Gm-C filter integrated in a 0.18μm standard digital CMOS process," in Proc. *ESSCIRC*, 2002, pp. 647-650.

[5] S. Pavan, et al., "Widely programmable high frequency continuous-time filters in digital CMOS technology," IEEE JSSC, vol. 35, no. 4, pp. 503–511, Apr. 2000.

Analysis and Design of an LC Oscillator-based Injection-Locked Frequency Divider

Saeid Daneshgar and Michael Peter Kennedy
Department of Microelectronic Engineering and Tyndall National Institute
University College Cork, Ireland
Email: saeidd@ue.ucc.ie

Abstract—**For high frequencies of operation, lowering power supplies and shrinking device sizes, prescalers prove to be a major source of power consumption in modern frequency synthesizers. The use of analog injection-locked frequency dividers (ILFDs) has been considered rather than conventional digital prescalers in order to tackle this problem. In recent years, numerous articles have proposed diverse and sometimes contradictory algorithms for estimating the locking range in these dividers. In this paper, a typical (LC oscillator-based) injection-locked frequency divider is considered. By resorting to nonlinear analysis combined with optimization techniques, we propose a new method to predict accurately and improve (widen) the locking range. In order to support our analysis, we provide SPICE simulations in 0.35-μm CMOS technology.**

I. INTRODUCTION

RF phase locked loops (PLLs) are widely used in wireless applications such as frequency synthesizers and clock sources [1]. Injection-Locked frequency dividers (ILFDs) are analog circuit blocks that are useful in PLL-based frequency synthesizers, among others, because they can consume much less power than conventional digital dividers [2]. The main concerns for frequency divider are low power consumption and high frequency capability with a wide locking range[1].

A variety of ILFD design implementations has been reported with locking ranges from 12% in [4] up to 50% in [5] over a range of operating frequencies (up to 50 GHz in [6]), and with power consumption as low as 44 μW [7]. However, despite important analytical studies [2]-[3], there is still no robust design methodology to determine the oscillator locking range with respect to design parameters. A low frequency SPICE validation of a new idea [8] to predict the locking range is considered in this paper and our future work will focus on an integrated circuit implementation at microwave frequencies.

In the following section, by considering a popular topology, we review the circuit parameters and equations. By using an appropriate normalization, we achieve a closed form state equation for this topology. In Sec. III, we assume that the circuit equations have been solved by combining normal form theory [9] and numerical continuation techniques [10] and that the design parameter space is well known. We then present a design procedure for the circuit together with simulation results for both the real circuit and a simplified equivalent

[1]Defined as $L_R = \frac{f_M - f_m}{f_c}$ where f_m and f_M are the minimal and maximal frequencies for which the exact frequency division (locking) occurs, and $f_c \simeq \frac{f_M + f_m}{2}$

circuit. In Sec. IV, we discuss the simulation results. Finally, in Sec. V, we draw some conclusions.

II. CONSIDERED CIRCUIT AND DESIGN PARAMETERS

The circuit we consider is the complementary cross-coupled LC CMOS ILFD topology shown in Fig. 1(a) [11]. The input signal, $v_I(t) = V_{dc} + v_i(t) = V_{dc} + V_i \sin(\omega_i t + \phi)$, is injected through the gate of transistor M_5. The output signal, $v_o(t)$, which has a fundamental angular frequency ω_o, appears across the RLC tank. The division ratio of the divider in the locked region is defined as $\frac{\omega_i}{\omega_o}$.

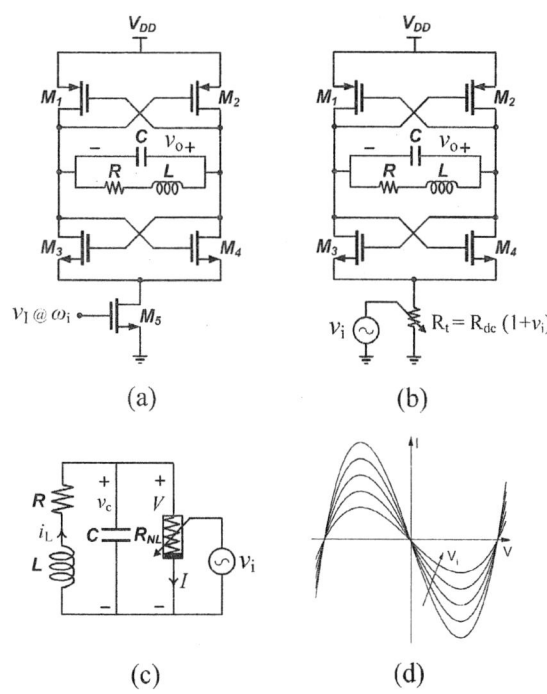

Fig. 1. ILFD oscillator: (a) Considered ILFD circuit schematic; (b) Equivalent circuit with VCR; (c) Simplified parametric model; (d) Qualitative $(I - V)$ characteristic of the controlled nonlinear resistor indicating its dependence on the control voltage V_i.

The injected signal v_i causes the tail transistor (M_5) to act as a modulated resistor. Therefore, replacing transistor M_5 by a voltage controlled resistor (VCR), we will model the circuit using the simplified equivalent circuit shown in Fig. 1(b). It can be shown that the simplified parametric model for this topology is similar to the circuit in Fig. 1(c) [8]. The simplified circuit consists of an RLC tank in parallel with a nonlinear resistor and has the following state equations:

$$\begin{cases} C\dfrac{dv_C}{dt} = i_L - f\left(v_C, v_i\right), \\[2mm] L\dfrac{di_L}{dt} = -R\,i_L - v_C, \end{cases} \quad (1)$$

where the nonlinear function $f\left(v_C, v_{in}\right)$ captures the behavior of all the nonlinearities. It can be shown that the I-V characteristic of the nonlinear resistor, which consists of the four cross coupled transistors and the voltage controlled resistor, can be approximated by a cubic of the form:

$$I_N = f\left(V_N, v_i\right) = a(v_i)\, V_N \left(1 - \left(\frac{V_N}{V_{DD}}\right)^2\right), \quad (2)$$

where V_{DD} is the supply voltage and a is the slope of the I-V curve in Fig. 1(d) at the zero crossing point. To simplify the analysis, we assumed that the modulation of the tail resistor (R_t) is mainly manifested as a variation of the parameter a. We will show that this assumption is valid for small amplitude forcing.

The next step of this analysis is to find a suitable normalization of parameters and to extract a general equation and design parameters. Applying the following linear transformation of the state variables and time:

$$x = \frac{v_c}{V_{DD}}, \quad y = \frac{R\,i_L}{V_{DD}}, \quad \tau = \frac{t}{\sqrt{LC}}, \quad (3)$$

and defining the following new parameters,

$$Q = \frac{1}{R}\sqrt{\frac{L}{C}}, \quad G = R \cdot a(V_{dc}), \quad \omega = \omega_i \sqrt{LC},$$

$$m_f = K \cdot V_i, \quad K = \frac{\left.\dfrac{\partial a\left(v_i\right)}{\partial v_i}\right|_{v_i = V_{dc}}}{a(V_{dc})} \quad (4)$$

we obtain canonical state equations as follows:

$$\begin{cases} \dot{x} = Q\left[-y + G\left(1 + m_f \sin\left(\omega\tau\right)\right) x\left(1 - x^2\right)\right], \\[2mm] \dot{y} = \dfrac{1}{Q}\left(x - y\right), \end{cases} \quad (5)$$

where the $\dot{}$ operator stands for $\frac{d}{d\tau}$ and Q is the quality factor of the tank. m_f is the relative variation of the parameter a for small variations of the input signal v_i; it indirectly quantifies the strength of injection. Finally, G is a parameter which depends on the value of the slope of the nonlinearity (a), and the resistance of the tank (R); qualitatively, it represents the gain of the oscillator at the origin.

In [8], we showed that the dynamics of the unforced circuit (free-running oscillator) and the forced circuit (locked oscillator) can be summarized using a simplified two parameter

(G and Q) bifurcation diagram. Furthermore, the valid ranges of G and Q for oscillation in the circuit can be calculated from this bifurcation diagram. We showed that the locking ranges in the locked oscillator depend in a complex manner on the values of G and Q. These two parameters are the main design parameters. By combining continuation techniques and optimization methods, we can optimize the width of the locking range with respect to them. The goal of this work is to show how the simplified qualitative approach based on Fig. 1(c) works when applied in the real circuit in Fig. 1(a).

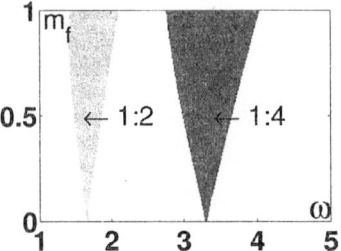

Fig. 2. Locking ranges for design parameters $Q = 20$ and $G = 0.1$ extracted from ODE Simulation in MATLAB (from [8]). The regions are not centered at 2 and 4 respectively, because the unforced oscillation frequency of the oscillator is not $\frac{1}{\sqrt{LC}}$.

III. DESIGN ANALYSIS AND SPICE SIMULATION

Using the results of [8], we have selected moderately wide divide-by-2 and divide-by-4 regions, as shown in Fig. 2(a). This bifurcation diagram shows the locked region with respect to ω (unit-less normalized input frequency) and m_f (input amplitude) for design parameters $Q = 20$ and $G = 0.1$. In order to produce this figure, simulations in MATLAB and AUTO have been performed for the range of $\omega \in [1, 5]$ and $m_f \in [0, 0.5]$ and then they have been verified with numerical continuation in MATCONT (a MATLAB package for numerical bifurcation analysis of ODEs) [10]. Focusing on the divide-by-4 region, the graph shows that the center frequency is $\omega = 3.3$ and the width of the locked region at $m_f = 1$ is approximately 1.27. This represents a locking range of 40%.

The results obtained by these simulations depend strongly on the modeling hypothesis assumed. In practice, the real nonlinearity seen by the LC tank is not an *ideal cubic* and its slope at the origin (a) is not *linearly modulated* by the injected signal. Therefore, we have first analyzed and measured the locking range of the circuit with the VCR in the tail (Fig. 1(b)) whose tail resistance is linearly modulated by the injected signal. We then consider the real circuit (Fig. 1(a)) which has a nonlinear tail resistance. This approach will let us have a better sense of the effect of the current source and also the amplitude of the injected signal on the locking range.

A. Design and simulation of the transistor circuit with a VCR in the tail

The general rule to design the circuit using this method of analyzing is first to extract the $a - K - V_{dc}$ graph for the nonlinear resistor seen across the tank. This graph shows the variation of K with respect to a for small variations of the input ac signal in the neighborhood of the dc bias voltage (V_{dc}) of the current source. To calculate the values of a and K, the I-V characteristic of the nonlinear resistor has been extracted by SPICE simulation and then the best cubic graph has been fitted to it using MATLAB.

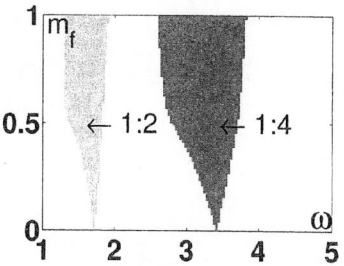

Fig. 4. Locking ranges for design parameters $Q = 20$ and $G = 0.1$: SPICE Simulation of equivalent circuit with VCR with device sizes of $(W/L)_{1,2} = 16\,\mu\text{m}/0.35\mu\text{m}$ and $(W/L)_{3,4} = 8\,\mu\text{m}/0.35\mu\text{m}$ (this work).

tail resistance is modulated linearly. Beyond $m_f = 0.5$, the approximation is no longer valid.

B. Design and simulation of the real transistor circuit with a current source in the tail

Next we replace the VCR by a real current source. We have analyzed the behavior of the real circuit with the current source in the tail in 0.35-μm CMOS technology using three sets of device sizes (small, medium and large) for the cross coupled transistors and over a wide range of current values and device sizes in the current source. Sweeping the dc bias voltage of transistor M_5 from 0.6 V up to 2.2 V, the $a-K-V_{dc}$ graph has been extracted and is shown in Figure 5 as an example for medium sizes of cross-coupled transistors and four different current source widths.

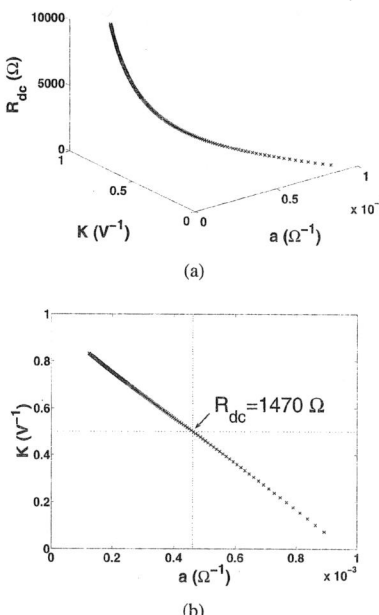

Fig. 3. (a) Extracted $a-K-V_{dc}$ diagram for the circuit with the VCR with device sizes of $(W/L)_{1,2} = 16\,\mu\text{m}/0.35\mu\text{m}$ and $(W/L)_{3,4} = 8\,\mu\text{m}/0.35\mu\text{m}$; (b) Projection of diagram (a) which shows a with respect to K.

In the VCR case, the $a-K-R_{dc}$ graph has been extracted and is shown in Fig. 3(a). This graph shows the variations of a and K as the dc resistance (DC operating point) in the tail is varied. Figure 3(b), which is a projection of Figure 3(a), illustrates the almost linear dependence of a on K. At the point $K = 0.5$, for which the value of R_{dc} is 1470 Ω, the slope of the nonlinearity is $a = 4.596e\text{-}4$.

Finding the value of a, and using equation (4), combined with the general equation for the operating frequency ($\omega_o = 1/2\pi\sqrt{LC}$), the RLC tank can be designed for the desired combination of operating frequency, Q, and G.

Figure 4 shows the simulation results for the locking regions of the circuit with a VCR in the tail designed in a 0.35-μm CMOS technology at an operating frequency of 1MHz with design parameters $Q = 20$ and $G = 0.1$. The related device sizes are $(W/L)_{1,2} = 16\,\mu\text{m}/0.35\mu\text{m}$ and $(W/L)_{3,4} = 8\,\mu\text{m}/0.35\mu\text{m}$.

Focusing on the divide-by-4 region in Fig. 4, we see that modeling the current source as a VCR is a valid assumption when the input signal is small ($m_f < 0.5$), in which case the

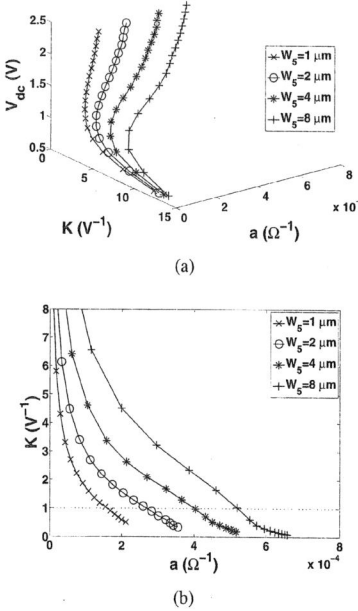

Fig. 5. (a) Extracted $a-K-V_{dc}$ diagram for the real circuit with current source with device sizes of $(W/L)_{1,2} = 16\,\mu\text{m}/0.35\mu\text{m}$ and $(W/L)_{3,4} = 8\,\mu\text{m}/0.35\mu\text{m}$; (b) Projection of diagram (a).

978-1-4244-1983-8/08/$25.00 ©2008 IEEE

Selecting $K = 1$ and the current source transistor size of $(W/L)_5 = 2\,\mu m/0.35\mu m$, the value of a is $2.713e-4$ and the tank can be designed with parameters of $Q = 20$ and $G = 0.1$. As the value of K has been selected to be 1, an input signal of $1\,V$ amplitude has been applied to the circuit in order to achive the maximum m_f of 1. Simulation results are shown in Fig. 6.

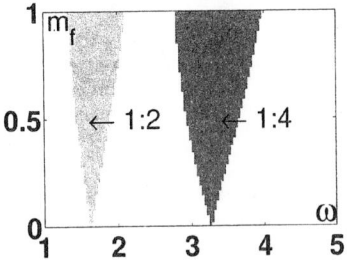

Fig. 6. Locking ranges for design parameters $Q = 20$ and $G = 0.1$ for the circuit with current source at $K = 1$ with device sizes of $(W/L)_{1,2} = 16\,\mu m/0.35\mu m$, $(W/L)_{3,4} = 8\,\mu m/0.35\mu m$ and $(W/L)_5 = 2\,\mu m/0.35\mu m$.

The center frequency of the divide-by-4 region is $\omega = 3.26$ and the width of the locked region at $m_f = 1$ is approximately 1.19, which represents a locking range of 36.5%.

IV. COMPARISON BETWEEN SIMULATION RESULTS

Simulation results for different transistor sizes have been analyzed from two points of view: (a) the maximum locking range at $m_f = 1$ and (b) the center frequency of the locked region. It can be seen that they match qualitatively with the predictions provided by the ideal cubic model of the I-V characteristic for the nonlinear resistance. For quantitative comparison, two examples are shown in Figs. 7(a) and (b). Figure 7(a) shows the maximum locking range at $m_f = 1$ with respect to K. For the divide-by-4 region, which is important from a designer's point of view, increasing the value of K will decrease the width of the locked region.

Note that the cubic approximation is valid only for $K < 1$. In this region, our simplified model predicts the center frequency and the locking range to within 10% of their simulated values.

The deviation of the center frequency of the locked region is shown in Fig. 7(b) for both the divide-by-2 and 4 regions. The graph shows that the deviations are not far from the center frequencies in Fig. 2(a), which are 3.3 and 1.65 for the divide-by-4 and 2 regions, respectively.

V. CONCLUSION

LC Injection-Locked Frequency Dividers can operate at high frequencies with low power dissipation. In this paper a new approach ([8]) for analyzing these dividers has been validated by SPICE simulations in 0.35-μm CMOS technology at an operating frequency of 1MHz. This approach is capable of predicting the locking range and the center frequency. We aim to validate this methodology by implementing the circuit in silicon.

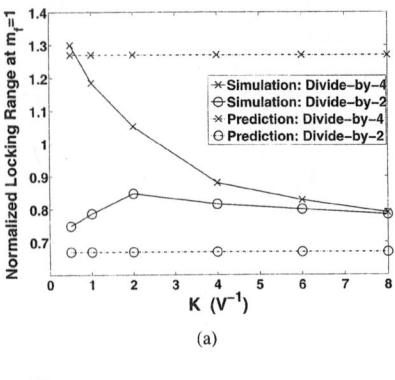

(a)

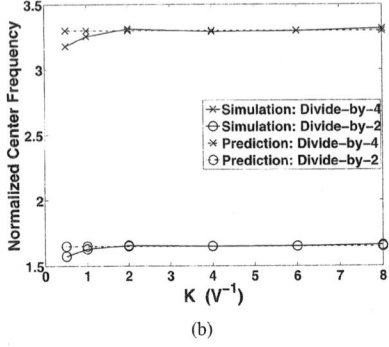

(b)

Fig. 7. SPICE simulation results for the circuit with device sizes of $(W/L)_{1,2} = 16\,\mu m/0.35\mu m$, $(W/L)_{3,4} = 8\,\mu m/0.35\mu m$ and $(W/L)_5 = 2\,\mu m/0.35\mu m$. (a) Locking range at $m_f = 1$ with respect to K; (b) center frequency with respect to K.

REFERENCES

[1] J. Craninckx and M. Steyaer, *Wireless CMOS Frequency Synthesizer Design.* London, U.K.: Kluwer, 1998.

[2] S. Verma, H. R. Rategh and T. H. Lee, "A Unified Model for Injection-Locked Frequency Dividers," *IEEE Journal of Solid State Circuits*, vol. 38, pp. 1015-1027, June 2003.

[3] B. Razavi,"A Study of Injection Locking and Pulling in Oscillators," *IEEE Journal of Solid State Circuits*, vol. 39, pp. 1415-1424, Sep. 2004.

[4] H. R. Rategh and T. H. Lee, "Superharmonic Injection Locked Oscillators as Low Power Frequency Dividers," *Symposium on VLSI Circuits Digest*, pp. 132-135, 1998.

[5] S. Cheng, H. Tong, J. Silva-Martinez and A. I. Karsilayan, "A Fully Differential Low-Power Divide-by-8 Injection-Locked Frequency Divider Up to 18 GHz," *IEEE Journal of Solid State Circuits*, vol 42, NO. 3, pp. 583-591, March 2007.

[6] M. Tiebout,"A CMOS Direct Injection-Locked Oscillator Topology as High-Frequency Low-Power Frequency Divider," *IEEE Journal of Solid State Circuits*, vol 39, NO. 7, pp. 1170-1174, July 2004.

[7] K. Yamamoto, and M. Fujishima,"A 44-μ W 4.3-GHz Injection-Locked Frequency Divider with 2.3-GHz Locking Range," *IEEE Journal of Solid State Circuits*, vol 40, NO. 3, pp. 671-676, March 2005.

[8] M. M. Ghahramani, S. Daneshgar, M. P. Kennedy, and O. De Feo, "Optimizing the design of an Injection-Locked Frequency Divider by means of nonlinear analysis," *European Conference on Circuit Theory and Design (ECCTD)*, pp. 571-574, August 2007.

[9] Y. A. Kuznetsov,*Elements of Applied Bifurcation Theory*, 3rd ed. New York: Springer-Verlag, 2004.

[10] H. Dhooge, W. Govaerts, and Y. A. Kuznetsov, "MATCONT: A MATLAB package for numerical bifurcation analysis of ODEs," *ACM Transactions on Mathematical Software*, vol. 29, pp. 141-164, 2003.

[11] E. Hegazi, J. Rael, and A. Abidi, *The Designer's Guide to High-Purity Oscillators.* United States of America: Kluwer, 2005.

Analyses and Design of Low Power Clock Generators for RFID TAGs

Christian Klapf, Albert Missoni, Wolfgang Pribyl
Institute of Electronics
Graz University of Technology
Graz, Austria
Klapf@TUGraz.at

Gerald Holweg, Günter Hofer
Infineon AG, Design Center Graz
Graz, Austria
Gerald.Holweg@Infineon.com

Abstract — **This paper introduces a new clock generation concept with a PLL for HF RFID systems. Low power consumption of 1.9µW and a good decoupling against power supply and bias variations are necessary to reach HF RFID timing and energy performance requirements. All presented oscillator topologies can be used in UHF EPCglobal class1 gen2 RFID systems as local oscillator with a minimum frequency of 1.92MHz. For all oscillators the PSR, power consumption and temperature drift are simulated and partly measured. In the CTS[1] project a new VCO and local oscillator concept was developed and manufactured on an Infineon 120nm CMOS test-chip. The PLL is simulated with the same process technology.**

I. INTRODUCTION

In RFID systems the clock for the digital state machine, decoder and encoder is extracted either from the carrier or it is generated by a local oscillator. Timing constraints of dedicated RFID specifications require a precise absolute oscillator frequency or only a quite constant clock frequency during a time period. Power supply voltage and bias current variations during RxD and TxD phase will proof the PSR of the design. Spectral considerations like phase noise or the time equivalent jitter which are typically one of the most important oscillator characteristics are less important in RFID systems. Typically simple peak detection ASK - demodulators don't need any frequency constant sources like IQ demodulators which are often the reason for oscillators with low phase noise. A contactless system gets the energy from the magnetic or electromagnetic field and is limited when big interrogator to TAG distances should be reached. This power limitation affects also the clock extraction or generation module and causes a typical power consumption of about 500nW to 3µW.

HF RFID systems extract the clock from the contactless field and convert the sinusoidal coil voltage to a 13.56MHz rectangular signal. A local low power frequency divider reduces the clock speed additionally. But during 100% ASK

---
[1] CTS - Comprehensive Transponder Systems (funded by FFG)

RxD the coil voltage drops down below 300mVpeak and is too small for even a very sensitive clock recovery. In this phase field clock losses are handled by the state machine and will not change the RxD - timing. Clock losses during very deep TxD load modulation phase will modify the subcarrier load modulation frequency. A deep modulation of the coil voltage is necessary to increase sideband voltage in the interrogator and is essential for a competitive TAG system. Even in sub micron CMOS technologies the optimization of the coil-voltage dynamic gets more important and leads to a deepest possible modulation depth of less than 50mV. No clock recovery can extract from such a small voltage a constant 13.56MHz clock. In this case a low power PLL could provide the internal clock and will guarantee the correct subcarrier timing. A very important characteristic of this PLL is the frequency stability in the hold-mode (charge pump of the PLL is not active) versus voltage and bias variation. To fulfill this requirement, a low power VCO with good PSR is essential.

UHF RFID systems at carrier frequencies of 869MHz and 2.45MHz anyway use a local oscillator (LO). Clock extraction from the field carrier consumes too much current even in a modern Si technology. Timing constraints in UHF RFID are relative (dynamic) timings and a tight tolerance is allowed additionally. Nevertheless a short time (few 100µs) constant LO is necessary with small frequency dependency against power supply and bias current variations.

II. OSCILLATOR TOPOLOGIES

Four different oscillator concepts and designs are shown in this section. All of them are low power low voltage ring or relaxation oscillators. In UHF EPCglobal class1 gen2 systems the clock frequency has to be higher than 1.92MHz [1]. In the proposed HF PLL system the field clock is already divided by a factor of 8. Phase discriminator, PLL charge pump and VCO are oscillating tight around the center frequency of 1.68MHz. Because of the similar characteristics of HF and UHF systems, the target frequency for all following topologies is set to 2.2MHz typically.

A. RC delay ring oscillator

A single RC stage acts as delay and is the dominant part which defines the frequency. The first inverter stage defines the operation point of VDD/2. The high gain of the 3 inverter stages result in a big voltage step at the transistor pair P3 and N3. The step size depends very much on the power supply voltage VDD and additionally the RC delay time increases. Big resistor and capacitance values are necessary to reach the low oscillation frequency requires and this will cause once again a big silicon area. The advantage of this architecture is the low voltage characteristic. Only little more than $1V_{GS}$ is necessary to start this module.

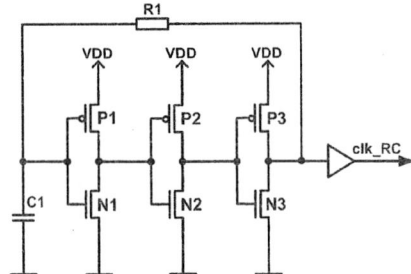

Figure 1. Common RC Oscillator

B. Current controlled ring oscillator

Several ring oscillator concepts are used as local oscillators in UHF RFID chips. One or two current sources can be used to reduce the speed. Only gate and parasitic capacitors are charged by the PMOS current sources P2 - P4 and linearly discharged via the NMOS transistors N3 - N5. This triangular voltage is greatly independent of power supply voltage changes but varies much with bias current variations.

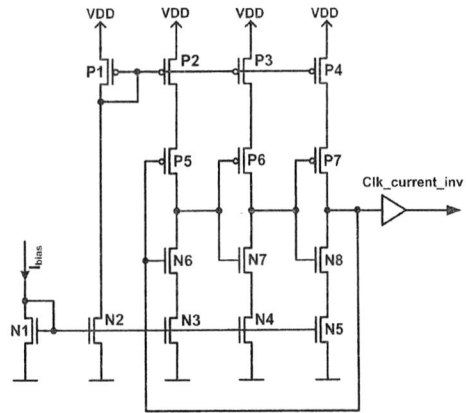

Figure 2. Current controlled Ring Oscillator

C. Ring oscillator with differential stages

Similar to the previous inverter architecture, differential stages are used as gain stages. Current sources and sinks guarantee once again a good power supply rejection.

Because of the differential design, phase noise and jitter is reduced but it has anyway no negative impact in UHF and HF RFID systems. Minimum voltage headroom is increased by $1V_{GS} + 2V_{DS}$.

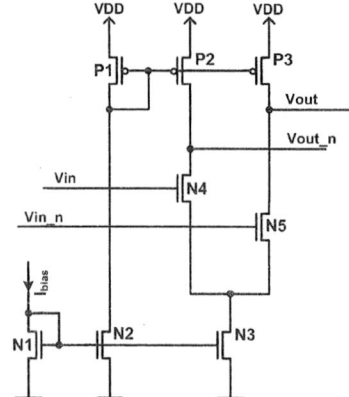

Figure 3. One differnetial stage from a ring oscillator

D. Proposed CTS relaxation oscillator

The goal was to develop an oscillator which has a reduced dependence on power supply and bias variations. Furthermore the frequency should be defined by passive elements. The result is a relaxation oscillator [2] where a resistor and a capacitor define the frequency. To use the oscillator also as VCO, I_{vco} has to be separated from I_{Bias}. The minimum supply voltage is $U_{R1} + U_{GS} + V_{DS}$.

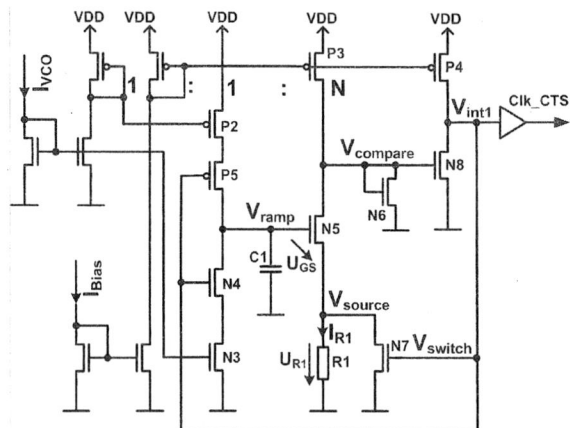

Figure 4. CTS VCO - Relaxations oscillator

Oscillator design ($I_{VCO}=I_{Bias}$) - Let's start with the assumption, that the oscillator output Clk_CTS and the node V_{switch} are LOW. N_7 is nonconductive and the inverter P_5/N_4 charges C_1 via the current source P_2. The voltage at the source of N_5 will follow proportional to the voltage at C_1 - V_{ramp}. This provokes a voltage drop over the resistor R_1 and results in a current I_{R1}. A current comparator at node $C_{compare}$ weights the current I_{R1} which rises linearly (proportional to

978-1-4244-1983-8/08/$25.00 ©2008 IEEE

V_{ramp}) and the constant sourced current delivered by P_3. V_{ramp} will rise until the current I_{R1} is bigger than the sourced current. $V_{compare}$ will drop to VSS and the weak inverter P_4/N_8 toggles from LOW to HIGH. The switch N_7 discharges U_{R1} to VSS but the gate voltage is still at a very high potential. The resulted current trough N_5 is much higher than the I_{P3} and the oscillator is in a stable phase where V_{ramp} is reduced linearly by the current sink N_3. The next switching point is reached, when the current source P_3 is stronger than the I_{DS} of N_5.

Correlation between oscillator period T and passive components:

$$\frac{T}{2} = \frac{\Delta Q_{C1}}{I_{Bias}} = \frac{C_1 \cdot \Delta V_{ramp}}{I_{Bias}} = C_1 \cdot R_1 \cdot N$$

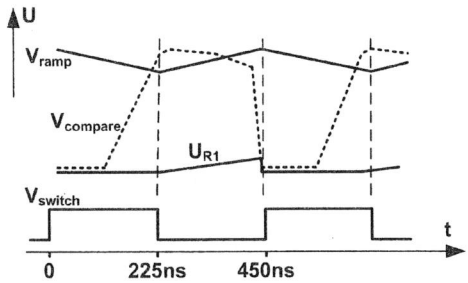

Figure 5.   CTS oscillator - Timing of internal notes

III.   SIMULATION & MEASUREMENT

RFID relevant electrical characteristics for all 4 oscillators are simulated and two of them are compared with measurement results.

A.   Power supply voltage variation

During field modulation or backscatter, the power supply varies in 120nm CMOS from 1.6V down to 0.8V! This is the big difference compared to other low power applications with e.g. battery energy source. The lower end of the dynamic range is dominated by the power on/down reset, the upper limit by the technology limit of 1.6V.

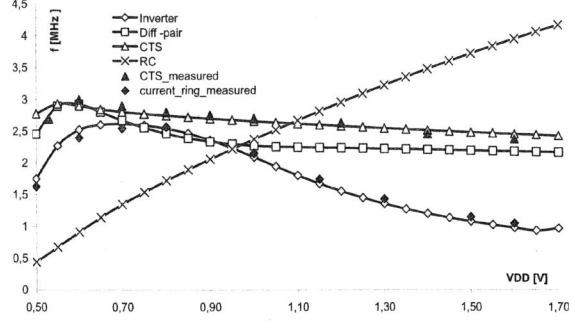

Figure 6.   Oscillator frequency over VDD variation

CTS and the differential pair oscillators show the most constant frequency versus VDD. The ring oscillator characteristic can be improved by reducing the bias current to 1.6MHz at 1.7V.

B.   Temperature variation

RFID is most often used in consumer electronics at a temperature range of 0°C - 50°C. If a TAG is close to the interrogator, the upper temperature will increase at ambient temperature of 50°C to 100°C silicon temperature! But a typical communication time interval takes less than 10ms. Only during this time the frequency should be constant.

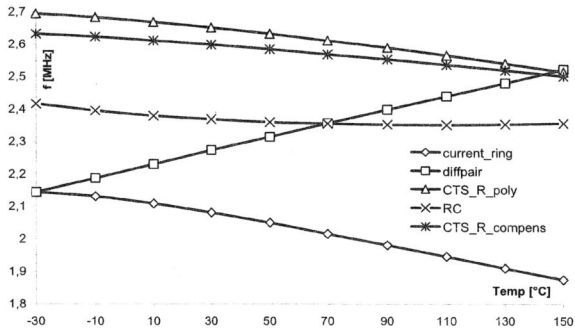

Figure 7.   Oscillator frequency over temperature variation

The temperature drift of the resistor impacts the characteristic (see CTS_R_compens).

C.   Bias variation

Bias generators in RFID are often very simple beta-multiplier circuits with bad power supply rejection. A nominal bias current of 150nA will typically vary from 100nA - 230nA.

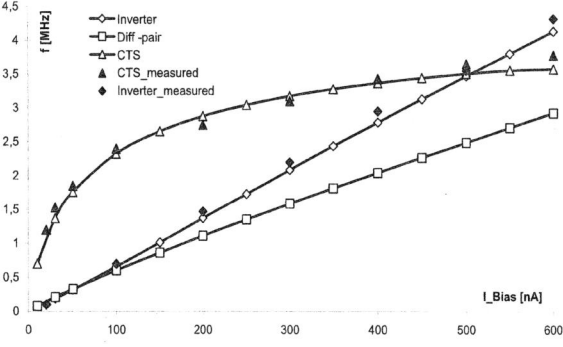

Figure 8.   Frequency variation over bias current variation

For the CTS oscillator R and C are the frequency dominant components. Parasitic capacitors at node $V_{compare}$ reduce the resistivity against current variation.

TABLE I.   POWER CONSUMPTION COMPARISON

	RC	Inverter	Diff-pair	CTS
VDD **[V]**	**P** **[µW]**	**P** **[µW]**	**P** **[µW]**	**P** **[µW]**
0.70	0.567	0.361	0.720	0.731
1.00	1.245	0.452	0.855	0.789
1.50	2.871	0.526	0.958	0.861

IV.   PROPOSED PLL SUPPORTED CLOCK GENERATION

A.   Clock Generation System

The proposed CTS oscillator is now used as CTS VCO in a new low power PLL system optimized for HF RFID applications.

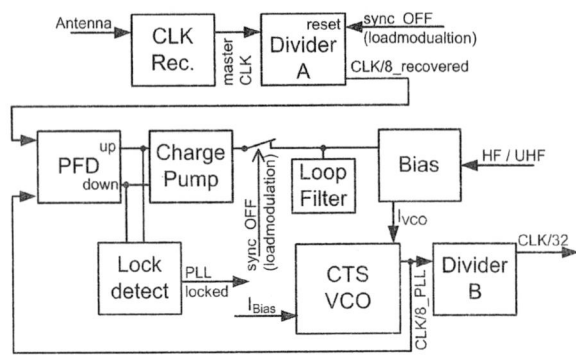

Figure 9.   Comprehensive clock generation system – overview

Figure 9 shows a clock generation system which provides a constant system clock without clock losses, even the antenna voltage is too low for clock extraction (during very deep HF load modulation - e.g. short of the two antenna pads) [4]. The 13.56MHz clock recovery followed by the low power Divider-A extracts the chip reference clock from the sinusoidal antenna voltage and divides it by the factor 8. A D-FF phase discriminator and a glitch-compensated charge pump generates with the following pole/zero loop filter a bias current for the input-stage of the CTS VCO. In Figure 4, the capacitor C_1 of the CTS oscillator is sourced by the variable bias current. The current comparator path N_5 and R_1 gets still the constant bias current. The charge pump is disconnected from the loop filter (sync_OFF) when the antenna voltage is shorted (during load modulation). Now the VCO frequency is stable (autonomic oscillator) and the Divider-A is switched OFF (to save power and to get a defined startup). ISO15693-2 compliant, Manchester coding with a subcarrier frequency of 424kHz is used for load modulation. The longest continuous modulation time is 56.64µs (SOF, EOF). In the worst case the longest free running phase are 96 oscillator clock cycles when no resynchronization takes place.

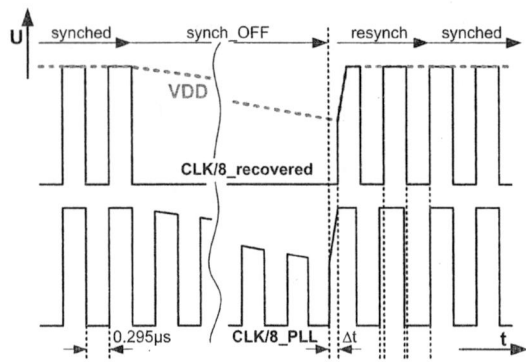

Figure 10.   PLL clock generation

Resynchronization during load modulation when the antenna-voltage gets to the high phase back again (t_{HIGH} = 1.18µs) requires a fast settling. Simulations of this PLL have shown, that a resynchronization after the longer time of 56.64µs generates less interference than the fast sync each 1.18µs. After load modulation phase (56.64µs) the phase-difference of CLK/8_PLL and CLK/8_recovered must not exceed 0.295µs, otherwise clock cycles could get lost. (Figure 10). At 1V power supply voltage the whole PLL system draws 2.1µW (oscillator, divider, bias generation and VCO - 120nm CMOS simulation result).

V.   CONCLUSION

We present four oscillator architectures for UHF RFID TAGs and include the proposed CTS oscillator as VCO into a new low power PLL system for HF RFID. The new oscillator concept draws 0.8µW at 1V and is robust against power supply and bias current variations. The most important RFID relevant electrical characteristics are compared for all oscillator types and summarized in Table II. The PLL concept consumes 2.1µW and is functional at a minimum supply voltage of 0.8V.

TABLE II.   OSCILLATOR COMPARISON

	RC	Inverter	Diff-pair	CTS
power cons.	-	+	O	O
frequ. stability	-	O	O	+
low volt. op.	+	-	-	O
chip area	-	+	O	+

REFERENCES

[1] Impinj, "Gen 2 tag clock rate – What You Need to Know", Impinj®, Inc.

[2] A. Missoni, Ch. Klapf, W. Pribyl, G. Hofer and G. Holweg, "A Triple Band Passive RFID Tag", ISSCC 2008 – Session 15.2

[3] J. Curty, M. Declercq, C. Dehollain and N. Joehl, "Design and Optimization of Passive UHF RFID Systems", Springer 2007

[4] D. Berger, "CONTACTLESS DATA CARRIER", Patent WO 2004/032040 A1, Infineon Technologies AG, 2004

A 24 GHz, 18 dBm Fully Integrated Power Amplifier in a 0.13µm SiGe HBT Technology

Nejdat Demirel, Eric Kerhervé
IMS Laboratory, COFI Department
Microwave Team
33405 Talence Cedex, France
nejdat.demirel@ims-bordeaux.fr

Denis Pache
STMicroelectronics
R&D
Crolles, France
denis.pache@st.com

Robert Plana
LAAS-CNRS, University of
Toulouse, France
plana@laas.fr

Abstract—A 24GHz, +18.0dBm fully-integrated power amplifier (PA) with 50Ω input and output matching is designed in 0.13µm SiGe BiCMOS process. The power amplifier features a peak power gain of 7.8dB with 15.89 dBm output power at 1dB compression and a maximum single-ended output power of +18.0dBm with 25.9% of power-added efficiency (PAE). The power amplifier uses a single 1.8 V supply and was fully integrated (including matching elements and bias circuit). The matching networks use inductors and MIM capacitors for high integration purpose, the circuit occupies a small area of 0.3mm² (including pads and matching networks).

I. INTRODUCTION

Safety and security issues in transport are gaining more and more importance that motivates a lot of research to explore solutions for improving existing solutions in term of resolution, sensitivity, functionalities, integration density and cost. Different frequency band have been allocated for two types of automotive radar referred to as Short Range Radar (SRR) and Long Range Radar (LRR) that are in the millimeter-wave range (i.e 24 GHz and 79 GHz).

Automotive radar performs different functions (e.g., obstacle and turn detection, collision anticipation, adaptative cruise control), and uses different operating principles (e.g., pulse radar, frequency-modulated continuous-wave [FMCW] radar, microwave impulse radar). Due to their simplicity, ultra-wideband pulse radio systems will gain importance for low-power, short-range sensors and communication devices. As an example, the SRR devices planned to be used for automotive applications have the following specifications [1]: antenna gain = 30-35 dBi, E.I.R.P. (Equivalent Isotropically Radiated Power) = 16-20 dBW, and the corresponding output power of the amplifier (P_{out}) = 15-20 dBm. For LRR devices, the requirements for the emitter path will be even stronger. From these specifications, it is understood that the power amplifier will be a key component that will drive the overall architecture of the system integration. As previously stated, the choice of the

technology will have an impact on both, the architecture performance and manufacturing cost of the equipment. It has to be outline that today the conventional radar architectures are limited by two drawbacks: a low integration density and a high cost. These two drawbacks could be overcome by silicon based technologies and more precisely the Silicon Germanium (SiGe) Heterojunction Bipolar technologies (HBT) that have already demonstrated impressive performances in term of cut-off frequency and maximum oscillation frequency. Additionally, the passive quality factor on silicon have made significant progress due to the multiplication of metal level and the use of appropriate design rules to get rid from the eddy current and ohmic losses. It has to be outline that moving to higher frequency range will result in a reduced dimension for the passive that will minimize the losses impact.

It has to be worth mentioned that the SiGe technology is compatible with CMOS process that make possible the integration of the base band circuits onto the same chip and that allow also to introduce some programmability that will relax the system architecture constraints. However, there are still some issues to solve or to overcome with the SiGe based technologies. The main issue deals with the avalanche breakdown voltages (BV_{CEO} BV_{CBO}) which are further and further reduced. This will drastically impact the architecture and the power capabilities of this technology and there is a strong need to develop power amplifier featuring high power, high efficiency with moderate or even low supply voltage. This is why in the literature, we find a lot of SiGe based circuit demonstration of LNA, VCO, Divider, mixer but the power amplifier demonstration are more rare. The best results to our knowledge are monolithic microwave integrated circuit (MMIC) power amplifiers featuring 13% of power added efficiency (PAE) for 21dBm of output power (P_{out}) at 24 GHz [2], and 3.5 % of PAE for 12.5 dBm of P_{out} at 77 GHz [3].

This paper proposes to explore the design of a fully integrated power amplifier at 24 GHz optimized with respect go the PAE as it seems to be a crucial parameter for the

978-1-4244-1983-8/08/$25.00 ©2008 IEEE

system performance. The paper will be organized as follow. Section II will briefly describe the technology used and the design methodology that has been implemented. Section III will address the simulation results that have been obtained and compared with the state of the art when conclusions are outlined in the last section.

II. TECHNOLOGY DESCRIPTION and DESIGN METHODOLOGY

The power amplifier under discussion has been designed in advanced bipolar technology BiCMOS9 from STMicroelectronics [4]. It is a npn 0.13μm SiGe HBT technology with a transit frequency f_T = 160GHz and a maximum oscillation frequency f_{max} = 175GHz, high quality passives and a 6-level copper back-end. The device characteristics show a maximum current density J_{cmax} of 2mA/μm, and a collector–emitter breakdown voltage BV_{CE0} of 1.8V and a collector-base breakdown voltage BV_{CB0} of 5.5V. The MMICs are realized on 375 μm-thick silicon substrate with a medium-resistivity of 10-15 Ω-cm.

Figure 1 shows the schematic of the single-ended common-emitter amplifier which consists of an active device (Q0) between integrated input and output matching networks. At the input, a matching network, consisting of a shunt capacitor (C3) and a serie inductor (L1), is used as pre-match for the active device. A type of current mirror biasing circuit has been used to provide an optimum condition for the power device as function of temperature and output power [5]. Output bias is provided by a choke inductor (L0) and bypass capacitors (C5 and C6) in the collector circuit. Input and output are DC isolated by series capacitors (C0 and C1 respectively).

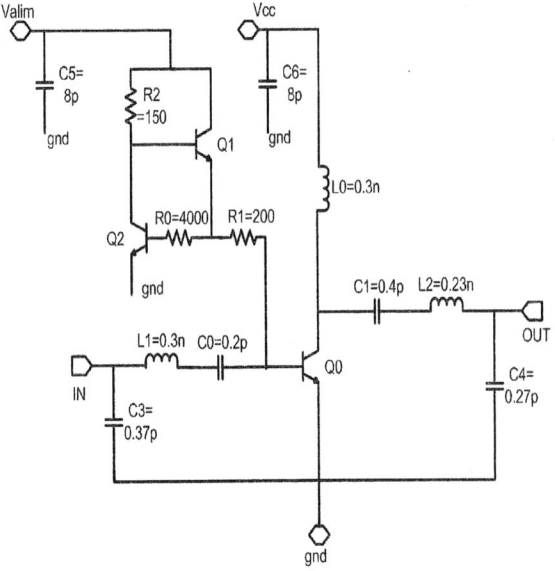

Figure 1. Circuit schematic of the single-ended common-emitter power amplifier.

A. Design methodology

The design methodology that has been implemented mainly takes into account the fact that this technology features breakdown voltage limitations. As the output transistor voltage cannot exceed twice the supply voltage Vcc. It is known that the output voltage is mostly limited to BV_{CE0} defined at I_B = 0 (open base). Then the low breakdown voltage BV_{CE0} = 1.8V would limit the supply voltage to 0.9V. The limitation of the output voltage is driven by impact ionization effect (which generates electron-hole pair in the base current) [6]-[7] which will results to reliability problems for the final circuit.

To achieve the maximum output power, the output transistors have been driven near to their practical limits, given by high-current effects and avalanche breakdown. The required optimum load-line resistance R_{opt} can be calculated from the device characteristics, i.e maximum collector current I_{cmax} and breakdown voltage BV_{CE0} as follows:

$$R_{opt} \approx \frac{BV_{CE0}}{I_{cmax}} \qquad (1)$$

The load-line impedance for maximum output power at the fundamental frequency is calculated by equation (1) to be equal to R_{opt} = 10 Ω for Vcc = 1.8V. The parasitic are extracted from the 6*15 μm HBT and included in the external matching network. The use of class-AB operation for the output stage provides a trade off between linearity and efficiency.

B. Matching circuit

It has been proved that the lumped element matching circuit is better than distributed element matching at these millimeter frequencies because the small size of lumped capacitors and wound inductors are significantly small when integrated on silicon. This particular lumped matching circuit has a low pass and high pass topologies (input and output respectively). If necessary to achieve a good match over the required bandwidth, additional sections may be added. Usually, the number of sections is minimized, in order to reduce both circuit complexity and cost.

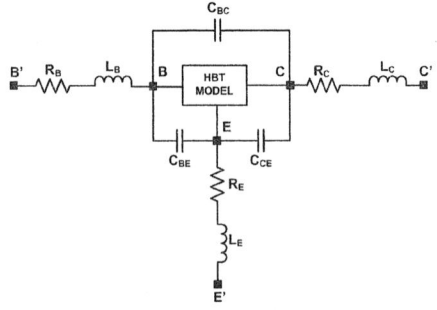

Figure 2. Parasitics extraction of the transistor accesses.

At 24GHz, λg/10 (where λg is the guided wavelength) is about 625μm. The lines between the matching components are modeled by inductance and resistance series. Input and output pad are modeled by shunt capacitor of 120fF. Figure 2 shows the extracted parasitic of the transistor accesses. These added elements are resistances (R_B, R_C and R_E), inductances (L_B, L_C and L_E) and capacitances between base, collector and emitter (C_{BE}, C_{CE} and C_{BE}). The HBT device use HICUM transistor model.

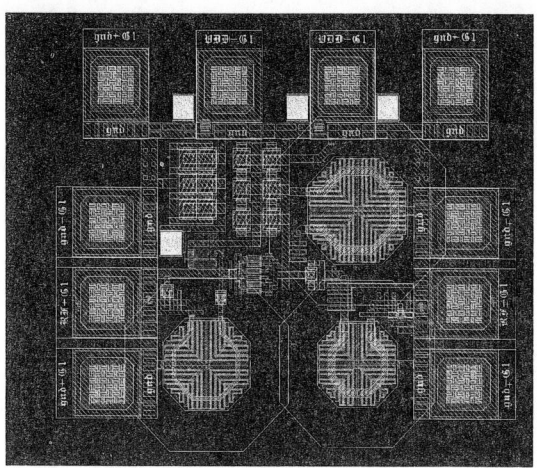

Figure 3. Layout of the integrated low-voltage 24 GHz power amplifier.

The layout of the designed MMIC common-emitter PA is shown in Figure 3. The total chip area (active device + integrated matching networks) is 0.3 mm² (600μm × 500μm).

III. SIMULATION RESULTS

Post layout circuit simulations, including small signal S-parameters and large signal analysis were performed using Cadence simulator.

A. Small-signal

Small-signal S-parameters of the power amplifier at Vcc= 1.8 V and Ice= 30 mA are plotted in Figure 4 from 20 up to 28 GHz. Simulations show a maximum power gain of 7.8 dB at 24 GHz. The input of the amplifier is optimized for gain and input match when the output is optimized with respect to output power (i.e R_{opt}).

A stability study has been carried out that has shown that the amplifier is unconditionally stable over a broadband frequency range. At 24GHz, the input return loss is -22dB and the output return loss is -20dB. The reverse isolation at 24GHz is -21dB.

B. Large-signal

Simulations were performed in SpectreRF. All active devices use the HICUM transistor model.

Simulated output power as function of input power at a quiescent operating point of Vcc= 1.8 V and Ice= 30 mA is plotted in Figure 5. The large-signal measurements show that the circuit presents an output power Pout of 18 dBm (63mW). The output power at 1dB compression is 15.89dBm. The power added efficiency (PAE) is defined by the equation (3).

$$PAE = \frac{Pout - Pin}{Pdc} \qquad (3)$$

where P_{out} and P_{in} are the output and input power respectively, and P_{dc} is the DC power consumption. The simulated power gain and PAE are shown in Figure 6. The simulations show that the circuit features a power gain of 7.8 dB. The power added efficiency at 1 dB compression is 23.3% and reaches a maximum value in the range of 25.9% at the maximum output power of 18 dBm.

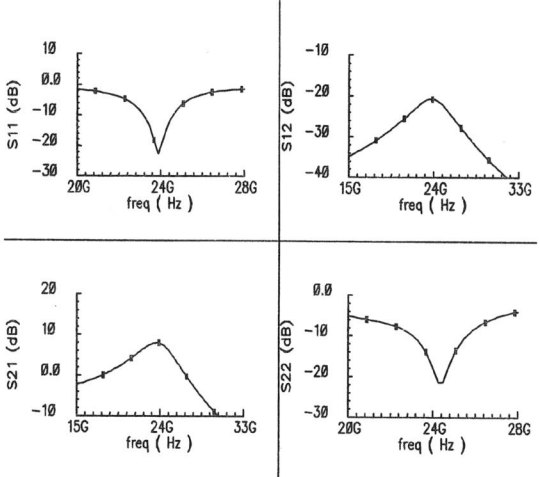

Figure 4. S-parameters of the 24 GHz power amplifier biased at Vcc= 1.8V and Ice= 30mA.

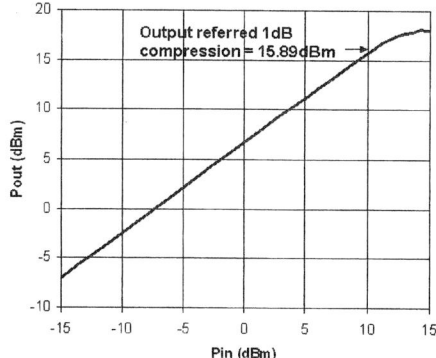

Figure 5. Simulated (post layout) output power vs. input power of the power amplifier at 24 GHz, Vcc=1.8V, and Ice= 30mA.

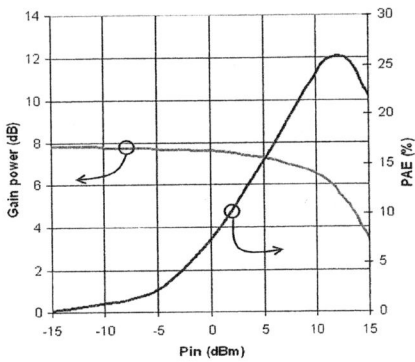

Figure 6. Simulated (post layout) power gain and PAE vs. input power of the power amplifier at 24 GHz, Vcc= 1.8V, and Ice= 30mA.

TABLE I.   COMPARISON WITH STATE OF THE ART

Freq. [GHz]	Technology	Pout [dBm]	PAE max [%]	area [mm²]	Vcc [V]	Ref.
24	0.13µm SiGe	18	25.9	0.3	1.8	This work (simulated)
24	0.18µm SiGe	21	13	6	1.8	[2]
22	0.18µm SiGe	23	19.7	6	1.8	[2]
24	0.18µm CMOS	14.5	11	1.26	2.8	[8]
17.2	0.13µm CMOS	17.8	15.6	0.81	1.5	[9]

A comparison of the power amplifier performances obtained in this work and previous work on power amplifier in silicon is presented in Table I. Promising results in high frequency performance of SiGe technology may lead to the development of low-cost automotive radars.

IV.   CONCLUSIONS

This paper reports on the design methodology that has been implemented to design a power amplifier at 24 GHz featuring outstanding performances at reduced supply voltage. The performance of the amplifier, occupying a chip area of 0.6mm×0.5mm, shows a maximum power gain of 7.8 dB. The peak power added efficiency of 25.9% was recorded at 24 GHz for 18 dBm of output power. A 24 GHz Si-based MMIC PA for low-cost SRR has been realized with high integration using lumped passives to minimize chip area. The chip demonstrates high level of mm-wave integration achievable in today's production silicon technology and feasibility of low-cost mm-wave systems for sensor and radio applications. The layout has been taped out and we should have the measurements for the conference.

REFERENCES

[1]   ETSI TR 102 263 V1.1.2 (2004-02) System Reference Document for automotive collision warning Short Range Radar.

[2]   Tak Shun Dickson Cheung and John R. Long, "A 21–26-GHz SiGe Bipolar Power Amplifier MMIC," IEEE JOURNAL OF SOLID-STATE CIRCUITS, VOL. 40, NO. 12, DECEMBER 2005.

[3]   U. R. Pfeiffer, S. K. Reynolds, and B. A. Floyd, "A 77 GHz SiGe power amplifier for potential applications in automotive radar systems," in *IEEE RFIC Symp. Dig. Papers*, Jun. 2004, pp. 91–94.

[4]   M. Laurens, B. Martinet, 0. Kermarrec, Y. Campidelli, F. Deleglise, D. Dutartre, G. Troillard, D. Gloria, J. Bonnouvrier, R. Beerkens, V. Rousset, F. Leverd, A. Chantre and A. Monroy, "A 15OGHz fT/fmax 0. 13µm SiGe:C BiCMOS technology," Bipolar/BiCMOS Circuits and Technology Meeting, 2003. Proceedings of the Volume , Issue , 28-30 Sept. 2003 Page(s): 199 - 202.

[5]   Esko Jarvinen, Sami Kalajo, and Mikko Matilainen, "Bias Circuits for GaAs HBT Power Amplifiers," Microwave Symposium Digest, 2001 IEEE MTT-S International, VOL 1, pages 507-510.

[6]   Matthias Rickelt and Hans-Martin Rein, "A Novel Transistor Model for Simulating Avalanche-Breakdown Effects in Si Bipolar Circuits," IEEE JOURNAL OF SOLID-STATE CIRCUITS, VOL. 37, NO. 9, SEPTEMBER 2002.

[7]   Matthias Rickelt , Hans-Martin Rein, and Eduard Rose, "Influence of Impact-Ionization-Induced Instabilities on the Maximum Usable Output Voltage of Si-Bipolar Transistors," IEEE TRANSACTIONS ON ELECTRON DEVICES, VOL. 48, NO. 4, APRIL 2001.

[8]   A. Komijani and A. Hajimiri, "A 24 GHz, +14:5 dBm fully-integrated power amplifier in 0.18 m CMOS," in *Proc. IEEE 2004 Custom Integrated Circuits Conf.*, Oct. 2004, pp. 561–564.

[9]   Andriy V. Vasylyev, Peter Weger, Winfried Bakalski, and Werner Simbuerger, "17-GHz 50–60 mW Power Amplifiers in 0.13-µm Standard CMOS," IEEE MICROWAVE AND WIRELESS COMPONENTS LETTERS, VOL. 16, NO. 1, JANUARY 2006.

978-1-4244-1983-8/08/$25.00 ©2008 IEEE

Noise analysis in Super-regenerative receiver systems

Prakash.E.Thoppay, Catherine Dehollain and Michel J.Declercq

Electronics laboratory(LEG1), Ecole Polytechnique Fédérale de Lausanne (EPFL), Switzerland

Email: prakash.thoppayegambaram@epfl.ch

Abstract—**The recent increase in the need for energy efficient wireless nodes have led to the study of various low power consumption receiver and transmitter architectures. The super-regenerative receiver architecture is one of the possible candidates for such a low power application. In this paper a detailed study on the noise analysis for such a kind of receiver is carried out. A closed form representation of the output signal to noise ratio for both narrow-band and wide-band communication is derived. The result indicates for a narrow-band communication the output signal to noise ratio cannot be better than a normal tuned amplifier. In the wide-band mode the SNR of super-regenerative receiver in linear regime is similar to a tuned amplifier.**

I. INTRODUCTION

There has been an increase in the study of wireless sensor network systems for its promising application such as body area networks(BAN) etc. Such a wireless sensor network consists of wireless nodes which are usually battery operated and hence an energy efficient wireless communication is required. The requirement for having an energy efficient wireless node has motivated the research community to study various low power consumption receiver architectures. The super-regenerative receiver architecture is one of the promising candidates for such a kind of low power application [1], [2].

The super-regenerative receiver architecture was originally conceptualized by E.H.Armstrong in 1922 [3]. A detailed explanation on the operating principle of the super-regenerative system is described in the section-II. The section-III gives an analysis on the noise added by the receiver circuitry and a closed form expression on the signal to noise ratio of the super-regenerative receiver system.

II. SUPER-REGENERATIVE RECEIVER: OPERATION PRINCIPLE

The main building blocks of a super-regenerative receiver system are the isolation amplifier/low noise amplifier, the core oscillator, the quench oscillator and the envelope detector as shown in Figure-1. The isolation/low noise amplifier has two purposes, it converts the received excitation voltage into current and injects into the main oscillator resonant circuit. Secondly, it prevents radiation of the voltage generated at the main oscillator back to the antenna thus acting as an isolation amplifier.

The core oscillator in Figure-1 consists of a LC resonant element in parallel with a negative resistance circuit. The negative resistance is controlled by the oscillator bias current. The quench signal periodically changes the oscillator bias

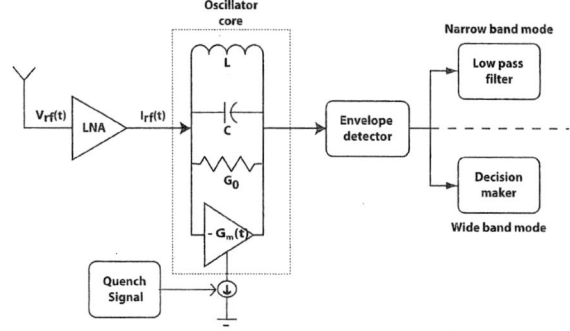

Fig. 1. Schematic of the super-regenerative receiver

current and thus the negative resistance seen by the resonant element. By increasing the bias current above the critical current, the core oscillator is allowed to start oscillation. When an excitation signal is injected into the resonant element around the time when the total conductance becomes negative, the oscillation starts in advance in comparison to the absence of the excitation signal. The difference in the start-up time of the oscillator is captured through an envelope detector. The output of the envelope detector is fed to a low pass filter or to a decision maker depending upon the operation mode. The explanation on the different operation modes are described in the later section.

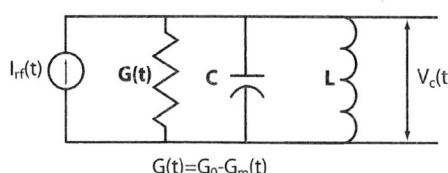

$$G(t)=G_0-G_m(t)$$

Fig. 2. Schematic of the resonant element

To understand the different parameters that affect the gain and bandwidth of the super-regenerative system, a short outline to the derivation of the output voltage across the resonator element is shown. The core oscillator resonant element can be modeled as shown in Figure-2. The effective conductance g(t) is periodically varied by the quench signal. The differential equation representing the voltage across the resonator is given by equation-1.

$$C\frac{d^2v_c}{dt^2} + g(t)\frac{v_c(t)}{dt} + \left(\frac{1}{L} + \frac{d(g(t))}{dt}\right)v_c(t) = \frac{di_{rf}(t)}{dt} \quad (1)$$

978-1-4244-1983-8/08/$25.00 ©2008 IEEE

By solving the above differential equation one can obtain the voltage across the resonator element. The rigorous derivation can be found in the reference [4], [5]. In this paper, the equations that are needed to understand the noise phenomenon in the super-regenerative receiver are mentioned. The output voltage across the resonator consists of terms due to free oscillation and forced oscillation. The forced oscillation is due to the input excitation signal. The equation-2 represent the super-regenerative system's output voltage when the input is $V_{in}(t) = V_0 p_c(t) \cos(2\pi f_c t)$, where $p_c(t)$ refers to the envelope of the input signal.

$$V_c(t) = V_0 K_r K_s p_e(t) \cos(2\pi f_c t) \qquad (2)$$

$$K_r = \zeta_0 \omega_0 \int_0^{t_2} p_c(t)s(t)dt; K_s = e^{-\omega_0 \int_{t_1}^{t_2} \zeta(t)dt} \qquad (3)$$

$$s(t) = e^{\omega_0 \int_{t_1}^{t} \zeta(t)dt}; p_e(t) = e^{-\omega_0 \int_{t_2}^{t} \zeta(t)dt} \qquad (4)$$

The super-regenerative system gain is given by the product of two terms namely super-regenerative(K_s) and regenerative gain(K_r) as shown in equation-3. The limits in the integration are as shown in Figure-4. The regenerative period refers to the region for which $G(t) > 0$ but less than G_0 or from time t=0 to t_1 as shown in Figure-4 and the gain during this period is called the regenerative gain. The super-regenerative period refers to the region for which $G(t) < 0$ and the gain during this period is called the super-regenerative gain.

Sensitivity and Selectivity: The term s(t) in equation-4 indicates the super-regenerative receiver's sensitivity. It should not be confused with the "minimum signal level" as defined in other standard receivers. Here the term sensitivity indicates the duration for which the super-regenerative receiver is influenced by the input excitation signal. Since the super-regenerative receiver encodes the message signal on the variation in the oscillation build-up, the receiver is sensitive only around the region when the effective conductance G(t) goes from positive to negative. The frequency response of the super-regenerative receiver is given by the selectivity curve. It is obtained by the Fourier transform of the sensitivity curve s(t).

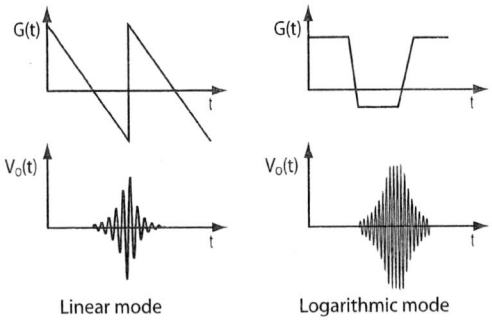

Fig. 3. Graph showing the various operation modes

Linear and Logarithmic mode: There are two operating modes in a super-regenerative receiver namely the linear and logarithmic mode. The two modes are distinguished based on

the output voltage level of the core oscillator. The output voltage level depends on the duration for which effective conductance across the resonator core is negative. If the output voltage level doesn't reach the saturation voltage before the effective conductance becomes positive then it is said to operate in a linear mode. When the output voltage reaches the saturation voltage it is said to operate in the logarithmic mode. The operating modes are shown in Figure-3. In a linear mode the output voltage is proportional to the input voltage as shown in equation-2.

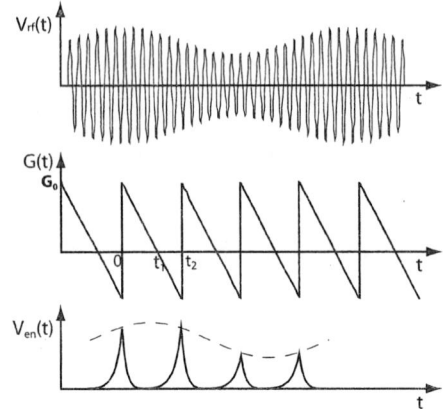

Fig. 4. Graph showing the narrow band operation mode

III. NOISE ANALYSIS IN SUPER-REGENERATIVE SYSTEMS

The noise analysis in the super-regenerative system is different from normal tuned amplifier due to its rather complicated demodulation/detection process. Lately the super-regenerative receiver is used for pulse detection apart from normal amplitude or OOK demodulation. When the super-regenerative receiver system samples the incoming signal it is said to operate in a narrow-band mode. When it is used to detect the presence of a pulse, it is said to operate in a wide-band mode. The noise analysis in both the modes are explained and comparisons are made with the normal tuned amplifier systems

A. Narrow-band mode:

Due to the fact that the super-regenerative receiver is sensitive to the excitation signal around the time when the effective conductance goes from positive to negative it can be thought of as a sampling process. Since the excitation signal is sampled it is necessary that the sampling frequency should be twice the message frequency. In the super-regenerative receiver the quench signal acts as the sampling signal. Thus for proper demodulation of the message signal the quench frequency should be atleast twice the message frequency.

The demodulation process for a narrow-band signal $[V_{rf}(t)]$ using a super-regenerative receiver is shown in Figure-4. The first row in Figure-4 indicates the incoming amplitude modulated RF signal $V_{rf}(t)$. The amplitude modulated excitation signal is fed into the resonant circuitry of the core oscillator. The second row in Figure-4 shows the quench signal. As the effective conductance goes from positive to negative the excitation signal is sampled and envelope variation is seen depending on the input excitation amplitude. The third row in Figure-4 shows output of the envelope detector $V_{en}(t)$. The output of the envelope detector is passed through a low pass filter where the message signal is demodulated shown as the dotted lines in the third row of Figure-4.

In a normal sampling system the signal is passed through an anti-aliasing filter and then to the sampler. The anti-aliasing filter restricts the incoming noise bandwidth. The sampling frequency is selected in such a way to respect the Nyquist criterion. In scenarios where the anti-aliasing filter is not present the incoming noise bandwidth which is usually higher than the signal bandwidth undergoes aliasing. Due to the aliasing process the out-of band noise is aliased into the desired bandwidth leading to the degradation in the signal to noise ratio in sampled systems. The noise aliasing phenomenon when the sampling frequency is less than the noise bandwidth is shown in Figure-5. In Figure-5 the sampling frequency is twice less than the noise bandwidth. The excess noise in the desired bandwidth is given by the equation-5. The factor two is due to the fact double sided power spectral density is assumed.

$$\text{Excess noise factor (ENF)} = 2 \left(\frac{BW_{noise}}{2 * f_s} \right) - 1 \quad (5)$$

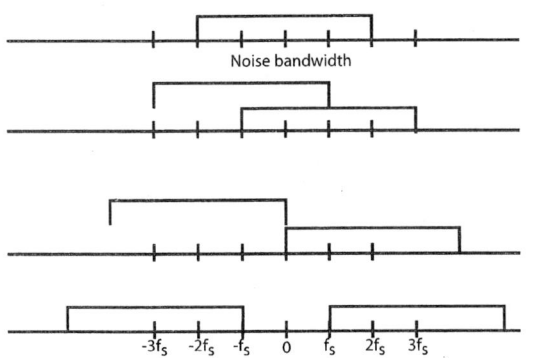

Fig. 5. Graph showing the noise aliasing phenomenon

The super-regenerative system is modeled as a sampler circuit providing gain as shown in Figure-6. The core oscillator along with the quench signal gives the functionality of the sampling circuitry. The representation is valid when the super-regenerative system is operating in the linear mode, since the gain is constant only in the linear regime. Also in the linear mode the noise variance at the output of the gain block is same as the input. The frequency selectivity of the super-regenerative receiver is due to the band-pass filter as shown in Figure-6.

As shown in the analysis on sampling circuits, when the noise bandwidth is larger than the sampling frequency, the sampling process leads to noise aliasing. This phenomenon is also called as noise folding. The super-regenerative receiver

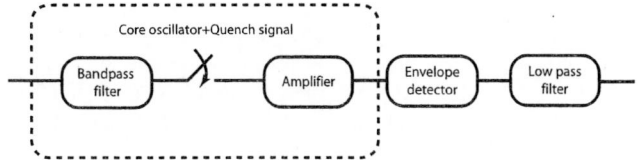

Fig. 6. Schematic of the super-regenerative system- Sampling model

system is modeled as a sampling system as shown in Figure-6. Henceforth noise folding occurs in such a system when the system bandwidth is greater than the received signal bandwidth. It can be shown that the super-regenerative receiver bandwidth is always greater than the received signal bandwidth. Therefore the super-regenerative receiver system has more noise power in the signal bandwidth due to noise folding process. Thus a normal tuned amplifier(Amp) has a better signal to noise ratio in comparison to a super-regenerative(SR) system for the same gain. The noise degradation in a super-regenerative system is by a factor $\frac{BW}{f_q}(> 1)$ as given by equation-6. The signal power is given by S and the bandwidth is given by BW. The noise power without folding is given by $N = \frac{N_0}{2} * BW$.

$$\text{Noise power after folding}(N_a) = \frac{N_0}{2} * BW * \frac{BW}{f_q}$$

$$\left(\frac{S}{N_a} \right)_{SR} = \left(\frac{S * f_q}{N * BW} \right), SNR_{amp} = \left(\frac{S}{N} \right)$$

$$SNR_{SR} = \left(\frac{SNR_{amp} * f_q}{BW} \right) \quad (6)$$

In a sampling system the noise folding phenomenon can be prevented by sampling at a higher sampling rate or by restricting the input signal bandwidth using an anti-aliasing filter. On contrary, in a super-regenerative receiver the noise folding phenomenon cannot be prevented simply because the bandwidth of the bandpass filter is depended on the quench frequency [2]. Hence the noise folding will occur in such systems degrading the SNR. Therefore the SNR of a super-regenerative receiver cannot be better than the normal tuned amplifiers.

B. Wide band mode:

When the super-regenerative receiver is used to detect the presence of a pulse it is said to operate in a wide-band mode. The principle is similar to the narrow-band mode, the incoming pulse influences the oscillation build-up of the core oscillator. The output of the core oscillator is passed through an envelope detector. The peak value of the envelope detector is sampled and compared with a threshold value. If the peak value is greater than the threshold then a pulse is detected or vice-versa. The difference between the narrow-band mode and the wide-band mode is, in the narrow-band mode the output is passed through a low pass filter whereas in the wide-band it is compared with a threshold value. Thus in the wide-band mode the core oscillator acts as a normal gain stage when operated in the linear mode. The advantage of using super-regenerative

978-1-4244-1983-8/08/$25.00 ©2008 IEEE

receiver for pulse detection is, the gain to current ratio is higher in comparison to a normal tuned amplifier due to the positive feedback in the oscillator. To compute the output signal to noise ratio, the output noise power $E[|n_o(t)|^2]$ need to be calculated. The output noise voltage for a noise waveform n(t) can be derived by solving equation-1 and is shown in equation-7 [4].

$$n_o(t) = 2K_s p_e(t)\zeta_0 \int_{-\infty}^{\infty} \dot{n}(\tau)s(\tau)\sin\omega_0(t-\tau)d\tau \quad (7)$$

Since we are interested around the center frequency (ω_0), the derivative of the noise term can be approximated as shown in equation-8.

$$\dot{n}(\tau) \approx \omega_0 n(\tau) \quad (8)$$

By using equation-8 in equation-7,

$$n_o(t) = 2K_s p_e(t)\zeta_0\omega_0 \int_{-\infty}^{\infty} n(\tau)s(\tau)\sin\omega_0(t-\tau)d\tau \quad (9)$$

The output noise variance is computed in equation-10, the noise term n(t) is assumed to be AWGN and hence the auto-correlation has a peak only when $\tau = \tau'$, in other words the second moment about origin is given by, $E[n(\tau)n(\tau')] = \frac{N_0}{2}\delta(\tau-\tau')$. N_0 indicates the double sided power spectral density. Similarly, the output signal voltage for an input signal $V_{in}(t) = V_a p_c(t)\cos(\omega_0 t)$ is given by equation-11.

$$V_0(t) = \left(\zeta_0\omega_0 V_a K_s p(t) \int_{-\infty}^{\infty} p_c(\tau)s(\tau)d\tau\right)\cos\omega_0 t \quad (11)$$

To calculate the output signal power the mean square value is computed and it is given in equation-12.

$$\overline{V_0^2(t)} = \frac{1}{2}\left(V_a\zeta_0\omega_0\overline{p_e(t)}K_s \int_{-\infty}^{\infty} p_c(\tau)s(\tau)d\tau\right)^2 \quad (12)$$

The output signal to noise ratio is calculated by dividing the output signal power and the noise power. In calculating the noise power and the signal power the mean square value is calculated across a unit resistance.

$$SNR = \frac{\overline{V_0^2(t)}}{E[|n_o(t)|^2]}$$
$$= \frac{V_a^2(\int_{-\infty}^{\infty} p_c(\tau)s(\tau)d\tau)^2}{2N_0 \int_{-\infty}^{\infty} s^2(\tau)d\tau} \quad (13)$$

The output signal to noise ratio can be rearranged in terms of the energy per bit, the input signal energy can be computed from $\overline{V_{in}^2(t)}$.

$$\overline{V_{in}^2(t)} = E_b = \frac{1}{2}V_a^2 \int_{-\infty}^{\infty} p_c^2(\tau)d\tau$$

$$SNR = \frac{E_b(\int_{-\infty}^{\infty} p_c(\tau)s(\tau)d\tau)^2}{N_0 \int_{-\infty}^{\infty} p_c^2(\tau)d\tau \int_{-\infty}^{\infty} s^2(\tau)d\tau} \quad (14)$$

The equation-14 gives the SNR in terms of the E_b and from this equation it is clear that if we want to maximize the SNR then $p_c(t) = s(t)$ [by applying schwartz inequality: $(\int_{-\infty}^{\infty} f_1(t)f_2(t))^2 dt = \int_{-\infty}^{\infty} f_1^2(t)dt \int_{-\infty}^{\infty} f_2^2(t)dt$ when $f_1(t) = f_2(t)$]. Thus in a super-regenerative receiver while operating in a wide-band mode, to achieve maximum signal to noise ratio, the quench signal shape is to be properly selected.

IV. CONCLUSION

In this paper the narrow-band and wide-band modes are explained. The influence of noise in both the regimes are described. It is shown in section IIIa that in narrow band, the noise figure of super-regenerative receiver cannot be better than a normal tuned amplifier due to the dependency of sampling frequency and noise bandwidth. In wide-band mode, the output signal to noise ratio is equivalent to a normal tuned amplifier and the importance of quench signal in maximizing the output SNR is shown. In both the analysis it is assumed that the super-regenerative receiver is operating in linear regime.

V. ACKNOWLEDGEMENT

This work is performed under MICS framework funded by the Swiss National Science Foundation.

REFERENCES

[1] B.Otis et al.,*A 400μW-RX, 1.6mW-TX Super- Regenerative Transceiver for Wireless Sensor Networks*,ISSCC, pp.396-397, Feb-2005.
[2] A.Vouilloz et al, *A Low-Power CMOS Super-Regenerative Receiver at 1 GHz*, IEEE Journal of Solid State Circuits, VOL. 36, pp. 440-451, March 2001.
[3] E.H.Armstrong,*Some recent developments of regenerative circuits*, Proc IRE, pp. p. 244-260., Aug. 1922,10,.
[4] F. Xavier Moncunill-Geniz et al, *A Generic Approach to the Theory of Superregenerative Reception*, IEEE Transactions on Circuits and Systems-I, VOL. 52, NO. 1,pp. 54-70., JANUARY 2005.
[5] Whitehead.J.R, *Super-Regenerative Receivers*, Cambridge Univ. Press, 1950.

$$E[n_o(t)n_o^*(t)] = (2K_s p_e(t)\zeta_0\omega_0)^2 \int_{-\infty}^{\infty}\int_{-\infty}^{\infty} E[n(\tau)n^*(\tau')]s(\tau)s^*(\tau')\sin\omega_0(t-\tau)\sin^*\omega_0(t-\tau')d\tau d\tau'$$

$$= (2K_s p_e(t)\zeta_0\omega_0)^2\frac{N_0}{2} \int_{-\infty}^{\infty}\int_{-\infty}^{\infty} \delta(\tau-\tau')s(\tau)s(\tau')\sin\omega_0(t-\tau)\sin\omega_0(t-\tau')d\tau d\tau'$$

$$= (2K_s p_e(t)\zeta_0\omega_0)^2\frac{N_0}{2} \int_{-\infty}^{\infty} s^2(\tau)\sin^2\omega_0(t-\tau)d\tau = (2K_s p_e(t)\zeta_0\omega_0)^2\frac{N_0}{2} \int_{-\infty}^{\infty} s^2(\tau)\sin^2(\omega_0\tau-\phi)d\tau$$

$$\approx (K_s p_e(t)\zeta_0\omega_0)^2 N_0 \int_{-\infty}^{\infty} s^2(\tau)d\tau \quad (10)$$

978-1-4244-1983-8/08/$25.00 ©2008 IEEE

A 150µW-11b Readout Circuit for Lab-on-a-Chip Applications

P. Delizia[1], S. D'Amico[1], A. Baschirotto[1,2]

[1]Department of Innovation Engineering, University of Salento, Italy
[2]Department of Physics, University of Milano Bicocca, Italy

Abstract - **An integrated readout circuit for Lab-on-a-Chip applications is presented. The overall system consist of a 640×480 array of capacitor sensors and actuators. Sensors detect dielectric permittivity variation thanks to dielectrophoresis (DEP) process. Usually for this kind of applications an off-chip analog-to-digital converter is used. As a consequence, the noise floor increases and high signal-to-noise ratios are difficulty achieved. On the other hand, these applications requires a stringent noise floor specification (>10b) and a relaxed linearity (≈8b). In this design, the noise coupled to the signal at the chip pad is reduced by using an on-chip analog-to-digital converter. The complete sensor readout channel is composed by two main blocks: a pre-amplifier with programmable gain and an algorithmic analog-to-digital converter with a 1.5-bit/stage architecture. Each one is realized with fully differential switched capacitor technique. In order to save chip area and power consumption a time sharing technique has been taken into account using a single operational amplifier for the pre-amplification stage and the conversion stage. The proposed A/D converter has 11b resolution, a sampling rate of about 100ksample/s and an input full-scale range of 1.2Vp-p differential. Simulation results show a SNR=65.7 dB and an ENOB value of 10.6b. Its power consumption is about 150µW. Readout chain is implemented in 0.35µm CMOS technology with a 3.3V supply voltage.**

I. INTRODUCTION

Significant research efforts in biology, chemistry and engineering have been recently aimed to pursuing the advantages of miniaturization for cheaper, better and faster sample analysis. Micro Total Analysis Systems (µTASs) were envisioned in the late 1980s as miniaturized, highly integrated chemical analysis systems. The advent of DNA microarrays, propelled by genomic research, captured the attention of researchers and investors alike. Although the field was generally indicated as that of biochips, the word "Lab-On-A-Chip" (LOAC) entered the jargon to differentiate between passive microarrays and microanalytical systems sporting some degree of integration, programmability or microfluidic capabilities.

LOAC systems are typically based on a capacitor sensor array. The system under development consists of a CMOS chip covered by a conductive glass lid which is separated by the chip by about 100µm [1]. The aim is to detect variations in dielectric permittivity due to the presence of particles in the region above superficial electrodes, which affects the coupling capacitance with the lid. This is possible thanks to

dielectrophoresis (DEP). In the system under development the elementary unit is replicated to form a 640×480 array. A block diagram of the chip is shown in Fig.I, each microsite is addressed by specific control signals generated by the row and column decoders through row and column circuits. The pitch of each site is 20µm (fixing also the spatial resolution). Each microsite is supplied by an actuator circuit that senses the dielectric permittivity variation and provides an output voltage, V_{oarr}.

This paper describes the design of the electronic read-out channel. Each column has a proper read-out channel in order to obtain the maximum flexibility, 640 identical channels are needed. The main requirements regard the signal-to-noise level (in excess of 10b in order to detect any small event) and the area occupancy (to be minimized because of the large number of read-out channels, 640). The linearity specification is less critical since the read-out input signal is provided by a source follower operating at low bias current, which is limiting the linearity performance. With respect to the previous realizations [1], the innovation of this design consists in realizing on-chip the A/D conversion. In the previous design the chip output signal was analog and then additional noise was collected at pad and board level. In this design, at the cost of design complexity and die area, the A/D conversion is on-chip and this guarantees significantly larger noise immunity at pad and board level.

The paper is organized as follow: in Section II the readout circuit architecture is reported and the time sharing technique is introduced; in Section III the pre-amplifier is described; in Section IV the A/D converter is reported, all blocks are described and noise considerations are reported. Section V reports simulation results and Section VI draws conclusions.

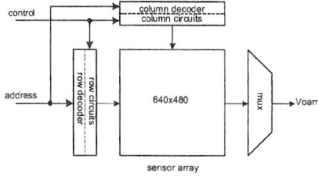

Fig. I - Block diagram of the chip

II. READOUT CHAIN ARCHITECTURE

The readout chain architecture is made up of a preamplifier and an A/D converter. The overall scheme is shown in Fig.II. A fully-differential philosophy was adopted

978-1-4244-1983-8/08/$25.00 ©2008 IEEE   193

due to the mixed-mode chip nature of this design. Fully-differential circuits are less susceptible than their single-ended counterparts to common-mode noise, such as noise on the power supplies that is generated by digital circuits integrated on the same substrate as the analog circuits.

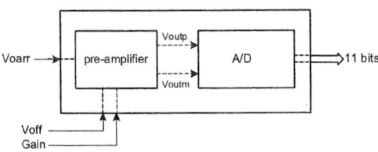

Fig. II - Readout architecture.

The pre-amplifier amplifies the single-ended input voltage V_{oarr} produced by the sensor array and provides a differential output voltage to the analog-to-digital converter. The sensor input signal may be unipolar. For this reason, input signals control V_{off} and $Gain$ allow pre-amplifier to obtain more sensibility and to optimally allocate the unipolar signal within the ADC input range. The analog-to-digital converter provides a resolution of 11 bits, it is realized by a fully-differential structure such as the pre-amplifier. The sampling frequency is about 100kHz. The input full-scale range of the A/D converter is about 1.2Vp-p differential.

The pre-amplifier stage and the conversion stage are performed in different time interval and both use the same time-shared operational amplifier, Fig.III. This choice gives chip area and power savings. Fig.III shows also the switching network for the time sharing technique. During the first time interval it behaves as pre-amplifier of the input signal, after which it realizes the multiply-by-2 feature of the conversion.

Phases $PHIXPA$ are obtained by a logic and between phases $PHIX$ and PA whereas $PHIXAD$ are obtained by a login and between phases $PHIX$ and AD. Phases $SIGXPA$ are obtained by a login and between phases $SIGX$ and PA. The phases provided from the outside are $SIG1, SIG2, PA, AD$ and a clock at the frequency of about 12MHz. A classic non-overlapping clocks circuit has been implemented on-chip to obtain $PHI1$ and $PHI2$ phases.

The pre-amplifier stage is performed during PA phase and the conversion stage is performed during AD phase.

III. PRE-AMPLIFIER.

The pre-amplifier stage is active only during PA phase. It is realized by a fully-differential switched-capacitor structure and uses the correlated-double-sampling technique for low frequency noise reduction [2]. During $PHI1PA$ phase the opamp input offset voltage is sampled across input capacitances C_i and C_{off}. After this, during $PHI2PA$ phase, the offset voltage is subtracted from the signal voltage by appropriate switching of the capacitors. It provides an accurate transfer function due to the accurate integrated capacitance matching. All the switches are CMOS pass-gate with $W_{NMOS}=0.8\mu m$, $W_{PMOS}=0.8\mu m$ and with $L=0.35\mu m$ for both transistors. The input capacitance C_i is fixed to be four times a unit capacitor $C_u \cong 270fF$ while the feedback capacitance C_f and the capacitance on the branch of V_{off} (C_{off})

are implemented as a bank of four unit capacitors each in series with CMOS pass-gate. The switches $S2PA, S3PA, S4PA$ are set by using the G_{ain} input signals to fix variable gains in the set of X={1, 4/3, 2, 4}. Then, in the following C_f will be

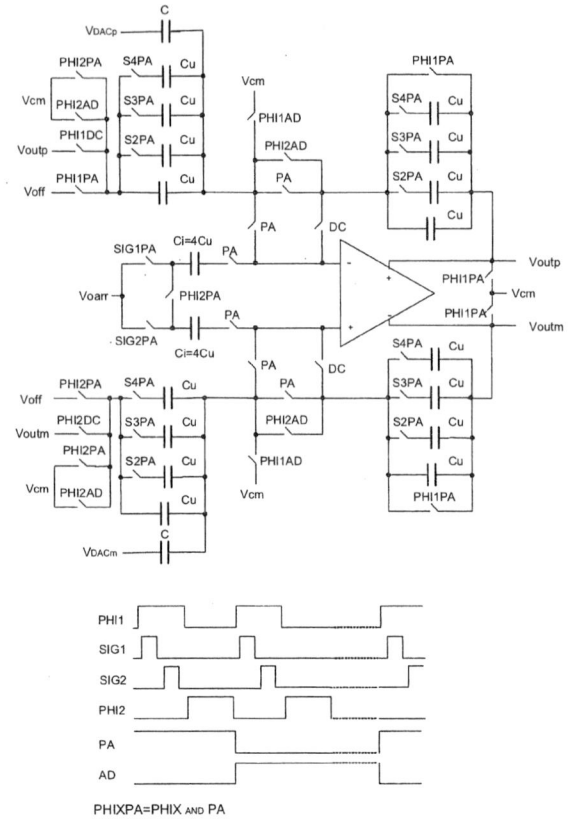

Fig. III - Switching network for time-sharing technique.

represented by XC_u. During $SIG1$ the input voltage V_{aorr} is sampled across C_i, it represents the result of the dielectrophoresis process of the addressed microsite. While $PHI1$ is still high a $RESET$ signal is activated to the same microsite and the reset voltage is sampled during $SIG2$ on the other C_i. The output differential voltage is provided during $PHI2$ according to the following relationship:

Eq.1
$$V_{outp} - V_{outm} =$$
$$= \frac{C_i}{C_f} \cdot \left(V_{oarr} @ SIG2 - V_{oarr} @ SIG1 \right) - \frac{V_{off}}{2}$$

V_{off} allows to modify the unipolar input signal common mode voltage and allocate it within the ADC input range.

To achieve the target signal-to-noise ratio > 10b, an accurate noise analysis of the circuit is proposed. At first only kT/C noise is taken into account and then the opamp noise contribution is analysed. During $PHI1$ the noise charge stored on C_i, C_{off} and C_f is:

Eq.2
$$Q^2_{n,Ci} = kT \cdot C_i$$
$$Q^2_{n,Coff} = kT \cdot C_{off}$$
$$Q^2_{n,Cf} = kT \cdot C_f$$

During $PHI2$ the output noise due to the previous charge noise and switches active in this phase is about:

Eq.3
$$V_n^2 = \frac{kT}{C_f} + \frac{kT \cdot C_i}{C_f^2} + \frac{kT \cdot C_{off}}{C_f^2} +$$
$$+ \frac{kT}{C_i} \cdot \left(\frac{C_i}{C_f}\right)^2 + \frac{kT}{C_{off}} \cdot \left(\frac{C_{off}}{C_f}\right)^2$$

The opamp noise contribution is about:

Eq.4
$$V_n^2 = V_{n,opamp}^2 \cdot \left(1 + \frac{C_i + C_{off}}{C_u}\right)^2$$

Therefore, substituting the capacitance values and supposing uncorrelated noise sources the total output noise of the pre-amplifier is about:

Eq.5
$$V_n^2 = 2 \cdot \left(\frac{kT}{X \cdot C_u} \cdot \left(3 + \frac{8}{X}\right) + V_{n,opamp}^2 \cdot \left(2 + \frac{4}{X}\right)^2\right)$$

In the previous formula the factor 2 is due to the fully differential structure.

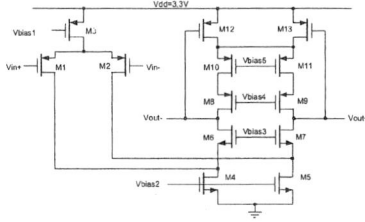

Fig. IV - Fully-differential folded cascode schematic.

The OTA design in the pre-amplifier schematic is a fully differential folded cascode with continuous time common mode feedback using triode devices, as shown in Fig.IV. This choice is principally due to its minimum occupied area. Its performance is as follow: bias current=20μA, dc-gain ≈ 80dB, unity-gain-bandwidth≈7MHz and phase margin ≈ 88° (C_{load}=1pF). These features allow obtaining an 11 bits settling precision in half period time.

IV. ANALOG TO DIGITAL CONVERTER ARCHITECTURE.

The essential point of this work is to design the electronic interface and the sensor array on the same chip. For this reason an algorithmic A/D converter was chosen where the number of components is drastically reduced in comparison with other structures [3]. Furthermore, a 1.5-bit/stage architecture, shown in Fig.V, has been chosen for relaxed component requirements. The internal clock frequency needed is about 1.2MHz to obtain a sampling frequency of 100ksample/s. A switched capacitor implementation was chosen.

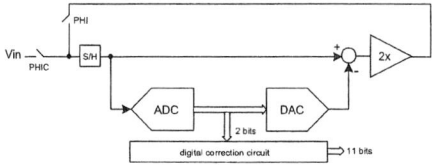

Fig. V - Algorithmic A/D converter with 1.5-bit/stage architecture.

The main blocks are a sample&hold (S/H), a multiply-by-two amplifier, a sub-ADC, a DAC and a digital correction

circuit. For this architecture the accuracy requirements of the sub-ADC are greatly reduced, in this case a maximum offset of $V_{ref}/4$ can be tolerated before bits error occurs. This allows the converter to neglect comparator offset performance. Otherwise any offset cancellation (CDS or large device area) would increase the ADC area occupancy. The A/D operation are described in [3].

The two-bits ADC block is implemented by two fully-differential clocked comparators and a switched network; each one schematic is depicted in Fig.VI. The reference voltages in this case are represented by the supply voltages, then an active charge redistribution is performed during *PHI2* phase. All switches designed are CMOS pass-gate with W_{NMOS}=0.8μm, W_{PMOS}=0.8μm and L=0.35μm for both transistors. The capacitor C_1 is set to 0.1pF, capacitor C_2 is set to 5fF. In this case a matching error between the two capacitors is not critical because this can be associated to an offset error of the comparator. This architecture can tolerate a maximum offset of $V_{ref}/4$ before bits error occurs. Simulation results show a correct behavior for C_2< 10fF. The comparator circuit (shown in Fig.VII) is based on a circuit reported by [4]. When *PHIC* is high the comparator is in the reset phase whereas when it is low the comparison starts. The bias current is set to 5μA.

The two bits DAC implementation is based on the circuit shown in Fig.VIII. It is realized by two different single-ended structures due to the fully differential philosophy of the entire interface. A logic circuit (not shown) processes the comparators outputs and generates the two bits of the stage and three phases (*PHI1*, *PHI2*, *PHI3*) to drive the reference voltages. Only the supply voltages and the common mode voltage are taken into account, therefore an active charge redistribution takes place. This solution allows saving chip area and power consumption in comparison with standard R-string DAC implementation.

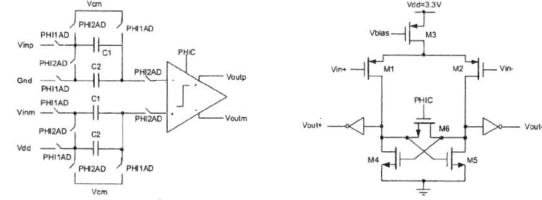

Fig. VI - Half sub-ADC schematic.   Fig. VII - Comparator schematic.

Fig. VIII - D/A converter circuit.

The multiply-by-two is obtained by circuit in Fig.3 when *AD* phase is active. The input branches V_{oarr} and V_{off} are disconnected during conversion and the only active capacitors are C_u without series switch and C (≈49fF). The circuit works as follow. During *PHI1AD* phase the opamp output voltage is stored on C_u input capacitance then, during PHI2AD and thanks to the active charge redistribution, the opamp output voltage become:

Eq.6
$$V_o = V_{i+1} =$$
$$= \begin{cases} 2 \cdot Vi - Vref, Vi > Vref/4, b_1 b_0 = 10 \\ 2 \cdot Vi, -Vref/4 < Vi < Vref/4, b_1 b_0 = 01 \\ 2 \cdot Vi + Vref, Vi < -Vref4, b_1 b_0 = 00 \end{cases}$$

About noise consideration, the total input referred noise of the algorithmic stage using the above circuit is given by:

Eq.7
$$V_n^2 = V_{n,1}^2 + \frac{V_{n,2}^2}{4} + \frac{V_{n,3}^2}{16} + \frac{V_{n,3}^2}{64} + \dots$$

The noise contributions of each cycle are due to kT/C noise and opamp *IRN*; the capacitance values and the IRN value of the designed opamp do not compromise the noise performance of the converter.

V. SIMULATION RESULTS.

Fig.IX shows the output spectra of the ADC. The input signal voltage is set to 1.2Vp-p i.e. the full-scale range, while the input frequency signal is about 20kHz which is about the maximum input signal frequency. The sampling frequency is 100ksamples/s. The number of samples is 512. The SNR value obtained is about 65.7 dB and the relative ENOB value obtained is 10.6 bits.

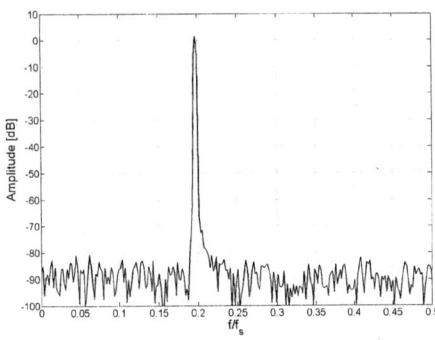

Fig. IX - ADC output spectra for f_s=100ks/s, $f_{in} \approx$20kHz and V_{in}=1.2Vp-pdiff

MATLAB simulations show that capacitance mismatch affects only linearity performance but not noise floor. This result respects the signal-to-noise ratio constraint that is a must for this design. Table I summarizes the performance obtained for this A/D converter design.

The layout of the readout channel has to be designed in a strip of 20μm because of the pitch site size. Since only three different metal layers can be used (the other two metal layers are used for power routing of the capacitor array) a digital approach has been adopted. These layers have been longitudinally and parallel orientated towards the layout. The layout of the readout channel occupies an area of about 2100×20μm², Fig.X. Bias circuits and phase generator occupy an area of about 150×200μm². Because of the fully-differential design philosophy, the layout has vertical symmetry axes.

Fig. X - Read-out channel layout.

TABLE I - A/D Converter Performance Summary.

Process	0.35μm CMOS
Supply	3.3V
Sampling rate	100ksample/s
Resolution	11b
Full-Scale Input	1.2Vp-p (differential)
Power Consumption	150μW
SNR	65.7dB
SNDR	65.6dB
ENOB	10.6b

VI. CONCLUSION

An integrated readout circuit for Lab-on-a-Chip applications has been described. On board A/D conversion allows readout circuit to obtain better performance in term of signal-to-noise ratio with respect to the off-chip solution. The readout chain is designed with the constraint of minimal occupied area and good performance at the same time. The main constraint of the design is the achievement of a >10b signal-to-noise ratio with a small die area at a 100kHz sampling rate. A fully-differential SC architecture is adopted. It is made up of a pre-amplifier circuit with programmable gain in the set X={1, 4/3, 2, 4}, and an algorithmic 1.5-bit/stage architecture A/D converter. This choice allows designing relaxed components requirements and saves die area. The achieved results show a good resolution for the readout circuit for any gain value and good A/D converter performances: SNR=65.7dB, SNDR=65.6dB, ENOB=10.6b. The power consumption is about 150μW.

The authors acknowledge Silicon Bio Systems s.r.l., for the collaboration and technical informations on sensor array. This project is partially funded by the Regione Puglia Explorative Project "Integrated microsystem for rapid biological cell detection (Lab-on-a-Chip)".

REFERENCES.

1. N. Manaresi, A. Romani, G. Medoro, L. Altomare, A. Leonardi, M. Tartagni, R. Guerrieri, "A CMOS Chip for Individual Cell Manipulation and Detection," *IEEE J. Solid-State Circuits*, Vol.38, No.12, December 2003, pp. 2297-2305.

2. G. C. Temes, and C. Enz, "Autozeroing, Correlated Double Sampling, and Chopper Stabilization," *IEEE Procedings*, to be published, 1996.

3. R. Van De Plassche, "Integrated Analog-to-Digital and Digital to Analog Converters ", 1994 Kluwer Academic Publichers.

4. V. Peluso, P. Vancorenland, A. M. Marques, M. S. J. Steyaert, and W. Sansen, " A 900mV low-power δΣ A/D converter with 77-dB dynamic range," *IEEE J. Solid-State Circuits*, Vol.33, December 1998, pp. 1887-1897.

An Interface Circuit for Temperature Control and Read-Out of Metal Oxide Gas Sensors

A. Lombardi, M. Grassi, L. Bruno, P. Malcovati
Department of Electrical Engineering,
University of Pavia,
Via Ferrata, 27100 Pavia, Italy
[andrea.lombardi, marco.grassi, luca.bruno,
piero.malcovati]@unipv.it

A. Baschirotto
Department of Physics,
University of Milano-Bicocca
Piazza Scienza, 20126 Milano, Italy
andrea.baschirotto@unimib.it

Abstract—**This paper presents a complete gas-sensing system. The latter consists of a high-efficiency temperature control loop with a switching heater based on a custom digital control logic and of a wide-dynamic-range interface circuit able to operate without calibration. The temperature control loop is driven by a digital set-point and the simulations results show that it controls the temperature of the sensor over a range of 250 °C with an accuracy better than 0.5 °C. The temperature control logic is reconfigurable and can be used to control the temperature of sensors with different baseline resistance value and different sensitivity. The gas-sensing interface circuit is based on a resistance to frequency conversion and achieves, without calibration, a precision in resistance measurements of 0.5% over a range of 5 decades (dynamic range, DR=146 dB). All voltage references are reconfigurable and on chip buffered.**

I. INTRODUCTION

Gas-sensing systems are getting more and more important in order to keep under control the air pollution and to avoid the human exposure to dangerous gases. Micromachined resistive gas-sensor now can be fabricated small enough to reach the operating temperature (around 300-400 °C) in few tens of milliseconds, exploiting a heater/thermometer embedded in the sensor itself, which consumes only a small amount of power even less than 100 mW in steady state [1,2,3]. For these reasons, handheld gas-sensing systems are getting more and more popular: manufacturers are dropping expensive ad hoc instruments and large PCB solutions, while academic institutions and government-funded consortia are working in the research field in order to develop devices based on the microsystem or micromodule approach [4,5].

A gas-sensing microsystem consists of an array of gas-sensors, a temperature control circuit, an electronic read-out block and a data processor. The most used sensors nowadays are the chemoresistive ones. This kind of devices are based on metal-oxide thin films (MOX), usually realized with the sol-gel technique and deposited by spin coating on silicon micromachined substrates, equipped with an integrated platinum heater-termometer and interdigitated electrodes. Since these types of gas-sensors are based on direct analyte

adsorption and charge process between the gas molecules and the MOX surface, which cause an electrical resistance variation of the gas-sensing element, they behave electrically as ohmic devices. Considering the state of the art of the manifacturing, the sensor resistance value may vary across several decades, being the combination of three variable components: the nominal baseline, the deviation from this nominal baseline due to ageing and working temperature and the resistance variation due to gas concentration [6]. Since each contribution to the resistance variation is of the order of one-two decades and the sensor resistance measurement accuracy required to measure gas concentrations of the order of tens of parts per million is about 1%, a dynamic range of about 140 dB is demanded [7].

Significant power saving can be obtained by using dynamic feature extraction techniques, that are able to extract information also from the dynamic sensor response, thus requiring a shorter array query time for achieving the same performance. Moreover, by using specific patterns to modulate the temperature of the sensor a higher selectivity with respect to different analytes can be obtained. In order to achieve the required accuracy in the chemical measurement, the temperature of the sensor has to be accurately controlled (within ±5 °C), thus requiring a dedicated temperature control loop [8]. Typically the chemoresistive sensors are used over a temperature range from 200 °C to 450 °C, with best performance in the range from 300 °C to 400 °C.

II. TEMPERATURE CONTROL SYSTEM STRUCTURE

The proposed temperature control system, as shown in Figure 1, is a closed loop circuit driven by a digital set-point. The temperature is controlled by means of two platinum resistors. The first resistor, called R_{HEATER} has one terminal connected to ground and the other connected either to V_{DD} or to ground. R_{HEATER} is embedded in the gas-sensor membrane, thus when it is connected to V_{DD}, the sensor is heated by Joule effect. Another platinum resistor, R_{SENS}, that is also embedded in the sensor membrane, is used as a thermometer.

Prototype fabrication has been funded by PRIN 2005092937 IT Project

978-1-4244-1983-8/08/$25.00 ©2008 IEEE

The value of R_{SENS}, processed by a suitable signal conditioning network, is sampled by an ADC.

Table I shows the relationship between the power dissipated by R_{HEATER} and the temperature reached by the gas-sensor, measured in an open-loop configuration. Instead of providing to R_{HEATER} the exact amount of power required to decrease the difference between the set-point and the actual temperature, a switching control strategy has been adopted. When it is necessary to increase the temperature of the gas-sensor, R_{HEATER} is directly connected to V_{DD}, thus providing the maximum power and hence maximizing the power efficiency. By contrast, R_{HEATER} is connected to ground when the temperature of the gas-sensor is higher than desired. The temperature control system has been designed for a gas-sensor with an R_{HEATER} whose value at room temperature is about 30 Ω and α, the thermal coefficient, is $2.2 \cdot 10^{-3}$ °C^{-1}. Therefore, the value of R_{HEATER} at 450 °C is less than 60 Ω. Even when R_{HEATER} is doubled by the effect of the temperature, the provided power is by far enough to allow the gas-sensor to reach the maximum temperature, which means also that the circuit can work with sensors characterized by an R_{HEATER} with a higher value at room temperature or with a higher thermal coefficient.

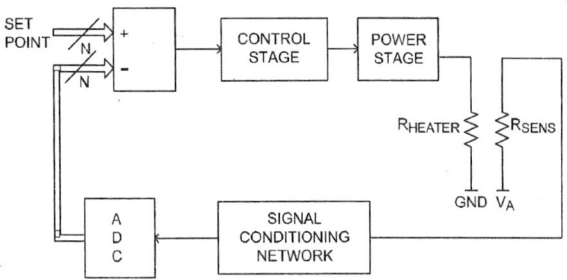

Figure 1. Temperature control system block diagram.

TABLE I. OPEN-LOOP POWER-TEMPERATURE RELATIONSHIP

POWER	23 mW	40 mW	60 mW	72 mW
TEMPERATURE	200 °C	300 °C	400 °C	450 °C

III. TEMPERATURE CONTROL SYSTEM CIRCUIT DETAILS

The forward path of the temperature control chain is reported in Figure 2. The 9 bits digital set-point allows the desired input pattern to be easily obtained. The value of the set-point is compared with the output of the ADC: when the set-point is larger than the ADC code the output will be 1, otherwise the output will be 0. The aim of the finite state machine placed after the digital comparator block is to force a minimum frequency to the control signal, higher than the frequency of the signal that needs to be delivered on the sensor for temperature modulation. This finite state machine has a single input and a single output and is characterized by three states called A, B and C, as explained in Table II. State B and state C occur once every 1024 clock cycles. The effect of the finite state machine is illustrated in Figure 3 on three example of the control signal. Forcing the output signal to assume the value "0", during state A, and "1", during the state B, every 1024 clock cycles with any input signal, a minimum frequency of the output is guaranteed.

TABLE II. FINITE STATE MACHINE

STATE	OUT
A	NOT(IN)
B	0
C	1

The p-MOS switch that connects R_{HEATER} to V_{DD}, needs a large W/L ratio, in order to achieve a low resistance and hence limit the power loss. A p-MOS switch characterized by W=10000 µm and L=0.5 µm achieves an impedance that is about 3 Ω, thus the power dissipated by the switch is about 10% of the power dissipated by R_{HEATER}, when its temperature is 25 °C and decreases when the temperature of the sensor increases, since R_{HEATER} increases with temperature. For example at 300 °C, the power dissipated by the switch is about 6% of the power dissipated by R_{HEATER}.

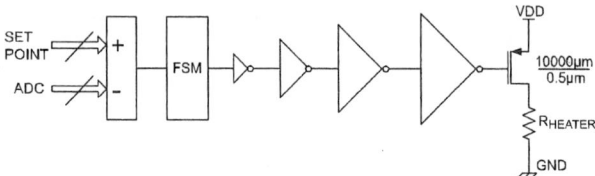

Figure 2. Forward path of the control chain.

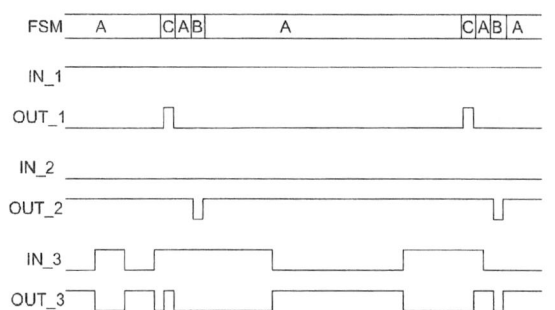

Figure 3. Finite state machine behavior.

Figure 4 shows the feedback path of the control system, that consists of a signal conditioning network and an A/D converter. The platinum resistance R_{SENS} is biased by means of a digitally programmable current generator, in order to allow the adjustment of the feedback loop gain. The selectable output currents are multiple of 0.2 mA with a maximum value of 3 mA. Of course in typical conditions with typical sensors, currents below 1.5 mA will be exploited to have higher efficiency.

Varying the output current of the digitally programmable current generator I_{DC} and the voltage level V_A, it is possible to use sensors that have different resistance values at room temperature or to vary the temperature operating range as shown in Table III.

978-1-4244-1983-8/08/$25.00 ©2008 IEEE

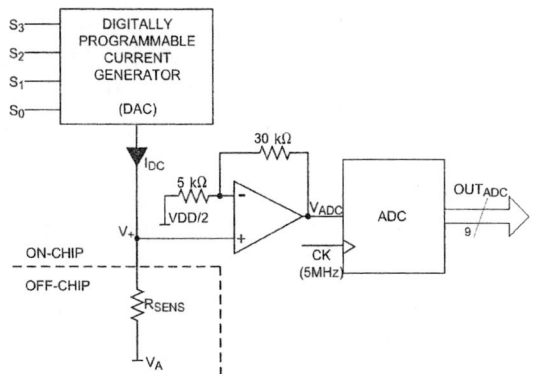

Figure 4. Feedback path of the control chain.

TABLE III. EXAMPLE OF OPERATING CONFIGURATIONS

First configuration	Second configuration	Third configuration
R_{sens}=75 Ω	R_{sens}=75 Ω	R_{sens}=150 Ω
I_{DC}=1.4 mA	I_{DC}=3 mA	I_{DC}=1.4 mA
α=0.0022 Ω/°C	α=0.0022 Ω/°C	α=0.0022 Ω/°C
V_A=1.5 V	V_A=1.25 V	V_A=1.3 V
V_+(25 °C)=1.605 V	V_+(250 °C)=1.586 V	V_+(225 °C)=1.602 V
V_+(450 °C)=1.703 V	V_+(450 °C)=1.685 V	V_+(450 °C)=1.684 V
V_{ADC}(25 °C)=1.335 V	V_{ADC}(250 °C)=1.205 V	V_{ADC}(225 °C)=1.31 V
V_{ADC}(450 °C)=2.02 V	V_{ADC}(450 °C)=1.898 V	V_{ADC}(450 °C)=1.89 V
OUT_{ADC}(25 °C)=109	OUT_{ADC}(250 °C)=49	OUT_{ADC}(225 °C)=100
OUT_{ADC}(450 °C)=429	OUT_{ADC}(450 °C)=371	OUT_{ADC}(450 °C)=367
STEP=1.3 °C	STEP=0.76 °C	STEP=0.84 °C

The used A/D converter is structurally similar to a first order single-bit sigma-delta modulator, but the output of the comparator drives 32 counters. The A/D converter works with a clock frequency equal to 5 MHz, and the output of each counter, that represents the converted value, is updated every 512 clock cycles. When the reset is released, only the first counter starts to count, while the second one starts after 16 clock cycles, the third one starts after 32, and so on. The multiplexer that follows the counters chooses the latest updated output, updating the global output signal every 3 µs. The actual temperature reached by the gas-sensor is, therefore, measured often enough to avoid that it gets too far from the set-point between two updates of the heater control signal.

The power consumption of the control circuit, including A/D converter, the counter registers and conditioning network is about 10 mW, which is much lower with respect to the power necessary to control the temperature of the gas-sensor in open loop for high-temperature set point.

TABLE IV. POWER EFFICIENCY OF THE SYSTEM

TEMPERATURE	200 °C	300 °C	400 °C	450 °C
EFFICIENCY	66%	76%	82%	84%

In Table IV the system power efficiency is reported for the most significant sensor temperatures in a typical configuration (first configuration in Table III).

IV. SIMULATION RESULTS

The system has been simulated at transistor level, substituting the gas-sensor with a simple RC cell with the same time constant (τ = 35 ms). Figure 5 and Figure 6 show the transient response and the steady state of the system, respectively, when the set-point temperature is 60 °C. When the gas-sensor has reached the desired temperature the control circuit switches off the heater and the temperature decreases to the set-point temperature with the time constant of the sensor. It is evident that the ringing around the mean temperature value is negligible. This simulation has been done with a very low set-point temperature in order to lower the simulation time required to reach the set-point temperature starting from the environment temperature T_0=25 °C.

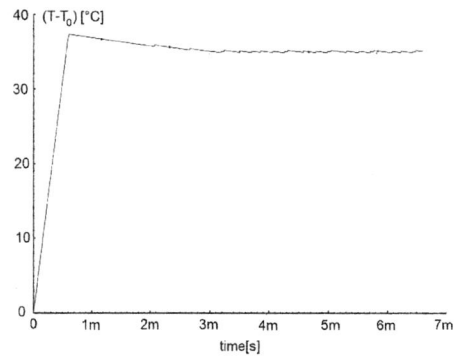

Figure 5. Transient temperature response waveform.

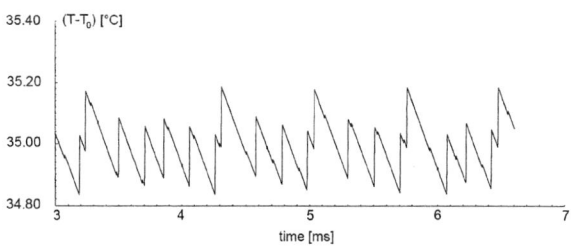

Figure 6. Steady state temperature response.

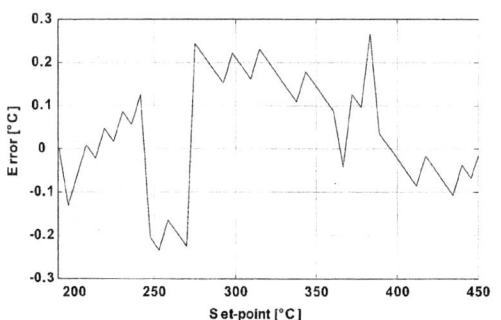

Figure 7. Temperature control circuit accuracy.

In order to verify the behavior of the system, several simulations have been performed in Simulink (in order to achieve an acceptable simulation time) and the results are reported in Figure 7, which shows the difference between the set-point and the mean temperature.

The linearity of the system is good since the error is always lower than 0.3 °C on the range 190 °C - 450 °C. The same results are also expected in the test chip measurement, because the linearity of the overall system is mainly due to the linearity of the A/D converter, that has been designed to meet higher specifications [7]. Figure 8 reports the accuracy performance of the chemical sensor read-out circuit over 5 decades range. The achieved precision in simulation is better than 0.5% over the entire range of interest (dynamic range, DR = 146 dB). The design details may be found in [6].

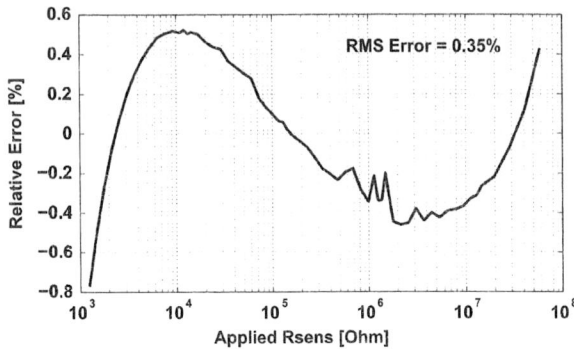

Figure 8.   Relative error in R_{sens} sweep simulation over [1 kΩ-100 MΩ].

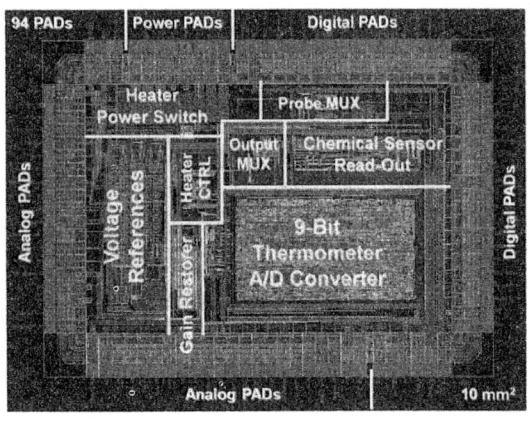

Figure 9.   Layout of the presented system.

The layout of the chip, realized in a 0.35 μm CMOS technology, is shown in Figure 9. The chip area is 10 mm^2 and 94 pads are used, divided in three groups: analog, digital and power pads. The block referred as "Voltage References" consists of several buffers and has the aim of generating the

voltage references that are necessary to bias the different blocks. The block "Gain Restorer" is the digitally programmable current generator. The block "Heater Control" is the block that compares the output of the A/D converter and the set-point and generates the control signal for the "Heater Power switch". By driving the "Output Mux", which is a 32 inputs, 16 outputs multiplexer, it is possible to read the 9 bits outputs of the thermometer and other control signals or the 16 output bits of the gas-sensor read-out circuit called "Chemical Sensor Read-Out". Exploiting the "Probe Mux", it is possible to measure the voltage level of some critical internal nodes, e.g. in order to allow fine tuning of the voltage references adding external resistors on apposite pins.

V.   CONCLUSIONS

In this paper we proposed a complete gas-sensing system consisting of a switching temperature control loop and of a wide-dynamic-range interface circuit, which does not require calibration. The proposed integrated circuit is able to control the temperature of the sensor, selected by means of a digital set-point, over a range of 250 °C with an accuracy better than 0.5 °C. Moreover, it is able to read-out the gas sensor resistance, exploiting a resistance to frequency conversion, with an equivalent dynamic range of 146 dB without calibration. Reconfigurability of the entire integrated circuit parameters allows the use of sensors with different features. The chip has been sent for fabrication and is presently under test.

REFERENCES

[1] S. M. Martin, T. Strong, R. B. Brown, "Design, implementation and verification of a CMOS-integrated chemical sensor system", IEEE ICMENS proceedings, pp. 336-342, 2004.

[2] G. Sberveglieri, "Recent Developments in Semiconducting Thin-Film Gas-Sensor", Sensors and Actuators B, vol. 23, pp. 103–109, 1995.

[3] S. Capone and P. Siciliano, "Gas-sensors from nanostructered metal oxides," Encyclopedia of Nanoscience and Nanotechnology, vol. 3, pp. 769–804, 2004.

[4] "Portable System for Ambient Gas Monitoring with Smart A/D Front-End Improving Sensor Resolution and Accuracy," PRIN 2003091427, funded by Italian Government, http://ims.unipv.it/prin03.

[5] "General Olfaction and Sensing on a European Level, GOSPEL", Network of Excellence funded by European Community, http://www.gospel-network.org.

[6] V. Ferragina, M. Ferri, M. Grassi, A. Rossini, P. Malcovati and A. Baschirotto, "A 12.4 ENOB Incremental A/D Converter for High-Linearity Sensors Read-Out Applications", Proceedings of IEEE International Symposium on Circuits and Systems (ISCAS '07), New Orleans, USA, pp. 3582–3585, 2007.

[7] M. Grassi, P. Malcovati and A. Baschirotto, "A 141-dB Dynamic Range CMOS Gas-Sensor Interface Circuit Without Calibration With 16-Bit Digital Output Word", IEEE Journal of Solid-State Circuits, vol. 42, pp. 1543–1554, 2007.

[8] G. Ferri, N. Guerini, V. Stornelli and C. Catalani, "A novel CMOS temperature control system for resistive gas sensor arrays", Proceedings of the 2005 European Conference on Circuit Theory and Design 2005, vol. 3, pp III/27-III/30.

Transimpedance amplifier for very high sensitivity current detection over 5MHz bandwidth

Giorgio Ferrari, Fabio Gozzini and Marco Sampietro

Dipartimento di Elettronica e Informazione

Politecnico di Milano, 20133 Milano, Italy

Email: fabio.gozzini@polimi.it

Abstract—The paper presents a transimpedance amplifier made on standard $0.35\mu m$ CMOS technology specifically conceived to measure the impedance of low-conductivity nano-bio devices. The circuit combines a bandwidth of 5MHz with an extremely low noise of 3fA/sqrt(Hz) by using an integrator-differentiator scheme. An innovative feedback network continuously discharges the integrator capacitance to ensure an unlimited measuring time irrespective of the dc input current up to 20nA. An highly linear 300GΩ active resistance, used in the feedback network to extend the lower limit of the bandwidth down to 100Hz, will be also described.

I. INTRODUCTION

Electrical characterization of single molecules and nanometer scaled devices requires special care because of their very small dimensions and correspondingly low current signals, often in the pA range in the nano-bio research field [1]–[4]. To measure these values, very low-noise and high-sensitivity front-end preamplifiers are needed. In addition, to be able to track the time evolution of a biological system the preamplifiers need fast settling times, i.e. relatively large bandwidths. For example, to detect the chemical identity of single molecules passing through an ion channel one should discriminate ionic current variations on the tens-of-microseconds time-scale, which corresponds to a bandwidth of 100kHz [5].

Wide bandwidth and high sensitivity in current detection can be simultaneously obtained by using the integrator-differentiator amplifier whose scheme is illustrated in Fig. 1. The absence of physical resistors at the input node ensures minimum noise. In many practical cases the bandwidth of the circuit is given by $GBP\frac{C_i}{C_i+C_s}$, where GBP is the gain-bandwidth product of the operational amplifier OP_{int} and C_s is the total capacitance at the input node of the integrator stage. A bandwidth in the MHz range can be obtained by choosing C_i of the same order as C_s. The disadvantage of this class of instrument is that the feedback capacitance of the integrator stage, C_i, must be discharged to prevent its saturation due to the input leakage current. Obviously a simple feedback resistor in parallel with C_i cannot be used in an integrated realization because of the large value required to minimize noise. This task is usually performed by a MOSFET switch placed in parallel with the integrator capacitor C_i, which discharges it when the output voltage reaches a defined threshold. This switch sets a limit to the measurement time available, which depends on both the input leakage current

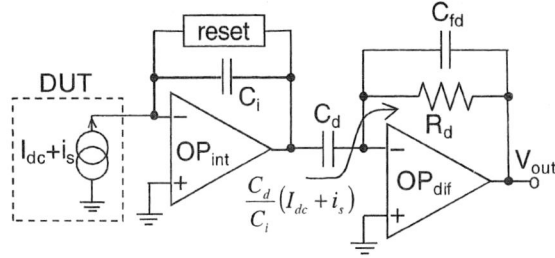

Fig. 1. Principle scheme of an integrator-differentiator transimpedance amplifier. The reset network prevents the saturation of the first stage due to the dc input current.

and C_i. In an integrated solution using a 100fF feedback capacitor, a 10nA leakage current would give a discharge period of only few tens of microseconds, unacceptable for many applications. This paper describes a fully integrated solution to this problem, consisting of a circuit network that provides a dc path to ground for the standing current of the DUT and for the charge accumulated on the feedback capacitance, leaving free the signal current to go through the integrator-differentiator stages. The result is a system that maintain an high sensitivity amplification of fast signal while ensuring an unlimited measuring time opportunity.

II. DESCRIPTION OF WORKING PRINCIPLE

The proposed circuit is shown in Fig. 2 and is based on the topology reported in [6]. The amplifier H is designed with an high gain in the dc frequency range but a strong attenuation in the signal bandwidth. In this way the dc variation of the integrator output, due to the dc current integration on the feedback capacitance, is amplified to node B producing, through the active transconductor G_f, the current necessary to discharge C_i and to continuously drain the DUT standing current in G_f. At signal frequencies, the strong attenuation introduced by the amplifier H makes ineffective the loop and, consequently, the input signal, i_s, is collected in the capacitance C_i.

Concerning the stability of the integrator stage with this new feedback loop, let us consider its loop gain

$$G_{loop} = -\frac{1}{sC_i}H(s)G_f \qquad (1)$$

978-1-4244-1983-8/08/$25.00 ©2008 IEEE

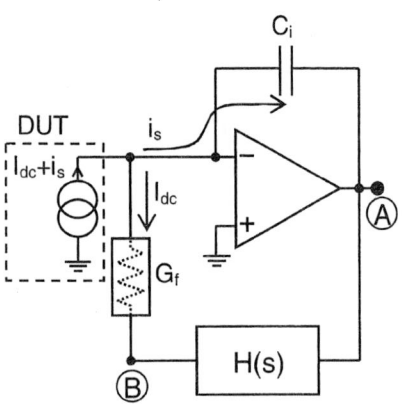

Fig. 2. Schematic of the integrator stage with the new feedback network to discharge the standing current from the device under test (DUT).

where we have assumed the integrator ideal. Considering that at low frequency the transconductor G_f has a flat transfer function and the integrator introduces a pole at zero frequency, stability is ensured if the amplifier $H(s)$ does not add more than 90° of phase shift at the frequency f_m at which $|G_{loop}(f_m)| = 1$ [7]. Our solution uses an amplifier $H(s)$ with one pair pole-zero:

$$H(s) = H_0 \frac{1 + s\tau_z}{1 + s\tau_p} \qquad (2)$$

where the frequency of the pole, f_p, must be lower than the frequency of the zero, f_z, to obtain an high frequency gain $H_h = H_0\tau_z/\tau_p$ less than 1. Stability with a phase margin greater than 45° is therefore obtained provided that $f_z < f_m$. In this condition the frequency f_m is given by:

$$f_m = \frac{H_h G_f}{2\pi C_i} \qquad (3)$$

Since for frequencies greater than f_m, where $|G_{loop}| < 1$, the feedback is inactive and the input current flows in the integrator, f_m indicates also the minimum frequency amplified by the circuit. The high frequency gain H_h of the amplifier H(s) is a free parameter that can be chosen to set the desired value of the minimum frequency. The choice of H_h does not affect any other amplifier characteristics (noise, maximum frequency, maximum input dc current), differently from the values of transconductor G_f and of the capacitor C_i which are fixed by noise considerations as will be explained later.

III. CMOS REALIZATION

A. Design of active transconductance G_f

The transconductor G_f injects its output noise directly to the input of the circuit. Therefore it cannot be integrated with a simple resistor because, even using the largest available (some hundreds of kΩ), the resulting equivalent current noise would be too high. So to make the integration possible, the transconductor G_f is implemented with a circuital system that uses a "matched-MOSFET" system as sketched in the

dashed box on left-side of Fig. 3 [8]. The two transistors T_{att} and T_{spill} have been made in the same N-well which is short circuited with their sources. Such connection makes the MOSFETs able to drive current on both direction: with negative V_{GS}, the device operates as a Mos-diode in the different bias regions depending on the current; with positive V_{GS}, the parasitic drain-well (p-n) junction is forward-biased, so the transistor acts as a p-n diode. Irrespective to the sign of the leakage current I_{dc} from the DUT, thanks to the virtual grounds V_1 e V_2, the two transistors are always biased with the same potential, thus the ratio of the currents flowing in T_{spill} and in T_{att} depends only on transistor sizes. Sizing T_{att} M-times larger than T_{spill}, the current flowing in the resistor R_{att} is injected in the input node reduced by a factor M, resulting in a linear transconductor of value $G_f = 1/(MR_{att})$. In our case we have obtained an equivalent resistance of about $45M\Omega$ by choosing $R_{att} = 300k\Omega$ and $M = 150$. From the noise point of view, note that the current noise of the resistor R_{att} is reduced by a factor M^2. Consequently, the noise injected to the input is equivalent to a very large resistance of 5GΩ. Such low noise condition is preserved for input current less than some tens of pA, i.e. in the most of bio/nanomeasurements. For I_{dc} currents greater than tens of pA, the shot noise $2qI_{dc}$ of the T_{spill} transistor operating in sub-threshold regime or as p-n diode, becomes the dominant effect. The flicker noise component has been made negligible in the signal band by using a non-minimal area of the MOSFET.

B. Amplifier H(s) design

The amplifier H(s) architecture is reported in the dotted box on the schematic in Fig.3 which is characterized by the ideal transfer function:

$$H(s) = \frac{1 + sC_2 R_a}{sR_a C_1} \qquad (4)$$

The very high dc gain of the amplifier H has been chosen in order to keep the output of the integrator close to zero irrespectively of the dc input current, ensuring the maximum input range for the signal and a high linearity of the integrator in any bias condition.

The main issue of this design is the placement of singularities. From Eq. 3 we can derive that, to obtain a low f_m of 100Hz, the $H_h = C_2/C_1$ factor must be equal to 400, leading to the choice of capacitor C_1 of 10pF and C_2 of 25fF. In order to set the zero frequency at one decade before f_m, the resistor R_a must be in the order of hundreds GΩ. Such large value, clearly impossible to achieve by an integrated physical resistor, has been implemented by cascading four current reducers similar to that discussed in the previous section. The measured I/V characteristic, sketched in the two graphs of the Fig.3, shows a very large equivalent resistance, of about 300Ω, with a very good linearity that assures a stable feedback loop in every condition. Note that the circuit can operate precisely at current levels as low as fA since the only leakage current, from substrate to N-well, is driven directly by the operational amplifiers controlling the transistor voltages without affecting the current in the transconductor.

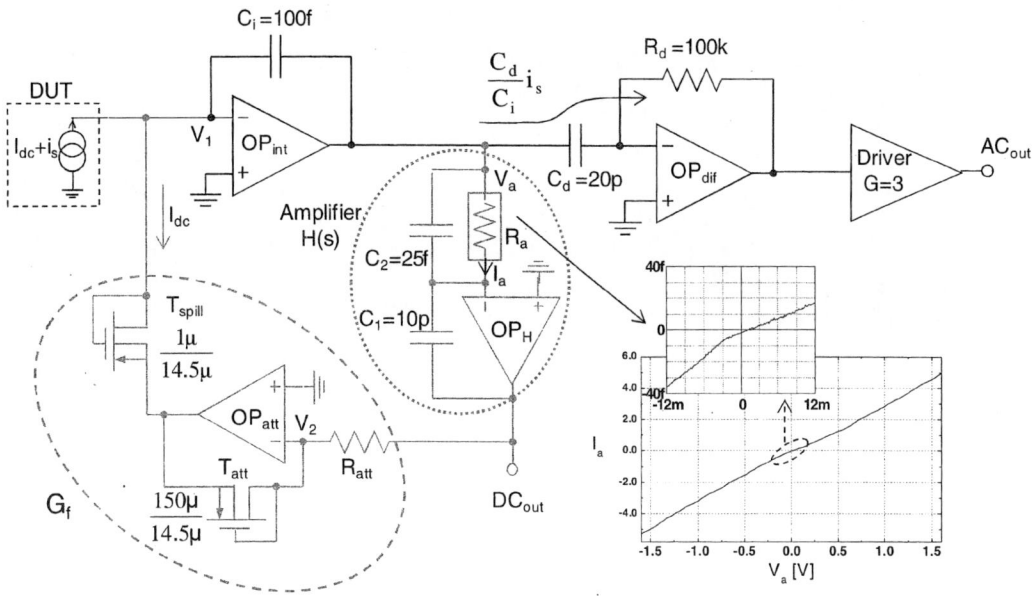

Fig. 3. Schematics of the transimpedance prototype with active network to draw the dc input current. The G_f transconductor is realized by the current reducer system (dashed box), while the amplifier H(s), (dotted box), is provided of a 300GΩ equivalent resistance R_a implemented by cascading 4 current reducer systems: the experimental I/V graph reported shows the high linearity and the wide current dynamic range of such active resistance

IV. NOISE ANALYSIS

The integrator-differentiator stage together with the feedback network has an equivalent input noise in the signal bandwidth, considering only the main terms, given by:

$$\overline{i_n^2} \cong \frac{4kT}{R_{att}M^2} + \frac{4kT}{R_d\left(\frac{C_d}{C_i}\right)^2} + \overline{i_{Tspill}^2} + \overline{e_{int}^2}\omega^2(C_i+C_s)^2 \quad (5)$$

where $\overline{i_{Tspil}^2}$ is the current noise of the MOSFET T_{spil}, $\overline{e_{int}^2}$ is the noise voltage source of the operational amplifier OP_{int}; $C_s = C_g + C_{DUT} + C_{stray}$ is the total capacitance at the input node of the integrator due to the operational amplifier (C_g), the device under test (C_{DUT}) and the stray due to the chip-DUT interconnection (C_{stray}).

By choosing $M = 150$ and a capacitor ratio $C_d/C_i = 200$, the thermal noise of the physical resistors R_{att} and R_d gives a contribution as low as $3fA/\sqrt{Hz}$. The current noise of MOSFET T_{spil} depends on the input dc current and it is negligible for currents smaller than 30pA as discussed in sec. III-A.

The last term in eq. 5 depends on frequency becoming dominant within the signal bandwidth. To minimize it, special care should be drawn in the design of the differential input transistors of OP_{int} that set both C_g and $\overline{e_{int}^2}$. An optimum is obtained for $C_g = C_i + C_{DUT} + C_{stray}$ [9], leading in our case to input transistor with length $L = 0.6\mu m$ and width $W = 220\mu m$. The differential input stage uses a pair of p-MOS transistors with a resistive load to minimize the voltage noise.

Finally, although eq. 5 would suggest to decrease the value of the capacitance C_i, one has to keep in mind that the value of C_i is also setting: i) the bandwidth of the integrator, equal to $GBP \cdot C_i/(C_i + C_s)$, where GBP is the gain-bandwidth product of the operational amplifier OP_{int}; ii) the ac dynamic range of the integrator so that small values of C_i correspondingly reduce the maximum ac input current before saturation.

V. RESULTS AND DISCUSSION

Fig. 3 shows the fully CMOS implementation of the integrator-differentiator amplifier. The circuit uses a dual supply voltage of ±1.5V and the total power consumption is 25mW. The equivalent input noise measured without DUT is reported in Fig. 4. The experimental result is in agreement with the theoretical prediction when in eq.5 is assumed the equivalent noise of the amplifier OP_{int} of $\overline{e_{int}^2} = (4nV)^2/Hz$, and a total input capacitance C_s of 700fF given by the operational amplifier and by the input bonding pad. The white noise in the low frequency region is equivalent to the thermal noise of a 2GΩ resistor. The white noise increases by increasing the input bias current above 30pA, as shown in the inset of Fig. 4. The increase follows a theoretical shot noise as expected for a transistor operating in sub-threshold regime or as a p-n diode. For large negative current the transistor T_{spill} operates in inversion regime and the noise is correspondingly less than the shot noise. The $1/f$ noise a frequencies lower than 1kHz is added by the differentiator stage and therefore is independent by the input bias.

The overall transfer function, measured in the frequency range $10Hz - 10MHz$, is reported in Fig. 5. The expected

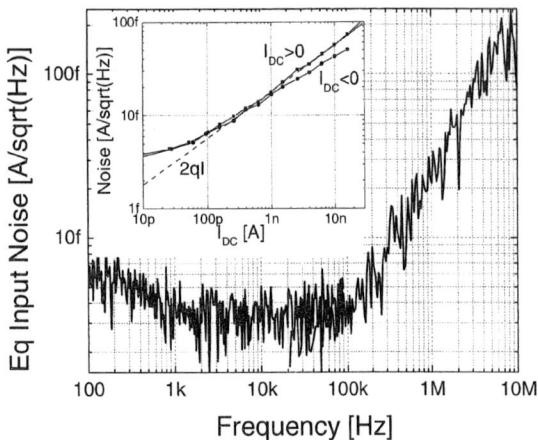

Fig. 4. Equivalent input current noise measured on the prototype shown in fig. 3, operating with low bias current. The raise in the spectrum at higher frequencies is due to the input stray capacitances. *Inset*: the white noise as a function of the input dc current.

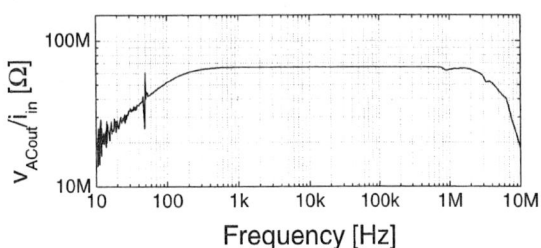

Fig. 5. Experimental frequency response of the integrator-differentiator.

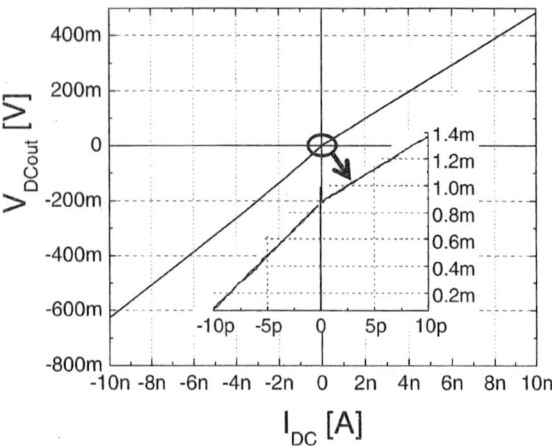

Fig. 6. Transconductance of the DC_{out}: a good linearity on a wide current range (1p-10nA) for both positive and negative currents is achieved. The slight asymmetry present, is due to the offset between the virtual ground V_1 and V_2 and to the mismatch on the MOSFETs, which work, in the negative branch, as MOS-diode and, in the positive branch, with the drain-well p-n junction

standard integrator-differentiator schemes and to commercially available amplifiers, it can be integrated on a single chip to address all applications in which size, portability and minimum dose of biological material to be measured are of primary interest. In addition, it provides extended bandwidth and continuous response capability to analyze long lasting events even in presence of large leakage current.

ACKNOWLEDGMENT

The authors acknowledge S. Masci for the bonding of the circuits, A.Molari for discussions on the circuit architecture. This work has been supported by the SPOT-NOSED European project and by italian MURST-PRIN 2005.

REFERENCES

[1] P. J. de Pablo, F. Moreno-Herrero, J. Colchero, J. G. Herrero, P. Herrero, A. M. B. P. O. J. Soler, and E. Artacho, "Absence of dc-conductivity in λ-dna," *Phys. Rev. Lett.*, vol. 85, pp. 4992–4995, 2000.
[2] L. Movileanu, S. Howorka, O. Braha, and H. Bayley, "Detecting protein analytes that modulate transmembrane movement of a polymer chain within a single protein pore," *Nat. Biotechnol.*, vol. 18, pp. 1091–1095, 2000.
[3] A. Stamouli, J. Frenken, T. Oosterkamp, R. Cogdell, and T. Aartsma, "The electron conduction of photosynthetic protein complexes embedded in a membrane," *FEBS Letters*, vol. 560, pp. 109–114, 2004.
[4] S. M. Iqbal, G. Balasundaram, S. Ghosh, D. E. Bergstrom, and R. Bashir, "Direct current electrical characterization of ds-dna in nanogap junctions," *Appl. Phys. Lett.*, vol. 86, p. 153901, 2005.
[5] C. Y. Kong and M. Muthukumar, "Simulations of stochastic sensing of proteins," *J. Am. Chem. Soc.*, vol. 127, pp. 18 252–18 261, 2005.
[6] M. Sampietro and G. Ferrari, "Wide bandwidth transimpedance amplifier for extremely high sensitivity continuous measurements," *Rev. Sci. Instr.*, vol. 78, p. 094703, 2007, and International patent, WO 2005/062061.
[7] P. Gray and R. Meyer, *Analysis and Design of Analog Integrated Circuits*, 3rd ed. John Wiley & Sons, Inc., 1993, ch. 9, p. 602.
[8] F. Gozzini, G. Ferrari, and M. Sampietro, "Linear transconductor with rail-to-rail input swing for very large time constant applications," *Electronics Letters*, vol. 42, no. 19, pp. 1069–1070, September 2006.
[9] L.Fasoli and M.Sampietro, "Criteria for setting the width of ccd front end transistor to reach minimum pixel noise," *IEEE Trans. Electron Devices*, vol. 43, no. 7, pp. 1073–1076, 1996.

bandwidth of the amplifier is fully confirmed by the experimental value: the minimum frequency is set by eq. 3 and the maximum frequency is limited only by the integrator stage in agreement with the GBP of OP_{int} equal to 100MHz and assuming a stray capacitance of C_{stray} =800fF, due to the input device.

Finally, note that the amplifier has actually two outputs: the signal output AC_{out} with bandwidth 100Hz-5MHz and a dc output DC_{out} that senses the voltage across R_{att} containing the frequency components lower than 100Hz. Even if the feedback network is made by non-linear elements, the I/V characteristic of the DC_{out}, sketched in Fig.6 shows a good linearity on a wide current range from 1pA to 10nA with only a slight asymmetry between positive and negative range. This is due to both the mismatch between the MOSFETs T_{spill}-T_{att} working as MOS-diode or as drain-Nwell diode and the offset between the virtual grounds V_1 and V_2 (see Fig.3). Despite these small asymmetry, the DC_{out} is well suited to monitor the bias condition of the DUT during the measurement and to track continuously the low frequency variations of the DUT.

VI. CONCLUSION

A current sensing amplifier with a new feedback continuous time reset network has been presented. Differently from

Design of CMOS Chopper Amplifiers for Thermal Sensor Interfacing

Michele Dei, Paolo Bruschi
Dipartimento di Ingegneria dell'Informazione
Università di Pisa
via G. Caruso 16, I-56122 Pisa, Italy
e-mail: michele.dei@iet.unipi.it

Massimo Piotto
IEIIT Pisa
CNR
via G. Caruso 16, I-56122 Pisa, Italy

Abstract—**An analytical approach to the design of compact CMOS chopper amplifiers for integrated thermoelectric sensors is presented. The impact of the high resistance and low signal bandwidth of thermopile sources on the design is illustrated. The proposed approach, regarding the precision vs noise tradeoff, is applied to the design of a practical prototype, using a commercial process. Accurate electrical simulations are provided to confirm the effectiveness of the proposed design methodology.**

I. INTRODUCTION

Thermal sensors play an important role in the world of Micro-Electro-Mechanical Systems (MEMS), since a large variety of physical and chemical quantities can be converted into a temperature difference [1]. Furthermore, temperature sensors can be easily integrated into silicon chips, using only the layers available in commercial microelectronic processes. Thermopiles represent an effective approach for the detection of temperature differences in MEMS: they offer intrinsically null offset, null power consumption and self heating and a sufficiently linear behavior. Thermopiles consist of a series connection of several thermocouples, which, in turn, can be obtained by connecting two different conductive layers. The use of n-polysilicon/p-polysilicon or p-polysilicon/aluminum couples is the typical choice in a standard CMOS process.

The d.c. signal (V_{th}) and rms value of the thermal noise (V_{nt}) produced by a thermopile, made of a series of n thermocouples are given by

$$V_{th} = n\alpha\Delta T \; ; \quad V_{nt} = \sqrt{4kTnR_TB} \; . \tag{1}$$

Here α and R_T are the sensitivity (Seebeck coefficient) and resistance of a single thermocouple, respectively, k is the Boltzmann constant, B is the bandwidth of the readout channel and ΔT is the temperature difference sensed by the thermopile. For a given bandwidth (dictated by application) the minimum detectable temperature difference is given by:

$$\min(\Delta T) = \frac{1}{\alpha}\sqrt{\frac{4kTR_TB}{n}} \; . \tag{2}$$

Therefore, to improve the resolution it is desirable to increase n. In practice, value of n up to a few tens, are customary in MEMS with dimensions of several hundreds microns. It should be observed that integrated thermopiles are based on polysilicon, which, differently from metals used in conventional thermocouples has sheet resistance in the range 10-100 Ω. As a result R_T, is typically of order of several kΩ and the total thermopile resistance (nR_T) falls around 100 kΩ and beyond. It is also important to observe that, in many thermopile based integrated sensors, such as flow meters [2] and bolometers, the temperature variations to be detected are as small as to require a resolution of the order of 1 μV on the output signal. This dictates the use of chopper stabilized (CHS) amplifiers to cancel the input offset and minimize the residual baseband noise. The mentioned high output resistance of integrated thermopiles relaxes the amplifier requirements in terms of input noise, but, at the same time, introduces a few peculiar problems.

In this paper, the way these issues affect the design of a chopper amplifier for integrated thermoelectric sensors is presented, obtaining a series of simple formulas useful for the design of compact CMOS readout channels for MEMS interfacing.

II. CHS AMPLIFIER FOR THERMOPILE BASED SENSORS

Figure 1 shows the block diagram of a chopper amplifier. The thermopile has been considered to be split into two identical parts of $n/2$ elements each, represented by the sources $V_d/2$ and $-V_d/2$. The central terminal is used to apply a common mode voltage to comply with the input range of the next stage, indicated with AMP. The latter is a CMOS, continuous time fully differential amplifier with gain A_D and an upper frequency limit much higher than the frequency of the input pole, deriving from the input

978-1-4244-1983-8/08/$25.00 ©2008 IEEE

capacitance C_{IN} and source resistance $R_S = nR_T$. The modulator S1 and demodulator S2, controlled by a clock signal of frequency f_{CH}, ideally introduce multiplication by 1 and -1 in the first and in the second half clock period, respectively. Due to the time constant $R_S C_{IN}$, the signal at the amplifier input (V_{IN}) is not an ideal square wave (see Fig. 1). As a result, the demodulated signal V_A is not a constant voltage but short pulses at frequency $2f_{CH}$ are present. The latter are suppressed by the low pass filter LP but the output d.c. value is altered by the amount ΔV, as schematically shown in Fig.1.

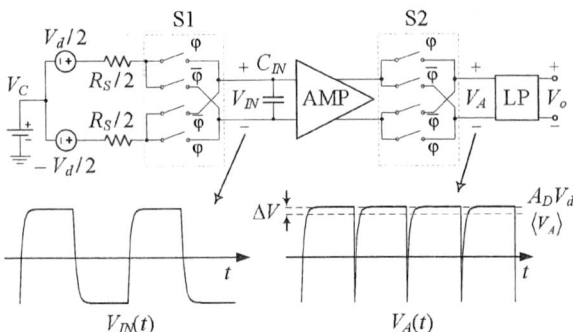

Figure 1. Block representation of a Chopper stabilized amplifier. Waveforms $V_{IN}(t)$ and $V_A(t)$ refer to the case $V_d = $ const.

It can be shown that, for a input d.c. signal this effect is equivalent to introduce a gain error, the relative value of which is given by:

$$\varepsilon_G = \frac{\Delta V}{A_D V_d} = 2 f_{CH} C_{IN} R_S \ , \qquad (3)$$

where $A_D V_d$ is the ideal output voltage. This error is particularly detrimental and has to be minimized, since it depends on the source resistance and amplifier input capacitance, affected by large temperature and/or process variations. Since the number n of thermocouples, and thus R_S, is fixed by the desired temperature resolution through Eq. (2), viable alternatives involve reducing either C_{IN} or f_{CH}. Unfortunately both options produce an increase of the residual equivalent noise spectral density (S_{Veq}), and should then be applied carefully. To understand this, we will introduce the following simplifying conditions: (i) the RTI (referred to the input) noise of block AMP can be completely ascribed to its input MOSFETs; (ii) the frequency f_{CH} is well below the flicker noise corner frequency, so we can neglect thermal noise; The first condition can be reached by proper design choices, as shown later, while the second is reasonable, due to the typical high flicker noise coefficients of MOS devices. We will also assume that the input capacitance C_{IN} can be written as:

$$C_{IN} = k_C C_{ox} WL , \qquad (4)$$

where W and L are the width and length of the input MOSFETs, respectively, while k_C is a constant factor,

depending on the amplifier topology. The intrinsic parasitic capacitance of the thermopile will be assumed to be negligible with respect to the amplifier input capacitance.

We will first find a general relationship, useful to estimate the best noise figure that can be achieved. With the above assumptions, the equivalent RTI noise of the chopper amplifier is given by [2]:

$$S_{Veq} \cong 0.85 \cdot \frac{S_{VF}(1)}{f_{CH}} \ , \qquad (5)$$

where $S_{VF}(1)$ is the RTI flicker noise density at 1 Hz for amplifier AMP. Using a simple noise model and recalling condition (i):

$$S_{VF}(1) = m \frac{k_F}{WL} , \qquad (6)$$

where k_F is the flicker noise coefficient, m the number of input devices with width W and length L. The noise figure F can be written as:

$$F = 1 + \frac{S_{Veq}}{4kTR_S} \ . \qquad (7)$$

Putting together Eqns. (3-7), we get the best noise figure F_{min}:

$$F_{min} = 1 + 0.85 \cdot \frac{k_F C_{ox}}{2kT} \cdot \frac{m k_C}{\varepsilon_G} \ . \qquad (8)$$

Equation (8) states that, once the acceptable gain error ε_G and the amplifier topology have been decided, the best noise figure depends only on process parameters or physical constants. Another important general expression is obtained relating the noise figure to the chopper frequency f_{CH} and the input transistor area WL. From Eqns. (4-7), we get:

$$f_{CH} = \frac{0.85 \cdot m \cdot k_F}{4kTR_S (F-1) WL} \ . \qquad (9)$$

Equation (9) gives the clock frequency necessary to obtain a given noise figure with a given input device gate area. It is important to observe that, in many thermal sensors, the required signal bandwidth is at most a few hundred Hertz, allowing the reduction of the clock frequency down to a few kHz with no consequences on the signal. Decreasing the clock frequency proportionally reduces the residual offset due to parasitic spikes introduced by the input modulator that should be otherwise mitigated with more complicated pass band amplifiers [4] or nested chopper configurations [5]. Equation (9) clearly states that a low clock frequency is paid in terms of device area; furthermore, for a given noise figure and chip area budget, the higher the source resistance, the lower the clock frequency. The application of these relationships to the design of a practical chopper amplifier will be illustrated in next section.

III. CASE STUDY

A. Instrumentation Amplifier

The instrumentation amplifier schematic, shown in Fig. 2, is based on a conventional folded cascode architecture. The use of the OTAs in the input stage enhances the voltage follower function of M1 and M2, resulting in improved linearity and lower temperature sensitivity of the amplifier. The gain is given by the ratio of the output differential resistor R_2 and the source degeneration resistor R_1, i.e. $A_D = R_2 / R_1$. Note that the output demodulator (indicated with S2 in Fig. 1) has been incorporated into the amplifier by means of chopper modulators S2a and S2b, shown in Fig 3(b). In this way, the output nodes are fed with the base-band signals, reducing the problems introduced by the bandwidth limitation due to the output port time constant [6]. Even if S2b does not have effect on the signal path, it is necessary in order to modulate also the M5-M6 offset and low frequency noise. The V_{o1}, V_{o2} common mode is fixed by a conventional CMFB circuit that senses the voltage at node n_C and acts on V_{cmfb}. The OTAs are simple p-type differential amplifiers shown in Fig. 3(a). All chopper modulators have been implemented by the means of complementary pass gates.

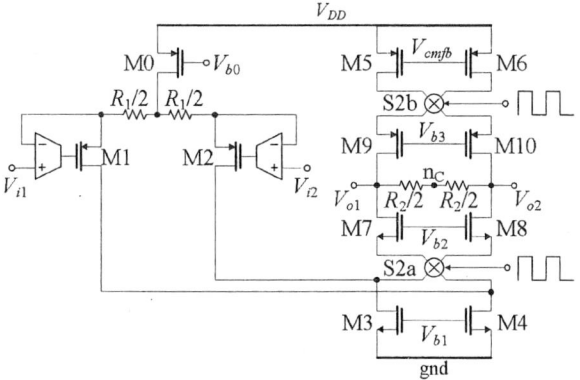

Figure 2. Schematic view of the instrumentation amplifier.

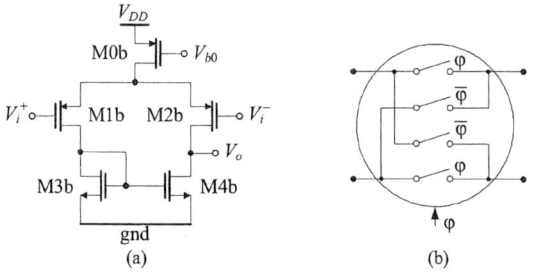

Figure 3.   (a) Differential amplifier with PMOS input devices. (b) Structure of chopper modulators.

The input referred noise spectrum of the amplifier of Fig. 2 can be calculated as

$$S_{V,rti} = 2S_{V,OTA} + 4KTR_1(1 + A_D^{-1}) + \\ + 2(R_1 g_{m3})^2 S_{V3} + 2(R_1 g_{m5})^2 S_{V5} \tag{10}$$

where $S_{V,OTA}$ and S_{Vi} indicate the PSD (power spectral density) of the OTA input referred noise and device Mi gate referred noise, respectively. Furthermore, the OTA noise can be calculated by [3]:

$$S_{V,OTA} = 2S_{V1b} + 2a^2 S_{V3b}, \tag{11}$$

where S_{V1b} and S_{V3b} are the gate referred noise PSDs of the OTA input and mirror devices, respectively, while a is a numerical factor given by

$$a = \frac{g_{m3b}}{g_{m1b}} \cong \frac{(V_{GS} - V_t)_{1b}}{(V_{GS} - V_t)_{3b}}. \tag{12}$$

Choosing a, $R_1 g_{m3}$ and $R_1 g_{m5}$ sufficiently small it is possible to reduce the noise contributions of all devices except the input MOSFETs, as required by condition (i) of previous section. Clearly, to obtain this, the gate areas of M3b, M4b, M3-6 should be similar to that of the input devices, to make their gate referred noise S_{Vi} comparable. For this reason, a specification of maximum area occupation for the whole cell can be easily converted into an area budget for the input devices.

B. Design of a prototype

A prototype has been designed using the 0.32 μm / 3.3 V CMOS devices of the Bipolar-CMOS-DMOS "BCD6s" process of STMicroelectronics. The gain A_D was fixed to 500, while the source resistance R_S was 200 kΩ. As an example of reasonable specification, we have chosen a target gain error of 0.5 and a maximum area for the input devices of 10000 μm². Considering that we have $m = 4$ input transistors (two for each OTA), the individual input device gate area is $WL = 2500$ μm². With the topology of Figs. 2 and 3, the k_C factor, determined by AC simulations, is nearly 0.1. First, we applied Eq. (8) with k_F and C_{ox} taken from the process documentation, obtaining a best noise figure of 1.12, so we choose this F value as feasible specification. Applying Eq. (9), we found a chopper frequency about 4 kHz. The main component parameters in Figs. 2 and 3 are reported in Table I.

TABLE I.   ASPECT RATIOS OF THE MORE IMPORTANT TRANSISTORS

M1b-M2b	1000/2.5	M3-M4	720/40
M3b-M4b	160/10	M5-M6	216/10
R_1	400 Ω	R_2	200 kΩ

The simulated RTI noise voltage spectral density of the amplifier (block AMP) is shown in Fig. 4 where the value at the chosen chopper frequency is shown. It is possible to observe that, in compliance with the hypothesis, noise flicker dominates at f_{CH}.

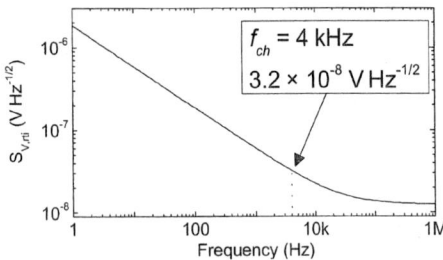

Figure 4. RTI noise spectrum of the amplifier AMP.

The gain error has been estimated by means of transient simulations performed with different d.c input voltages V_d. The output d.c. value is calculated by extracting the mean value from the amplifier output (signal V_A in Fig. 1). The resulting transfer characteristic is shown in Fig. 5. The actual gain of the chopper amplifier was estimated from the slope of a linear fit calculated for $V_d < 200\ \mu V$. A gain error of 0.56 % has been estimated.

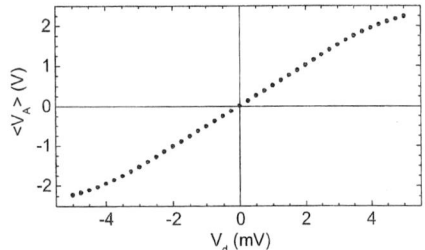

Figure 5. Simulated input/output characteristic of the amplifier.

The equivalent S_{Veq} has been estimated from the result of NOISETRAN simulations performed over a total time interval of 30 ms. Noise data have been extracted at the output of the low pass filter in Fig. 1. To this aim, the filter was implemented with ideal components (R, L, C) to obtain a second order Butterworth low pass transfer function with unity d.c gain. The filter cutoff frequency was set to 200 Hz.

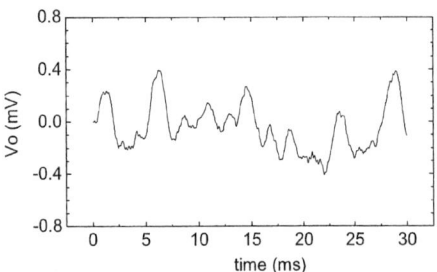

Figure 6. Output noise waveform with noiseless input source resistance R_S.

The spectral density was estimated by dividing the output rms value by the filter bandwidth. The result is reported in Table II together with other main performance data.

TABLE II.   DESIGN SPECIFICATIONS AND SIMULATED PERFORMACE

	Design specification	Simulated
Power consumption @ 3.3 V		5.78 mW
Input device area	2500 μm^2	
Noise Figure	1.12	1.2
Gain error	0.5 %	0.56 %
Gain temperature sensitivity 0÷100 °C		93.5 ppm/°C

IV.   CONCLUSIONS

The design of chopper amplifiers for integrated thermoelectric sensors is marked by a few peculiar aspects, mainly related to the low signal level, high source resistance and narrow signal bandwidth. An approach to simplify the design of compact CMOS chopper amplifier has been presented taking into account noise figure, gain error, area budget and chopper frequency. Application to a practical prototype cell showed that the performances estimated by detailed electrical simulations are close to the initial specifications. The slightly higher noise figure can be ascribed to having neglected all noise sources except the input MOSFETs.

ACKNOWLEDGMENT

The authors would like to thanks the R & D group of the STMicroelectronics of Cornaredo (MI, Italy) for providing the design kit of the BCD6s process.

REFERENCES

[1] H. Baltes, O. Paul, O. Brand, "Micromachined Thermally Based CMOS Microsensors", Proceedings Of The IEEE, Vol. 86, No. 8, August 1998.

[2] P. Bruschi, D. Navarrini, M. Piotto, "A Closed-Loop Mass Flow Controller Based on Static Solid-State Devices", J. Microelectromechanical Systems, Vol. 15, No. 3, June 2006.

[3] C. C. Enz, G. C. Temes, "Circuit Techniques for Reducing the Effects of Op-Amp Imperfections: Auotzeroing, Correlated Double Sampling, and Chopper Stabilization," Proceedings Of The IEEE, Vol. 84, No. 11, November 1996.

[4] C. Menolfi, Q. Huang, "A Fully Integrated CMOS Istrumentation Amplifier with Submicrovolt Offset," IEEE J. Solid-State Circuits, pp. 415-420, March 1999.

[5] A. Bakker, K. Thiele, J. H. Huijsing, "A CMOS Nested-Chopper Instrumentation Amplifier with 100-nV Offset," IEEE J. Solid-State Circuits, Vol. 35, No. 12, pp. 1877-1883, December 2000.

[6] Y. Christoforou, "A Chopper Based CMOS Current Sense Instrumentation Amplifier," IEEE Instrumentation and Measurement Technology Conference Anchorage, AK, USA, 21-23 May 2002, pp. 271-273.

[7] H. Gray, L. Meyer, "Analysis and design of analog integrated circuits", John Wiley & Sons, 2001, p. 792.

978-1-4244-1983-8/08/$25.00 ©2008 IEEE

Slewing Investigation and Improved Design Rules for SC Circuits Employing Two-Stage Amplifiers with Current-Buffer Miller Compensation

F. A. Amoroso, A. Pugliese, G. Cappuccino, G. Cocorullo

Department of Electronics, Computer Science and Systems
University of Calabria
Via P. Bucci, 42C, 87036-Rende (CS), Italy
{f.amoroso, a.pugliese, g.cappuccino, g.cocorullo}@deis.unical.it

Abstract— **The slewing behavior of two-stage operational amplifiers (op-amps) with current-buffer Miller compensation (CBMC) is examined. It is shown that the current buffer used in the compensation network significantly affects the slewing characteristics of the amplifier in typical switched-capacitor (SC) applications. On the basis of the amplifier behavior investigation, improved early-design rules are suggested to overcome the limitation of the conventional approach that neglects the current buffer impact on the slewing circuit performance. To demonstrate the validity of the proposed rules, a design example of a CBMC op-amp in a commercial 0.35 μm CMOS technology is also presented.**

I. INTRODUCTION

Two-stage operational amplifiers (op-amp) with current-buffer Miller compensation (CBMC) are widely employed in switched-capacitor (SC) applications, such as integrators and multiplying digital-to-analog converters (MDACs), owing to advantages arising from connecting a current buffer in series to the Miller capacitance. The current buffer can be implemented as a separate element from the two amplifier stages, or by exploiting the common-gate transistors of the first cascoded stage [1]-[8].

One of the most important design issues for a CBMC amplifier used in SC circuits is the slewing performance, which is mainly controlled by the bias currents, the compensation and the load capacitances. These parameters also influence other important op-amp design aspects such as the DC gain, the output swing, the linear settling time, the noise and the power consumption. The straight correlation among all circuit characteristics makes the op-amp design then a hugely complicated multidimensional problem. Moreover, the intrinsic non-linearity of the slewing makes a rigorous mathematical analysis of the phenomenon very complex or even impractical.

For this reason, simple rules are extensively exploited to cope with the amplifier design, at least in the early-design phase. These rules are derived by approximating the op-amp response in the slewing period by a straight line whose slope is the amplifier slew rate [9]-[11]. However, this may prove to be an excessively rough approach for CBMC amplifiers used in SC circuits. In these cases, in fact, the feedforward signal path introduced by the external capacitive feedback network causes an initial transient in which the op-amp response drifts away from its steady-state value [12], [13]. This behavior significantly impacts on the operation of the current buffer, causing the slewing to be remarkably different from a straight line. Although stricter mathematical slewing analyses are available in the literature [14]-[15], they refer to two-stage op-amps in voltage follower configurations and thus they cannot be directly extended to SC applications.

The slewing behavior of two-stage CBMC op-amps is investigated in this paper. The conventional design approach is revised in Section II, and applied to design an amplifier in 0.35 μm CMOS technology in Section III. As will be shown, important phenomena which are usually neglected affect the step response of CBMC amplifiers in real SC circuits. These phenomena have to be taken into account in order to achieve the desired slewing performances. To this aim, improved early-design rules are suggested and proved by the design example presented in Section IV. Finally, some conclusions are reported in Section V.

II. CONVENTIONAL SR-ORIENTED DESIGN APPROACH

The typical fully-differential schematic of a two-stage CBMC amplifier is depicted in Fig. 1. The differential and the common-source stages are implemented by MOSFETs *M1-M5* and *M6-M7/M8-M9*, respectively. The frequency compensation is achieved by the capacitances C_C and the current buffers. The latter are implemented by the transistors *M10* and *M11*, which are biased by the current I_b. C_1 and C_L are the output capacitances of the first and second stage, respectively, also including the possible loading caused by common-mode feedback circuits [6],[16].

978-1-4244-1983-8/08/$25.00 ©2008 IEEE

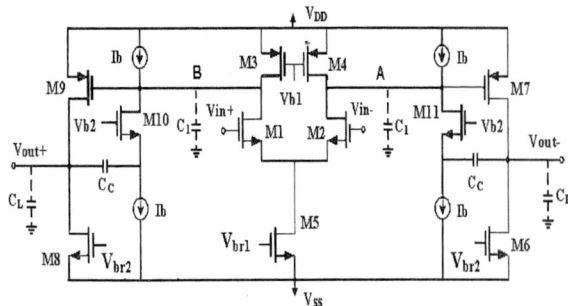

Figure 1. Fully-differential two-stage CBMC amplifier

The slewing behavior of the amplifier is examined by considering the circuit of Fig. 2. It describes the SC integrator operation during the integration phase as well as the MDACs operation during the amplifying phase, when the switch resistances and the input op-amp capacitance are neglected. Conventionally, the output voltage of the circuit in Fig. 2 during the slewing phase is roughly modeled by considering a straight line which starts from a non-null initial condition and whose slope is the slew rate. By indicating with $V_{out} = V_{out+} - V_{out-}$ the op-amp differential output voltage, its initial value is given by [12], [13]:

$$V_{out}(0) = -\frac{C_F C_I}{(C_I + C_F)C_L} A_0, \qquad (1)$$

A_0 being the step amplitude at the differential input voltage $V_i = V_{i+} - V_{i-}$. Clearly, $V_{out+}(0) = -V_{out-}(0) = \frac{1}{2}V_{out}(0)$.

The slew rate is determined by the maximum speed with which the first stage of the amplifier in Fig. 1 charges/discharges the internal capacitances at nodes A and B (namely the internal slew rate), and by the maximum speed with which the second stage charges/discharges the external capacitances at nodes V_{out+} and V_{out-} (namely the external slew rate). The single-ended internal (SR_{INT}) and external (SR_{EXT}) op-amp slew rate are normally expressed as [9]-[11]:

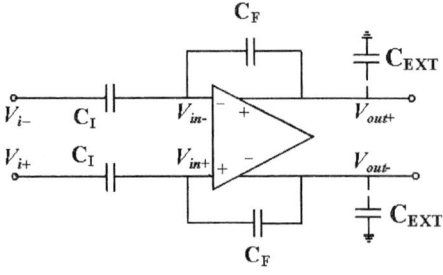

Figure 2. SC circuit in typical operations

$$SR_{INT} = \frac{I_1}{C_C}, \qquad (2)$$

$$SR_{EXT} = \frac{I_2}{C_C + C_L}, \qquad (3)$$

where I_1 and I_2 are the bias currents of M1-M2 and M6-M9, respectively. For the circuit in Fig. 2, C_L is [12]:

$$C_L = C_{EXT} + \frac{C_F C_I}{C_F + C_I} \qquad (4)$$

The op-amp single-ended slew rate (SR) is then $SR=min(SR_{INT}, SR_{EXT})$ [9]-[11]. In many practical cases, a reasonable early-design choice can be $SR=SR_{INT} = SR_{EXT}$ [17] i.e.,

$$\frac{I_1}{C_C} = \frac{I_2}{C_C + C_L}. \qquad (5)$$

III. INVESTIGATION OF THE ACTUAL AMPLIFIER SLEWING BEHAVIOUR

To study the actual amplifier slewing behavior, the circuit of Fig. 2 has been designed according to (5) in a commercial 0.35µm CMOS technology with a 3.3V supply voltage. C_I=2pF, C_F=2pF and C_{EXT}=4pF were assumed and the first stage bias current I_1 was set equal to 7.5μA. The compensation network was sized to guarantee an open loop amplifier phase margin of about 70°. In particular, I_b=14.5μA and C_C=0.6 pF were chosen. In correspondence to these parameter values, I_2=70μA and SR= 12.5$V/\mu s$ derive from (4) and (5). HSPICE simulation results related to the 1V-amplitude step responses of the two op-amp branches are depicted in Fig. 3. To allow a straightforward graphical comparison of the slewing behaviors at the two output nodes,$-V_{out-}$ is plotted instead of V_{out-}.As it appears, the slewing behavior of one op-amp branch (in the case under examination, V_{out-}) is well-described by a straight line of slope SR which starts from the initial condition derivable from (1), i.e. 100 mV. Unexpectedly, the other branch (V_{out+}) is instead characterized by a completely different slewing behavior. In fact, the V_{out+} output presents a faster response at the beginning of the transient with respect to the V_{out-} output. After the initial phase, the V_{out+} slope decreases, becoming less than SR, whereas it becomes about equal to SR in the final phase of the slewing. This asymmetric behavior of V_{out+} and V_{out-} can be explained by considering the different phenomena which control the slewing of the two op-amp branches. In fact, V_{out+} and A nodes are subject to a positive voltage step, whereas V_{out-} and B nodes are subject to a negative voltage step. The growing of the voltage at node A tends to turn-off M7. As a consequence, the bias current I_2 of M6

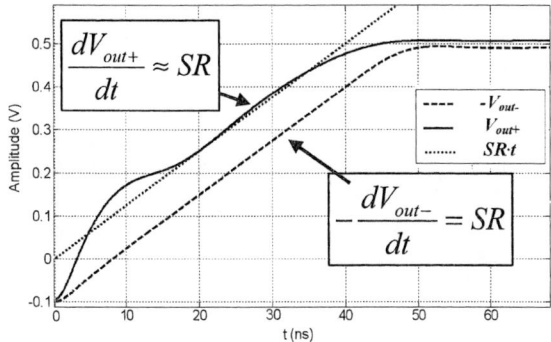

Figure 3. 1V-step responses of the two amplifier outputs (V_{out+} and $-V_{out-}$) when the conventional design rules are used

determines the speed of discharge of the total capacitance C_C+C_L at the node V_{out-}, according to (3). Referring to the complementary branch, $M9$ is ON because the voltage at the node B decreases. The slope of V_{out+} is then limited by the internal slew rate which is not well-predicted by (2). In fact, the slewing of V_{out+} is a complex phenomenon which is not simply related to the discharge of the compensation capacitance C_C through the first stage bias current I_1. In this scenario, the drain current of $M10$ (I_{D10}) plays a fundamental role in determining the slewing behavior of V_{out+}. The trend of I_{D10} during the entire op-amp response is depicted in Fig. 4. As shown, I_{D10} is much greater than its bias value I_b in the initial part of the response. The increasing of I_{D10} is related to the initial undershoot in V_{out+} (Fig. 3), which also leads to an undershoot in the source voltage of $M10$. As a consequence, the gate-source voltage and the drain current of the transistor rise. The first part of the slewing is then characterized by the discharge of the capacitance C_I at the node B through the total current $I_1+I_{D10}-I_b$. This causes the slope of V_{out+} to be significantly greater than SR_{INT} (2). Afterwards, I_{D10} becomes less than I_b, C_C starts to discharge and the slope of V_{out+} decreases. Finally, $I_{D10}-I_b \approx I_1$, i.e. $I_{D10} \approx 7\mu A$. In this case, the current available for the discharge of C_C is about I_1, and the slope of V_{out+} equals about SR_{INT} (2). As proved by the above design, the

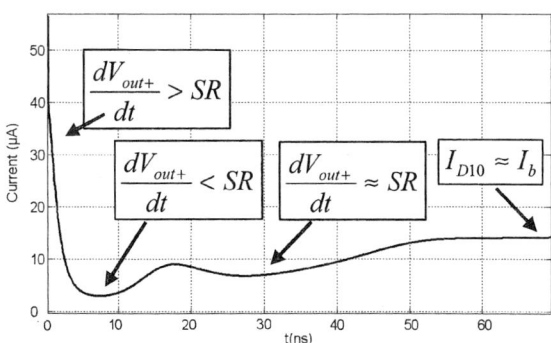

Figure 4. Drain current of $M10$ (I_{D10})

conventional description of the slewing behavior of two-stage op-amps is not well-founded for CBMC amplifiers employed in SC circuits.

IV. PROPOSED APPROACH AND DESIGN EXAMPLE

A complex non-linear analysis should be carried out to describe the above discussed phenomena. It is however worth pointing out that the behavior of V_{out+} (Fig. 3) at the end of the slewing phase can be anyway approximated by a straight line of slope SR_{INT} (2), but starting from a null initial condition, as has been experimentally verified by several designs carried out. This allows the definition of quite simple early-design rules to manage the CBMC op-amp slewing behavior. In order to identify a general analytical formulation to describe the slewing in the presence of both positive and negative input steps, the op-amp output voltages are indicated with $V_{out1}(t)$ and $V_{out2}(t)$ in the following. If the amplitude A_0 of the input step is positive, then $V_{out1}(t)=V_{out+}(t)$ and $V_{out2}(t)=V_{out-}(t)$, otherwise $V_{out1}(t)=V_{out-}(t)$ and $V_{out2}(t)=V_{out+}(t)$. By taking these positions into account, the previous analysis of the slewing phenomenon suggests to consider the following relationships as useful rules of thumb in order to evaluate the two op-amp output voltages at the final instant of the slewing period (namely t_{LS}):

$$V_{out1}(t_{LS}) = \frac{I_1}{C_C} t_{LS} \qquad (6)$$

$$V_{out2}(t_{LS}) = -[\frac{I_2}{C_C+C_L} t_{LS} - \frac{V_{out}(0)}{2}] \qquad (7)$$

t_{LS} is the amount of time required by the amplifier input voltage to reach the value $\sqrt{2}(V_{GS1,2}-V_{TH1,2})$, $V_{GS1,2}$ and $V_{TH1,2}$ being the gate-source voltage in the bias point and the threshold voltage of MOSFETs $M1,2$, respectively [14]. By applying Kirchoff's laws to the SC circuit in Fig. 2, the following equation can be easily found:

$$\frac{C_I|A_0|}{C_I+C_F} - \frac{C_F|V_{out}(t_{LS})|}{C_I+C_F} = \sqrt{2}(V_{GS1,2}-V_{TH1,2}), \quad (8)$$

where $|V_{out}(t_{LS})| = |V_{out1}(t_{LS}) - V_{out2}(t_{LS})|$. Equations (6)-(8) can be then exploited to fix the op-amp parameters in order to achieve the desired value for t_{LS}. Moreover, these equations allow useful design hints to be defined in order to limit the asymmetry of the two amplifier branch responses by properly balancing the stage bias currents. In fact, by imposing $V_{out1} = -V_{out2}$ (i.e., $V_{out+} = -V_{out-}$) at the end of the slewing period, the following equation can be written according to (1), (6) and (7):

$$\frac{I_1}{C_C} t_{LS} = \frac{I_2}{C_C+C_L} t_{LS} - \frac{C_F C_I}{2(C_I+C_F)C_L}|A_0| \qquad (9)$$

If (9) is satisfied, $|V_{out}(t_{LS})| = 2(I_1/C_C)t_{LS}$. From (8), the slewing period is then given by:

978-1-4244-1983-8/08/$25.00 ©2008 IEEE

$$t_{LS} = \frac{C_I C_C |A_0| - \sqrt{2}(V_{GS1,2} - V_{TH1,2})C_C(C_I + C_F)}{2I_1 C_F} \quad (10)$$

To demonstrate the validity of the proposed rules, the circuit in Fig. 2 was redesigned accordingly. The same values of I_1, I_b, C_I, C_F, C_C and C_{EXT} used for the design of Section II were considered, i.e. $I_1 = 7.5\mu A$, $C_I = 2pF$, $C_F = 2pF$, $C_{EXT} = 4pF$, $I_b = 14.5\mu A$ and $C_C = 0.6pF$. The transistors $M1$ and $M2$ were sized to have an overdrive voltage $(V_{GS1,2} - V_{T1,2})$ equal to 100 mV. Therefore, $\sqrt{2}(V_{GS1,2} - V_{TH1,2}) \approx 140mV$. The design was carried out by considering an input step of amplitude $A_0 = 1V$. On the basis of the chosen amplifier parameters, $t_{LS} = 30$ ns results from (10). From (9), $I_2 = 90\mu A$ is then the value of the second stage bias current required to achieve symmetrical slewing behaviors of the two op-amp output voltages, instead of $I_2 = 70\mu A$ resulting from the conventional design rule (5). HSPICE simulation results related to $V_{out1}(t) = V_{out+}(t)$ and $-V_{out2}(t) = -V_{out-}(t)$ are depicted in Fig. 5. In the figure, the difference between the positive and the negative input of the amplifier (namely $V_{in,op}$) is also reported. It appears that $V_{in,op}$ reaches the threshold of 140 mV in a period of time of about 30 ns, according to the value predicted by (10). As shown, by balancing the stage bias currents according to rules (9) and (10), the unavoidable asymmetry between the responses of the two op-amp branches is significantly reduced. The above example proves that the proposed design approach is very effective to control the slewing behavior of the two-stage CBMC op-amp, when it is used in typical SC circuits.

V. CONCLUSIONS

The slewing behavior of two-stage CBMC op-amps in SC circuits has been investigated in this paper. It has been shown that the current buffer used in the compensation network strongly affects the op-amp response, causing the slewing behavior to deviate remarkably from the conventional straight-line approximation. Improved early-design rules have been proposed in order to control the actual CBMC amplifier slewing performance.

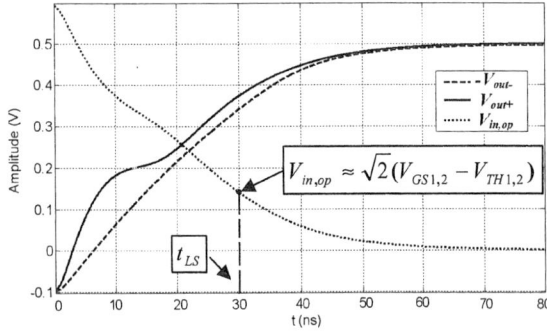

Figure 5. 1V-step responses of V_{out+}, $-V_{out-}$ and $V_{in,op}$ when the proposed design rules are used

REFERENCES

[1] M. Waltari, and K.A.I. Halonen, "1-V 9-Bit Pipelined Switched-Opamp ADC," IEEE Journal of Solid-State Circuits, vol. 36, No. 1, pp. 129- 134, January 2001.

[2] J. Li, Gil-Cho Ahn, Dong-Young Chang, and Un-Ku Moon, "A 0.9-V 12-mW 5-MSPS Algorithmic ADC With 77-dB SFDR," IEEE Journal of Solid-State Circuits, vol. 40, No. 4, pp. 960-969, April 2005.

[3] Hung-Chih Liu, Zwei-Mei Lee, and Jieh-Tsorng Wu, "A 15-b 40-MS/s CMOS Pipelined Analog-to-Digital Converter With Digital Background Calibration," IEEE Journal of Solid-State Circuits, vol. 40, No. 5, pp. 1047-1056, May 2005.

[4] Jiang Yu, and F. Maloberti, "A Low-Power Multi-Bit ΣΔ Modulator in 90-nm Digital CMOS Without DEM," IEEE Journal of Solid-State Circuits, vol. 40, No. 12, pp. 2428-2436, December 2005.

[5] Kye-Shin Lee, S. Kwon, and F. Maloberti, "A Power-Efficient Two-Channel Time-Interleaved ΣΔ Modulator for Broadband Applications," IEEE Journal of Solid-State Circuits, vol. 42, No. 6, pp. 1206-1215, June 2007.

[6] P.J. Hurst, S.H. Lewis, J. P. Keane, F. Aram, and K.C. Dyer, "Miller Compensation Using Current Buffers in Fully-Differential CMOS Two-Stage Operational Amplifiers," IEEE Trans. on Circ. and Syst. I, vol. 51, No. 2, pp. 275-285, February 2004.

[7] G. Palmisano, and G. Palumbo, "A Compensation Strategy for Two-Stage CMOS Opamps Based on Current Buffer," IEEE Trans. On Circ. And Syst. I, vol. 44, No. 3, pp. 257-262, March 1997.

[8] A. Pugliese, F.A. Amoroso, G. Cappuccino, and G. Cocorullo, "Settling time optimisation for two-stage CMOS amplifiers with current-buffer Miller compensation," Electronics Letters, Vol. 43, No. 23, pp. 1257-1258, November 2007.

[9] S. Rabii, and B. A. Wooley, "A 1.8-V Digital-Audio Sigma-Delta Modulator in 0.8-μm CMOS," IEEE Journal of Solid-State Circuits, VOL. 32, No. 6, pp. 783-796, June 1997.

[10] H. Aminzadeh, M. Danaie, and R. Lotfi, "Design of Two-Stage Miller-Compensated Amplifiers Based on an Optimized Settling Model," IEEE Conference on VLSI Design 2007.

[11] R. Lotfi, M. Taherzadeh-Sani, M.Yaser Azizi, and O. Shoaei, "Low-power design techniques for low-voltage fast-settling operational amplifiers in switched-capacitor applications," Integration, the VLSI Journal, Elsevier, vol. 36, pp. 175-189, August 2003.

[12] R. Naiknaware, and T.S. Fiez, "Process-Insensitive Low-Power Design of Switched-Capacitor Integrators," IEEE Trans. On Circ. and Syst. I, vol. 51, NO. 10, pp. 1940-1952, October 2004.

[13] R. del Rio, F. Medeiro, B. Perez-Verdù, and A. Rodriguez-Vazquez, "Reliable analysis of settling errors in SC integrators: application to ΣΔ modulators," Electronics Letters, vol. 36, No. 6, pp. 503-504, March 2000.

[14] F. Wang, and R. Harjani, "An Improved Model for the Slewing Behavior of Opamps," IEEE Trans. On Circ. And Syst. II, vol. 42, NO.10, pp. 679-681, October 1995.

[15] M. Yavari, N. Maghari, and O. Shoaei, "An Accurate Analysis of Slew Rate for Two-Stage CMOS Opamps," IEEE Trans. On Circ. And Syst. II, vol. 52, No. 3, pp. 164-167, March 2005.

[16] O. Choksi, and L. R. Carley, "Analysis of Switched-Capacitor Common-Mode Feedback Circuit," IEEE Trans. On Circ. And Syst. II, vol. 50, No. 12, pp. 906-917, December 2003.

[17] G. Palmisano, G. Palumbo, and S. Pennisi, "Design Procedure for Two-Stage CMOS Transconductance Operational Amplifiers: A Tutorial," Analog Integrated Circuits and Signal Processing, Kluwer, vol. 27, pp. 179-189, 2001.

A CMOS 90nm 4-mW 15dBm-IIP3 Base-Band Programmable Gain Amplifier for UWB Receivers

A.Cito[1,4], M. De Matteis[1], S. D'Amico[1], P.Delizia[1],
A.Baschirotto[1,3]
[1]Dep. of Innovation Engineering, University of Salento– Italy
[3]Dep. of Physics - University of Milano Bicocca, Italy
[4]Dep. Of Electric Engineering, University of Pavia, Italy

C. Azeredo-Leme[2], Ricardo Reis[2], Wen-Hu Zhao[2],
A. Tavares[2]
[2]Chipidea Microelectronica - Portugal

Abstract – In this paper the design of a Programmable Gain Amplifier (0dB÷15dB, 3dB per step) embedded in the analog base-band chain for UWB Receivers in a standard 90nm CMOS technology is presented. The design has been realized using a proper Matlab procedure. Due to the large bandwidth required in the UWB Receivers, the procedure takes into account the second order effects related with the opamp parasitic capacitance and finite bandwidth. In this way the model guarantees to satisfy the IRN and transfer function accuracy requirements and to avoid linearity performance degradation. Thanks to the results obtained by the Matlab procedure, no capacitance are used in the PGA schematic. From a single 1.2V supply voltage, the PGA total current consumption is 3.5mA, the in-band IIP3 is 15dBm and out-of-band IIP3 is 23dBm, the maximum IRN is 16nV/√Hz at 15dB gain.

I. INTRODUCTION

Ultra Wide Band (UWB) systems have recently received a great deal of interest due to their potential for high speed wireless communication. UWB technology is characterized by using a large portion of the radio spectrum with a limited output power level in a way that doesn't interfere with other narrow band technology with which shares the same spectrum. For the extremely low emission level, UWB is prevalently used for short-range and indoors application (called Wireless Personal Area Network (WPAN) systems). At present, both direct-sequence impulse communication and multi-band OFDM (MB-OFDM) UWB systems are under consideration to realize a short-range high data rate communication link.In this paper an application of the last system is presented. This MB-OFDM UWB system divides the UWB spectrum into sub-bands of 528MHz [1]. A typical MB-OFDM Receiver in zero-IF architecture is shown in block diagram of Fig. 1. Compared to narrowband receivers, MB-OFDM poses a new set of challenges. The low signal power at the Receiver input requires low-noise Filter and PGA. Coexistence with other bands gives rise to new linearity constraints. These bands feature strong out-of-band interferers for UWB system that needs a high linearity. The Receiver needs a channel-select filter with a high cut-off frequency of 264M. Out-band interferers

give rise to in-band intermodulation tones. Therefore high frequency response accuracy is required to avoid in-band distortion and in-band signal degradations.

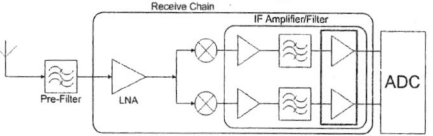

Fig. 1 - UWB receiver zero-IF architecture

In this paper a design procedure for UWB active cells is presented. In particular the procedure gives the design parameters of the Programmable Gain Amplifier (PGA) embedded in analog base-band section of the UWB zero-IF receiver shown in Fig. 1. The challenges first described are more stringent with reference to poorer performance (larger parasitic capacitance and lower gain transistor) of analog devices in a so scaled technology. In fact, due to the large input signal bandwidth the design of the UWB PGA must take into account the 2nd-order effects in terms of transistors parasitic capacitance and opamp finite bandwidth. These two effects can affect critically the transfer function accuracy and the stability of the circuit. The Matlab procedure, as described in the section III, starts from the overall PGA specifications. A Matlab model of the PGA is determined, taking into account the opamp input and output stage parasitic capacitance, and its unity gain bandwidth, UGBW. The output results of the procedure are the following:

- PGA resistances values.
- the opamp specifications (IRN, dc gain A_o, UGBW f_u, Phase Margin).

II. PGA

A. PGA Circuital Topology

The general specifications for the UWB Receiver base-band chain are presented in the Table I.

Table I - UWB Receivers Baseband Chain Specifications

	Baseband Chain	PGA
Cut-off frequency (f_o)	264MHz	>264MHz
Dc-gain (G)	0dB÷60dB	0dB÷15dB
Attenuation at 660MHz	-40dB	-
Maximum Ripple	2dB	1dB
In-band IIP3	12dBm	>12dBm
Out-of-band IIP3	20dBm	>20dBm
IRN (for G>15dB)	10nV/√Hz	-

978-1-4244-1983-8/08/$25.00 ©2008 IEEE   213

Starting from the overall UWB Receiver requirements the PGA specifications are obtained. They are also reported in Table I.

In order to select one of the MB-OFDM sub-band the base-band chain must have a cut-off frequency of 264MHz. The f@-3dB frequency is typically guaranteed by the previous filter, so that the PGA features a maximum flat frequency response at 264MHz. In this way significant in-band signal drop are avoided. The overall UWB Receiver chain features a dc gain of 60dB. The PGA dc gain is 15dB, the rest of the gain is given by the filter. The out-of-band attenuation is performed by the filter and for the PGA a minimum 20dB/decade is sufficient. Stringent linearity performance is required to the Receiver chain. The PGA is the last block of the chain and so it is the most critical for the linearity. The IIP3 of 12dBm and 20dBm is required for in-band and out-band linearity test respectively. The IRN requirement for the PGA is not so critical because the low IRN must be guaranteed for large gain – G>15dB typically – where the noise contribution of the PGA can be neglected. So that to satisfy the stringent linearity specifications, a closed loop configuration has been selected. The topology used for this cell is shown in the single ended version in Fig. 2.

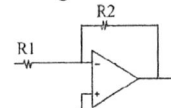

Fig. 2 – PGA single-ended schematic

The PGA programmability has been realized by the configuration shown in Fig. 3[2]. In this structure, the current signal doesn't flow through the switches avoiding linearity degradation and transfer function modifications.

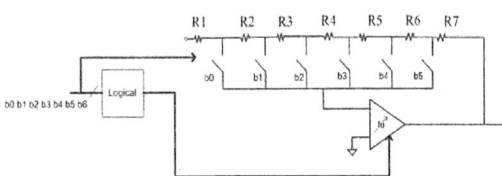

Fig. 3 – PGA gain programmability strategy

PGA must have a variable gain from 0dB to 15dB with 3dB step. Programmability is realized switching the resistances. The opamp bandwidth is regulated for each gain, by the control logic. A large opamp gain can be achieved by adopting two stage structures, therefore a Class A Miller opamp has been designed [3].

A low opamp dc gain can critically affect the PGA dc gain. Simulation results show that to maintain the maximum PGA gain drop below the 3% of the nominal value an opamp dc gain higher than 45dB is necessary. To guarantee a minimum opamp dc gain the input stage transistors – M1, M2 – should be very large. Large transistors in CMOS 90nm technology introduce relevant

parasitic capacitance. This effect is more and more important in the scaled technologies. So that, the design of the PGA operating in the UWB Receivers is critical for the combined presence of the following no-idealities:

- the opamp finite unity gain bandwidth (f_u)
- the input stage opamp parasitic capacitance (C_p).

In particular the parasitic capacitance and the input resistance R_1 creates one pole whose frequency is placed close to the opamp UGBW. This can generates complex poles with a proper Q and f_o. If the complex poles are not controlled the maximum flat shaping of the PGA frequency response can be affected. There are two typical problems related with the no-controlled PGA complex poles:

- High Q frequency response – Q>1.
- 264MHz drop higher than -3dB.

In both cases it appears very difficult to comply with the maximum ripple requirement – 2dB – indicated in the Table I. Furthermore the high Q frequency response degrades the in-band and out-band linearity performance increasing the power of the intermodulation harmonics falling down into the band.

B. Matlab Procedure

The starting point of the Matlab procedure is to design the PGA model considering the second order effects as shown in Fig. 4 and the selected circuital topology.

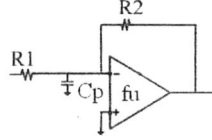

Fig. 4– PGA Matlab model

This model of the PGA exploits the parasitic capacitance C_p and finite opamp f_u. The PGA parameters depend on the f_u and C_p. The basic formulas are reported in (1).

$$(1)\quad G = \frac{R_2}{R_1}, \quad \omega_o = \sqrt{\frac{2\pi f_u}{C_p \cdot R_2}}, \quad Q = \frac{1}{1+G}\sqrt{2\pi f_u \cdot C_p \cdot R_2}$$

Without considering the C_p and f_u contribution in the overall PGA frequency response plotting, high Q poles can be generated. In fact the opamp bandwidth should be obtained to comply with the minimum gain bandwidth product requirement at 15dB – f_u>2GHz. The R_1 of 1kΩ allows maintaining good noise performance for low chain gains. But basing on the previous formulas, and considering a maximum C_p of 200fF, the frequency response features a quality factor higher than 5dB. This value is evidently relevant for PGA that must have a limited gain flatness, as just explained.

In order to obtain a flat frequency response for the PGA the most common solution is to place a feedback capacitance in parallel with the R_2 resistance in order to control the stability [4]. This is not feasible in this design for the following reasons:

978-1-4244-1983-8/08/$25.00 ©2008 IEEE

- The strategy used for the programmability gain allows that the PGA gain varies changing the feedback resistance R_2. So to control the PGA pole the feedback capacitance should be adequately programmed. This increases the switches in the schematic and in general the complexity of the PGA degrading the linearity performance.
- The feedback capacitance decreases the frequency of the opamp second pole, degrading the opamp UGBW and the Phase Margin.

So the strategy adopted in this work is to find a trade-off between the opamp f_u and the opamp maximum input stage C_p. In this way high Q poles or drops at 264MHz larger than -3dB, can be avoided. Furthermore the absence of the capacitance reduce the area of the definitive integrated circuit.

III. PGA UWB DESIGN PROCEDURE

The Matlab procedure is based on the equivalent model of PGA that takes into account the opamp input stage parasitic capacitance C_p and the opamp UGBW f_u. The equivalent circuit of the PGA used in the Matlab model is shown in Fig. . As just explained, this cell can synthesize two complex poles, with a proper f_o and Q. If a large opamp f_u is selected, the complex poles features high quality factor, generating stability or linearity problems. If the opamp f_u is too low the drop at 264MHz can degrade the in-band signal. The goal of the Matlab procedure is to find a trade-off between the opamp f_u and the maximum C_p due to the opamp input stage.

A preliminary analysis has shown that f_o and Q must have values in the range [400MHz;1GHz] and [0.707;0.9] respectively, to avoid high peaks and in-band drop. These ranges individuate an area AR_{PGA} of the Gauss plane. The Matlab procedure selects the PGA complex poles that fall into this area for each gain. The selected complex pole pair guarantees a minimum 10% accuracy of the PGA frequency response.

As example in the Fig. 5 and Fig. 6 the complex poles position in the Gauss plane and the correspondent frequency response are respectively reported, without using the Matlab design procedure. The opamp UGBW is 2GHz and the opamp input stage parasitic capacitance is 200fF. In particular in Fig. are reported the real part (on the x-axis) and the imaginary part (on the y-axis) of the PGA poles. The bold area fixes the AR_{PGA}.

The PGA transfer function features three poles. The first pair depends on the just commented complex poles. The other poles in the left side of the Fig. are placed at high frequency, and they are related with the opamp second pole. Their influence is not so significant in this analysis due to the large frequency.

For the case shown in Fig. , only one of the couple of pole fall into AR_{PGA}. The complex poles can fall out the delimited area because the quality factor is too high or the f_o is placed too close to the 264MHz frequency.

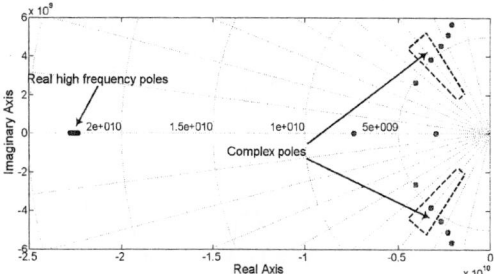

Fig. 5 – Position poles into Gauss plane

The corresponding frequency response, plotted in the Fig. 6, features peaks at high frequency for G=0dB÷6dB, and in-band signal attenuation higher than 3dB for G=12dB÷15dB, while satisfies the f_o and Q requirements for G=9dB, as expected considering the poles position in the Fig. .

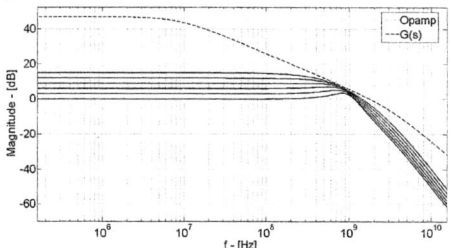

Fig. 6 – PGA frequency response with fixed f_u and C_p

So that regulation of the opamp f_u is needed for each gain. In fact the Matlab procedure adjusts the f_o and Q of the complex poles finding the right opamp f_u for each gain. The basic steps of the Matlab procedure are presented in the flow-chart of the Fig. 7. The principal steps of the Matlab procedure are here explained.

Step 1-2-3. The cell parameters (G, IRN) and the opamp parameter (A_o, PM, maximum C_p) are fixed. The resistance values are calculated.

Step 4,5,6,7,8,10,11. The algorithm loop starts from G=0dB and f_u=3GHz. Fixing the gain the frequency response is calculated at f_u=3GHz. If the PGA poles fall into AR_{PGA} the correspondent opamp f_u is maintained and the G is increased of 3dB. Otherwise the opamp f_u is reduced of 500MHz and the frequency response is calculated.

Step 9. When the f_u=500MHz and there were no poles falling into the AR_{PGA}, for the current gain, the procedure is repeated adjusting the PGA IRN. In fact, reducing the IRN, the PGA input resistance decreases and the parasitic pole effect is less significant, as in the formulas (1).

Step 12. The PGA and respective opamp frequency responses are plotted.

The PGA resistance values calculated with this procedure are indicated in the Table II. The f_u and C_p values are also indicated for each PGA gain in the Table III. The opamp f_u is adjusted by regulating the miller capacitance C_c.

978-1-4244-1983-8/08/$25.00 ©2008 IEEE   215

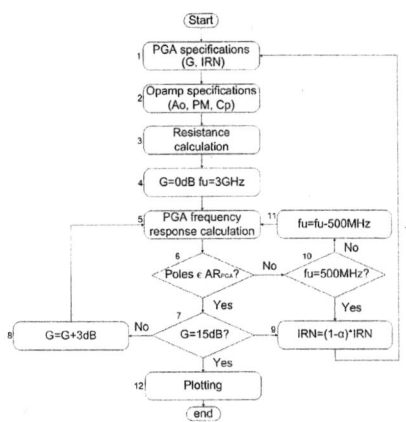

Fig. 7– Matlab Procedure Alogorithm

Table II – PGA resistance

R_1	R_2	R_3	R_4	R_5	R_6	R_7
951Ω	313Ω	385Ω	453Ω	508Ω	539Ω	3.15kΩ

Table III – Output results of the Matlab procedure

PGA Gain	Opamp fu	Cc
0dB 3dB	500MHz	1.7pF
6dB 9dB	800MHz	1.3pF
12dB 15dB	1GHz	500fF

IV. SIMULATION RESULTS

Starting from the results produced by the Matlab models the design of the PGA has been The simulation results for different gain levels are reported in Fig. 8 and Fig. 9, frequency response and IRN. In Fig. 10 the IIP3 for the unwanted UWB signal and for the WLAN interferes case are calculated. The input signal is a dual tone signal at 280MHz and 300MHz frequency, for the unwanted UWB signal, and at 480MHz and 810MHz frequency, for the WLAN interferes case. The Filter Gain is 0dB. The simulation results resume is reported in the following table.

V. CONCLUSION

A design procedure for the PGA embedded in the analog base-band section of the UWB receiver has been described in this paper. The procedure has been developed using a Matlab model. The output results of the Matlab procedure allows using a large bandwidth stable closed-loop configuration without feedback capacitance. The PGA has been developed in 90nm CMOS technology with a 1.2V supply voltage. Thanks to the Malab model Active configuration is used and no capacitance are used. The maximum gain is 14.7dB. The maximum attenuation at 264MHz and 660 MHz of -1dB and -10dB respectively guarantee the in-band and out-band attenuation required for the whole chain. The IIP3 is 15dBm for unwanted UWB and 23dBm for WLAN interferer. The total current consumption is 3.5mA.

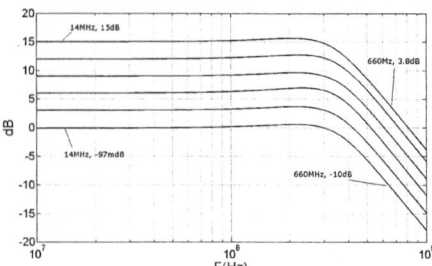

Fig. 8– PGA Frequency Response for each gain

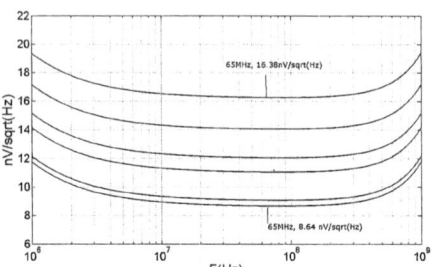

Fig. 9– PGA IRN for each gain

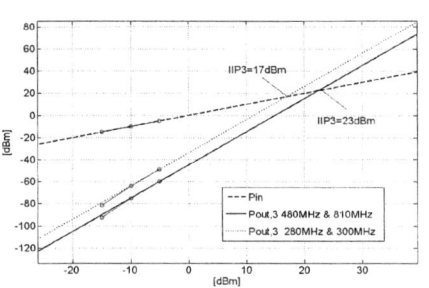

Fig. 10– IIP3 for out-band and in-band test

TABLE IV – UWB BASEBAND CHAIN PERFORMANCE SUMMARY

Parameter	Sim Results
Technology	90nm CMOS
VDD	1.2 V
Gmax	14.7dB
Max Attenuation at 264MHz	-1dB
Min Attenuation at 660MHz	-10dB
IRN	16.3nV√Hz@0dB 8.6nV√Hz@15dB
Out-of-band IIP3 - (480MHz&810MHz)	23dBm
In-band IIP3 - (280MHz&300MHz)	15dBm
I_{tot}^* – Current Consumption (*considering the Bias and the CMFB)	3.5mA
Output Load	5kΩ//500fF

VI. REFERENCES

[1] "Multi-band OFDM Physical Layer Proposal for IEEE 802.15" Task Group 3a", IEEE P802.15, March 2004.

[2] D'Amico, et. all, "A 5nV/√Hz 78dB-Gain-Range 82dB-DR Multistandard Baseband Chain for Bluetooth, UMTS and WLAN" Munich, ESSCIRC 2007

[3] J. Mahattanakul et.all, "Design Procedure for Two-Stage CMOS OpampWith Flexible Noise-Power Balancing Scheme", IEEE Transaction on circuits and systems

[4] R. Rooverset.all, "An Interference-Robust Receiver for Ultra-Wideband Radio in SiGe BiCMOS Technology", IEEE Journal of solid-state circuits, NO. 12, DECEMBER 2005

Wideband Self-Biased CMOS CCII

Emre ARSLAN

Department of Electrical and Electronics Engineering
Bogazici University, 34342, Bebek
Istanbul, Turkey
emre.arslan@boun.edu.tr

Avni MORGUL

Department of Electrical and Electronics Engineering
Bogazici University, 34342, Bebek
Istanbul, Turkey
morgul@boun.edu.tr

Abstract—**Wideband, self-biased, second generation current conveyor (CCII) is proposed. It is shown that the proposed CCII exhibits superior performance compared to its previous counterparts in terms of bandwidth, parasitic resistance and voltage swing on port *X*. Also, the proposed CCII uses no additional biasing voltage or current sources other than the two supply rails.**

I. INTRODUCTION

High-frequency analog circuits are used for numerous applications in telecommunication, video and hard disk areas. The current-mode circuits are being widely used in high frequency circuit design applications [1-6]. The second generation current conveyor (CCII) is the most widely used active element in signal processing applications. Recently, a great deal of study has been done in order to design high performance, wideband, low X port parasitic resistance CMOS and BiCMOS CCIIs [2-9]. The CCII circuits based on the translinear loop have excellent wideband current following behavior, but the voltage following behavior is poor and there is a large offset voltage. Also, using translinear-loop based circuits increases the slew rate [4-5].

In an analog circuit, different active devices should be properly biased in order to get high performance. For example, most of the transistors in a CMOS circuit have to be biased in a saturation region to work properly. Almost all of the CMOS CCII circuits need additional voltage or current source in order to bias the transistors. Using independent biasing sources results in numerous drawbacks, namely, an area and power overhead, and high sensitivity of the bias point to process variations. The main disadvantage of using independent biasing sources is that, the wires used between the supply voltage and the CMOS circuit are susceptible to noise and crosstalk. [1-2]

II. THE CCII

The CCII is a three terminal active block as shown in Fig. 1. It ensures two functionalities between its terminals:

- A current follower between ports X and Z

- A voltage follower between ports Y and X

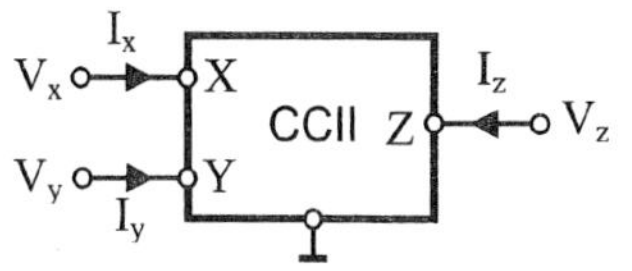

Figure 1. Second generation current conveyor

In mathematical terms, the input-output characteristics of the CCII can be described by the following matrix equation:

$$\begin{bmatrix} I_y \\ V_x \\ V_z \end{bmatrix} = \begin{bmatrix} 0 & 0 & 0 \\ \beta & 0 & 0 \\ 0 & \pm\alpha & 0 \end{bmatrix} \begin{bmatrix} V_y \\ I_x \\ V_z \end{bmatrix} \tag{1}$$

where ideally $\alpha=1$ and $\beta=1$ which represent the current and voltage transfer ratios of the current conveyor, respectively. The sign of α, plus or minus, denotes positive or negative type current conveyor respectively.

III. THE PROPOSED CCII

In this paper, CMOS realization of the self-biased, wideband, low parasitic resistance and large voltage swing CCII based on the long tail differential pair is proposed. The proposed circuit is realized as shown in Fig. 2. It is possible to manage input signals from positive supply voltage down to negative supply voltage, rail to rail operation, by using both the n-type and p-type differential pairs. P-type differential pair (M_7-M_8) operates when the input voltage on port Y is low and n-type differential pair (M_6-M_9) operates when the input voltage is high. Joining these two differential pairs in parallel allows implementing "rail to rail input stage" based CCII [12]. Also the proposed CCII is free from

978-1-4244-1983-8/08/$25.00 ©2008 IEEE

the drawbacks of using additional biasing sources as mentioned in the introduction part.

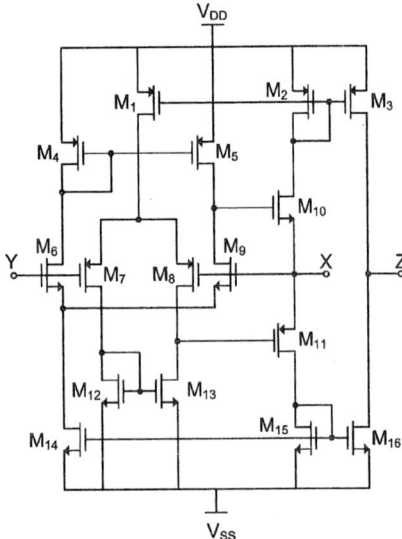

Figure 2. Self-biased CMOS CCII circuit

In addition to rail to rail input stage, two source followers, M_{10} and M_{11}, are used in order to preserve low parasitic resistance on port X. These transistors also establish the biasing current of the circuit. This biasing current is carried to the rail to rail stage by using the mirror transistors M_2 and M_{15}. Also, the biasing current flowing through the source follower transistors forms the port X current, so the mirror transistors M_2 and M_{15} copy this current to port Z.

If we consider the differential pair having the same small signal parameters (equal r_o's), the parasitic resistance on port X can be evaluated as:

$$R_x \cong \left(g_{m10} \left(1 + \frac{r_o}{2} g_{m9} \right) \right)^{-1} // \left(g_{m11} \left(1 + \frac{r_o}{2} g_{m8} \right) \right)^{-1} \quad (2)$$

The value for the output resistance seen at port Z is:

$$R_z \cong \frac{r_{o3} r_{o16}}{r_{o3} + r_{o16}} \quad (3)$$

IV. SIMULATION RESULTS

The characteristics of the self-biased current conveyor have been determined from HSPICE simulations by using UMC 0.18 μm process parameters. The current and voltage gains are 1.01 and 0.96, respectively. Aspect ratios of the transistors and simulation results of the proposed circuit are listed in the tables below.

TABLE I.   ASPECT RATIOS

Transistor	Aspect Ratio, W/L
M_1, M_2, M_3, M_7, M_8	30/0.36 (μm)
M_6, M_9, M_{14}, M_{15}, M_{16}	10/0.36 (μm)
M_4, M_5	45/0.36 (μm)
M_{12}, M_{13}	15/0.36 (μm)
M_{10}	40/0.36 (μm)
M_{11}	120/0.36 (μm)

TABLE II.   SIMULATION RESULTS OF THE PROPOSED CMOS CCII

Simulation Results	
Power supply	±1.65 V
Power dissipation	3.1 mW
Voltage gain, V_x / V_y	0.96
Voltage transfer BW	990 MHz
Current gain, I_z / I_x	1.01
Current transfer BW	700 MHz
Port X resistance	0.018 Ω
Voltage swing on port X	±1.44 V

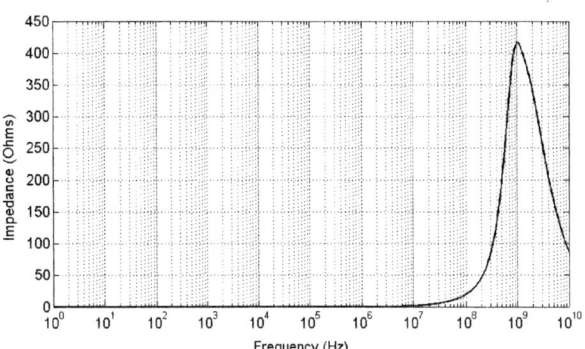

Figure 3. Port X impedance variation with frequency

The variation of the port X intrinsic impedance with frequency is shown in Fig. 3. As it can be seen from the figure, it has very small values up to very high frequencies. The value for this intrinsic impedance is about 1.9Ω for 10 MHz and 20Ω for 100MHz. So, the proposed circuit is useful to be used in high frequency ranges. With using the rail to rail input stage, it is possible to have port X voltage swing very close to the supply voltages as shown in the figure below.

978-1-4244-1983-8/08/$25.00 ©2008 IEEE

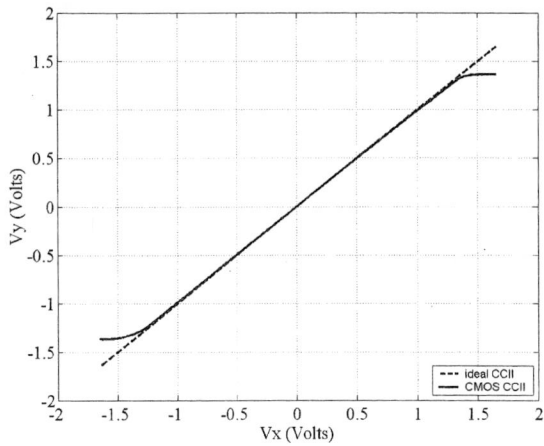

Figure 4.　Voltage transfer characteristics from port Y to port X

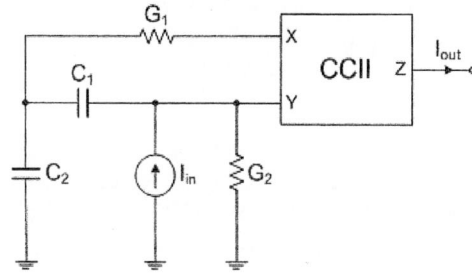

Figure 5.　Band-pass filter circuit employing the proposed CMOS CCII

As it can be seen from the simulation result shown in Fig. 6, the response of the proposed circuit is very close to the ideal filter response.

V.　APPLICATION EXAMPLE

As an application example, a wideband, high output impedance, current mode band-pass (BP) filter, which employs the proposed CMOS CCII, is demonstrated in this section [13]. The topology of the filter is shown in Fig. 5. It is based on a single CCII block with the transfer function as:

$$\frac{i_{out}}{i_{in}} = \frac{C_2 G_1 s}{G_1 G_2 + (C_1 + C_2)G_2 s + C_1 C_2 s^2} \qquad (4)$$

BP filter condition for this transfer function is:

$$G_1 = \left(1 + \frac{C_1}{C_2}\right)G_2 \qquad (5)$$

After applying the condition, the transfer function can be obtained as:

$$\frac{i_{out}}{i_{in}} = \frac{(2G/C)s}{s^2 + (2G/C)s + (2G^2/C^2)} \qquad (6)$$

The centre frequency and the quality factor for the BP filter can be obtained as:

$$w_o = \frac{\sqrt{2}G}{C}, \; Q = \frac{\sqrt{2}}{2} \qquad (7)$$

The functionality of the filter circuit is illustrated with the element values of R_1=0.5kΩ, R_2=1kΩ, C_1=2pF, C_2=2pF, which determine the center frequency as f_0 = 112.5 MHz.

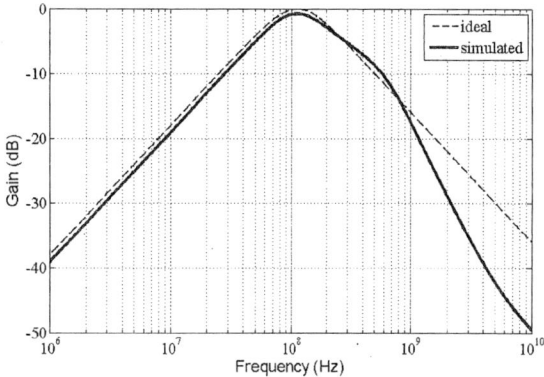

Figure 6.　Simulation results of the BP filter response

VI.　CONCLUSION

In this paper, self-biased, wideband, low parasitic impedance current conveyor with rail to rail input stage is proposed. The proposed CCII uses no additional biasing voltage or current sources other than the two supply rails. 990MHz voltage transfer bandwidth and 700MHz current transfer bandwidth are obtained with 3.1mW power consumption. The value of the parasitic resistance is 18mΩ at DC and it has very small values up to high frequencies; so the circuit may be useful to be used in high frequency applications. In addition to that, the most important advantage of the proposed circuit is the increase of the dynamic range. The proposed circuit operates linearly in the range of ±1.4 V. A wideband, high output impedance band-pass filter which uses the proposed CMOS CCII is given as an application example. The simulation results are in good agreement with the theoretical one.

978-1-4244-1983-8/08/$25.00 ©2008 IEEE

ACKNOWLEDGMENT

This work is sponsored by Bogazici University Scientific Research Projects Fund: 05HA201.

REFERENCES

[1] M. A. Ibrahim, H. Kuntman and O. Cicekoglu, "A very high-frequency CMOS self-biasing complementary folded cascode differential difference current conveyor with application examples", Midwest Symposium on Circuits and Systems, vol. 1, pp. 279-282, 2002.

[2] P. Mandal and V. Visvanathan, "A self-biased high performance folded cascode CMOS op-amp", 10th International Conference on VLSI Design, pp. 429-434, 1997.

[3] A. Fabre, O. Saaid, F. Wiest and C. Boucheron, "High frequency applications based on a new current controlled conveyor", IEEE CAS-I, pp. 82-91, 1996.

[4] B. Calva, S. Celma, P. A. Martinez and M. T. Sanz, "High-speed high-precision CMOS current conveyor", Analog Integrated Circuits and Signal Processing, pp. 265-269, 2003.

[5] W. S. Hassanein, I. A. Awad and A. M. Soliman, "New wide band low power CMOS current conveyors", Analog Integrated Circuits and Signal Processing, pp. 91-97, 2004.

[6] S. B. Salem, M. Fakhfakh, D. S. Masmoudi, M. Loulou, P. Loumeu and N. Masmodi, "A high performances CMOS CCII and high frequency applications", Analog Integrated Circuits and Signal Processing, pp. 71-78, 2006.

[7] A. M. Ismail and A. M. Soliman, "Wideband CMOS current conveyor", Electronics Letters, pp. 2368-2369, 1998.

[8] A. Fabre, "On the frequency limitations of the circuits based on second generation current conveyors", Analog Integrated Circuits and Signal Processing, pp. 113-129, 1995.

[9] A. J. L. Martin, J. R. Angulo and R. Carjaval, "Low-voltage low-power wideband CMOS current conveyors based on the flipped voltage follower", IEEE, pp. 801-804, 2003.

[10] L. N. Alves and R. Aguiar, "Maximizing bandwidth in CCII for wireless optical applications", IEEE International Conference on Electronics, Circuits and Systems, vol. 3, pp. 1107-1110, 2001.

[11] H. W. Cha and K. Watanabe, "Wideband CMOS current conveyor", Electronics Letters, pp. 1245-1246, 1996.

[12] G. Ferri and N. C. Guerrini, "Low-voltage, low-power CMOS current conveyors", Kluwer Academic Publishers, 2003.

[13] S. Ozcan, O. Cicekoglu and H. Kuntman, "Multi-input single-output filter with reduced number of passive elements employing single current conveyor", Computers and Electrical Engineering, vol. 29, pp. 45-53, 2003.

An analog baseband chain for DS-UWB system

Tero Koivisto, Janne Maunu and Esa Tiiliharju
University of Turku
Department of Information Technology, Microelectronics Laboratory
Turku, FIN-20014
Email: tejuko@utu.fi

Abstract—**In this paper, we propose a mixed-mode baseband chain using a high-speed analog Viterbi decoder. The circuit system architecture is presented with a link budget calculations and the baseband circuitry is introduced. Our target performance is 125 Mbit/s and 1Gbit/s throughput at the distances of 10 and 1 m using a 1 GHz bandwidth dedicated to UWB systems between 3.1-4.9 GHz utilizing the analog Viterbi decoder. Proposed analog receiver architecture makes possible high-speed communication with decreased power consumption.**

I. INTRODUCTION

UWB wireless radios are capable of carrying extremely high data rates (>100 Mbit/s) for up to 10 meters with little transmit power. At present, the high-speed UWB systems can be divided into two categories: Direct-Sequence (DS), also called impulse radio and Multiband (MB-OFDM) [1]. The DS-UWB system proposed in [2] [3] uses an emission bandwidth of 1.8 GHz and can be used with center frequencies of 4, 7 and 9 GHz. The key property of this system is to spread each transmission data with a code to expand the signal over a large bandwidth. The spreading gain compensates for the low level of transmit power required to comply with the FCC power density spectrum mask. Therefore, the DS-UWB transmitter operates at lower voltages as MB-OFDM and without a power amplifier. Furthermore, the transmitted wideband signal suffers no Rayleigh fading [4]. The MB-OFDM is based on the subdivision of large available bandwidth (3.1-10.6 GHz) in subbands of 528 MHz [5]. However, each 528 MHz band consists of 128 channels of 4.125 MHz and therefore the system is narrowband in nature and suffers Rayleigh fading. The MB-OFDM exploits frequency diversity through band hopping, but this results in a tight settling time specification of 10ns. It is difficult to accommodate such fast band switching with a phase-locked loop, and therefore this issue complicates the transceiver design.

In this paper, we propose an 500 MHz baseband chain for DS-UWB receiver using an analog Viterbi decoder instead of a high speed AD-converter and a digital Viterbi decoder. The analog Viterbi decoder is proposed so as to enable stronger coding with a lower power consumption and to omit the high-speed power hungry 1 Gs/s 4-bit A/D-converters and the digital Viterbi decoder. Our ultimate target is the CMOS implementation of the whole baseband chain.

II. THE DS-UWB RECEIVER

The architecture of the proposed receiver is shown in Fig. 1. It consists of a RF front-end, low-pass filter (LPF), variable gain amplifier (VGA), 1-bit AD-converter (comparator) and the analog Viterbi decoder. The main difference as compared to standard wireless baseband architectures is, that by utilizing a direct analog approach in Viterbi decoding, an 4-bit A/D converter is no longer required and therefore the output of the amplifiers are directly connected to the analog Viterbi decoder. The receiver operates as follows: The RF front-end downconverts the received UWB signal to baseband I and Q channels. These downconverted pulses are bandlimited by the LPFs with a cut-off frequency of 500 MHz. The low-pass filtered baseband pulse is further amplified by the VGA and the output of the VGA is fed to the analog Viterbi decoder, which performs the decoding process. Under high signal-to-noise (SNR) conditions, the Viterbi decoder is switched off and the signal is by-passed to the 1-bit AD-converter through a buffer amplifier. In figure 1, the receiver is operating under low SNR conditions, and therefore the analog Viterbi decoder is switched on and the 1-bit AD-converter switched off.

The DS-UWB system proposed in [2] uses the lower band between 3.1-4.9 GHz. The DS employs the transmission of short duration pulses (in the subnanosecond range) by using $\pi/2$ BPSK modulation with a chip rate of 1 GHz. Alternating 2 ns pulses on I and Q channels yields a signal with a constant envelope (stable characteristics) easing CMOS implementation. A flexible forward error correction (FEC) scheme uses convolutional (r=1/2, K=3 or 7) and Reed Solomon (240, 224, 16) codes either individually or concatenated. The data rate adaptation between 125 Mbit/s to 1 Gbit/s (58 Mbit/s to 466 Mbit/s with redundancy) is accomplished by varying spreading factor N_{ss} between 1-8. The radio architecture proposed in this work uses a analog Viterbi decoder with an r=1/2, and therefore the effective data rate is 62.5 Mbit/s to 500 Mbit/s with FEC redundancy. However, the effective data rate can be doubled with the 1-bit AD-converter as high SNR conditions allow(small distance) the analog Viterbi decoder to be switched off and the signal is by-passed to the converter. Therofore, the actual data rate varies between 62.5 Mbit/s to 1 Gbit/s. The output of the proposed architecture is further processed in the digital baseband [2]. The link budget of the analog Viterbi decoder based DS-UWB radio receiver is shown in table 1. The average transmit power is calculated from the FCC part 15 regulation, which limits the transmission to an EIRP of -41.3 dBm/MHz. The transmitted power is calculated by multiplying this limit by the bandwidth of the DS-UWB signal, which is 1 GHz. Therefore average Tx power is -

978-1-4244-1983-8/08/$25.00 ©2008 IEEE

TABLE I
THE DS-UWB LINK BUDGET

Parameter	Value	Value
Distance	10m	1m
Data Rate R_b	125 Mbs	1 Gbs
Average Tx power	-11.3 dBm	-11.3 dBm
Tx, Rx antenna gain	0 dBi	0 dBi
Path loss at	-64.2 dB	- 44.2 dB
Rx power	-75.5 dBm	-55.8 dBm
N	-93 dBm	-84 dBm
Rx NF	6.6 dB	6.6 dB
Viterbi E_b/N_0	5 dB	8 dB (QPSK)
Implementation loss	1 dB	1 dB
Link margin	4.9 dB	12.6 dB
Rx sensitivity	-80.5 dBm	-63.8 dBm

41.3dBm/MHz·1000MHz = -41.3dBm + 10log(1000) dB = -41.3dBm + 30dB = -11.3dBm. The path loss of a wireless channel is given by:

$$PL = 20log\left(\frac{4\pi d f_c}{c}\right) \qquad (1)$$

where f_c is 4 GHz, a geometric mean of the high and low -10 dB frequencies, d is the range and c is the speed of light in free space. Assuming a 0 dBi gain for transmitter and receiver antennas, the Rx power = Average Tx power + Path loss at 10m = -75.5 dBm. The average noise power N per bit is

$$N = -174dBm + 10log(R_b) = -93dBm. \qquad (2)$$

To operate correctly the analog Viterbi decoder demands a SNR of 5 dB, which with an implementation loss of 1 dB and a receiver noise figure of 6.6 dB, the link margin (LM) is

$$LM = -75.5dBm + 93dBm - 6.6dB - 1dB - 5dB = 4.9dB \qquad (3)$$

The sensitivity is -75.5 dBm - 5 dB = -80.5 dBm, which corresponds to a 30 μV signal at the input of the receiver. The corresponding values at the distance of 1 m are also given in table 1. In this case the input signal is fed to the 1-bit A/D converter.

III. THE ANALOG DS-UWB RECEIVER BUILDING BLOCKS

A. The analog baseband chain

The analog baseband chain consists of LPFs and VGAs. Although the nominal bandwidth of I and Q channels is reduced from 1 GHz to 500 MHz by using direct-conversion architecture, the required bandwidth is still large and therefore is the main baseband design challenge [5]. On the other hand, this large bandwidth makes it possible to realize high-order passive on-chip filters, because the required components are physically small [6]. However, our aim is to realize the whole receiver using a 0.13-μm digital CMOS process (no inductors) and therefore an active filter realization is chosen. Assuming a 0.2 m distance to the wanted UWB signal, the received signal power is around -40 dBm, or 3 mV. Whereas the smallest signal at the input of the receiver is 30 μV. Therefore, the

wanted UWB signal varies between 30 μV to 3 mV. The signal strength required at the Viterbi decoder input is 25 mV, and the total required maximum gain (RF + analog baseband) is $20\cdot log(25mV/30\mu V) = 59$ dB. The minimum required gain 18 dB. Assuming a RF front-end gain of 25 dB, the maximum gain required by the analog baseband circuits is 34 dB.

The overall filtering requirements can be estimated assuming the closest 802.11a interferer located only 1.15 GHz away from the carrier frequency of 4 GHz (5.15-4 GHz) at a distance of 0.2 m while the wanted UWB signal is at 10 m distance. The filter has to provide more than 35 dB of attenuation relative to DC at 1.15 GHz offset [5]. The filtering requirements can be accomplished with a 4th order Sallen-Key filter assuming a 30 dB attenuation in the RF pre-filter. Furthermore, this filter can be merged with a mixer circuit [7]. The VGAs are realized as a chain of current reuse CMOS amplifiers.

B. 1-bit AD-converter and the Differential Analog Viterbi Decoder

The 1-bit AD-converter is in fact a comparator and it is used under high SNR conditions. The attenuator is controlled by the digital baseband in such a way that the input signal to the comparator is 250 mV. In [8], a 5-bit comparator based 1 Gbit/s AD-converter was designed using 0.13-mum CMOS-process. The core analog circuitry consumed 42 mW, whereas the digital logic consumed 4 mW, a total of 46 mW. We apply similar techniques to design the 1 Gbit/s 1-bit AD-converter, which consists of a buffer stage to isolate the clocked comparator from the other circuitry and the actual comparator. The amplifier must be able to drive the analog Viterbi decoder input and the AD-converter input buffer. The decoder input consists of a four differential pairs, a total of five differential pairs.

A convolutional encoder is a finite-state machine and its code rate is the number of input bits to output bits. The constraint length of the encoder is a measure of the memory within a code. The example (2,1,7) convolutional encoder is shown in Figure 2. The encoder consists of two modulo-2 adders and a seven - bit shift register structure. The number of states is, in general, determined by $N = 2^K - 1$. Therefore, by defining the first bit as the input and the last six bit register as the state vector results in a 64-state finite-state machine. Hence, the trellis diagram corresponding the (2,1,7) convolutional code consists of 64 states.

In an analog implementation of the Viterbi decoder all the metric computations are performed and stored as analog signals by analog add-compare-select (ACS) circuits. By using a direct analog approach on Viterbi decoding a high-speed A/D converter is no longer needed. Traditionally, the decoding is performed by using the survivor path storage and trace-back procedure. In our approach [12] the two competitive paths corresponding the input bits 0 and 1 are processed by the differential add-compare-select (ACS) units. Thus, by comparing the minimum accumulated error metrics of these units the decoded output 0 or 1 is produced depending on whether the metric of the upper ACS unit is smaller or not.

978-1-4244-1983-8/08/$25.00 ©2008 IEEE

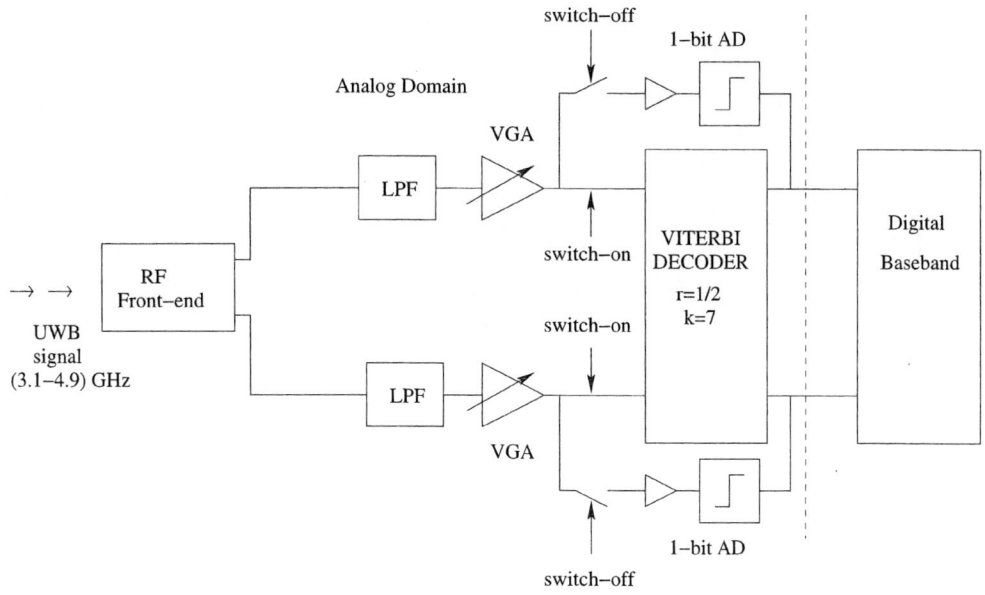

Fig. 1. The proposed DS-UWB receiver

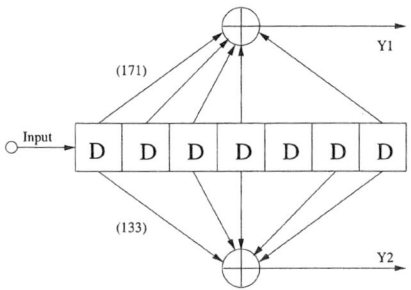

Fig. 2. The convolutional encoder for a (2,1,7) convolutional code

As a result, the path memory structure could be excluded in the differential approach.

The resulting block diagram of the Viterbi decoder with differential ACS units and the state-parallel approach is shown in Figure 3. The network block corresponds the connections between the states in the trellis diagram. Both differential processing units consist of 64 computation elements which perform a large proportion of the Viterbi algorithm functionality, namely the branch metric calculation (4 BMC), the ACS and path metric update. A single computation element simply corresponds one state in the trellis diagram. The sample/hold (S/H) circuits are used to store the calculated state metrics. Survivor path information is extracted by the decision unit, which consists of two 64-input minimum circuits. These minimum values correspond the accumulated error metrics for two competitive paths. Finally, the decoded output is produced by analyzing the difference of these currents with minimum error metric at the current comparator output.

During one computation cycle, 64 decision bits are gener-

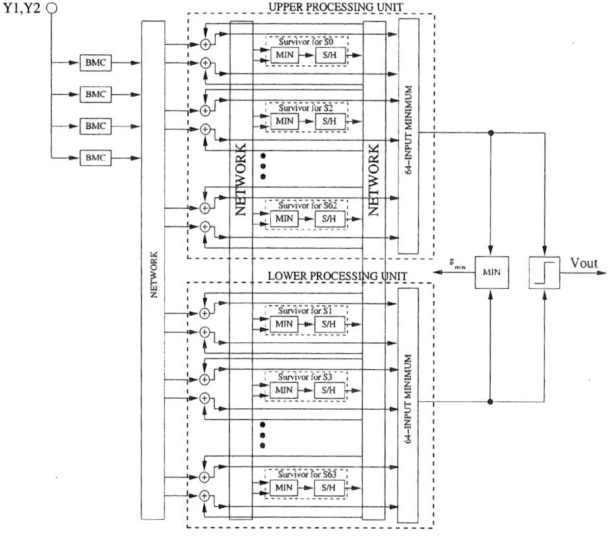

Fig. 3. Simplified block diagram of the proposed Viterbi architecture

ated in both differential processing units. Consequently, the 128 times faster operation may be achieved in comparison to the state-sequential architecture, where only one ACS processor is implemented. As presented in [13] [14] the differential structure also enables on-line decoding after initial loading stage, and thus the decoder can operate at even data rates up to hundreds of megabits per second as required by the DS-UWB system. However, the faster operation is obtained at the expense of the 128 times larger area. Nevertheless, a compact architecture may still be attained with the aid of analog implementation, since the large path memory with its

978-1-4244-1983-8/08/$25.00 ©2008 IEEE

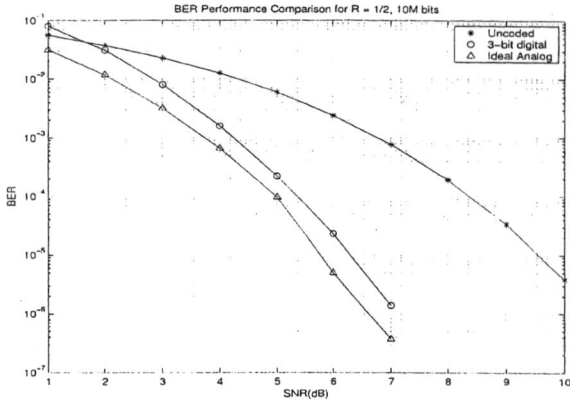

Fig. 4. BER comparison

additional circuitry is not required.

The bit error rate (BER) performance simulations in Figure 4 are accomplished by applying a random binary test sequence of length 10^7 to the convolutional encoder. Therefore, the actual test input of the decoder is the output of the binary analog channel. For the three-bit soft decision digital decoder the trace-back path memory of length seven is used, which corresponds the capability of the proposed implementation. For the decoders, the BER is calculated for SNR from one to ten decibels.

IV. DISCUSSION

The proposed UWB receiver using the analog Viterbi decoder can potentially save a lot of power, because of the elimination of the power-hungry wideband A/D-converter and the digital Viterbi decoder. The power consumption of the state of the art 4-bit 1 Gbit/s AD-converter can be estimated from [8], since power consumption of the flash converter comparators is proportional to $1/2^n$ (n is number of the converter bits). Therefore, the power comsumption can be estimated as $1/2 \cdot 42mW = 21mW$, a total of 25 mW including digital logic. The corresponding digital Viterbi decoder consumes 200 mW, a total of 225 mW [1] [15]. Our preliminary estimation of the analog Viterbi decoder power consumption is 20 mW, whereas the 1-bit AD-converter consumes at most 2 mW. This is based especially on ACS circuit simulation and to the other researchers work on the same field [16].

V. CONCLUSION

In this paper, a novel system/circuit architecture for DS-UWB radio baseband processing was proposed. The architecture is based on the analog Viterbi decoder with the 1-bit AD-converter under high SNR conditions. The power consuming 4-bit AD-converter and especially the digital Viterbi decoder are eliminated. We estimate that by the elimination of the 4-bit A/D-conversion and the digital Viterbi decoder, we can save a considerable amount of power. Therefore, on the phrases of these promising results, individual circuit blocks will be

implemented in 2008, and the whole system integration will be made on 2009.

REFERENCES

[1] Payam Heydari, 'A Study of Low-Power Ultra Wideband Radio Tranceiver Architectures ', WCNC, 2005.

[2] Sachi Iida, Katsuyuki Tanaka et.al, 'A 3.1 to 5 GHz CMOS DSSS UWB Transceiver for WPANs ', ISSCC 2005.

[3] Fujita, et.al, http://grouper.ieee.org/groups/802/15/pub/2003/May03, 03138r2P802-15.

[4] John McCorkle, 'Ultra Wide Bandwidth (UWB): Gigabit Wireless Communications for Battery Operated Consumer Applications ', Symposium on VLSI Circuits, 2005.

[5] D.M.W. Leenaerts, 'Transceiver Design for Multiband OFDM UWB ', EURASIP Journal on Wireless Communications and Networking, 2006.

[6] Aly Ismail and Asad A. Abidi, 'A 3.1-8.2-GHz Zero-IF Receiver and Direct Frequency Synthesizer in 0.18-/mum SiGe BiCMOS for Mode-2 MB-OFDM UWB Communication ', IEEE Journal of Solid-State Circuits, vol 40, No.12, December 2005.

[7] Behzad Razavi et.al, 'A UWB CMOS Transceiver ', IEEE Journal of Solid-State Circuits, vol 40, No.12, December 2005.

[8] Olli Viitala, Saska Lindfors and Kari Halonen, 'A 5-bit 1-GS/s Flash-ADC in 0.13-/mum CMOS Using Active Interpolation', ESSCIRC 2006.

[9] Martin Vogels and Georges Gielen, 'Architectural selection of AD-converters', DAC 2003.

[10] M. H. Shakiba, D. A. Johns, K. W. Martin, 'An Integrated 200-MHz 3.3-V BiCMOS CLASS-IV Partial Response Analog Viterbi Decoder', IEEE Journal of Solid-State Circuits, Vol. 33, Jan. 1998, pp.61 - 75.

[11] K. He., G. Cauwenberghs, 'An Integrated 64-state Parallel Analog Viterbi Decoder', proceedings of the IEEE International Symposium on Circuits and Systems, May 2000, pp. IV.761-IV-764.

[12] Janne Maunu, Tero Koivisto, Mika Laiho and Ari Paasio, 'An Analog Viterbi Decoder Array for DS-UWB receiver', International Conference on Ultra-Wideband', 2006.

[13] J. Maunu, M. Laiho, A.Paasio, 'A Differential Approach to Analog Viterbi Decoding', proceedings of MWSCAS06, Aug. 2006.

[14] J. Maunu, M. Laiho, A. Paasio, 'A Differential Architecture for an Online Analog Viterbi Decoder', accepted to IEEE Transactions on Circuits and Systems- I, vol. 55, no.4, May 2008.

[15] Tobias Gemmeke, Michale Gansen and Tobias G. Noll, 'Implementation of Scalable Power and Area Efficient High-Throughput Viterbi Decoders ', IEEE Journal of Solid-State Circuits, vol 37, No.7, July 2002.

[16] A. Demosthenous, J. Taylor, 'A 100-Mb/s 2.8-V CMOS Current-Mode Analog Viterbi Decoder ', IEEE Journal of Solid-State Circuits, vol 37, No.7, pp.904-910, July 2002.

978-1-4244-1983-8/08/$25.00 ©2008 IEEE

Online SNR Detection for Dynamic Power Management in Wireless *Ad-Hoc* Networks

Erdem S. ERDOGAN, Sule OZEV and Leslie M. COLLINS
Duke University
ECE Department Durham, NC 27708-0291 USA
Contact Author: Erdem S. Erdogan (ese@ee.duke.edu)

Abstract—Power consumption is a critical factor that determines the lifetime of a wireless network node. Substantial portion of the power is consumed by the transceiver in the transmit phase. This paper presents a new SNR detection technique for dynamic power management of *ad-hoc* wireless network nodes. The technique is based on the spectral analysis together with a predetermined spectral mask of the received signal. The peak power of the transmitting party can be adjusted according to the SNR information computed at the receiving end. In this way, each wireless network node transmits at the minimum power level avoiding dropped packages. Measurements on sensor network nodes, we have determined 50% potential savings in the average power consumption of the overall node. Experiments on a small network show that the power management scheme provides 65% power savings.

Index Terms—Ad-hoc sensor networks, online SNR measurement, dynamic power management

I. INTRODUCTION

Recently, *ad-hoc* wireless networks with peer-to-peer communication capability [1] have attracted much attention. The mobility and adjustability of *ad-hoc* wireless networks make them attractive for many applications, such as sensor networks for homeland security or safety monitoring.

Wireless networks are usually composed of battery-operated nodes such that many nodes may transmit or receive packets at the same time. Network structure and conditions, may change over time due to the mobility and lifetime limitations of the network nodes. One critical factor that determines the lifetime and availability of each network node is the battery life. In order to extend the lifetime of each node, minimizing its power consumption is essential.

Many factors impact the level of power consumption in a node, including the amount of digital processing, sensing, transmission and reception of signals and voltage regulation. There is an inherent trade-off between the amount of processing that needs to be done in the node and the amount of information that needs to be transmitted. However, once the amount of signal processing and bit-level transmission are determined, power optimizations are still possible at the hardware and software levels.

The power consumption in a network node has traditionally been estimated through the number of software computations. As a result, most of the research on power optimization in wireless nodes has concentrated on software-based optimizations [2]–[5]. However, an important portion of the node power is consumed in the transceiver during the transmit phase. Hardware-level power management techniques that

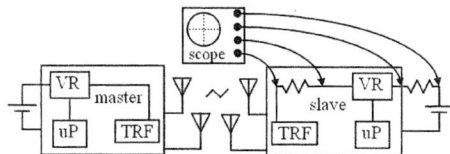

Fig. 1. Block diagram of the experimental setup

compliment software-based techniques can take advantage of potential reductions in the transmit power based on environmental conditions.

In this paper, we propose a new transmit power management technique based on detecting the signal to noise ratio (SNR) of the incoming signal. Our main goal is to reduce the power consumption of the wireless network node by continuously adjusting the transmit power level. Our online SNR detection technique is based on the frequency spectrum analysis of the received signal. The information on the communication scheme of the system enables reducing the amount of additional signal processing needed for the SNR calculations. Available SNR detection techniques [6]–[8] are not suitable for small battery-operated devices due to the required extensive computations.

II. TRANSCEIVER POWER IN A WIRELESS SENSOR

The peak transmit power significantly affects the overall power consumption in a transceiver. The transmit power needs to be high enough to reach the receiving party at the worst-case distance under the worst-case interference and noise scenarios. Transceivers typically have various power modes where the peak transmit power can be adjusted. Since environmental conditions as well as the distance of the receiving party may significantly vary over time, adjusting the transmit power based on such information may provide potential power savings as well as reduce the rate of dropped packages.

In order to evaluate the potential power savings, we perform power measurements with two identical *ad-hoc* sensor nodes placed apart, which are programmed to communicate with each other. The nodes are configured as master and slave. We modify the slave node board slightly by adding series 10Ω sense resistors to the main power input of the board as well as the power input of the transceiver. Since the resistors are small, the noise introduced by adding a series resistor to the transceiver power supply input is negligible [9]. The measurement setup and its picture are given in Figures 1 and 2 respectively.

978-1-4244-1983-8/08/$25.00 ©2008 IEEE 225

We set the slave node transmit power mode to $0dB$ and then $20dB$ attenuation for the high and low power measurements respectively. The set of four voltage waveform data stored by a digital scope suffices to calculate the total power consumed in the slave board, by the voltage regulators (VR), and by the transceiver. Table I tabulates the distribution of the average power consumption for the slave node in the two-abovementioned power modes. These measurements indicate a more than 50% difference in the power consumption of the transceiver between its two power modes.

The SNR of the received signal for a wireless network node is impacted by environmental conditions, such as the distance between the communicating parties, the communication channel noise, and the co-channel and adjacent channel interference from other devices. In an *ad-hoc* wireless network node, all the parameters continuously change as nodes move around or go on and off-line. The peak power of the received signal, which also affects its SNR, is determined by the transmit power level and the distance between the communicating nodes. Since the communicating parties have no control over the noise and interference, the SNR can only be controlled by changing the peak transmit power.

Detecting the SNR of the incoming signal at the receiving end and communicating this information to the transmitting end provides the sender node a chance to adapt itself to the channel and environmental conditions. Continuously adapting increases the efficiency of the wireless network by decreasing the number of lost packets and the overall power consumption of the wireless network.

III. Online SNR Detection

In order to determine the SNR, we need to extract both the noise and the signal power from the incoming signal. The signal processing technique to be used for the SNR measurements will impose additional computational load on the system. Fortunately, some invariant information about the communication system can be used to reduce the number of computations for the SNR calculations.

TABLE I
AVERAGE POWER CONSUMPTION OF THE EXPERIMENTAL WIRELESS NETWORK NODE (TRF = TRANSCEIVER VR = VOLTAGE REGULATOR

(mW)	Total	TRF	VR	Others
High Power Mode	291	128	167	1
Low Power Mode	160	63	97	1
Difference (%)	46	51	42	0

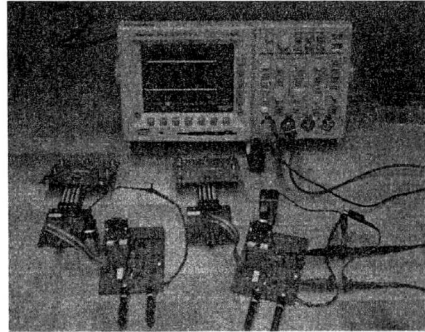

Fig. 2. Picture of the experimental setup

The online SNR detection technique that we propose is based on a limited-band spectral analysis. Here, we use the information on the intermediate frequency (IF) as well as the modulation scheme of the digital communication system to reduce the number of additional computations. The signal is sampled in the IF stage of the receiver before the demodulator module. The block level diagram of the system together with the added computational blocks is shown in Figure 3.

While the exact power spectral density (PSD) of the signal depends on the bit pattern, an approximate spectral mask for the desired signal can be obtained *a priori* using the information on the modulation scheme, the IF frequency and/or using some experimental data of the incoming signal. The dashed arrows in Figure 3 indicate two possible ways of calculating the spectral mask. In one case, an average spectral mask can be used for any bit pattern eliminating the need to compute the desired signal spectrum from the incoming bits. In another case, the decoded bit pattern is plugged into an analytical expression to estimate the desired signal spectrum within the band of interest.

The analytical expression for the spectral mask with using the modulation scheme of Frequency Shift Keying (FSK), can be calculated starting from the time domain representation of the ideal received signal:

$$y(t) = \cos[(\omega_{IF} + b(t)\omega_d)t] , \quad (1)$$

where ω_{IF} is the IF frequency, ω_d is the frequency shift and $b(t)$ is the receiver output which is composed of pulses with an amplitude of 1 or -1. For a single pulse i the spectrum of FSK modulated signal is:

$$
\begin{aligned}
Y_i(f) &= 0.5[G(\alpha)e^{-j2\pi iT(\alpha)} + G(\beta)e^{-j2\pi iT(\beta)}] \quad (2)\\
\alpha &= (f - f_{IF} - b_i f_d), \ \beta = (f + f_{IF} + b_i f_d)\\
G(f) &= Te^{-j\pi Tf}sinc(\pi Tf) .
\end{aligned}
$$

The spectral mask of the FSK modulated pulse stream of any size can be calculated by using (2). Since there is a band-pass filter, only the pass-band calculation is needed for the spectral mask. Since the signal power is appreciably high near the IF signal, it will dominate the PSD calculations in a large band. The modulated signal power decreases as the frequency offset from the center of the channel increases. For the extraction of the noise power, we concentrate on small bands near the corners of the channel. We use the spectral mask and the peak signal information to determine the expected and desired signal powers in these bands of interest. The difference of the two gives the noise power within that

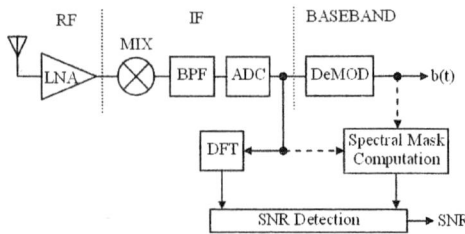

Fig. 3. Block diagram of the detection system

TABLE II
RMS ERROR IN SNR DETECTION FOR RANDOM NOISE AND ADJACENT
CHANNEL INTERFERENCE

(dB)	NO Interferer			One Side Interferer	
Pulse	Error	Actual SNR	Meas. SNR	Average Spectral Mask PSD	Analytical Spectral Mask PSD
10	4.7	25.1	22.4		
20	3.9	24.6	24.4		
40	3.3	23.1	21.3	6.0	5.4

frequency band, which we then extrapolate over the complete communication channel. This band-limited measurement technique is illustrated in Figure 4.

The number of samples and the sampling frequency determine the frequency resolution of the FT. Since we concentrate on small bands near the edge of the communication channel, the number of frequency bins to be analyzed in terms of PSD is small.

IV. ACCURACY OF THE ONLINE SNR DETECTION

We evaluate our SNR detection technique with MATLAB simulations by implementing the DFT, the spectral mask computation and the SNR detection blocks in Figure 3.

We assume a low-IF system with an incorporated ADC. The IF frequency, ω_{IF}, is chosen as $51.175MHz$ and the frequency offset, ω_d, for the FSK modulation is set as $50KHz$. We set the sampling frequency for the time domain samples at four times the IF frequency ($200MHz$). We compute the spectral mask using two methods: first by using the analytical expression 2, second by averaging the FTs of random pulse streams. For the pulse stream, 10 and 40 pulses with pulse durations of $1\mu s$ are used. The communication bandwidth is chosen as $1MHz$ and the bandwidth for noise analysis is set as $200kHz$. The DFT frequency resolution of $100kHz$ results in only 4 frequency bins to analyze.

With the above configuration, for a $10 - pulse$ wide pulse stream, the SNR can be estimated with a $4.7dB$ error. Left hand side of Table II shows the error in estimating the SNR with pulse streams of various durations.

To determine the system response in the case of a strong interferer from adjacent channels together with random noise, we perform simulations for an incoming signal modulated by a $40 - pulse$ wide pulse stream. We also impose a strong interferer at the adjacent channel. Right hand side of Table II shows the RMS error in SNR detection for the strong interferer scenario using the two spectral mask (SM) calculation methods.

The accuracy of the SNR detection depends on the size of the pulse stream, the number of time samples, and the ADC resolution. The SNR detection error will decrease as the number of samples increase. However, the increase in these parameters increases the power consumption and detection time. Therefore, the parameters should be tuned for an acceptable accuracy.

V. THE POWER MANAGEMENT SCHEME AND EXPERIMENTAL RESULTS

It is not essential to know the exact SNR of the incoming signal for the power management measuring quality of the received signal. The goal of the management scheme is to keep the received signal SNR within an ideal window where the peak transmit power and the bit error rate (BER) are minimized. Each system has an acceptable BER and a minimum receive SNR is specified to satisfy this rate. Thus, one needs to only detect the SNR level that is above a certain margin from the minimum acceptable SNR. The goal then is to keep the SNR level within this window. After the SNR computation at the receive end, the desired transmit level information is coded and added to the transmit package.

In order to evaluate the dynamic adoption capability of the proposed technique, we simulate on a network model, which consists of 4 nodes. As in many proposed network architectures the nodes are mobile and two separate transmissions may occur at adjacent channels based on *ad-hoc* routing [10], [11]. We compare our transmit power management technique with a fixed transmit power scheme in terms of average power consumption and the number of dropped packages.

The network model is implemented in MATLAB with the proposed online SNR detection technique. Figure 5 shows the distance relations between the communicating and interfering nodes. In the experiment, Node 1 transmits signals to Node 2 and Node 3 transmits signals to Node 4 in adjacent channels. The packet length is $40 - bits$ and the bit length is $1ms$. The communication band is around $900MHz$ with a channel band width of $1MHz$.

The communication between Node 1 and Node 2 is disturbed by environmental conditions as well as by the adjacent channel signal transmitted by Node 3. Same conditions are valid for the communication between Node 3 and Node 4. Initial distances are assigned between nodes to simulate the path loss effect on the proposed technique.

A packet-based communication is established between the transmitting and receiving nodes. After the each received packet, the receiving party sends an acknowledgement about the transmission to the transmitter. In our power management scheme, the receiver embeds information on whether

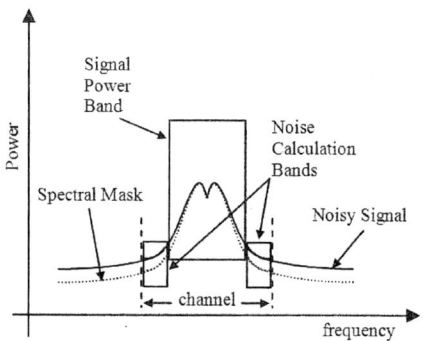

Fig. 4.   SNR detection technique

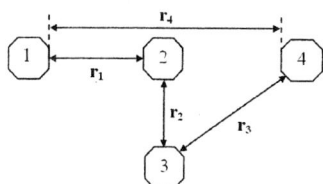

Fig. 5.   Experimental network model

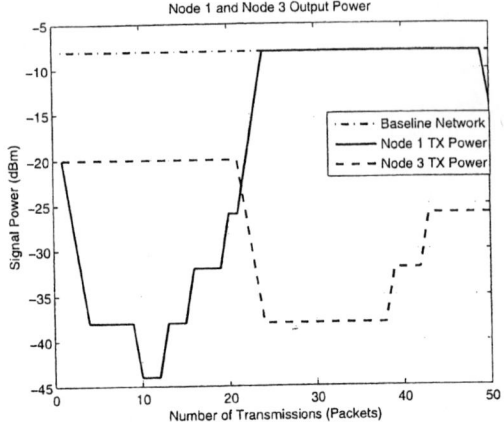

Fig. 6.   Transmit powers for Node 1 and Node 3

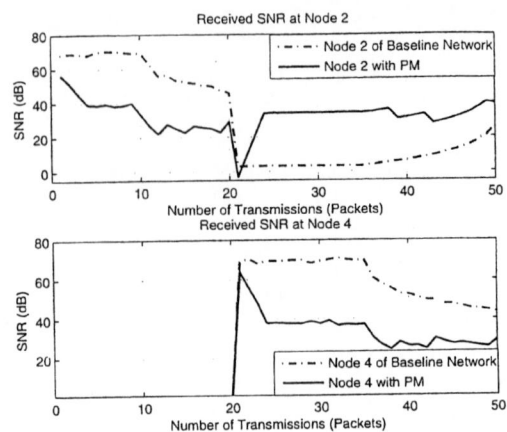

Fig. 7.   Received SNR for Node 2 and Node 4

to increase/decrease or retain transmit power in to the acknowledgement. This action is decided based on the received signal SNR together with the preset SNR window for proper communication. A power step value of $6dB$ is assigned to the transmitters.

For the experiment, we define a SNR window of $(25, 40)dB$ for the power management scheme and a dropped packet threshold of $20dB$. The transmitters initial power level is set as $-20dBm$. Initial distances between nodes are assigned as $r_1 = 80m, r_2 = 120m, r_3 = 60m$ and $r_4 = 160m$ as defined in Figure 5. To evaluate the power savings as a result of the proposed power management technique, a baseline network is established for the experiment without any power management and maximum transmit power ($-8dBm$) is assigned to each wireless node.

In the experiment, we keep Node 2 and Node 3 stationary while the other two are mobile. At some time Node 3 (initially silent) starts communicating with Node 4 which is moving away from the other nodes. In Figure 6, the transmit signal power levels for Node 1 and Node 3 together with the baseline network are shown. Also in the plots of Figure 7, the received signal SNR values at Node 2 and Node 4 are shown in comparison with the base line network. As can be seen, with the help of power management (PM), SNR values are tried to be kept between the preset SNR window by continuously adjusting transmit signal power level. The upper SNR plot in Figure 7 also shows that the network with power management lost only two packets when Node 3 starts transmitting in the adjacent channel. For the baseline network, the number of dropped packets is 28 due to the excessive transmit power of Node 3. Average power savings for Node 1 and Node 3 are $13.3mW$ and $9.8mW$ per package respectively. The baseline network power consumption per package is increased to $19.6mW$ due to the dropped packets.

VI. CONCLUSION

In this paper, we propose a new online SNR detection technique based on a limited-band frequency spectrum analysis. We also present a dynamic power management scheme for wireless battery operated network nodes. Measurements on an example wireless sensor network node indicate that appreciable savings in the average power consumption can be attained by reducing the peak transmit power.

In the proposed method, SNR information is computed at the receiving end and is used to capture the collective environmental conditions including the communication distance, noise, and interference. This information is communicated back to the transmitting party to enable adjustments in the power level. Simulation results show that the online SNR detection provides adequate accuracy to adjust the peak transmit power level. The proposed technique is capable of saving power as well as reducing the number of dropped packages by maintaining an optimum received SNR value.

REFERENCES

[1] G. J. Pottie and W. Kaiser, "Wireless Sensor Networks," *Communications of the ACM*, 2000.

[2] E. Chung, L. Benini, and G. Micheli, "Dynamic Power Management Using Adaptive Learning Tree," in *IEEE Int. Conference on Computer-Aided Design*, Nov 1999, pp. 274–279.

[3] A. Sinha and A. Chandrakasan., "Dynamic Power Management in Wireless Sensor Networks," *IEEE Design & Test of Computers*, vol. 18, no. 2, pp. 62–74, 2001.

[4] R. Passos, C. Coelho, A. Loureiro, and R. Mini, "Dynamic Power Management in Wireless Sensor Networks: An Application-Driven Approach," in *Annual Conference on Wireless On-demand Network Systems and Services*, Jan 2005, pp. 109–118.

[5] M. Neugebauer and K. Kabitzsch, "Proposal and Application Aware Analysis of A Wireless Sensor Network Protocol," in *Telecommunications Network Strategy and Planning Symposium*, Jun 2004, pp. 237–242.

[6] M. Toner and G. Roberts, "A BIST Scheme for A SNR, Gain Tracking, and Frequency Response Test of A Sigma-Delta ADC," *IEEE Tran. on Circuits and Systems-II: Analog and Digital Signal Processing*, vol. 42, no. 1, pp. 1–15, Jan 1995.

[7] "IEEE Trial-Use Standard for Digitizing Waveform Recorders," *IEEE Std 1057*, Jul 1989.

[8] M. Padmanabhan and K. Martin, "Filter banks for Time-Recursive Implementation of Transforms," *IEEE Tran. on Circuits and Systems-II: Analog and Digital Signal Processing*, vol. 40, no. 1, pp. 41–50, Jan 1993.

[9] A. Soldo, A. Gopalan, P. Mukund, and M. Margala, "A Current Sensor for On-chip, Non-Intrusive Testing of RF Systems," in *IEEE Int. Conference on VLSI Design*, 2004.

[10] G. Pei, M. Gerla, and T. Chen, "Fisheye State Routing: A Routing Scheme for Ad-hoc Wireless Networks," in *IEEE Int. Conference on Communications*, vol. 1, Jun 2000, pp. 70–74.

[11] M. Ng and I. Lu, "Peer-to-Peer Zone-Based Two-Level Link State Routing for Mobile Ad Hoc Networks," *IEEE Journal on Selected Areas in Communications*, vol. 17, no. 8, pp. 1415–1425, Aug 1999.

978-1-4244-1983-8/08/$25.00 ©2008 IEEE

Improvement of Power Efficiency of Inductive Links for Implantable Devices

Kanber Mithat Silay, Catherine Dehollain, Michel Declercq

Institute of Electrical Engineering
Ecole Polytechnique Fédérale de Lausanne (EPFL)
CH-1015, Lausanne, Switzerland
e–mail: kanbermithat.silay@epfl.ch

Abstract— **This paper presents the analysis of inductive links for remote powering of implantable devices and a method to improve the power link efficiency by modifying the geometrical parameters of planar spiral inductors. The analysis of the inductive links includes a model for the inductors, which is more accurate for larger bandwidths. The corresponding equations for the power and voltage transfer functions in the inductive links are solved by using MATLAB's Symbolic Math Toolbox. These equations are verified in Agilent ADS, and the results perfectly match. Besides the analysis, a method to improve the power efficiency by changing the geometrical parameters of the spiral inductors is proposed.**

I. INTRODUCTION

Transferring power via an inductive link is the most commonly used method for powering implantable biomedical devices [1]–[6]. This method eliminates the need for transcutaneous wires or implantable batteries. Moreover, energy harvesting via inductive link is not only used in biomedical devices but can also be used in RF identification tags [2].

The inductive link presented in this article is going to be used in a cortical implant for monitoring neural activity of the cortex. In such an application, for patient mobility and comfort, it is important to use small and light–weight batteries at the external reader, which reads the data from the implant. To be able to use small batteries for long durations, the remote powering of the implant should be as efficient as possible. Besides, the efficiency of the link is also important in terms of safety regulations as the emitted fields are absorbed in the tissues. Hence, the design of the inductive link is one of the most important challenges in the implantable devices.

Previous studies have shown that the geometric design of the inductors affects the performance of the inductive link [5], [6]. It has been presented that the coupling factor can be increased by distributing the turns of a spiral coil to improve efficiency [6]. However, the analysis and simulations presented in this article show that the geometry of the inductors for optimum coupling factor does not imply optimum efficiency for power transfer.

In this paper, we present a detailed analysis of inductive links, by including a model which is more accurate for larger bandwidths. Then, a method to design the inductive link is proposed for improving the efficiency by modifying the geometrical parameters of spiral inductors. The simulations

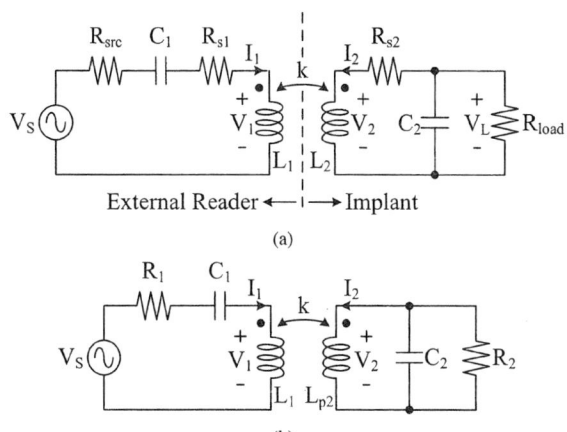

Fig. 1. (a) Simplified model of the inductive link for remote powering of the implantable devices and (b) model of the inductive link with narrowband approximation.

are done with MATLAB and FastHenry2, and the accuracies of the derived equations are verified with Agilent ADS.

II. INDUCTIVE LINK ANALYSIS

Inductive links for data telemetry and/or remote powering of the implantable devices are generally designed with a series resonance circuit in the external reader and a parallel–tuned tank circuit in the implant [1]–[5]. The series resonance circuit in the reader loads the transmitter with a low–impedance, whereas, the parallel tank in the implant improves the driving performance for non–linear loads, in this case, the rectifier [1]. Fig. 1(a) shows the simplified model of the inductive link for remote powering of the implantable devices. The power amplifier driving the primary coil is modeled as a voltage source with a series resistance of R_{src}, for simplicity of the calculations. The resistances R_{s1} and R_{s2} are the series resistances of the inductors L_1 and L_2, respectively. The load is modeled as a resistor, whose resistance can be found from $R_{load} = V_{L,pp}^2/8P_{load}$, where, $V_{L,pp}$ is the peak–to–peak voltage at the load and P_{load} is the power delivered to the load. The resistances of the capacitors are ignored in this analysis, since the quality factors of the capacitors used are much larger than the quality factors of the inductors.

978-1-4244-1983-8/08/$25.00 ©2008 IEEE

$$A_P(\omega) = \frac{\omega^2 k^2 L_1 L_{p2} R_2}{\omega^2 \left(L_{p2}^2 R_1 + k^2 L_1 L_{p2} R_2\right) + R_1 R_2^2 \left(1 - \omega^2 L_{p2} C_2\right)^2} \frac{R_{p2}}{R_{p2} + R_{load}} \tag{2}$$

The series resistance in the secondary coil complicates the analysis of the system. To have a simpler analysis, the series resistance R_{s2} can be converted to its parallel equivalent R_{p2} with narrowband approximation [7]. Fig. 1(b) shows the model of the inductive link with narrowband approximation. In this figure, R_1 is equal to $R_{src} + R_{s1}$ and R_2 is the parallel combination of R_{p2} and R_{load}. Note that, L_{p2} in Fig. 1(b) is the parallel equivalent of the inductance L_2 in Fig. 1(a), as defined in [7]. The analysis of Fig. 1(b) is much simpler than the analysis of Fig. 1(a). However, this narrowband approximation is valid only around the resonance frequency. Therefore, to obtain a more realistic solution, the circuit in Fig. 1(a) should also be analyzed. Both of these circuits are analyzed by using MATLAB's Symbolic Math Toolbox.

A. Analysis of the Inductive Link with Narrowband Approximation

The currents I_1 and I_2 are calculated in the Laplace domain by solving Kirchoff's voltage law in the primary and secondary sides. Then, the power transfer function from source to load is calculated by:

$$A_P(s) = \frac{P_{load}}{P_{src}} = \frac{\text{Re}\left\{V_L I_{R_{load}}^*\right\}}{\text{Re}\left\{V_s I_1^*\right\}} \tag{1}$$

From (1), the power transfer function for Fig. 1(b) can be found as in (2). The frequency, where the efficiency is maximum, can be obtained by differentiating (2) w.r.t. ω and equating the result to zero. This "peak frequency" is equal to:

$$\omega_p = \frac{1}{\sqrt{L_{p2} C_2}} \tag{3}$$

which is the resonance frequency of tank circuit on the secondary side. By inserting (3) into (2), the maximum efficiency of the circuit can be calculated as:

$$\eta_{max} = A_P(\omega_p) = \frac{1}{1 + \frac{1}{k^2 Q_1' Q_2'}} \frac{R_{p2}}{R_{p2} + R_{load}} \tag{4}$$

where, $Q_1' = \omega L_1 / R_1$ and $Q_2' = R_2 / \omega L_{p2}$. This result is very similar to the equations obtained in [4], [6].

The bandwidth of the link can also be found by solving (2) at half of its peak value and calculating the corner frequencies. The equations for the corner frequencies are not given in this paper, however, they can be easily calculated by using MATLAB's Symbolic Math Toolbox.

B. Analysis of the Inductive Link without Narrowband Approximation

The same analysis can also be done for the circuit presented in Fig. 1(a) by following the same procedure. The power transfer function for this case is not given in this paper

since the equation is much more complex than (2). The peak frequency for this case can be found as:

$$\omega_p = \frac{\sqrt{R_{s2} + R_{load}} \sqrt[4]{R_{src} + R_{s1}}}{\sqrt{C_2 R_{load}} \sqrt[4]{L_2^2 \left(R_{src} + R_{s1}\right) + k^2 L_1 L_2 R_{s2}}} \tag{5}$$

It can be seen from this equation that the peak frequency actually depends on the coupling factor and the resistances.[1]

C. Effect of C_1 on the Inductive Link Performance

It can be seen from (2) that the efficiency is not a function of the series tuning capacitor in the reader. It can also be proven that the power transfer function for Fig. 1(a) is also independent from C_1. Therefore, there is no constraint for C_1 for the efficiency. However, the voltage transfer function from the source to the load, which is defined as $A_V(\omega) = |V_2/V_S|$, is a function of C_1. It can be found that there is an optimum value of C_1 for maximum voltage transfer and this C_1 value can be calculated at ω_p for narrowband approximation from:

$$L_1 C_{1,n} \left(1 - k^2\right) = L_{p2} C_2 \tag{6}$$

III. INDUCTANCE CALCULATIONS

The inductors that are going to be used for remote powering of the implanted devices will be fabricated on printed circuit boards (PCB), because the sizes of the coils for efficient power transfer (e.g. 10 mm) are quite large to be placed on–chip.

The inductors are designed as spiral inductors in order to improve the coupling factor by distributing the turns on the radius, as discussed in [6]. In the literature, it is possible to obtain some approximate equations for self and mutual inductance of the spiral shaped coupled coils [4]–[7]. Similar to [5], instead of using these equations, we developed a MATLAB code to generate coordinates of multiple rectangular and/or circular spiral shaped inductors for FastHenry2.[2] The results from FastHenry2 is processed in MATLAB by using the equations given in the previous section to get the power transfer function, peak frequency, maximum efficiency, bandwidth, and values of the resonance capacitors. Besides, MATLAB code is further developed to sweep the geometric parameters of the inductors (e.g. number of turns, outer dimension, etc.) to be able to find the optimum parameters of the inductors to operate with maximum efficiency.

IV. DESIGN METHODOLOGY

In this paper, we propose a design methodology for improving the efficiency of the inductive links modeled in Fig. 1. The inductors are designed as rectangular spirals given in Fig. 2. The path of the spiral is generated in MATLAB with the width and height of the conductor for FastHenry2.

---

[1]It can be easily proven that (3) can be obtained from (5) by making the assumptions from Section II-A.

[2]FastHenry2 is capable of calculating self and mutual inductances and parasitic resistances of 3–D conductors and can be obtained from [8].

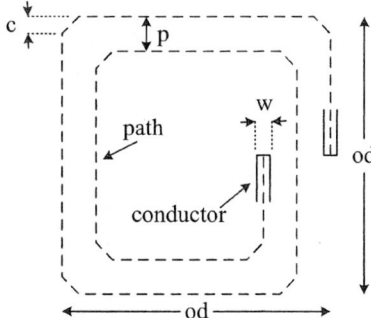

Fig. 2. Geometric definitions for the rectangular spiral inductor. od stands for outer dimension, p stands for pitch, w stands for width, and c stands for chamfer. h is the height of the conductor, which is not shown in this figure.

For an inductor, the thickness of the conductor (h), minimum width of the conductor (w_{min}), and minimum spacing between two conductors ($s = p - w$) are determined by the PCB fabrication capabilities. Moreover, the outer dimension (od) of the implanted inductor is limited by the size of the implant. Therefore, it is not possible to increase od_2 to improve the power efficiency. Hence, the design of the implanted inductor in the secondary is not as flexible as the design of the primary inductor. Although some of the parameters are not under designer's control, there are still so many parameters to sweep simultaneously for optimization of the inductive link.

In this paper, to simplify the geometric design of the inductors, we propose a methodology to improve power efficiency. As a starting point, we assume that both inductors are identical and has a fixed outer dimension equal to the outer dimension of the implanted inductor. After the power efficiency is optimized for identical inductors, the primary inductor is redesigned.

The steps of the proposed methodology can be done as:[3]

1) Sweep the pitch (p) and number of turns (n) of the inductors. Choose p and n for maximum efficiency considering a reasonable resonance capacitor.
2) Sweep c to tune the inductance with the capacitor C_2 at the operation frequency f_0.
3) As the secondary inductor design is finalized, sweep the outer dimension of the primary (od_1) for different n_1 and p_1 values. Find od_1, p_1, and n_1 for maximum efficiency.
4) Tune the primary inductance by sweeping c_1 to obtain a reasonable value of C_1 (see Section II-C).

V. DESIGN EXAMPLE: REMOTE POWERING OF A CORTICAL IMPLANT

In this section, we give a case study for an inductive link for remote powering of a cortical implant.[4] The system parameters for the inductive link are given in Table I.

[3]Simulations have shown that maximum coupling and efficiency is obtained at maximum allowed width (i.e. minimum spacing) for a fixed pitch. Therefore, minimum spacing (s_{min}) is used in all of the inductor designs.

[4]For this example, narrowband approximation is not used, except for comparison at Table III.

TABLE I
SYSTEM PARAMETERS FOR THE INDUCTIVE LINK

Parameter	Value	Explanation
f_0	1 MHz	Operation frequency
P_{load}	10 mW	Load power
$V_{L,pp}$	3.3 V	Load voltage
R_{src}	3 Ω	Source (amplifier) resistance
h	40 μm	Conductor thickness
s_{min}	150 μm	Minimum conductor spacing
w_{min}	150 μm	Minimum conductor width

TABLE II
GEOMETRY OF THE INDUCTORS, COMPONENT VALUES, AND SYSTEM
PERFORMANCE FOR OPTIMUM EFFICIENCY AND OPTIMUM COUPLING

Param.	Unit	Optimum Efficiency		Optimum Coupling	
		Primary	Secondary	Primary	Secondary
n	—	50	15	12	5
od	mm	43	10	19	10
p	mm	0.4	0.3	0.7	0.6
w	mm	0.25	0.15	0.55	0.45
c	mm	1.1	0.3	0.7	0.7
L	μH	51.986	1.221	1.526	0.250
R_s	Ω	8.089	0.965	0.460	0.146
C	nF	0.488	20.74	16.61	101.2
k	—	0.1251		0.1846	
$A_P(f_0)$	—	0.2213		0.0523	

The distance between the coils is determined by the thickness of the tissues in between. Considering a cortical implant with a secondary inductor underneath the skin, the minimum distance is taken as 5 mm. The outer dimension of the implanted inductor is chosen as 10 mm.

By following the procedure proposed in Section IV, the inductors are designed for the inductive link given in Fig. 1(a) with the system parameters of Table I, to obtain maximum efficiency with reasonable component values. Furthermore, the same method is re–run to have maximum coupling coefficient between inductors, instead of considering maximum efficiency.

Fig. 3 shows one sample set of simulation results obtained in the third step of the proposed method for optimum efficiency analysis. These graphs do not show the optimum efficiency or coupling for the system, but are given to demonstrate that the maximum efficiency and coupling may even occur at different values of a geometric parameter while the others are constant.

Table II gives the geometry of the inductors, component values, and system performance for optimum efficiency and optimum coupling cases. It can be seen that the inductor designs for optimum efficiency and optimum coupling are quite different, i.e. optimizing coupling factor does not imply optimum efficiency. For this case, the efficiency is increased from 5% to 22% by considering optimum efficiency instead of optimum coupling. Table III summarizes the performance of the designed inductive link for optimum efficiency case with and without narrowband approximation.

Since the method proposed is based on a successive ap-

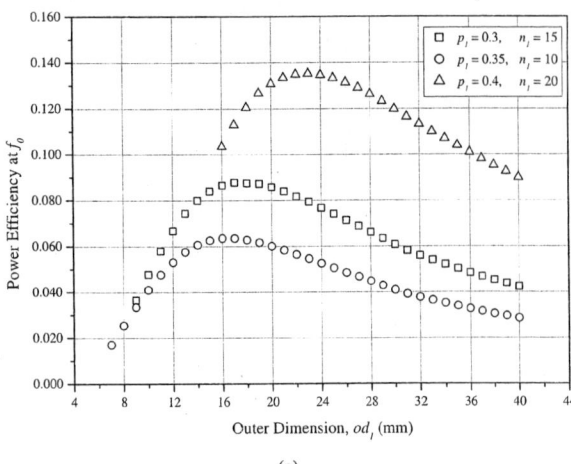

(a)

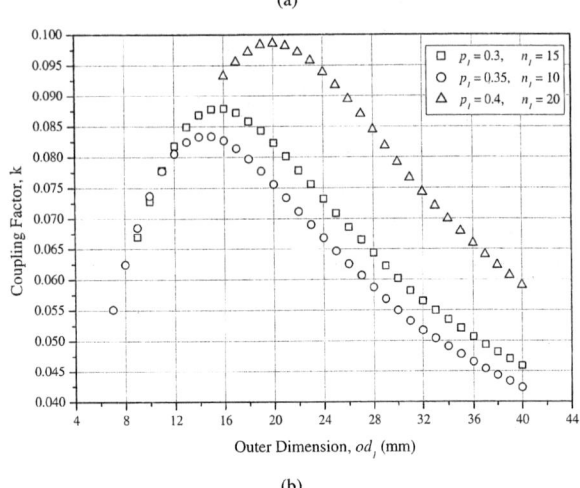

(b)

Fig. 3. (a) Power efficiency at operation frequency (f_0) and (b) coupling factor (k) vs. outer dimension of the primary inductor.

proach, it cannot always guarantee that the inductors have the optimum geometry. However, it helps the designer to design the system near maximum efficiency operation.

The model in Fig. 1(a) is simulated in Agilent ADS with the components given in Table II (optimum efficiency case). Fig. 4 gives the power and voltage transfer functions for the designed inductive link from MATLAB model of Fig. 1(a) and from ADS. As seen from Fig. 4, the simulation results of the model are very accurate.

VI. CONCLUSION

In this paper, inductive links for remote powering of implantable devices are analyzed and a method to improve the power efficiency is proposed. The inductive link is analyzed with and without making narrowband approximation. The power and voltage transfer functions are calculated with MATLAB and FastHenry2, and their accuracies are verified in Agilent ADS. A method to improve the power efficiency is proposed by changing the geometry of rectangular spiral

TABLE III
PERFORMANCE OF THE DESIGNED INDUCTIVE LINK

Parameter	W/ narrowband approximation	W/o narrowband approximation	Error due to approx. (%)
η_{max}	0.2238	0.2222	0.75
f_p	0.9922 MHz	0.9895 MHz	0.27
$A_P(f_0)$	0.2234	0.2213	0.94
BW	337.9 kHz	332.4 kHz	1.67

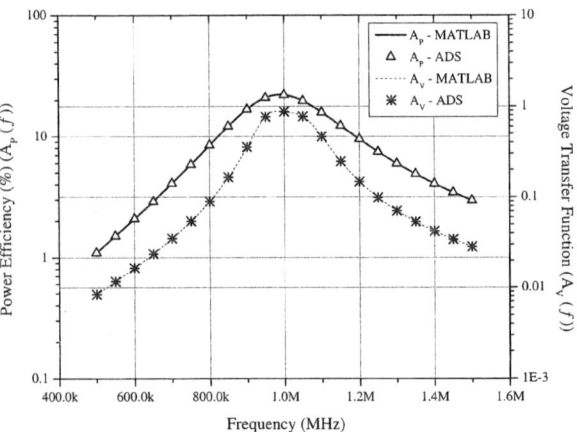

Fig. 4. Power and voltage transfer functions for the designed inductive link from the model in Fig. 1(a).

inductors. For the design example given, the efficiency of the link is improved from 5% to 22% by following the proposed method for optimum efficiency instead of optimum coupling. Experimental measurements for verification of the link model are currently under progress.

ACKNOWLEDGMENT

The *NEURO–IC* project is supported by the Swiss National Funding (SNF). The authors would like to thank Altug Oz for his feedbacks.

REFERENCES

[1] M. Sawan, Y. Hu, and J. Coulombe, "Wireless smart implants dedicated to multichannel monitoring and microstimulation," *IEEE Circuits Syst. Mag.*, vol. 5, no. 1, pp. 21–39, 2005.
[2] C. Sauer, M. Stanacevic, G. Cauwenberghs, and N. Thakor, "Power harvesting and telemetry in CMOS for implanted devices," *IEEE Trans. Circuits Syst. I*, vol. 52, no. 12, pp. 2605–2613, 2005.
[3] G. A. Kendir, W. T. Liu, G. X. Wang, M. Sivaprakasam, R. Bashirullah, M. S. Humayun, and J. D. Weiland, "An optimal design methodology for inductive power link with class-E amplifier," *IEEE Trans. Circuits Syst. I*, vol. 52, no. 5, pp. 857–866, 2005.
[4] R. R. Harrison, "Designing efficient inductive power links for implantable devices," *Proc. ISCAS'07*, pp. 2080–2083, 2007.
[5] S. Atluri and M. Ghovanloo, "Design of a wideband power–efficient inductive wireless link for implantable biomedical devices using multiple carriers," *Proc. 2nd IEEE–EMBS Conf. Neural Eng.*, pp. 533–537, 2005.
[6] C. M. Zierhofer and E. S. Hochmair, "Geometric approach for coupling enhancement of magnetically coupled coils," *IEEE Trans. Biomed. Eng.*, vol. 43, no. 7, pp. 708–714, 1996.
[7] T. H. Lee, *The Design of CMOS Radio–Frequency Integrated Circuits*. New York: Cambridge, 2004.
[8] (2004). [Online]. Available: http://www.fastfieldsolvers.com.

Measurements of Impulsive Noise in Broad-band Wireless Communication Channels

Okan Z. Batur, Mutlu Koca, Gunhan Dundar
Electrical and Electronics Engineering Department
Bogazici University Istanbul, Turkey
Email: okan.batur, mutlu.koca, dundar@boun.edu.tr

Abstract— Electrical and mechanical equipments can be major sources of interference for wireless systems. This paper presents the effects of both noise and interference in a wide frequency band between 100 kHz and 3GHz. Electromagnetic field (EMF) strength measurements of the impulsive noise taken at various locations such as computer labs, TV stations and hospitals are presented. Measurement results are fitted into a mathematical model and it is shown that the amplitude probability distribution (APD) functions derived from the measurements can be described by the Middleton Class-A model.

I. INTRODUCTION

Wideband and ultra-wideband (UWB) systems are popular research topics in recent years. In the previous wideband and UWB transciever designs, the transmission channel is assumed to have Additive White Guassian Noise (AWGN) behaviour. Effect of impulsive noise is ignored for UWB systems. The traditional approach considers just the thermal noise. However, The thermal noise is modelled as a stationary and memoryless Gaussian random process which does not agree with relevant field measurements. It has been reported in the indoor measurements [1] that there is performance degradation seen, when the UWB system subject to the impulsive noise produced by the office machines such as printers, fax machines and photocopiers.

Impulsive noise consists of sudden step-like transitions between two or more levels (non-Gaussian), as high as several hundred microvolts, at random and unpredictable times. Electrical and mechanical machinery can produce non-Gaussian impulsive noise bursts in wireless receivers. Effects of impulsive noise can be seen in the populated metropolitan areas. Devices with high voltages such as the ignition motors, microwave owens, hairdryers, blenders, photocopiers and printers are significant sources of impulsive noise in office and retail environments. In crowded cities, impulsive noise is most probably caused by the large number of old vehicles that have been used for more than 10 years and also due to the electromagnetic noise generated by industrial equipments and household appliances [2]-[4].

Our aim is to characterise and present the impulsive behaviour of the wideband and UWB communication channels. In this work, taking into account impulsive noise and interference, Middleton's Class-A mathematical noise models are constructed from the EMF measurements. By presenting the channel which has impulsive behaviour even at higher frequencies, we can design new wideband transciever which can work

Fig. 1. Rohde and Schwarz FSH-6 EMF measurement device

under impulsive noisy and interfering communication channel as well as AWGN. This work shows the importance of impulsive noise aware design of an UWB system. The noise measurements are taken at various indoor and outdoor locations to show the presence of the impulsive noise. Measurements are taken with the portable spectrum analyser which is shown in Figure 1. The Spectrum analyser (Rohde-Schwarz FSH6) with tri-axis isotropic antenna is capable of taking measurements from 100 kHz to 3 GHz. Measurements are modelled with Middleton's Class-A noise model. Measurement results fit quite well to the mathematical models.

978-1-4244-1983-8/08/$25.00 ©2008 IEEE

II. FIELD TESTS

EMF measurements were taken in indoor and outdoor locations to observe the impusive noise in different frequencies. These locations are presented in Table I with their respective total field strength over the range 100 kHz to 3 GHz.

At hospital, measurements are taken at the technical room of the MR service. Computer lab of the university is used to measure the noise. For the measurements in the house, blenders, aspirator and other household devices are used. At the bus terminal, vehicle ignition noise was condsidered. Noise of the electronic devices are measured at the TV stations.

TABLE I

MEASUREMENTS

Locations	Field Strength (V/m)
HOSPITAL	0,8796
MEMS LAB	0,8808
COMPUTER LAB	0,9384
HOUSE	0,9475
HOUSE 2	0,9483
BUS TERMINAL	1,6032
TV STATION	4,1227
TV STATION 2	7,2493
TV STATION 3	12,1032

In the following section detailed results are given with figures.

A. Measurements

Locations where impulsive noise dominates are as follows.

1) House: Two measurements are taken. At the first measurement all household devices were left working. Blender, aspirator, vacuum cleaner, and dishwasher were used while taking the measurements. In the second measurement, switching activity of aspirator only is recorded.

In Figure 2, the measurement taken between 900 MHz and 1 GHz. Note that there is also a GSM signal at around 950MHz. It can be seen that GSM signal is affected by the impulsive noise.

2) Bus Terminal: Measurements are taken at a crowded bus terminal. Vehicles in the terminal were very old. Therefore, vehicle ignition noise is dominating as impulsive noise.

In Figure 3, The ignition noise can also be observed around 1.4 GHz. Vehicle ignition noise is also measured at lower frequencies.

3) Computer Lab: Measurement are taken at the computer lab with 17 computers. A task was given to the computers to run them at full speed. Measurements do not show big noise impulses and it is very hard to detect them.

The results are shown in Figure 4 at between 1.8 Ghz and 1.9 Ghz range. It can be seen that small impulsive noise can be detected near GSM signal.

4) TV Station: TV station measurements were taken at three different locations. Measurements in Figure 5 were taken at the TV Station.

TV broadcasting signal and GSM signals are dominating but it can be observed from Figure 5 that small impulses above

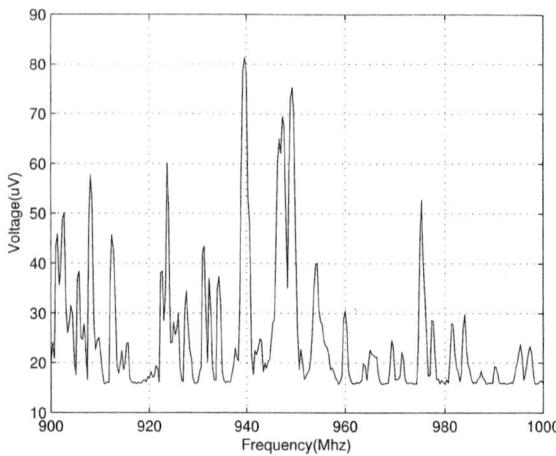

Fig. 2. Indoor EMF measurement taken in the presence of household appliances

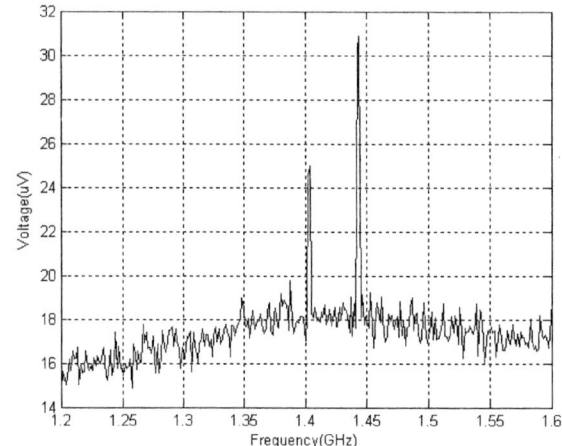

Fig. 3. Outdoor EMF measurement taken at a bus terminal

the noise floor are present. Since the other signals are very strong, impulsive noise appears to be small.

III. IMPULSIVE NOISE MODEL

Noise and distortion are the main limiting factors in communication and measurement systems. Therefore, the modelling and removal of the effects of noise and distortion have been at the core of the theory and practice of communications and signal processing [5]. Middleton's Class-A noise model is a well known statistical-physical model for man-made impulsive radio noise [6],[7]. The model is constructed from Gaussian mixtures with different variances and means. Therefore, the Gaussian mixture well fits to the impulsive noise shape.

On the other hand, Additive White Gaussian Noise (AWGN) does not well represent the communication channel, because the impulse bursts can happen anywhere at any time. From the measurements, it can be observed that the impulsive noise has

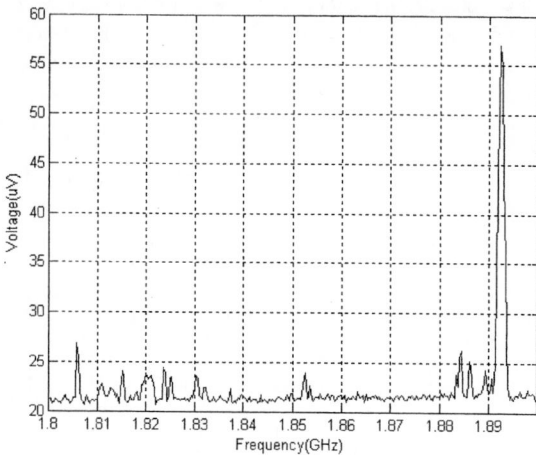

Fig. 4.   Indoor EMF measurement taken at a computer lab

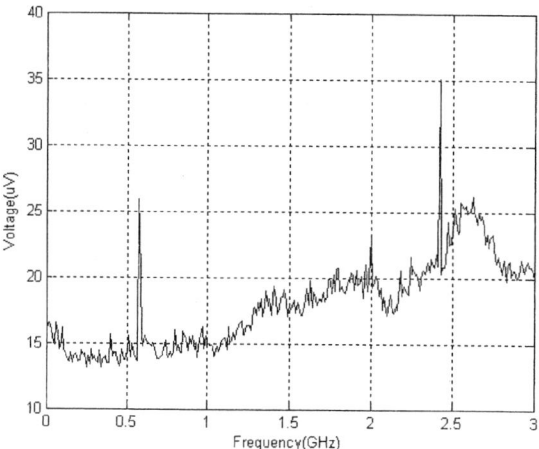

Fig. 5.   Indoor EMF measurement taken at a TV station

been found at high frequencies. Middleton's Class-A impulsive noise model is used. Class-A noise model is given in equation 1.

$$p\left(x\right) = \exp\left(-A\right) \sum_{m=0}^{\infty} \frac{A^m}{\sqrt{2\pi\sigma_m{}^2}} exp\left(-\frac{x^2}{2\sigma_m{}^2}\right) \quad (1)$$

where

$$\sigma_m{}^2 = \frac{m/A + \Gamma'}{1 + \Gamma'} \quad (2)$$

In equation 2, Gamma is defined as the ratio of the Gaussian component of noise power to the non-Gaussian component of the noise power.

$$\Gamma' = \frac{2e_2\left(e_6 + 12e_2{}^3 - 9e_2e_4\right)}{3(e_4 - 2e_2{}^2)^2} - 1 \quad (3)$$

In the model equations, A shows the impulsive behaviour of the model. Impulsiveness increases with parameter A.

$$A = \frac{9\left(e_4 - 2e_2{}^2\right)^3}{2(e_6 + 12e_2{}^3 - 9e_2e_4)^2} \quad (4)$$

By using these two parameters, class-A noise model for a given system is constructed. These parameters can be estimated by finding the envelope of the measured noise. From the envelope second, fourth and sixth order of moments which are given in equation 1 are calculated and the parameters are found.

$$e_2 = E\left[e^2\right], e_4 = E\left[e^4\right], e_6 = E\left[e^6\right] \quad (5)$$

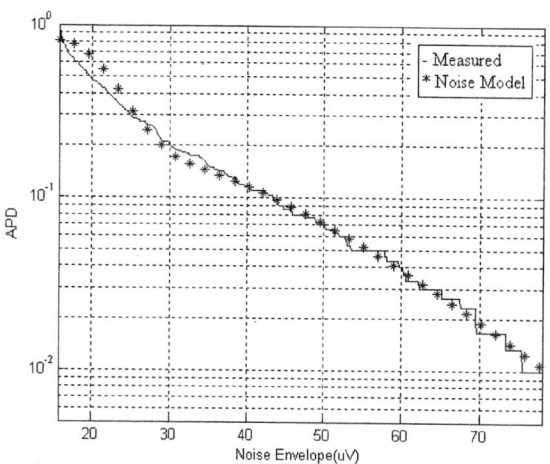

Fig. 6.   Class-A noise model for signals in Figure 2

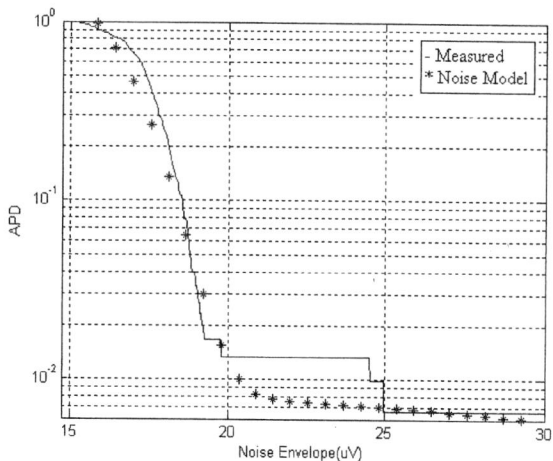

Fig. 7.   Class-A noise model for signals in Figure 3

To compare the impulsive noise and its model, the amplitude probability distribution (APD) is used. APD is defined as the

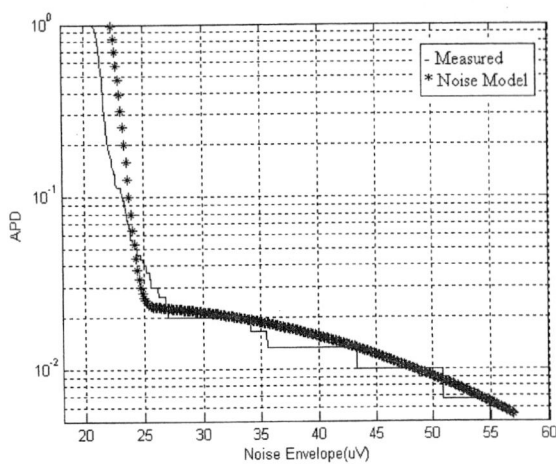

Fig. 8. Class-A noise model for signals in Figure 4

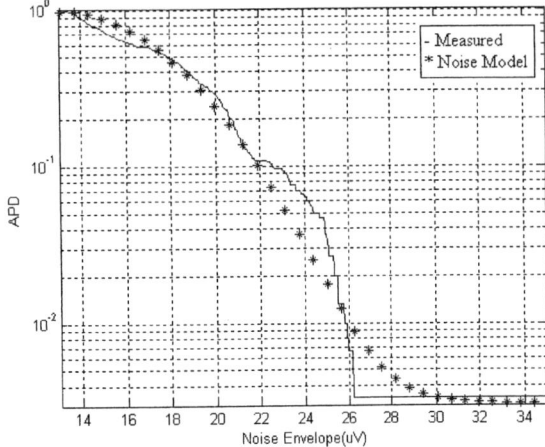

Fig. 9. Class-A noise model for signals in Figure 5

haviour. AWGN noise model does not hold at impulsive noise enviroments. Therefore, effect of impulsive noise over the channel must be considered, while designing a communication hardware for broad-band wireless systems.

ACKNOWLEDGMENT

This research is funded by TUBITAK (Turkish Research and Scientific Council) EEEAG under Grant 105E077.

REFERENCES

[1] K.L. Blackard, T.S. Rappaport and C.W. Bostian "Measurements and Models of Radio Frequency Impulsive Noise for Indoor Wireless Communications", *IEEE Journal on Selected Areas in Communication*, vol.11, no:7, September 1993.
[2] G. Bedicks, C.E.S. Dantas, F. Sukys, F. Yamada, L.T.M. Raunheitte and C. Akamine, "Digital Signal Disturbed by Impulsive Noise", *IEEE Transactions on Broadcasting*, vol.51, no:3, September 2005.
[3] G.L. Maxam, H.P. Hsu, and P.W. Wood, "Radiated Ignition Noise Due to the Individual Cylinders of an Automobile Engine", *IEEE Bans. Vehic. Technol.*, vol. 25, May 1976.
[4] R.A. Shepherd, "Measurements of Amplitude Probability Distributions and Power of Automobile Ignition Noise at HF", *IEEE Bans. Vehic. Technol.*, vol. 23, August 1974.
[5] J.Y.C. Chia, "Interference Characteristics of Microwave Ovens in Indoor Radio Communications", *Tech. Rep., IEEE 802.11 Tech. Commit. Pub.*, 11/91-52, May 1991.
[6] D. Middleton, "Statistical-Physical Models of Electromagnetic Interference", *IEEE Trans. Electromagn. Compat.*, vol 19, no 3, pp.106-126, Aug.1977.
[7] H. Kanemoto, S. Miyamoto, N. Morinaga, "Statistical Model of Microwave Oven Interference and Optimum Reception", *IEEE International Conference on Communications*, vol. 3, pp. 1660-1664, Jun 1998.
[8] T.K. Blakenship, D.M. Krizman, and T.S. Rappaport, "Measurements and Simulation of Radio Frequency Impulsive Noise in Hospitals and Clinics", *IEEE Vehicular Technology Conference*, vol. 3, pp. 1942-1946, May 1997.

probability that noise power exceeds a defined noise level [8]. In the case of impulsive noise, this level is the background noise. Figure 2 to 5 shows the indoor and outdoor EMF measurement results. Figure 6 to 9 shows the respective APD results and Middleton's Class-A noise model. It can be seen that, measurement results are in good agreement with the class-A impulsive noise model.

IV. CONCLUSION

This paper shows the analysis and measurements of impulsive noise for a wide band frequency band between 100 kHz and 3 GHz. It has been shown that impulsive noise can be observed even at high frequencies. Indoor and outdoor EMF measurements are taken and Middleton's Class-A noise models are constructed. Class-A mathematical models and measurement results are in good correspondence. Middleton's Gaussian Mixture model better represents the channel be-

978-1-4244-1983-8/08/$25.00 ©2008 IEEE

Architecture of Automatic Monitoring System for Fresh Food Quality using Wireless Sensor Network

Risang G. Yudanto, Danilo Burdese, Marco Mulassano, Leonardo Reyneri

Dipartimento Elettronica, Politecnico di Torino

Corso Duca degli Abruzzi 24, Torino, Italy 10129

Phone: +39 011 5644170, Email: risang.yudanto@polito.it, leonardo.reyneri@polito.it

Abstract—An automatic monitoring system for fresh food quality is being developed. The system consists of several sensor nodes that measure the temperature of the product samples and the surrounding environment and transmit the result via wireless in 2.4 GHz frequency to the central station. It is developed in order to have fully automatic monitoring system in which the operator will be able to remotely monitor the product's quality. The low-cost, small-sized and battery-powered sensor nodes make them easy to be distributed. This easiness will increase the scalability of the products that can be monitored and reduce the overall operational cost with respect to the manual measurement using wired sensor devices.

I. INTRODUCTION

Maintaining high quality products is always been a priority in fresh food industries. Generally, in order to maintain the quality of the products, the temperature must be kept within some certain of limits. Several operators are also required to monitor the temperature by frequently taking the temperature data of several samples in which temperature sensor or thermometer is attached. However, during some other processes, like delivery from the factory to selling points for example, the temperature of the products may change especially when the temperature of the surrounding environment is changing. This condition can be even worse especially in the case when the temperature of the surrounding environment is not homogeneous, in which it would be very difficult to estimate the temperature of each product item.

Electronic Department of Politecnico di Torino and Centrale del Latte di Torino, Italy, have been working together to build an automatic monitoring system based on Wireless Sensor Network (WSN) that is able to give information of the actual temperature of the monitored fresh-food products. The concept in this development is to built an automatic monitoring system based on distributed-sensors and wireless-networking that composed by commercial off the shelf (COTS) components, especially low-cost and low-power-consumption microcontrollers, sensors and transceiver chips that are highly available in the market nowadays. Wireless network gives a lot of advantages regarding the flexibility in the nodes placement and the network scalability since no wires are needed. On the other hand, the use of distributed sensor nodes also gives advantage regarding the robustness of the system against node failures. Thanks to the availability of COTS components in the market, the prototype development can be realized in relatively short time without spending too much costs.

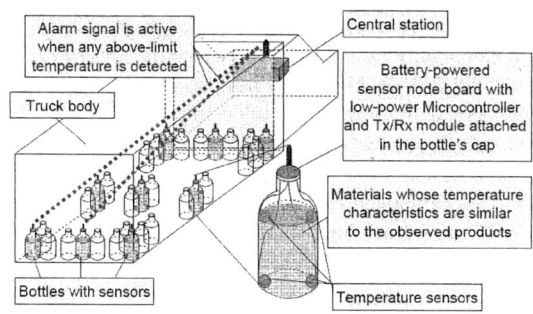

Figure 1. The complete overview of the monitoring system

II. SYSTEM ARCHITECTURE

The monitoring system will be implemented in every truck that deliver fresh milk from the factory to the selling points. It is composed by one central station in the driver's cabin and several sensor nodes distributed inside the truck's body.

Placing the sensor node and measuring the temperature in every product item will cost too much, since in that case we have to provide the number of sensor nodes as much as the number of bottles inside the truck even though it will guarantee the temperature of each. Fortunately, it is possible to estimate the temperature of several identical product items which are put in the same place and close to each other by measuring the actual temperature of one of them. Therefore, only some bottles will have sensor nodes attached.

The sensor nodes will be attached on several marked bottles (8-12 bottles per truck) filled with product sample or other liquid whose temperature characteristic is similar with milk and sense directly the temperature of the liquid inside. These marked bottles are used as part of the measurement tool to estimate the fresh-milk temperature in the neighboring bottles, therefore they are not going to be delivered to the customers.

Since every node will use battery power for its operation, the use of application with algorithm that consider the power management become critical. In this application, each node will sleep most of the time, it will only wake up every period of one minute to measure the temperature, store the result in the memory and then go back to sleep. Right after 20 measurements for example, the node will have to send all the results to the central and then back to sleep. However, when the temperature of the liquid is out of the limitation (the limit

978-1-4244-1983-8/08/$25.00 ©2008 IEEE

for fresh milk is between 4-6°C), the corresponding node will have to send an alarm signal and the measurement result to the central station immediately after the measurement. The overall monitoring system overview is depicted in Figure 1.

A. Sensor Nodes

The sensor nodes development is one of the critical part of the design, because the nodes should be able to perform temperature measurement that represent the actual temperature of the sample. Hence, the temperature sensor should directly sense the sample. The radio frequency (RF) part of the nodes, however, will not optimally work if they are put inside the sample, which is liquid. Therefore, the temperature sensors will not be totally mounted on the nodes mainboard but they will be put directly inside the sample and connect them to the mainboard by wire. The nodes mainboard itself will be placed in the bottle cap, therefore the size of the board is also important.

The main components in the developed sensor nodes are low power RF transceiver, microcontroller and temperature sensor. At this moment, we are developing two different platforms of sensor node. One platform uses MRF24J40 RF transceiver and PIC18LF2620 microcontroller from Microchip, and the alternative node uses CC2430 from Texas Instruments which is RF transceiver and microcontroller that was built in a single chip.

1) RF Transceiver: The RF transceiver is an important component in the sensor nodes since it is the component that performs signal transmission and reception. We decided to use component that complies to the IEEE 802.15.4 radio standard [1], [2] since this standard has been ratified for simple wireless applications that do not require high power operations. The advantages given by this standard are mainly due to its modulation technique, which is Direct Sequence Spread Spectrum (DSSS) and the low data rate, which are 20 kbps in 868 MHz, 40 kbps in 915 MHz and 250 kbps in 2.4 GHz. For sending sensor readings, which are typically a few tens of bytes, high bandwidth is not necessary and the low bandwidth given by this standard helps it fulfill its goals of low power, low cost, and robustness.

Operating frequency in 2.4 GHz has been chosen since this frequency is accepted worldwide and working in this frequency would require less dimension of the antenna, which is important in our application. Working in higher frequency will reduce the maximum range of line-of-sight communication but for this application where the distance between central station and the sensor nodes is never above 10 meters this losses can be neglected.

The two transceivers used in the two platforms that are being evaluated, which are CC2430 [6] from Texas Instruments and MRF24J40 [7] from Microchip, are working in the 2.4 GHz band, a wideband radio with O-QPSK modulation with DSSS at 250kbps. The main difference between the two is that CC2430 has also an 8-bit microcontroller 8051 integrated inside the chip that gives advantage of no need to put another microcontroller especially for small applications.

	CC2430	MRF24J40
Frequency	2.4 GHz	2.4 GHz
PHY & MAC Layer	IEEE 802.15.4	IEEE 802.15.4
Sleep Mode Current Consumption	190 μA * 0.5 μA **	2 μA
Receive Mode	26.7 mA	18 mA
Transmit Mode, 0 dBm	26.9 mA	22 mA
Receiver Sensitivity	-92 dBm	-91 dBm

* digital voltage regulator is on
** digital voltage regulator is off

Table I
RF CHARACTERISTICS COMPARISON BETWEEN CC2430 AND MRF24J40

Table I shows some characteristics between the two, and it is evident that MRF24J40 performs slightly better than CC2430. However, the availability of microcontroller and RF transceiver in a single chip as offered by CC2430 still make it as an attractive solution alternative for this application.

2) Microcontroller: Using a powerful and high performance microcontroller for this platform is not our main priority, since it will consume a lot of power and there is no need to process data that require processor with high computational ability for this application. We have chosen to use PIC18LF2620 [8] low power 8-bit microcontroller from Microchip as the pair for MRF24J40 transceiver, which can be connected by simple SPI connection. This microcontroller has 64 kB flash with 4 kB data memory that divided into 3 kB SRAM and 1 kB EEPROM, which is enough for this application. It has also capability to manage the power, for example by turning off the CPU with peripherals still on (idle mode) the current down to 2.5 μA. Turning off both CPU and peripherals (also called sleep mode) will achieve current down to 100 nA typically. It has also up to 13 channels of 10 bit Analog-to-Digital Converter (ADC) with possibility of conversion even during sleep mode.

On the other hand, the microcontroller provided by CC2430 also gives a lot of interesting features. It is built with an enhanced version of 8051 core CPU with 128 kB flash and 8 kB RAM and also support up to 8 channels of 12-bit ADC inputs. There are also power management modes, with the lowest current consumption down to 300 nA by turning off the digital voltage regulator and all the peripherals.

3) Temperature Sensor: Various types of sensors are used to measure temperature including thermistor, or temperature-sensitive resistor. The simplicity of the circuit required to measure temperature using thermistor make it a good option to be used in this application. It is possible to use a simple voltage divider circuit as depicted in Figure 2 to estimate the temperature inside the liquid as required in this application. However, of all passive temperature measurement sensors, thermistors have the highest sensitivity (resistance change per degree of temperature change) and they do not have a linear temperature/resistance curve. In this case, the parameter to

978-1-4244-1983-8/08/$25.00 ©2008 IEEE

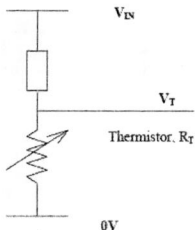

Figure 2. Simple temperature measurement using thermistor

measure the temperature using thermistor must be provided by the thermistor manufacturer. For Negative Temperature Coefficient (NTC) thermistor, accuracy ranges from 3% to 6% depends on the ambient temperature [5].

Another option is to use digital temperature sensor. We have chosen TC72 [9] temperature sensor from Microchip to be evaluated with our platforms since it does not require any additional components in order to measure temperature apart of one decoupling capacitor. This device features a four-wire serial interface that is fully compatible with the SPI specification and, therefore, allows simple communications with common microcontrollers and processors.

The TC72 consists of a band-gap type temperature sensor, a 10-bit Sigma Delta Analog-to-Digital Converter (ADC), an internal conversion oscillator and a double buffer digital output port. The 10-bit ADC is scaled from -128°C to +127°C; therefore, the resolution is 0.25°C per bit. The ambient temperature operating range of the TC72 is specified from -55°C to +125°C.

The TC72 can be used either in a Continuous Temperature Conversion mode or a One-Shot Conversion mode. The Continuous Conversion mode measures the temperature approximately every 150 ms and stores the data in the temperature registers, while the One-Shot mode performs a single temperature measurement and returns to the power-saving shut down mode. This mode is especially useful for low power applications and will be used also in this application. Since the temperature inside the liquid will not change drastically in a short time, a measurement of every one minute would be enough.

B. Central Station

The central station is basically composed by two parts, which are RF Communication Module and Data Acquisition Module.

1) RF Communication Module: Basically is the same architecture as the nodes, which is composed by the same RF transceiver and microcontroller as the nodes have. The only differences are that there is no need to put sensor and there is additional USB or RS232 connector to communicate with Data Acquisition Module.

2) Data Acquisition Module: This module use ARM926EJ-S RISC processor [10], which is a member of ARM9 family of general-purpose microprocessors targeted at multi-tasking applications where full memory management, high performance, low die size, and low power are all important. Some features of this module including: optimized 10/100 Ethernet MAC, LCD controller, USB Ports v.2.0 full speed (12 Mbps) and low speed (1.5 Mbps), 2 Serial Ports, I2C Port, 1284 Parallel Peripheral-to-Host Port and System Bus DMA and Peripheral Bus DMA. It also has 32 MB NAND Flash, 4 MB NOR Flash (up to 8 MB) and 8+8 MB SDRAM (up to 64+64) in the memory part and Altera MAX II CPLD. This module can be operated using Linux operating system.

C. Network Topology

There are several network topologies in WSN, such as star, mesh and clustered hierarchical topology. Each of them has its own pros and cons under the specific working environment. For this application, where the distance between nodes and central station is never above 10 meters, the star topology would be the best due to its simplicity and energy efficiency [3]. This conclusion has been verified by the electromagnetic fields measurement with the presence of the milk bottles/packs inside the truck that will be explained in the next section.

III. PRELIMINARY TESTS AND VERIFICATIONS

A. Verifying Sensing Applications

Several preliminary tests have been performed in order to verify some sensing algorithms that are applied to perform direct sensing of liquid device. The experiment has been performed by using Tmote Sky module (or also called Telos-b) [4] with additional temperature sensor that located inside a test bottle that filled with water in order to perform measurement of the water temperature. Tmote Sky is chosen because of its features, which are very similar with our developed platforms. Tmote Sky uses MSP430F1611 low power microcontroller and CC2420 RF transceiver chip, which is IEEE 802.15.4 compliant that operates in 2.4 GHz.

The test was performed using two motes that acted as the sensor nodes and one node as a base station. In the sensor nodes, a circuit as depicted in Figure 2 was created with a NTC thermistor (R_T) was put inside a test bottle that filled with water, and connect V_{IN}, GND, and V_T to VCC, GND and one of ADC inputs of the Tmote Sky expansion pins. Meanwhile, the base station mote was connected to computer. The setup of the sensor node is depicted in Figure 3.

By heating up or cooling down the temperature of the surrounding environment, the temperature of the water slowly changed. In this case, the regular potential divider equations [5] can be applied and the standard function can be expressed as:

$$\frac{V_T}{V_{IN}} = \frac{R_T}{R_l + R_T}$$

Since R_l is constant, as the temperature changed, also did V_T, and by knowing V_T, R_T can be determined. The value of V_T itself can be determined from the output of the ADC. Once the value of R_T is known, the actual temperature can be

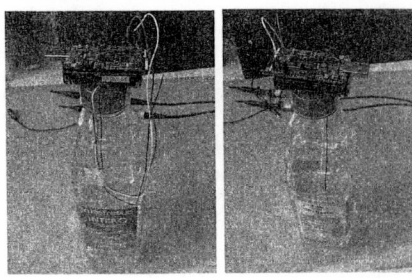

Figure 3. Test setup using Tmote Sky mote connected with thermistor that directly measure the temperature of the water inside the bottle.

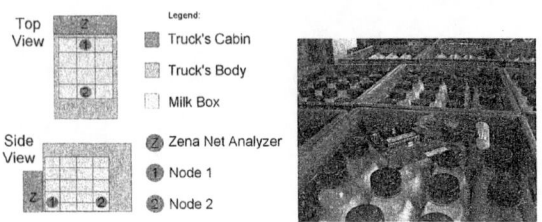

Figure 4. Scheme of the test inside the truck with the presence of the milk (left) and the photograph of the milk boxes and the test device tool from Microchip (right)

estimated from the curve or table R_T vs Temperature provided by thermistor manufacturer.

We have performed different tests using sensing applications that have been developed under TinyOS [11] and Mantis OS [12], which are open-source operating systems for WSN. The tests have verified the possibility to use those applications in order to perform direct measurement of liquid device using external sensors and send the result via 2.4 GHz radio.

B. Verifying Network Topology

As explained before, star topology would be ideal for this application. However, the actual condition in the field is that, there is always possibility to have the truck full with bottles/packs with milk inside. In this condition, especially when the density is quite high, it is difficult to model the medium between the nodes and the central station, since there are huge amount of liquids around the nodes, but still there are some spaces even though these spaces are very narrow.

Therefore, we have performed several tests to measure the received signal strength inside the truck with and without the presence of milk. This measurement was performed using a demo kit from Microchip that composed by two nodes and one net analyzer that connected to computer. This measurement can be performed thanks to the Received Signal Strength Indicator (RSSI) feature that available on the MRF24J40 RF transceiver chip that also used in the demo kit.

The measurement was performed by placing one node inside the truck body just near the back door (Node 2), another node was put in the truck body but near the border with the cabin (Node 1), and the net analyzer (Node Z) was in the driver's cabin, as illustrated in Figure 4. The distance from Node Z to Node 2 is around 5 meters. From the RSSI measurement, we obtained minimum received signal strength of -27 dBm when the truck was empty (without the presence of the milk). When the truck was full of milk the minimum received signal strength was -40 dBm and no single packet was lost during the transmission. This value is still far above the theoretical receiver sensitivity limit of MRF24J40 RF transceiver which is -91 dBm. Based on our experiences, however, the minimum received signal strength required in order to establish a good communication is -60 dBm, and this value is also still below the minimum received signal strength that we obtained during our measurement in the truck under the worst possible condition. This measurement has verified that direct communication from each sensor node to the central station can be performed without the need to have relay station as in multi-hop communication.

IV. SUMMARY

An architecture of automatic monitoring system for fresh food quality has been presented. The use of wireless networking with distributed sensors system will improve the efficiency, fault tolerant, and scalability with respect to manual monitoring system using wired sensors and operators. By using COTS components, the prototype development has become easier without spending too much costs. Several tests to verify sensing applications and the chosen network topology have been performed. The development of sensor nodes is in the final stages and soon after the prototypes ready several tests will be performed to verify the design.

V. ACKNOWLEDGMENTS

We would like to thank to Denis Avanzi from Centrale del Latte di Torino, Danilo Demarchi and Fabrizio Vacca for the work in data acquisition module, and Alberto Vallan and Simone Corbellini for the advices in measurement system. This project is funded by Regione Piemonte.

REFERENCES

[1] Jon T. Adams, "An Introduction to IEEE STD 802.15.4", IEEE Aerospace Conference, 2006
[2] IEEE Computer Society, "IEEE Std 802.15.4™-2006 (Revision of IEEE Std 802.15.4-2003)"
[3] Shrestha, Akhilesh; Xing, Liudong; "A Performance Comparison of Different Topologies for Wireless Sensor Networks". *IEEE Conference on Homeland Security*, pages 280-285, 2007
[4] Polastre, J.; Szewczyk, R.; Culler, D.; "Telos: enabling ultra-low power wireless research", *4th International Symposium on Information Processing in Sensor Networks*, 2005.
[5] Pat Lyons; Phil Waterworth, "The use of NTC Thermistors as sensing devices for TEC controllers and temperature control Integrated Circuits", http://www.mrccomponents.de/
[6] CC2430 Data sheet
[7] MRF24J40 Data sheet
[8] PIC18LF2620 Data sheet
[9] TC72 Data sheet
[10] ARM926EJ-S Technical Reference Manual
[11] TinyOS Community Forum, An open-source OS for the networked sensor regime. http://www.tinyos.net/
[12] MANTIS, MultimodAl NeTwork of In-situ Sensors. http://mantis.cs.colorado.edu/index.php/tiki-index.php

978-1-4244-1983-8/08/$25.00 ©2008 IEEE

Hardware Reduction in Digital MASH Delta-Sigma Modulators via Error Masking

Zhipeng Ye and Michael Peter Kennedy

Department of Microelectronic Engineering and Tyndall National Institute
University College Cork, Ireland
Email: zhipeng.ye@ue.ucc.ie, peter.kennedy@ucc.ie

Abstract—**A reduced complexity (RC) digital Multi-stAge noise SHaping (MASH) delta–sigma modulator (DSM) was proposed in [1]. The sequence length is maximized by setting the LSB of the input to "1"; a long word is used for the first modulator in a MASH structure; shorter words are used in subsequent stages. Rules for selecting the wordlengths of each stage are presented in this paper. We show that an appropriate selection of the wordlength for each stage of the DSM can yield similar performance compared with a conventional MASH DSM, but with less hardware and lower power consumption.**

I. INTRODUCTION

Digital delta-sigma modulators (DDSMs) are often found in consumer communications and entertainment products. In a DDSM, a discrete-time input is requantized to produce a lower resolution output. The quantization error is shaped by a high pass filter and attenuated in the signal band. Ideally, the requantization process produces a random error. In practice, the randomization is often insufficient and the quantization error forms short and repeating patterns; this gives rise to strong unwanted tones in the output spectrum [2].

Two classes of techniques have been developed to whiten the quantization noise: stochastic and deterministic. The "stochastic" approach is to use a "random" dither sequence to disrupt periodic cycles. Dithering breaks up the cycles and increases the effective sequence length, resulting in smooth noise-shaped spectra. However, it inherently adds noise to the spectrum; care must be taken to minimize the contribution of this additional noise.

By contrast, the deterministic approach to whitening the quantization noise is to guarantee sequence lengths by design. Kozak and Kale [3] and Borkowski *et al.* [4] have shown that the MASH 111 DSM does not exhibit large spurs when the initial condition of the first stage is odd, due to the inherent whitening of the quantization error. Unlike dithering, the deterministic approach does not inherently add noise to the spectrum.

In earlier work [1], we proposed a RC MASH DDSM which maximizes the sequence length in a deterministic way. In this paper, we show how error masking can be used to reduce the hardware consumption of MASH DDSMs.

In Sec. II, we review the architecture of the RC MASH DDSM. In Sec. III, we explain the design methodology using the deterministic technique. In Sec. IV, a design example is shown. Finally, we draw some conclusions in Sec. V.

II. RC MASH ARCHITECTURE

We consider the MASH 111 DDSM in this paper. A conventional structure consists of three *identical* N-bit accumulators and an error cancellation network. The output of the MASH 111 DDSM can be expressed in the Z–domain as:

$$Y(z) = STF(z) \cdot X(z) + NTF(z) \cdot E_3(z), \qquad (1)$$

where $X(z)$, $E_3(z)$, $STF(z)$, and $NTF(z)$ are the z–transforms of the input, the quantizer error of the *third* stage, and the signal and noise transfer functions of the modulator, respectively.

In our RC MASH DDSM [1], only the first accumulator has a full-width N-bit word. The error signal of the first stage passes through an M-bit interstage quantizer Q_M before being fed forward to the next (L-bit) stage. The other accumulators use fewer bits to perform the noise shaping. This allows us to reduce the wordlengths in the following stages without reducing the sequence length and compromising the spectral performance. The block diagram of our RC MASH 111 DDSM [1], [6] is shown in Fig. 1. A linearized model which

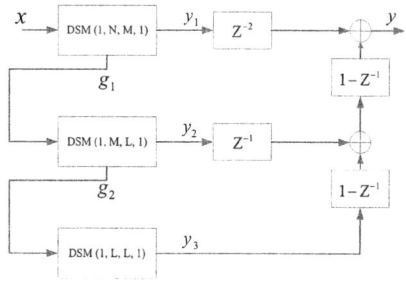

Fig. 1.   RC MASH 111 DDSM

illustrates all the quantization error sources is shown in Fig. 2; its output can be expressed in the Z–domain as:

$$
\begin{aligned}
Y(z) = {} & STF(z) \cdot X(z) + NTF(z) \cdot E_3(z) \\
& + z^{-1}(1 - z^{-1}) \cdot E_{12}(z) \\
& + (1 - z^{-1})^2 \cdot E_{23}(z). \qquad (2)
\end{aligned}
$$

$E_{12}(z)$ and $E_{23}(z)$ are the z-transforms of the quantizer errors introduced by the M-bit and L-bit interstage quantizers be-

978-1-4244-1983-8/08/$25.00 ©2008 IEEE

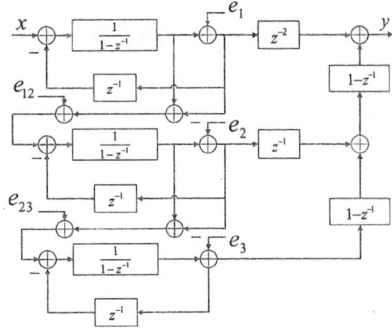

Fig. 2. Linearized model of the RC DDSM shown in Fig. 1.

tween the first and second and second and third accumulators, respectively.

Note that (2) differs qualitatively from (1) in that it contains two additional shaped noise terms resulting from the errors introduced by the interstage quantizers. By choosing the values of M and L appropriately, these contributions can be masked by the shaped E_3 term. Guidelines for choosing M and L are presented in Sec. III.

III. DESIGN METHODOLOGY

Guaranteed minimum sequence lengths for MASH DDSMs with identical first order stages have been determined [4], [5]. For a third-order DDSM, the result is 2^{N+1}, where N is the wordlength of the accumulator in the delta-sigma modulator, provided that the initial condition is odd [4], [5].

The quantization noise power is spread over a number of tones that is determined by the sequence length, resulting in a tone spacing of $\Delta f = \frac{f_s}{L_s}$, where f_s is the sampling frequency and L_s is the sequence length. The locations of these tones are given by

$$f[k] = k\Delta f, \qquad k = 1, 2, ..., \frac{L_s}{2}, \qquad (3)$$

where k is the index of the tone.

Assuming a cycle of length L_s and additive uniformly distributed white quantization noise, the idealized power spectrum of the shaped noise $NTF(z) \cdot E_3(z)$ is given by

$$N(f[k]) = \frac{1}{12L_s}|NTF(z)|^2\big|_{z=e^{j2\pi k/L_s}}, \quad k = 1, 2, ..., \frac{L_s}{2}. \quad (4)$$

For notational convenience, we rewrite (2) as follows:

$$Y(z) = STF(z)X(z) + N_3(z) + N_{12}(z) + N_{23}(z), \quad (5)$$

where $N_3(z) = NTF(z) \cdot E_3(z)$ is the filtered contribution from the quantizer in the third accumulator, $N_{12}(z) = z^{-1}(1 - z^{-1})E_{12}(z)$ is the filtered contribution of the first interstage quantizer error, and $N_{23}(z) = (1 - z^{-1})^2 E_{23}(z)$ is the filtered contribution of the second interstage quantizer error.

Assuming that all quantization noise terms can be modeled by independent white sources, we can express the idealized power spectrum of $N_3(z)$ as

$$N_3(f[k]) = \frac{1}{12L_s}\left|(1 - z^{-1})^3\right|^2\big|_{z=e^{j2\pi k/L_s}}, \quad (6)$$

where L_s is the sequence length.

In the same manner, the idealized spectra N_{12} and N_{23} resulting from e_{12} and e_{23} can be expressed as follows:

$$N_{12}(f[k]) = \frac{\Delta_{12}^2}{12L_{12}}\left|z^{-1}(1 - z^{-1})\right|^2\big|_{z=e^{j2\pi k/L_{12}}}, \quad (7)$$

$$N_{23}(f[k]) = \frac{\Delta_{23}^2}{12L_{23}}\left|(1 - z^{-1})^2\right|^2\big|_{z=e^{j2\pi k/L_{23}}}, \quad (8)$$

where Δ_{12} and Δ_{23} are the quantization intervals of the M- and L-bit interstage quantizers, which are $\frac{1}{2^M}$ and $\frac{1}{2^L}$, respectively. We denote by L_{12} and L_{23} the sequence lengths of the error signals from the interstage quantizers. We will show in the following how these lengths can be determined.

Let us express the DC input X to the DDSM in binary form, which is $x_N, x_{N-1}, ...x_1$, and separate it into its upper and lower pieces as follows:

$$\{x_N, x_{N-1}, ...x_1\} = X_{upper} + X_{lower} =$$
$$\{x_N, x_{N-1}, ...x_{N-M+1}, 0_{N-M}, ...0_1\} +$$
$$\{0_N, 0_{N-1}, ...0_{N-M+1}, x_{N-M}, x_{N-M-1}, ...x_1\} \quad (9)$$

Since the $(N - M)$ Least Significant Bits (LSBs) of X_{upper} are 0, they do not contribute to N_{12}. Therefore, X_{lower} alone determines N_{12}. Since the M Most Significant Bits (MSBs) are 0, the error output for X_{lower} feeding through the N-bit accumulator is the same as that feeding through an $(N - M)$-bit accumulator. Consequently, the sequence length of N_{12} is the same as the output sequence length of an $(N - M)$-bit word feeding through an $(N - M)$-bit accumulator. If we set the LSB of the input to "1", the sequence length L_{12} in this case is 2^{N-M} [5]. In the same manner, the sequence length L_{23} for N_{23} is 2^{N-L}. Thus,

$$L_3 = 2^N, L_{12} = 2^{N-M}, L_{23} = 2^{N-L}. \quad (10)$$

The correlation function γ, shown in Eq. (11), can be used to quantify the interdependencies of N_{12}, N_{23} and N_3:

$$\gamma = \frac{N\Sigma xy - (\Sigma x)(\Sigma y)}{\sqrt{[N\Sigma x^2 - (\Sigma x)^2][N\Sigma y^2 - (\Sigma y)^2]}}. \quad (11)$$

We have performed extensive calculations for different combinations of N, M and L, and the correlation results are below 0.02 in each case. Therefore, we conclude that N_{12}, N_{23} and N_3 can be made almost independent of each other.

By *assuming* independence, the total noise power at the output of the reduced complexity MASH DDSM can be expressed as:

$$N(f[k]) = N_3(f[k]) + N_{12}(f[k]) + N_{23}(f[k])$$
$$= \frac{1}{12L_s}\left|(1 - z^{-1})^3\right|^2\big|_{z=e^{j2\pi k/L_s}}$$
$$+ \frac{\Delta_{12}^2}{12L_{12}}\left|z^{-1}(1 - z^{-1})\right|^2\big|_{z=e^{j2\pi k/L_{12}}}$$
$$+ \frac{\Delta_{23}^2}{12L_{23}}\left|(1 - z^{-1})^2\right|^2\big|_{z=e^{j2\pi k/L_{23}}},$$
$$k = 1, 2, ..., \frac{L_s}{2}. \quad (12)$$

978-1-4244-1983-8/08/$25.00 ©2008 IEEE

The idea of our wordlength selection strategy [6] is to mask the contributions of the intermediate quantizers by hiding the noise components N_{12} and N_{23} below the N_3 component.

The idea is illustrated graphically in Fig. 3. The error spectra N_{12} and N_{23} due to the interstage quantizers should lie below the N_3 contribution. Since all are discrete spectra, the constraints apply at a finite number of points. In particular,

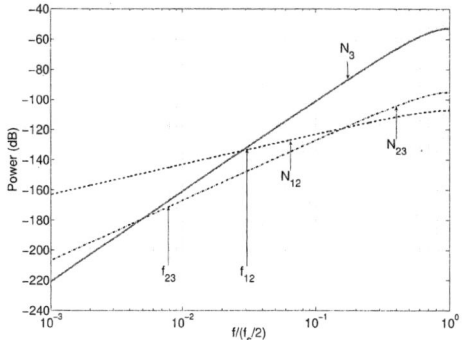

Fig. 3.   Masking N_{12} (dashed) and N_{23} (dotted) below N_3 (solid). The lowest frequency tone in N_{23} is at $f_{23}(= f_s/2^{N-L})$; the lowest frequency tone in N_{12} is at $f_{12}(= f_s/2^{N-M})$.

we require that:

$$N_{12} < N_3 \ @ \ f \ = \ f_s \cdot k/L_{12}, k = 1, 2, ..., L_{12}/2, \quad (13)$$
$$N_{23} < N_3 \ @ \ f \ = \ f_s \cdot k/L_{23}, k = 1, 2, ..., L_{23}/2. \quad (14)$$

Recall that for a DDSM with an output sequence length of L_s, the location of the lowest frequency tone is at f_s/L_s. Therefore, since the sequence lengths for N_{12} and N_{23} are 2^{N-M} and 2^{N-L}, the lowest frequency tones for N_{12} and N_{23} are at $f_s/2^{N-M}$ and $f_s/2^{N-L}$, respectively.

Additionally, at the output of the RC DDSM, since N_{12} and N_{23} are first- and second-order shaped, respectively, while N_3 is third-order shaped, if the levels of the lowest frequency tones for N_{12} and N_{23} are below that of N_3, the overall power of N_{12} and N_{23} should always be below the N_3 envelope. Based on this idea, the constraints can be rewritten as:

$$N_{12} < N_3 \ @ \ f \ = \ f_s/2^{N-M}, \quad (15)$$
$$N_{23} < N_3 \ @ \ f \ = \ f_s/2^{N-L}. \quad (16)$$

Since

$$\left|1 - z^{-1}\right|^2 = \left|1 - e^{-j2\pi f/f_s}\right|^2 = |2\sin(\pi f/f_s)|^2 \quad (17)$$

and

$$\sin(\pi f/f_s) \approx \pi f/f_s \quad \text{for} \quad f << f_s, \quad (18)$$

we can approximate N_{12}, N_{23} and N_3 at low frequencies by

$$N_{12} \approx \frac{\Delta_{12}^2}{12L_{12}} \cdot 2^2 (\pi f/f_s)^2, \quad (19)$$

$$N_{23} \approx \frac{\Delta_{23}^2}{12L_{23}} \cdot 2^4 (\pi f/f_s)^4, \quad (20)$$

$$N_3 \approx \frac{1}{12L_3} \cdot 2^6 (\pi f/f_s)^6. \quad (21)$$

Substituting Eqs. (19)–(21) into the constraints (15) and (16), we obtain

$$\frac{1}{2^{2M}} \cdot \frac{1}{12 \cdot 2^{N-M}} \cdot 2^2 \cdot \frac{\pi^2}{(2^{N-M})^2} < \frac{1}{12 \cdot 2^N} \frac{2^6 \pi^6}{(2^{N-M})^6} \quad (22)$$

and

$$\frac{1}{2^{2L}} \cdot \frac{1}{12 \cdot 2^{N-L}} \cdot 2^4 \cdot \frac{\pi^4}{(2^{N-L})^4} < \frac{1}{12 \cdot 2^N} \frac{2^6 \pi^6}{(2^{N-L})^6}, \quad (23)$$

which reduce to

$$4N - 5M - 4 < 4\log_2(\pi) \approx 6.6, \quad (24)$$
$$2N - 3L - 2 < 2\log_2(\pi) \approx 3.3. \quad (25)$$

Based on Eqs. (24) and (25), in order to design a reduced complexity N-M-L DDSM with the same sequence length and similar power spectrum as a conventional N_0-bit DDSM, the design procedure is:

1) Choose $N=N_0+1$ to ensure that the output sequence length of the reduced complexity DDSM is the same as that of the conventional N_0-bit DDSM.

2) Choose M=ceil($\frac{4N-10.6}{5}$) [from Eq. (24)] to ensure that the power of the first tone of N_{12} is less than N_3 at the frequency $f_s/2^{N-M}$, where ceil(x) means the smallest integer greater than x.

3) Choose L=ceil($\frac{2N-5.3}{3}$) [from Eq. (25)] to ensure that the power of the first tone of N_{23} is less than N_3 at the frequency $f_s/2^{N-L}$.

IV. Design Example

In order to verify the design methodology in Sec. III, we present a design example for a 19-bit MASH DDSM. The optimum wordlengths of the first, second and third stages of the RC DDSM are 20, 14 and 12, respectively.

Firstly, we simulate the 20-14-12 RC DDSM to show typical contributions of N_{12} and N_{23}. Figure 4 shows the power spectrum of N_{12}. The power spectrum based on the white noise approximation (7) is overlaid as well. As expected, the quantization powers are spread over $2^{20-14} = 2^6$ discrete tones, while the location of the lowest frequency tone is $f_s/2^6$. In addition, N_{12} is shaped by 20 dB/dec, which is the same as for a first order DDSM. Figure 5 shows the

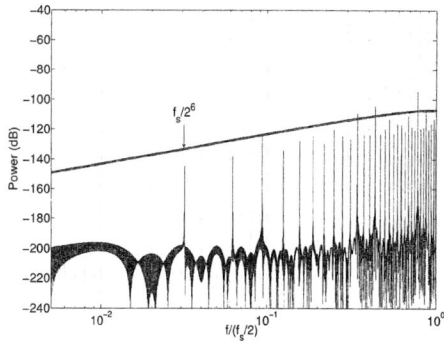

Fig. 4.   Simulated power for N_{12} when N=20, M=14, L=12; the input is 104857. The first spur is at $f_s/2^6$. The smooth curve is (7).

978-1-4244-1983-8/08/\$25.00 ©2008 IEEE

simulated power spectrum for N_{23} with the overlaid white noise prediction (8), from which we can see that N_{23} is shaped by 40 dB/dec, as expected. The lowest frequency tone is located at $f_s/2^{20-12}(=f_s/2^8)$. The output power for the

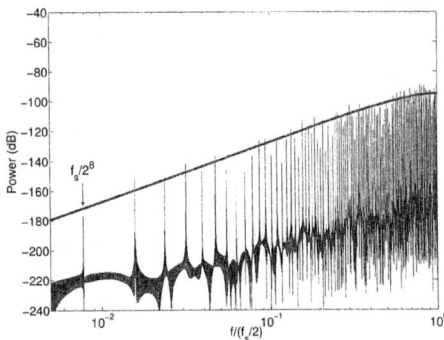

Fig. 5.   Simulated power for N_{23} when N=20, M=14, L=12; the input is 104857. The first spur is at $f_s/2^8$. The smooth curve is (8).

20-14-12 RC DDSM is shown in Fig. 6. Note that the N_{12} and N_{23} components lie below the spectral envelope of N_3 and are therefore masked by it, as expected. Consequently, N_{12} and N_{23} do not adversely affect the overall performance of the DDSM. To compare with the conventional MASH DDSM, the

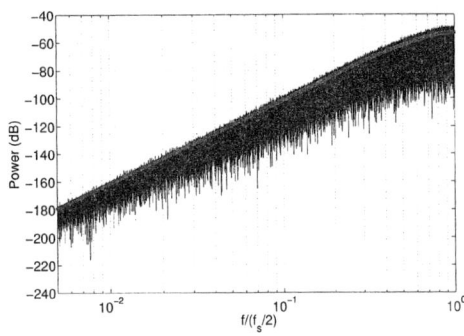

Fig. 6.   Simulated power for a RC 20-14-12-bit MASH 111 DDSM; the input is 104857. The smooth curve is (6).

simulated power of the conventional 19-bit DDSM is shown in Fig. 7. The spectrum for our RC MASH DDSM is smoother since the additional quantization error sources serve to whiten the error.

A. Hardware Consumption

The hardware requirements for a conventional 19-bit MASH 111 DDSM and the 20-14-12-bit RC-DDSM without dither are summarized in Table. I. The hardware consumption is reported as the number of flip–flops (FFs) and the number of 4–input look–up–tables (LUTs) which represent the synchronous logic and the asynchronous logic, respectively. The total equivalent gate (TEG) count for the design is given as well. These results are based on the map report from the Xilinx ISE program [7]; a full custom implementation could potentially

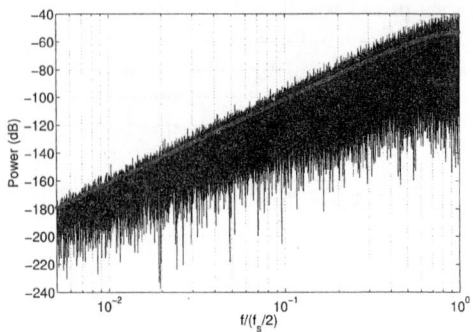

Fig. 7.   Simulated power for a conventional 19-bit MASH 111 DDSM; the input is 52429. The smooth curve is (6)(compared with Fig. 6.)

TABLE I

HARDWARE CONSUMPTION OF THE CONVENTIONAL 19-BIT MASH 111 DDSM AND THE 20-14-12-BIT RC-DDSM WITHOUT DITHER

MASH DDSM	Hardware Consumption		
	FFs	LUTs	TEGs
(1) 19 bit	113	68	1579
(2) 20-14-12 bit	90	57	1263
((2)/(1))%	79.6%	83.8%	80%

do better. The 20-14-12 reduced complexity MASH DDSM has marginally better spectral performance than the 19-bit conventional MASH DDSM, with 20% less hardware (TEG).

V.   CONCLUSIONS

In this paper, we have presented a design methodology for MASH DDSMs based on error masking. We have shown that, starting with a conventional DDSM, it is possible to find an optimized wordlength for each stage of the DDSM, which allows the hardware consumption to be reduced by 20% without degrading the spectral performance.

ACKNOWLEDGMENT

This work is supported in part by Science Foundation Ireland under Grant 02/IN.1/I45.

REFERENCES

[1] Z. Ye and M.P. Kennedy, "Reduced Complexity MASH delta-sigma modulator," *IEEE Trans. on Circuits and Systems II*, vol. 54, no. 8, pp. 725–729, Aug. 2007.

[2] V. Friedman, "The structure of the limit cycles in sigma delta modulation," *IEEE Trans. Commun.*, vol. 35, pp. 784–796, Jul. 1989.

[3] M. Kozak and I. Kale, "Rigorous analysis of delta–sigma modulator for fractional-N PLL frequency synthesis," *IEEE Transactions on Circuits and Systems I: Regular Papers*, vol. 51, no. 6, pp. 1148–1162, June 2004.

[4] M.J. Borkowski, T.A.D. Riley, J. Hakkinen and J. Kostamovaara, "A practical $\Delta\Sigma$ modulator design method based on periodical behavior analysis," *IEEE Transactions on Circuits and Systems II: Express Briefs*, vol. 52, no. 10, pp. 626–630, Oct. 2005.

[5] K. Hosseini and M.P. Kennedy, "Mathematical analysis of digital MASH delta-sigma modulator for fractional-N frequency synthesizers," in *Proc. of PRIME 2006.*, pp. 309–312, Otranto, Italy, Jun. 2006.

[6] M.P. Kennedy and Z. Ye, "A Delta Sigma Modulator," Irish Patent, application number 2006/0763, Oct, 17, 2006.

[7] ISE quick start tutorial, [online]. Available: *http://toolbox.xilinx.com /docsan/xilinx7/books/docs/qst/qst.pdf*, 2005.

978-1-4244-1983-8/08/$25.00 ©2008 IEEE

Design of an Ultra-Low Power Time Interleaved SAR Converter

Fabrizio Erario, Andrea Agnes, Edoardo Bonizzoni, and Franco Maloberti

Department of Electronics
University of Pavia
Via Ferrata, 1 - 27100 Pavia - ITALY
fabrizio.erario01@ateneopv.it, [andrea.agnes, edoardo.bonizzoni, franco.maloberti]@unipv.it

Abstract — **An ultra low-power SAR ADC is presented. The circuit is the interleaved version of an already designed SAR converter with improved performance. This design uses 7 interleaved converters and achieves a conversion rate of 700 kS/s. The converter has been simulated by using a 0.18-μm CMOS technology showing a power consumption as low as 40 μW which allows obtaining a state-of-the-art FoM equal to 37 fJ/conv.-step. The architectural study together with converter simulation results are presented.**

I. INTRODUCTION

The power consumption is important for micro-sensors wireless systems. After performing a first analog processing, many architectures use an A/D to move into the digital domain. The typical requirements for many sensor systems are medium resolution (~10–12 bit) and low speed (~500-700 kS/s). Therefore, the specs are not difficult but the required low power level is very challenging. Among the data converter architectures, the flash is a first generic option: it uses (2^n-1) comparators. However, the high number of comparators makes the architecture too power hungry even for low resolutions. Also the pipeline architecture is not a good approach for ultra low-power. As known, a pipeline converter reduces the power as it divides the conversion task into several consecutive steps; however, each stage uses an active gain element whose power equals the one of many comparators. Therefore, it becomes competitive for high resolution and, in general, requires too much power. Other architectures, like the sigma-delta and the time interleaved have similar limits because despite their use of speed or multiple paths to increase resolution or throughput, they use active power hungry elements. If the speed is low the most suitable algorithm is the successive approximation that uses a successive approximation register (SAR) to control a DAC in a feedback loop with a single comparator. The cost is that it requires $n+1$ comparison cycles to achieve n-bit resolution. The above considerations are summarized in Fig. 1, that shows the best data converter topologies as a function of the desired sampling rate and the required output bit resolution. The diagram depends on the technology and the used supply voltage; however, the SAR algorithm is preferable for signal bands up to one, two hundred of kHz.

As known, the SAR architecture has a circuit configuration like the one shown in Fig. 2, [1 - 4]. The comparator refers to

ground after the subtraction of the sampled-and-held input and its foreseen version generated by the DAC. The given topology increases the resolution by just increasing the number of cycles of the algorithm; therefore the power has the logarithmic dependence with the number of quantization steps.

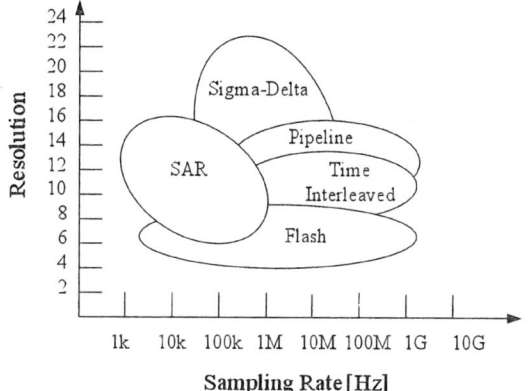

Figure 1. Tipical converter topologies as a function of the sampling rate and obtained resolution.

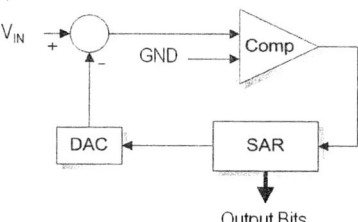

Figure 2. Conventional SAR converter architecture.

Starting form this favorable algorithm, this work studies architectural strategies to optimize the power consumption of a medium-speed medium-resolution SAR ADC. Namely, a careful design of a SAR architecture, [5], and the use of the time interleaved technique achieves ultra low power. With a 0.18-μm CMOS technology and a sampling rate of 700 kS/s, the expected power consumption is 40 μW, giving rise to a FoM as low as 37 fJ/conv.-step.

II. SAR CONVERTER BASIC BUILDING BLOCKS

The basic architecture of Fig. 2 has been realized with non conventional and custom designed circuits to minimize the consumed power. This Section discusses the design strategies and provides circuit details. The foreseen resolution is 12-bit (enough for many sensor systems) with a reference voltage of 1 V and supply voltage V_{DD}= 1.2 V.

A. DAC

Fig. 3 shows the circuit schematic of the DAC, the input S&H and subtractor. It consists of two identical 6-bits sub-array binary weighed capacitors with unary attenuation capacitor. The use of two sub-arrays is a common way to reduce the capacitor spread and, consequently the power due to the charging and discharging of capacitors during the conversion cycles. However, as discussed in [1], the capacitor used to bridge the two arrays and ensure the attenuation factor of 32 should be non unity. The solution of Fig. 3 uses a unity capacitor but the sub-array is made by 63 elements instead of 64. This causes a global gain error that is acceptable, [6]. The value of the unity capacitance is at least 100 fF and the power consumption turns out to be about 570 nW.

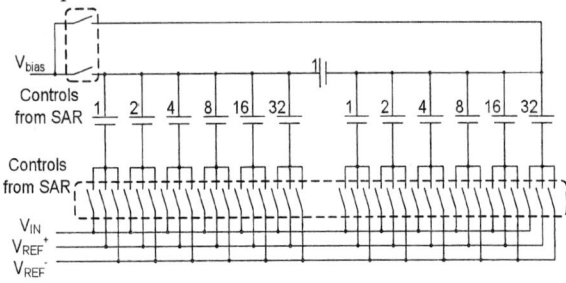

Figure 3. Binary capacitors arrays with attenuation capacitor.

The solution of Fig. 3 is the same used in [5], but the switches used for the input sampling are driven by a clock-boost, shows in Fig. 4. The switch transistor is MN1 and the others allow controlling the switch transistor with the same overdrive despite the input variations. The charge and discharge of bootstrap capacitors in the switches introduce a power consumption of about 850 nW.

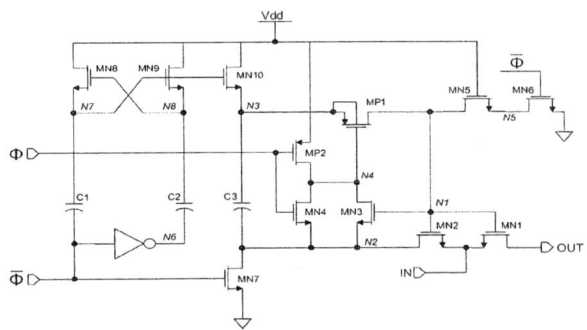

Figure 4. Schematic of a bootstrapping switch.

B. Time-Domain Comparator

Conventional SAR ADCs foresee the use of a voltage comparator, typically made of a preamplifier followed by a latch stage. Theoretical analysis and simulation results demonstrate that this approach is not the best in terms of power consumption, [6]. In order to minimize the power consumption, in this paper, a power effective time-domain comparator is used. Two identical voltage-to-time (V2T) converters make the core of the time-domain comparator, [5]. The V2T, depicted in Fig. 5, provides an output pulse which is delayed with respect to the reference clock, Φ_C, as a function of the input voltage, V_A. This process is based on a voltage-to-current conversion (i.e., the input voltage into the current flowing through R_D). This current is used to discharge capacitor C with a constant slope. When voltage V_C crosses the threshold of transistor M_5, the output voltage Out_{V2T} rises, driven by a tapered inverter chain.

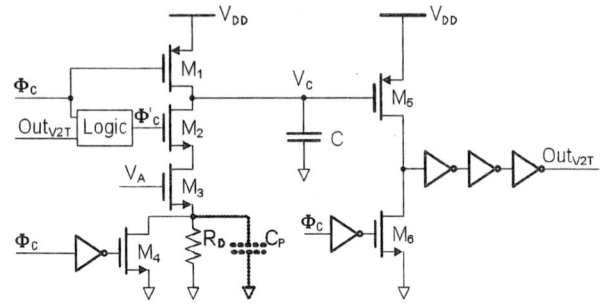

Figure 5. V2T schematic diagram.

A simple flip-flop delay (DFF) identifies the faster signal between the two V2T output pulses. The digital logic connected to the gate of transistor M_2 has the purpose to stop the discharge of the capacitor C after the commutation of the signal Out_{V2T}, as shown in Fig. 6.

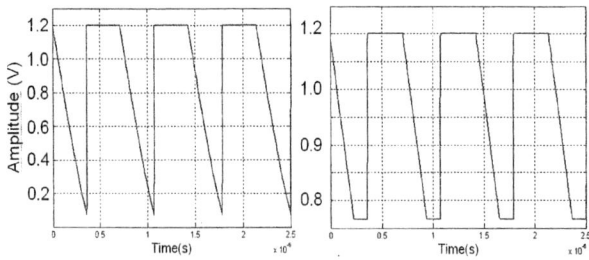

Figure 6. Voltage across C without and with the digital logic on gate of M2.

In this way the power consumption spent for a comparison step decreases of at least a factor two. The most important limit on the value of capacitance C is the thermal noise which imposes at least 0.8 pF. Therefore, the estimated power related to the comparator is about 880 nW.

When designing a time-domain comparator, the time error, conceptually described in Fig. 7, introduced by the DFF has to be taken into account.

978-1-4244-1983-8/08/$25.00 ©2008 IEEE   246

This error can be considered as an equivalent input voltage of

$$V_{IN,eq.err} = \frac{t_{ji}}{T} \cdot \left(V_{ref} - V_{GS3}\right) < \frac{V_{REF}}{\sqrt{2} \cdot 2^{N+1}} \qquad (1)$$

where the time t_{ji}, characteristic of the time comparator, is equal to 130 ps, V_{ref} is the input voltage of the V2T used as reference, and $V_{REF} = V_{REF+} - V_{REF-}$. Therefore, to obtain an equivalent input error less than a quarter of LSB, it is required a comparison time of at least 226 ns.

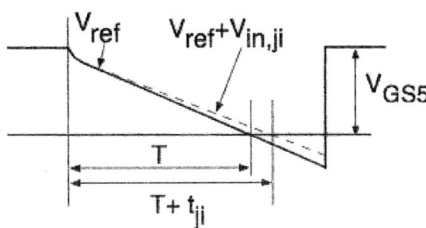

Figure 7. Considerations on time error.

C. SAR

As mentioned above, to meet the low power requirements, the successive approximation register is full custom designed, which implicates a power consumption of about 1.8 µW. It is worth to point out that the SAR logic generates also the A/D converter control phases. The SAR logic is optimized for minimum power consumption. The conversion requires 14 clock periods of the main clock: the first for the input sampling, 12 periods are for the successive approximation cycles and the last one for end of conversion and data transfer.

III. TIME INTERLEAVED SOLUTION CONSIDERATIONS

The SAR converter requires 14 clock periods to obtain a single conversion and this limited comparison speed of the time-domain comparator motivates the choice to consider a time interleaved approach to meet the required conversion rate. The number of different paths used in this approach must to be equal to an integer divisor of the number of clock periods needed for a conversion with a simple SAR. In particular, in this paper, an architecture with 7 parallel paths is analyzed.

In the single-path solution, as described above, two V2T converters are used in the time-domain comparator, one to process the input signal (signal V2T) and the other used as a reference (reference V2T). This choice ensures good matching performance between the two V2T blocks. When considering a 7 parallel paths solution, it is possible to use only one reference V2T for all 7 signal V2Ts, as shown in Fig. 8. This solution, obviously, allows strongly reducing the power consumption.

When approaching a time interleaved topology, issues related to gain and offset errors have to be carefully analyzed. In a time interleaved SAR ADC, the gain error of each converter depends mainly from the DAC. In this work, the DAC has been designed with analog output as a function of the reference voltages. It has to be pointed out that the reference voltages are the same for all DACs present on each path, thus leading to any gain error. The gain error depends also on DAC capacitors mismatch, but the DAC structure adopted in this design allows obtaining good matching performance, [6].

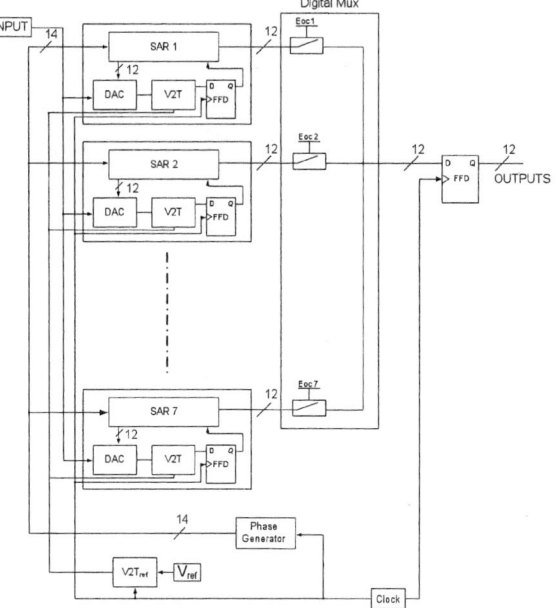

Figure 8. Time interleaved solution with SAR converter based on time-domain comparator.

The second critical issue to be considered when designing a time interleaved ADC is the offset error. In the case of a time interleaved SAR converter, the block that introduces this kind of error is the comparator. In our design, different sources of V2T input offset are present: mismatch among capacitors C or resistors R_D, transistor M_5 threshold variations, and FFD time errors. The basic idea in order to compensate for all these offset contributions is to trim the value of resistor R_D by means of an adequate digital trimming circuit. This leads to reduce the equivalent input offset below half LSB. To estimate the overall equivalent input offset of a reference V2T with comparison time equal to 230 ns, several Montecarlo simulations have been performed. As a result, it has been verified that the comparison time is in the range from 200 ns to 260 ns. Translating these time values in the voltage domain, it means that the possible equivalent input offset can assume values from -43 mV to +40 mV. Considering that half LSB is equal to 122 µV, it is apparent that the digital trimming of resistor R_D is not effective. The idea is to perform first a coarse offset reduction by trimming on each DAC the bias voltage, referred to as V_{bias} in Fig. 3, and then the fine digital trimming of resistor R_D. The manual course calibration precision is typically of about 10-15 mV. With regard to the fine digital trimming design, it is worth to point out that a V2T input offset error equal to half LSB can be compensated for by a R_D variation of 330 Ω. In this way, at least 100 330-Ω resistors are required to compensate for the residual offset of 10-15 mV. This leads to a digital trimming that foresees the use of 7 bits (i.e., 128 steps).

978-1-4244-1983-8/08/$25.00 ©2008 IEEE

Considering the technology mismatches on resistors values, a nominal resistor of 312 Ω is used. Considering mismatch, this resistor can vary from 294 to 330 Ω so that an overall input offset of 13.9-15.6 mV, respectively, can be corrected. The power consumption of the digital logic which manages the time interleaved structure is estimated to 10 μW.

IV. SIMULATION RESULTS

The proposed time interleaved SAR ADC has been realized at the transistor level and simulated by using a conventional 0.18-μm CMOS technology. The used power supply value is 1.2 V. Fig. 9 shows the simulated output spectrum (obtained with 1024 samples) of a SAR converter. The sampling rate is 100 kS/s and the sine wave input signal frequency is 29.8 kHz. It can be noted that the noise floor is almost flat. The simulated SNDR is equal to 65.6 dB.

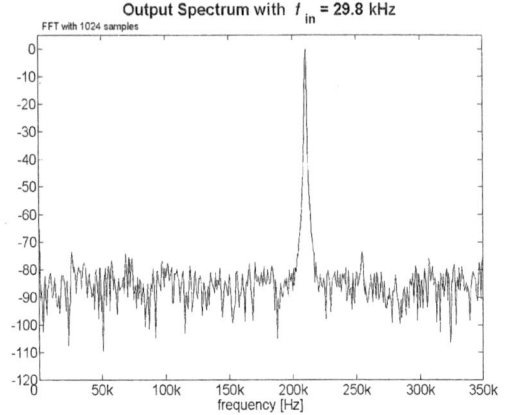

Figure 9. Single path SAR ADC output spectrum.

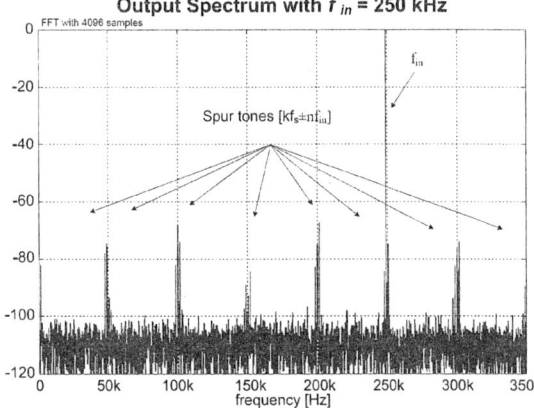

Figure 10. Effects of offset on output spectrum.

Fig. 10 shows the output spectrum simulated at the behavioral level in order to evaluate the effect of different offset into each path of the time interleaved structure, before the relative correction. It is apparent the presence of spur tones at 1/7 of the sampling frequency and its multiples.

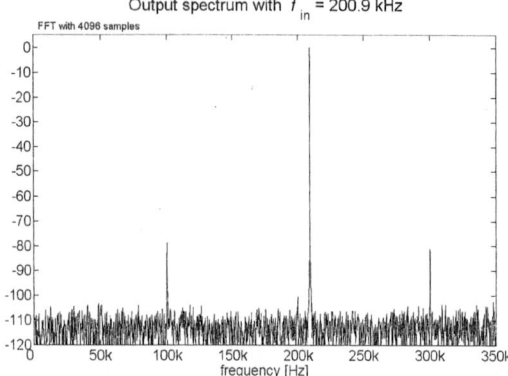

Figure 11. Simulated time interleaved SAR output spectrum.

Fig. 11 shows a behavioral dynamic simulation of the 7 parallel paths time interleaved SAR converter after the digital correction of the offset. The output spectrum depicted in Fig. 11 has been obtained including an offset error with standard deviation equal to 2 LSB. The frequency of the input sine wave is 200.9 kHz. It can be noted that the second and the third harmonics are present. The second harmonic tone, placed at about 300 kHz, is at -78 dB while the third harmonic tone, placed at about 100 kHz, is at -81 dB. The simulated power consumption of the time interleaved SAR ADC is about 40 μW. This value leads, considering a signal bandwidth of 350 kHz, to a FoM as low as about 37 fJ/conv.-step.

CONCLUSIONS

In this paper, an ultra low-power time interleaved SAR ADC is presented. This design uses 7 interleaved converters and achieves a conversion rate of 700 kS/s. Design considerations about time-domain comparator offset calibration have been drawn. The A/D converter has been simulated at the transistor level by using a 0.18-μm CMOS technology. The simulated power consumption is as low as 40 μW which allows obtaining a state-of-the-art FoM equal to 37 fJ/conv.-step.

REFERENCES

[1] F. Maloberti, "Data converters", Springer, 2007.

[2] M.D. Scott, B.E. Boser, and K.S.J. Pister, "An Ultralow-Energy ADC for Smart Dust" *IEEE Journal of Solid State Circuits*, vol. 38, pp. 1123-1129, Jul. 2003.

[3] N. Verma and A.C. Chandrakasan, "A 25μW 100 kS/s 12b ADC for wireless application", *IEEE International Solid-State Circuits Conference Dig. Tech. Papers (ISSCC)*, pp. 222-223, Feb. 2006.

[4] J.Sauerbrey, D. Schimitt-Landsiedel, R. Thewes, "A 0.5 V, 1μW Successive Approximation ADC", *IEEE European Solid State Circuits Conference (ESSCIRC)*, pp. 247-250, Sep. 2002.

[5] A. Agnes, E. Bonizzoni, P. Malcovati, and F. Maloberti, "A 9.4 ENoB, 1V, 3.8μW, 100kS/s SAR-ADC with Time-Domain Comparator", *IEEE International Solid-State Circuits Conference (ISSCC)*, pp. 246-247, Feb. 2008.

[6] A. Agnes, E. Bonizzoni, and F. Maloberti, "Design of an Ultra-Low Power SA-ADC with Medium/High Resolution and Speed", *IEEE International Symposium on Circuits and Systems*, May 2008.

A design methodology for asynchronous sigma-delta converters

Balkir Kayaalti
Boğazici University, Department of
Electrical and Electronics Engineering
Bebek,34342 Istanbul, Turkey
balkir.kayaalti@boun.edu.tr

Omer Cerid
Boğazici University, Department of
Electrical and Electronics Engineering
Bebek,34342 Istanbul, Turkey
cerid@siemens.ee.boun.edu.tr

Gunhan Dundar
Boğazici University, Department of
Electrical and Electronics Engineering
Bebek,34342 Istanbul, Turkey
dundar@boun.edu.tr

Abstract— **In this paper a design methodology to design asynchronous continuous time delta-sigma modulators (ASDM) is given. Performance equations relating circuit parameters and the input signal are derived. Asynchronous modulators eliminate the use of an external clock, and thereby jitter and give the opportunity to build faster modulators.**

I. Introduction

Delta-sigma modulation A/D conversion has received much attention since 80's. First examples were built with discrete time components using sampled signals. Today, discrete-time sigma delta (SD) modulators are still the most popular implementations. However, there is a trend to replace all the discrete-time blocks in traditional modulators with their continuous-time counterparts. Continuous-time synchronous modulators in which the signal is sampled before quantization have achieved high performances due to the better power-speed performance of continuous time filters as opposed to their discrete-time counterparts. Sampling is done by using an external clock signal, requiring an additional signal source and the modulator is subjected to possible performance penalties due to the clock jitter. In recent years there are a few examples utilizing asynchronous type modulators, which require no external clock. The amplitude-time transformation is done with a self generated limit-cycle. This type of modulator can be regarded as a full continuous-time modulator.

There is some related work which can be used to design such a circuit [1],[2] using graphical methods of describing functions; however no clear performance related design equations are given.

The method described in this paper consists of using a quasi-linear model of the quantizer and deriving some performance relations which are dependent on input and the circuit used. Similar work has been done for discrete time modulators in [3].

Stability of the system which is of great importance for the proper operation of the modulator is analyzed in the second section and some design parameter relations ensuring desired operation are derived.

II. Asynchronous Sigma Delta Modulators

A typical asynchronous delta-sigma modulator (ASDM) is shown in Fig.1. Loop filters H_1 and H_2 are simply continuous time integrators. The quantizer, which is a highly nonlinear element, is represented by a linear gain block followed by an adder representing quantization noise. The nonlinearity of the quantizer makes exact mathematical analysis difficult. To make the circuit operational, it is also necessary to establish the limit cycle condition. Nonlinearity of the quantizer and its degradation effect upon the performance and stability is mentioned in a related work [3]. The quasi-linear models for the quantizer used for synchronous modulators can also be adapted to ASDM's.

III. ASDM Performance Analysis

Two input random signal (Gaussian plus sine wave for signal and noise) analysis provides two associated gains K_n and K_x. The circuit now can be analyzed as two interlocked systems for signal and noise only. Fig. 2 shows the circuit for the input signal and Fig. 3 shows the circuit for noise [3].

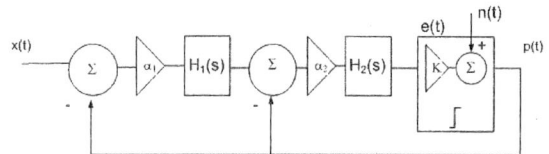

Figure 1. Second order ASDM with linear model of quantizer

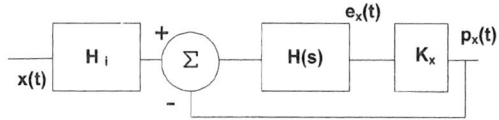

Figure 2. Signal path for the input signal

978-1-4244-1983-8/08/$25.00 ©2008 IEEE

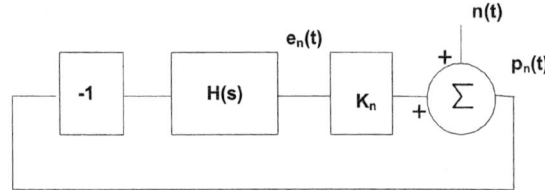

Figure 3. Signal path for noise

The quasi-linear gains are given in [4], K_n for noise and K_x for signal are dependent highly on the signal levels before the quantizer, as expected. 'M' in the equations stands for confluent hypergeometric functions [5] and ρ is the variance ratio of signal to noise at the input of quantizer.

$$K_x = \sqrt{\frac{2}{\pi}} \frac{\Delta}{\sigma_{en}} M(\frac{1}{2},2,-\rho^2) \qquad (1)$$

$$K_n = \sqrt{\frac{2}{\pi}} \frac{\Delta}{\sigma_{en}} M(\frac{1}{2},1,-\rho^2) \qquad (2)$$

To obtain linearized gains, noise deviation before the quantizer is needed. This can be calculated with the formula below, where N(s) is the Laplace transform of the noise added by the quantizer:

$$\sigma_{en}^2 = \frac{\sigma_n^2}{2\pi j} \int_{-j\infty}^{j\infty} \frac{N(s)N(-s)}{[1+K_n H(s)][1+K_n H(-s)]} ds \qquad (3)$$

However this formula is dependent on gain K, so (2),(3) should be solved iteratively, thus bringing mathematical complexity. To ease this calculation, the linearized gain K_n can be unified with the second filter function $H_2(s)$. Then, noise and signal powers at the output should be added up to Δ^2, where Δ is quantization level.

$$P_{yy} = P_{xx} + P_{nn}(\frac{1}{2\Delta f} \int_{-f}^{f} |H_n(jw,K)|^2 dw) = \Delta^2 \quad (4)$$

This kind of calculation was studied in [6] for DT implementations. The integral of the noise at the output which has s a highpass spectrum results in an infinite number quantity when all frequency range between minus and plus infinity is taken into account. However, there is a certain physical frequency range which the quantized signal spectrum covers. Detailed noise spectra analysis was done in [7], [8]. These studies give the spectrum is composed of odd tones of the carrier frequency and some frequency modulated components of the input signal.

In this work the limit 'f' of the integral is taken as the third harmonic of the quantizer square wave output, which is three times the limit cycle frequency 'f_c' of the ASDM. Assuming P_{nn}, the noise added by the quantizer is uniform with level $\Delta^2/3$ [3] the power of noise integral is defined as follows using (4) :

$$I_n(K_n) = \frac{1}{f} \int_0^f |H_p(jw,K_n(ax))|^2 dw = (\Delta^2 - \frac{a_x^2}{2})\frac{3}{\Delta^2} \quad (5)$$

Right hand side of (5) changes as input level changes. Left hand side of the equation shows how the uniform quantization noise is shaped and falls in the band of interest. Zero input will give $I_{n,max}= 3$; whereas full scale input (a_x) gives $I_{n,min}= 1.5$. $H_n(j\omega,K_n)$ is the noise transfer function.

$$I_n = \frac{1}{n \cdot f_c} \int_0^{n \cdot fc} \frac{1}{|1+H_n(jw,K_n)|^2} dw \qquad (6)$$

Clearly, location of the poles and zeros of the loop filter function determines the amount of noise added at the output by changing the value of the power integral 'I_n'. Considering a second order loop filter with a delayed integrator in the second integrator position, the following overall loop filter function is obtained:

$$H(s) = \frac{(K_n \alpha_2)e^{-Ts}}{s}(1+\frac{\alpha_1}{s}) \qquad (7)$$

Delay element is necessary to obtain a limit cycle pattern. In contrast; this behavior is undesired in synchronous delta sigma modulators (SDM). This topic is detailed in the next section.

The followed procedure for designing the ASDM is used: First determine a limit-cycle frequency; this can be done with frequency stability analysis. The systems frequency is where the instability starts. One can show the solution of the following equation gives the limit-cycle frequency of the second-order delayed system.

$$\tan(\omega_c \cdot T_d) = \frac{\omega_c}{\alpha_1} \qquad (8)$$

Start with normalized loop coefficients and then scale all loop coefficients for a desired high f_c. Note that filter cutoff frequency and limit cycle frequency can be different. To find the suitable value of K_n, $I_n(K_n)$ in (6) can be plotted and the intersection of this curve and horizontal lines $I_n=3$ and $I_n=1.5$ gives the design region of the modulator. The integral limits of (6) can be taken as 3 and 0 for the normalized case. The K values outside the region 1.5 and 3 lines are unfeasible gain values. Noise power integral versus K_n is given in Fig. 4.

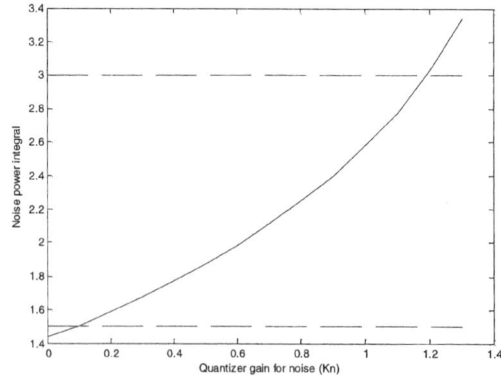

Figure 4. Power noise integral vs. K_n

978-1-4244-1983-8/08/$25.00 ©2008 IEEE

The delay value of the integrator T_d is 0.5. The feasible value of K is between 0.1 to 1.2. 'K_n' versus input is given in Fig. 5. If during frequency scaling, node voltages will be kept the same; according to (1),(2) K will be not scaled ,so normalized K can be used.

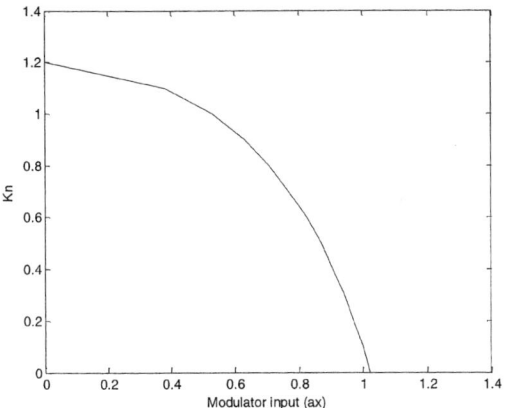

Figure 5. Kn versus input of the modulator

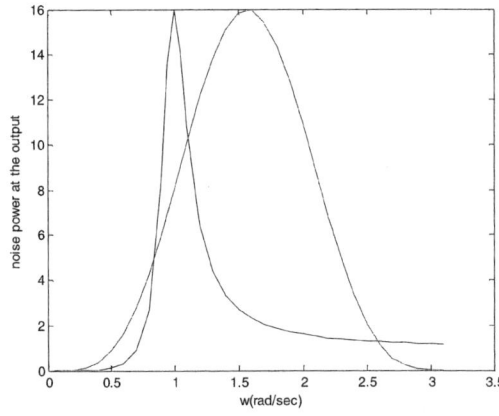

Figure 6. Comparasion of synchronous and asynchronous noise shaping

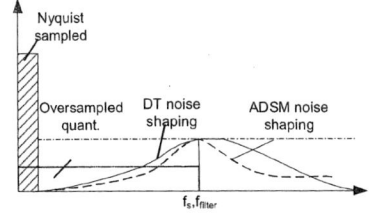

Figure 7. Effect of oversampling , noise-shaping in SDM and ASDM

Figures 6 and 7 show noise shaping action of the delta-sigma modulators. In conventional discrete time modulators,

the noise is suppressed in the in band at the expense of the total increase in the noise. Discrete time spectra repeats it self every $f_s/2$; however, noise is continuous in ASDM modulator; it goes to infinity after the f_c switching frequency; but after some harmonics the spectrum can be neglected.

Noise spectrum for different values of quantizer gain K_n is given in Fig. 8 with normalized frequency, $\omega=1$ rad/sec. It is seen that noise level increases with the decrease of K, which corresponds to a high input level. Therefore it can be concluded that, increase in the input level, also increases the distortion.

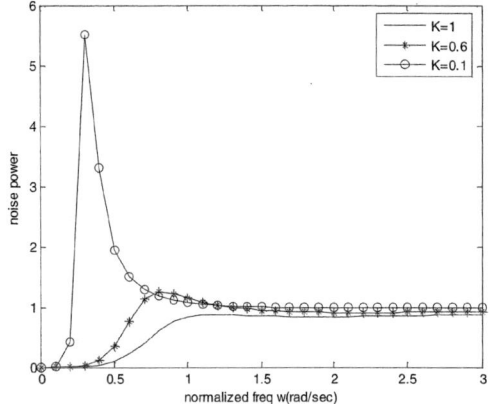

Figure 8. Effect of K on noise shaping.

A. SNR value

For a second order modulator noise spectrum around the signal band can be calculated approximately as follows:

$$S_{nn}(f) = P_n(jw)P_n(-jw) \approx \frac{\sigma_n^2}{fc} \cdot \frac{16\pi^4 f^4}{[(\alpha_1)^2(\alpha_2 K_n)^2]} \quad (9).$$

From (9) in band noise is:

$$\sigma_{nb}^2 = \int_0^{fb} S_{nn}(f)df = \frac{\sigma_n^2 16\pi^4 (f_b/f_c)^5}{5(\alpha_1')^2(\alpha_2' K_n)^2} \quad (10)$$

where integration time constant scaling of the integrators is done according to:

$$\alpha_{1,2} = \alpha_{1,2}' \cdot \frac{1}{\tau} = \alpha_{1,2}' \cdot f_c \quad (11)$$

Signal to noise ratio is calculated as follows, the value vs. input is depicted in Fig. 9 for the coefficients discussed before.

$$SNR = \frac{\sigma_x^2}{\sigma_{nb}^2} = \frac{15(\sigma_x^2)(\alpha_1')^2(\alpha_2' K_n)^2}{16\pi^4}(f_c/f_b)^5 \quad (12)$$

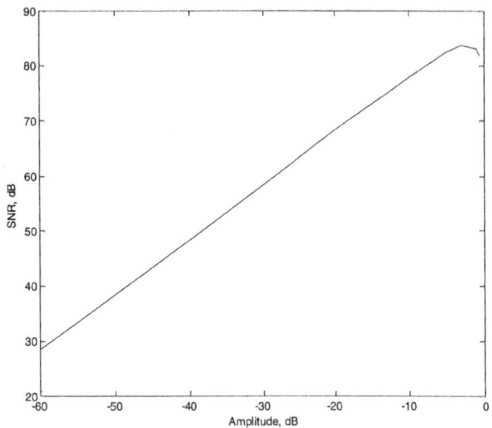

Figure 9. SNR vs. modulator input

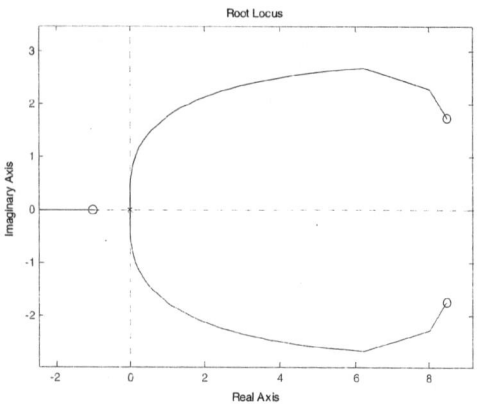

Figure 11. Root-loci for Td=1 sec.

V. CONCLUSION

A design methodology for ASDM's is given. Equations for noise spectrum and SNR for the asynchronous modulators are derived which are present for synchronous modulator designs.

B. Stability analysis

The ASDM modulator should follow a limit cycle pattern. Limit-cycle is not an exponentially unstable system, where node voltages grow to infinity; but rather the closed loop poles shift from their left hand position with the increase of K; pass the jω axis ,but with the decrease of K the poles go to left hand side. A root-locus plot analysis is necessary to see the behavior. The plots change with the value and poles oscillate between a point in the left hand plane and jω crossing.

There is a limit that T_d can be increased. After $T_d>1$, limit cycle breaks up and an unstable focus is observed. After that point, the node voltages grow exponentially, so T_d value should be selected to obtain root-locus plots which pass to the left hand plane. Selected delay values in design should be checked with Root-Locus plots. Figures 10 and 11 are plots for different T_d values for normalized frequency designs. Desired frequency scaling can be easily applied to T_d value to change f_c value.

REFERENCES

[1] E. Roza, "Analog-to-digital conversion via duty-cycle modulation," IEEE Trans. Circuits Syst. II, Analog Digit. Signal Process., vol. 44, no.11, pp. 907–914, Nov. 1997.

[2] S. Ouzounov, E. Roza, J. Hegt, G. van der Weide, "Analysis and Design of High-Performance Asynchronous Sigma-Delta Modulators With a Binary Quantizer," IEEE Journal Of Solid-State Circuits , vol..41, No. 3, March 2006

[3] S. H. Ardalan and J. J. Paulos. "An analysis of nonlinear behavior in delta-sigma modulators," IEEE Transactions on Circuits and Systems, vol. 33, pp. 287-301, Mar. 1987.

[4] D.P. Atherton, Nonlinear Control Engineering. London:VanNostrand, 1975

[5] M.Abromowitz and I.A. Stegun, Handbook of Mathematical Functions. New York: Dover Publications, 1970

[6] Y. Zhang, E. Hayahara, S. Hirano, N. Sakakibara, "Correlation of transfer function implementation on delta-sigma modulator stability analysis," IEICE Transaction on Fundementals of Electronics,Communucations and Computer Sciences ,vol. E83-A No.4 pp.733-739, April 2000.

[7] R.M. Gray "Quantization noise spectra", IEEE Trans. On Inform. Theory, vol.3, pp.1220-1244, Nov 1990.

[8] T.Green "Spectra of delta-sigma modulated inverters," IEEE Trans. on Power Electronics , vol. 7. No.4 ,Oct 1992.

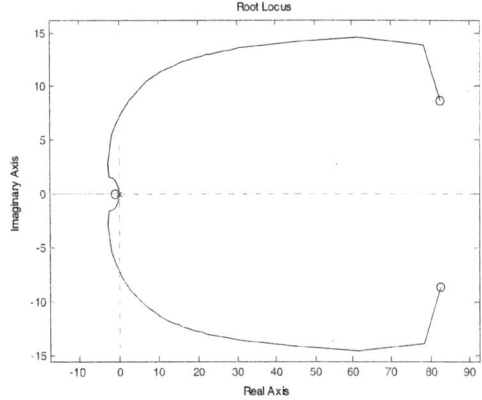

Figure 10. Root-loci for Td=0.25 sec.

978-1-4244-1983-8/08/$25.00 ©2008 IEEE

INNOVATIVE DEVICES IN BIOMEDICAL ELECTRONICS

Alessandro Convertino, Agostino Giorgio, Savino Giusto, Roberto Marani, Anna Gina Perri

CARDES Engineering S.r.l., Academic Spin-off, Electrotechnic and Electronic Department

Polytechnic of Bari

Bari, Italy

convertino@deemail.poliba.it; a.giorgio@poliba.it; robertom4@libero.it; perri@poliba.it

Abstract— **In this paper we present our principal projects in biomedical electronics, especially applied to telemedicine. The designed systems are characterized by originality and by plainness of use, as they planned with a very high level of automation (so called "intelligent" devices).**

I. INTRODUCTION

In this paper we present our principal projects in biomedical electronics, especially applied to telemedicine. The designed systems are characterized by originality, by plainness of use, as they planned with a very high level of automation (so called "intelligent" devices).

At this stage, they have been prototyped and tested but not still engineered even if a potential further reduction in dimensions is clearly obtainable using SMD package and optimizing the PCB design.

Firstly, we propose a system oriented to remote health monitoring named "Asclepius". It has been designed to be a wearable system and to allow real-time rescue in case of emergency without the necessity for data to be constantly monitored by a medical.

Also, we propose a low-cost, electronic medical device, named "Aglaia", designed for the non-invasive continuous real-time monitoring of breathing functions. It diagnoses respiratory pathologies by the electronic three dimensional (3-D) auscultation of lung sounds and performing a correlation between lung sounds and diseases.

At last we present a system for ECG transmission by Bluetooth and a digital cardioholter with multiple leads.

II. REMOTE MONITORING SYSTEM OF HEALTH (ASCLEPIUS)

It uses some technologies of GPRS/Bluetooth wireless telecommunication, satellite GPS more advanced technologies and internet. Asclepius allows real-time rescue in case of emergency without the necessity for data to be constantly monitored by a medical thanks to a reliable, real-time, automatic electrodiagnostic service and alarm system to alert medical staff, families or anyone able to arrange for rescue operations.

The system is able to monitor and to send via any wireless connection the most relevant health parameters and environmental factors affecting health, also leaving patients free to move and allowing prompt aid. Each monitored patient is given a case sheet on a Personal Computer (PC) functioning as a server (online doctor). Data can also be downloaded by any other PC, handheld or smartphone equipped with a browser. The system reliability rests on the use of a distributed server environment, which allows its functions non to depend on a single PC and gives more online doctors the chance to use them simultaneously.

The whole system consists of three hardware units and a management software properly developed. The units are:

- Elastic band: the sensors for the measurement of health parameters are embedded in an elastic band to be fastened round the patient's chest;
- Portable Unit (PU), having the following characteristics: wearable, wireless (GPRS/Bluetooth) and wireline (internet) connection, transmission, continuous or sampled or on demand, GPS satellite localization, real-time information devices, automatic diagnosing system, automatic alarm service, on board memory, USB port for data transfer, rechargeable battery
- Relocable Unit (RU): GPRS/Bluetooth Dongle;
- Management Software: GPS mapping, address and telephone number of nearest hospital, simultaneous monitoring of more than one patient, remote (computerized) medical visits and consultation service, creation and direct access to electronic case sheets (login and password).

In Fig. 1 there is a picture of the PU; the very small dimensions are remarkable, even if it is only a prototype and more reduction in dimensions is possible.

The monitorable parameters are: electrocardiogram, heart frequency, respiratory frequency, body kinetics, body temperature, oxygen saturation of hemoglobin, environmental pressure, temperature and humidity, position (GPS), arterial pressure. The system, in particular the PU, collects data continuously. These are stored in an on-board flash memory and then analysed real-time with an on-board automatic diagnosis program. Data can be sent to the local receiver, directly to the PC server (online doctor), or to an internet server, which allows anyone to download them by his/her own login and password.

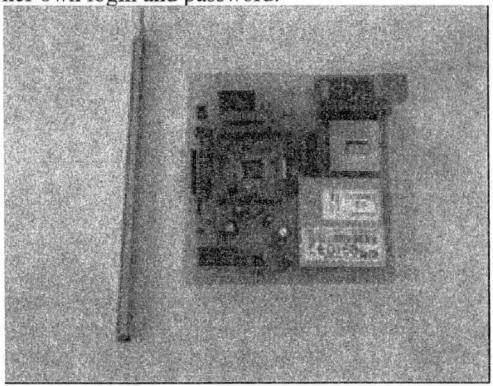

Figure 1. A picture of the Portable Unit (PU), the main part of the telemonitoring system

Data can be transmitted as follows:
1. real time continuously;
2. at programmable intervals (for 30 seconds every hour, for example);
3. automatically, when a danger is identified by the alarm system;
4. on demand, i.e. whenever required by the monitoring centre;
5. offline (not real-time), i.e. by downloading previously recorded (over 24 hours, for example) data to a PC.

In all cases patients do not need to do anything but supply power by simply switching on. When an emergency sign is detected through the real-time diagnosing system, the PU automatically sends a warning message, indicating also the diagnosis, to one person (or even more) who is able to verify the patient's health status and arrange for his/her rescue. In order to make rescue operations as prompt as possible, the PU by its GPS receiver also provides the patient's coordinates and the Management Software provides real-time a map indicating the position of the patient.

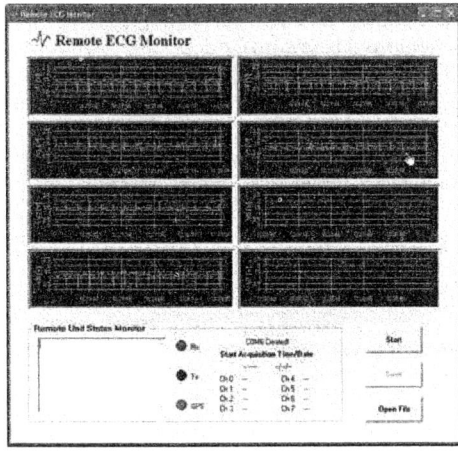

Figure 2. Example of acquisition by Bluetooth of an electrocardiogram

Fig. 2 shows a picture of an electrocardiogram transmitted by Bluetooth and plotted on a Personal Computer by the proper developed management software.

The sensors used are all commercially available and the signal processing algorithm is based on the time domain analysis of the QRS-wave of the electrocardiogram and a correlation between the ECG measurements and the data coming from the pulse oxymeter, by which also the breathing frequency is calculated, body kinetics and arterial pressure. A suspected pathology is confirmed by environmental parameters values.

Asclepius appears to be very innovative because, at the best of our knowledge, the only wearable medical device actually offered by the market and oriented to the a remote health monitoring is the electrocardiograph. Moreover, there are not "intelligent" devices, able to activate the rescue fully automatically.

III. MULTICHANNEL SYSTEM FOR THE ELECTRONIC AUSCULTATION OF THE PULMONARY SOUNDS (AGLAIA)

Aglaia is an electronic medical device intended to provide medical specialists with a totally non-invasive high-engineering device able to detect and analyse the widest number of data for the monitoring of the respiratory system. It performs the monitoring and the analysis of respiratory health by the simple recording and evaluation of lung sounds [1, 2]. It has been prototyped, as it can be seen in the picture of Fig.3. A specific management software program enables the user to process and save the acquired signals as well as to plot different graphs of them. Aglaia is able to plot the graph of a signal both while recording it (real time) and after saving it. The following options are available:
- temporal graph of a breathing sound;
- frequency graph of a breathing sound;
- spectrogram of a breathing sound;
- temporal graph of the airflow;
- measurement of both airflow and inspiratory/expiratory volume.

Figure3. The prototype: a double-sided printed circuit board.

Aglaia's configuration options also allow to choose the number of the channels to be displayed and those to be examined, as well as to preset a different acquiring time for

978-1-4244-1983-8/08/$25.00 ©2008 IEEE

each channel and an automatic detection procedure where the user is notified of pathologies by a warning message or an alarm.

Aglaia's main features are:
- extreme simplicity of use and very low cost;
- USB interface ;
- microcontroller based system ;
- ten on board acquisition channels;
- simultaneous acquisition;
- specific channel for the pneumotachograph;
- 16 bit resolution of the acquired signals;
- plug and play module for the sensors;

Recent research has pointed out the effectiveness of the frequency analysis of lung sounds for the diagnosis of pathologies [3].

A number of validation experiments show that computerized tomography (CT) results perfectly match those of a simple frequency analysis of previously recorded lung sounds. This is a reason for a great interest in Aglaia.

Many studies [4] have been carried out on the frequency analysis of lung sounds and researchers have set the threshold for the detection of pulmonary pathologies at 500Hz.

Spectrum components over that threshold (500 Hz) may be indicative of pulmonary disease.

It is widely known that in patients treated with mechanical ventilation a gradual PEEP increase (PEEP = positive end-expiratory pressure) results in a progressive re-expanding of alveoli which were previously collapsed due to a pathology.

The obtained experimental results shows that a gradual PEEP increase – from 5 to 20 – has effected a gradual reduction in lung damage, thereby leading to improvement in the patient's respiratory health.

The CT results perfectly match those of the frequency analysis.

Moreover, there are also research projects about pulmonary acoustic imaging for the diagnosis of respiratory diseases. In fact, the respiratory sounds contain mechanical and clinical pulmonary information. Many efforts have been devoted during the past decades to analysing, processing and visualising them [5]. We can now evaluate deterministic interpolating functions to generate surface respiratory acoustic thoracic images [6].

IV. SYSTEM FOR ECG TRANSMISSION BY BLUETOOTH

The system (Fig.4) uses the Bluetooth [7] to send electrocardiograms to 6 or 12 leads and it is also equipped with GPS module for the patient's location in real time.

It proves particularly useful indefinite places such as nursing homes and rest homes for elderly people. However by using a mobile phone the system also allows transmission within a long range by GPRS/GSM.

The microcontroller permits to implement a diagnostics algorithm and/or to download, in real time, the data by UDP channel.

The tracing can be also stored on flash cards legible with any PC equipped with a reader of flash memories.

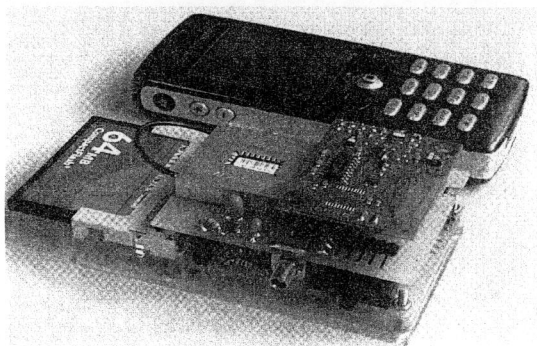

Figure 4. System for ECG transmission by Bluetooth

V. DIGITAL CARDIOHOLTER WITH MULTIPLE LEADS

Nowdays [8-10] the most used tape-recorder type electrocardiographs for the long term registration provide the acquisition of two or three channels thus allowing the detection of a limited number of pathologies and missing crucial details relevant to the morphology of the heart pulse and the related pathologies, given only by a static electrocardiogram (ECG) executed in the hospital or in medical centers.

We have designed a microcontroller-based device allowing data from up to 12 channels to be stored thus providing the diagnostic capabilities of the static ECG together with the wearability and the long term registration of the cardiac activity. Thanks to its specific sensors, embedded in a kind elastic band, it is possible to place on the thorax many electrodes without reducing the movement potentials. Moreover, the elastic band is provided with a wireless module (Bluetooth) to send the data to the recorder unit (shown in the photograph below). The storage support is a flash card. Therefore, the new system is miniaturised and results more comfortable than the commonly used tape-recorder type portable electrocardiographs.

In particular, the device (Fig.5) has been planned using SMD components and it is also made up of a proprietary software which allows the download of the recorded tracing and afterwards the processing of the same tracing thanks to the implementation of digital filters with "easy to use" interface.

The management software to data-download has been properly developed by the authors, being it custom for this application. It receives the data from the electrocardiograph and allows to store/plot them. The software also allows the creation of hospital files, the forwarding of the acquired tracing by e-mail as well as an automatic reading of the recorded result (automatic diagnosis).

The use of multiple leads gives diagnostic potentialities to the electrocardiogram which are not possible with 3 or 4 leads employed at present.

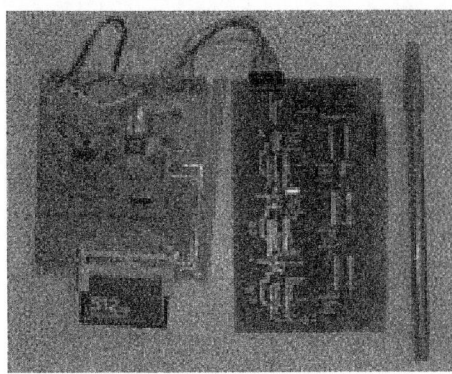

Figure 5. Digital cardioholter with multiple leads

CONCLUSIONS

In this paper we have presented some projects in biomedical electronics, especially applied to telemedicine. The designed systems are characterized by originality, by plainness of use, as they planned with a very high level of automation (so called "intelligent" devices).

The devices have to be engineered but appear very promising due to the innovation they introduce at reasonably low cost.

The systems are oriented to remote health monitoring by measuring and sending in real time more and more parameters vs other telemedicine systems that consist essentially in an electrocardiograph that stores and sends the only electrocardiogram.

Moreover, "Aglaia" is very promising because it can increase the diagnostic potentials of the lung sounds registration and analysis giving results comparable to those obtainable with the CT system.

REFERENCES

[1]   A.R.A. Sovijärvi, J. Vandershoot, J.E. Earis, Standardization of computerized respiratory sound analysis, European Respiratory Review, 2000, 10:77, p. 585.

[2]   J.E. Earis, B. M. G. Cheetham, Future perspectives for respiratory sound research, European Respiratory Review, 2000, 10:77, pp. 641-646.

[3]   Munakata M, Ukita H, Doi I, Ohtsuka Y, Masaki Y, Homma Y, Kawakami Y. Related Articles, Links Spectral and waveform characteristics of fine and coarse crackles. Thorax. 1991 Sep;46(9):651-7.

[4]   A.Vena, G. Perchiazzi, G. M. Insolera, R. Giuliani, T. Fiore, Computer analysis of acoustic respiratory signals, Modelling biomedical signals-Editors: Nardulli G. e Stramaglia S., World Scientific Publishing, pp. 60-66, Singapore, 2002.

[5]   Martin Kompis, MD, PhD, Hans Pasterkamp, MD, and George R. Wodicka, Acoustic Imaging Of The Human Chest, Chest 2001, 120:1309–1321.

[6]   S. Charleston-Villalobos, S. Cortés-Rubiano, R. González-Camarena, G. Chi-Lem, T. Aljama-Corrales, Respiratory acoustic thoracic imaging (RATHI): assesing deterministic interpolation techniques, - Medical & Biological Engineering & Computing, vol. 42, No. 5, pp:618-626, September 2004.

[7]   B. Senese,: "Implementing Wireless Communication in Hospital Environments with Bluetooth, 802.11b, and Other Technologies", Medical Device & Diagnostic Industry, July 2003

[8]   J.J. Carr and J.M. Brown, Introduction to Biomedical Equipment Technology, fourth edition, Prentice Hall (2001);

[9]   J.G. Webster, Bioinstrumentation, Jhon Wiley & Sons, Inc. (2004);

[10] E. Jovanov, P. Gelabert, R. Adhami, B. Wheelock, R. Adams, Real Time Holter Monitoring of Biomedical Signals DSP Technology and Education Conference DSPS'99, August 4-6, 1999, Houston, Texas.

Front-end IC Design for Intravascular Ultrasound Imaging

F. Yalcin Yamaner, Linga Reddy Cenkeramaddi[†], and Ayhan Bozkurt

Acoustic Group, Sabanci University, Orhanli Tuzla 34956 Istanbul Turkey

[†]Norwegian University of Science and Technology, NTNU, N-7491 Trondheim, Norway

Email: yalcin@su.sabanciuniv.edu

Abstract— Capacitive micromachined ultrasonic transducers (cMUT) technology is a new trend for intravascular ultrasound (IVUS) imaging. Large bandwidth, high sensitivity and compatibility to CMOS processes makes the cMUT a better choice compared to the conventional piezoelectric transducer. To exploit the merits of cMUT technology, an accurately designed front end circuit is required. The circuit functions as an output pulse driver for the generation of the acoustic signal and buffers the return echo. For an accurate evaluation before tape-out, the circuit has to be simulated using the post-layout extracted netlist of the IC with the electrical equivalent circuit that models the transducer pulse-echo behavior. In this paper, we present two different designs of front-end IC for 2D cMUT arrays that can be used for intravascular ultrasound imaging system. To simulate the response of the front-end circuit, we first developed a pulse-echo model for an array element using Mason Equivalent Circuit. The model is then combined with the front-end circuit using Cadence Spectre. The simulation results are verified by comparing them to experimental data obtained from the manufactured front-end IC. The results show that the front-end circuit tested with the equivalent circuit model of the cMUT elements is promising for the optimization of the overall system performance before manufacturing.

I. INTRODUCTION

Intravascular Ultrasound is an invasive diagnostic technique that provides the cardiovascular surgeon with useful information on the condition of vessels. Conventional IVUS consists of a side-looking 1D array to asses arteriosclerosis. Front-looking IVUS, on the other hand, is a newly emerging technology that uses 2D annular ring transducers placed at the tip of a catheter probe 1. The use of two dimensional arrays enables scanning in the transverse plane, resulting in a 3D (volumetric) image of the medium. The primary use of front-looking IVUS is flow measurement and guidance.

There are two major issues associated with 3D acoustic imaging: increased element count and the reduced signal strength of individual array elements due to limitations in the physical dimensions. These issues necessitate the design and use of specifically designed integrated circuits for driving and sensing purposes, and a thoroughly analysis of the system performance has to be done before the manufacturing of the IC 2. In this paper, we present two front-end IC designs with different amplifier topologies.. The first design is based on a trans-impedance amplifier while the second circuit uses a charge amplifier in the receiver section. Both designs include a high voltage pulse driver and a high voltage protection switch. The ICs were designed using the 0.35 m high-voltage CMOS

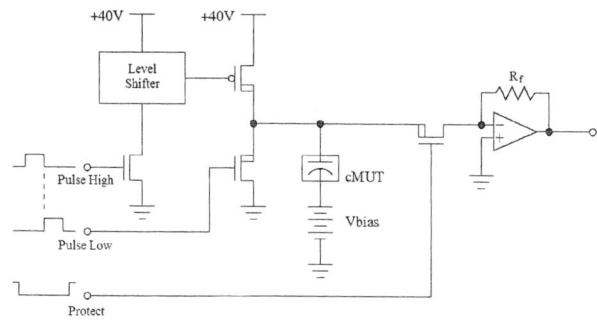

Fig. 1. The overall circuit topology: A high voltage pulser, a protection switch and a trans-impedance amplifier.

technology by AMS (AustriaMicroSystems AG, Graz, Austria). The designs have an overall layout area of approximately $170 \times 170 \ \mu m^2$. Extensive simulations have been run on the post-layout extracted netlist of the IC combined with a cMUT equivalent circuit model using Cadence Spectre. The analysis methodology is verified by experimental results.

II. CIRCUIT TOPOLOGY I: TRANSIMPEDANCE AMPLIFIER

Limitations on the dimensions of array elements used for IVUS result in reduced element sensitivity and relatively high radiation impedances. To detect small capacitance changes, the trans-impedance amplifier is a good choice for the cMUT front-end circuit [3]–[5]. The circuit is designed for a 2D cMUT array with 5×5 capacitive cells with 18 μm radius, 0.3 μm membrane thickness, and 0.15 μm air gap, the radiation impedance is approximately 130 kΩ for 100 volts bias voltage. The overall pulse-echo circuit consists of a pulser, a protection switch, and a read-out amplifier (Fig. 1).

The value of the feedback resistor R_f was chosen as identical to that of the radiation impedance of the transducer. The resulting trans-impedance amplifier acts as a unity gain buffer amplifier. Higher gains result in better noise performance, but limit the bandwidth. A simple two-stage op-amp with internal compensation and resistive bias was used in the circuit.

The high-voltage pulse driver in Fig. 1 is a simple push-pull stage. The TTL-level control signal for the PMOS driver needs to be shifted to the high voltage supply rail. In this design, there are two level shifter implementations: a simple

978-1-4244-1983-8/08/$25.00 ©2008 IEEE

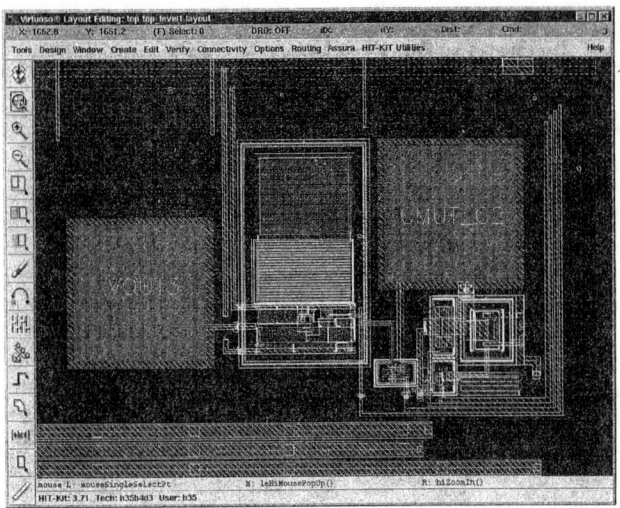

Fig. 2. Layout of the trans-impedance amplifier and the pulse driver.

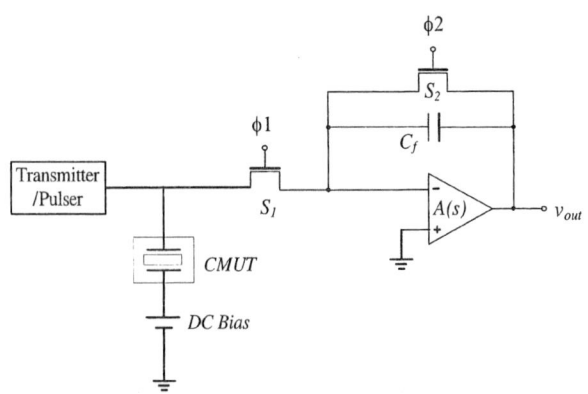

Fig. 3. The designed capacitive feedback analog front-end architecture for cMUT based IVUS imaging.

resistive load inverter and a flip-flop circuit known as a voltage mirror [6].

The physical layout in Fig. 2 contains the driver, buffer amplifier, protection switch and a bonding pad for the transducer element.

III. CIRCUIT TOPOLOGY II: CHARGE AMPLIFIER

The main problem associated with the resistive feedback circuit architectures is the noise contribution of the feedback resistor at the output of the amplifier. Capacitive feedback amplifiers (i.e., charge amplifiers) do not possess this noise problem. These circuits have been avoided due to the problem of undefined node voltages or floating gates [7].

Fig. 3, shows the designed capacitive feedback analog front end circuit architecture for amplifying the signals generated from the CMUTs. The circuit has two switches (S1 and S2) that control the operation during transmit and receive modes. During transmit mode switch S1 is switched off and the switch S2 is switched on. This situation makes the operational amplifier to be configured in a unity gain mode and the amplifier is electrically disconnected from the CMUTs. In this way switch S1 protects the amplifier from the high voltage pulses that are generated from transmit/pulser circuit. In this mode, input transistors of the operational amplifier will be biased to the DC voltage which is applied at the positive input of the operational amplifier. This DC voltage will be stored on the gate capacitance of the input transistors of the amplifier even if S2 is switched off. As the amplifier is electrically disconnected from the CMUTs, transmitter/pulser circuit is activated during this time so that the CMUTs are driven.

In receive mode, S1 is switched on and S2 is switched off. This configures the circuit to be as an inverting amplifier with CMUT as input capacitance and C_f as feedback capacitance. During this mode the circuit amplifies the signals generated by the CMUTs.

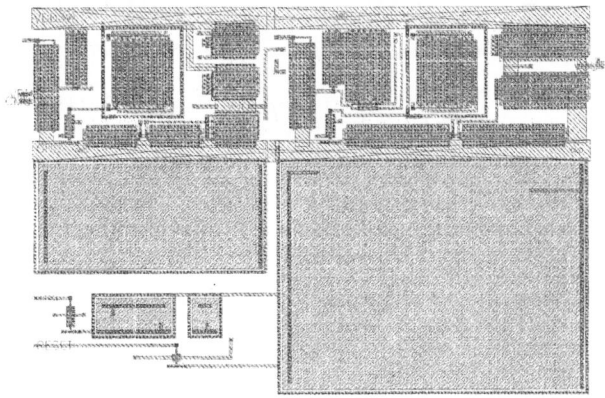

Fig. 4. Layout of charge amplifier.

For the operational amplifier in the proposed capacitive feed architecture, we have chosen a simple two stage amplifier topology with lead compensation. Post-layout simulations show that the in-band noise figure of the front-end is smaller that 2 dB. The layout of the circuit is shown in Fig. 4.

IV. EXPERIMENTAL SETUP

cMUT elements are wire bonded directly the amplifiers driving pad to test the ICs. The properties of the cMUT elements are given in Table I. The pulse echo environment is created using vegetable oil that filled over the chip carrier that encloses the IC and cMUTs (Fig. 5).

TABLE I
cMUT ELEMENT PARAMETERS

Element Size	150 μm x 150 μm
Number of cMUT cells	4 x 4 μm
Membrane Material	Silicon Nitride
Membrane Radius	15 μm
Membrane Thickness	0.4 μm
Gap Distance	0.2 μm

978-1-4244-1983-8/08/$25.00 ©2008 IEEE

Fig. 5. Picture of front-end IC combined with a cMUT array element.

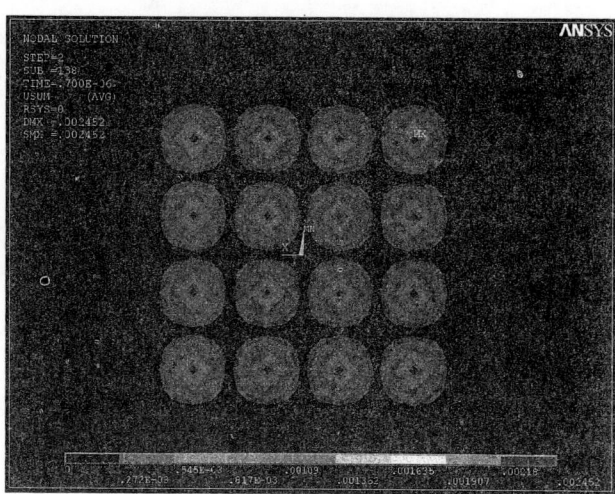

Fig. 7. FEM result showing membrane displacements of a cMUT array element.

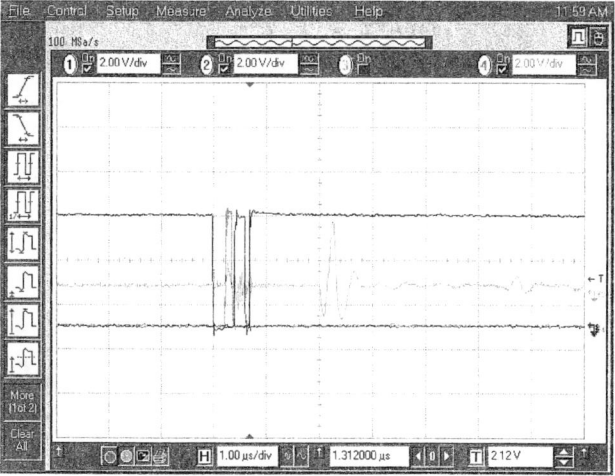

Fig. 6. The oscilloscope data obtained from hydrophone and pulser circuit.

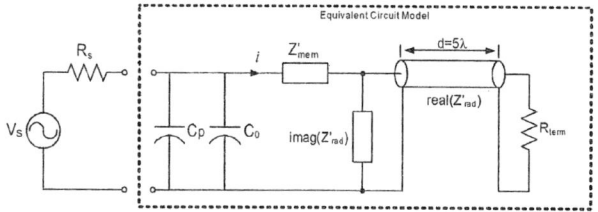

Fig. 8. The electrical equivalent circuit model.

Cadence Spectre circuit simulator. The calculated component values for the electrical side are given in Table II and the overall equivalent circuit is shown in Fig.8.

TABLE II
EQUIVALENT CIRCUIT PARAMETERS

C_0	336 fF
C_p	50 fF
n	9.6×10^{-5}
L_{mem}	4.1 mH
C_{mem}	810 fF
L_{rad}	2.3 mH
C_{rad}	2.4 fF
R_{rad}	140 kΩ
R_{term}	140.25 kΩ

The transducer element consists of 16 cells. cMUTs were driven with 15 V pulses and the DC bias was set to 45 V. Fig. 6 shows a hydrophone measurement obtained at a distance of 2 mm.

V. MODELING

For the electrical circuit model, the Mason Equivalent Circuit was modified and the radiation impedance term has been replaced by an RLC network to include the effects of finite transducer size and diffraction loss [10], [11]. The model has been verified by running transient FEA simulations using ANSYS. The length of the transmission line was set to the depth of the ail-oil interface. The transformer ratio was found as 9.6×10^{-6} Nt/Volt 9. The component "R_{term}" was used to mimic the lost on the reflected wave. The found mechanical components were transferred to the electrical side of the Mason equivalent circuit making the circuit suitable for

VI. RESULTS

The equivalent circuit was combined with the front end IC and transient analysis was done using extracted netlist of the design. System performance calculated by the model was then verified by experimental results. As shown in Fig. 9, the return echo signal amplitude and wave shape can be accurately calculated by the simulation setup.

VII. CONCLUSION

In this work, we presented the design and verification of a front-end circuit for IVUS imaging. A transducer model

978-1-4244-1983-8/08/$25.00 ©2008 IEEE   259

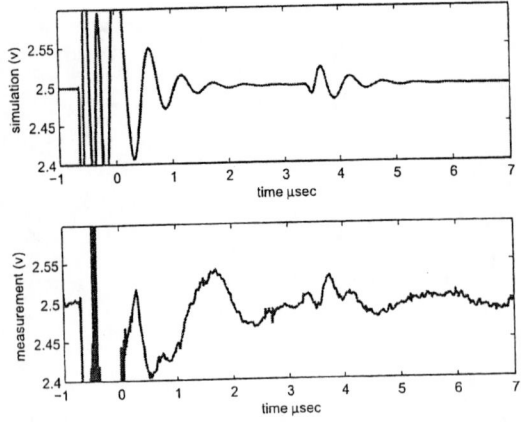

Fig. 9. Simulation results including the equivalent circuit model (top) and experimental results (bottom).

combined with the post-layout network of a front-end circuit produces an accurate prediction of the pulse-echo response. The transducer model includes the propagation medium and is derived using finite element method simulations. The proposed methodology can be used to thoroughly analyze and optimize pulse-echo system performance before IC tape-out, consequently reducing number of manufacturing iterations.

ACKNOWLEGMENTS

This work is supported by the Turkish Scientific and Technological Research Council TUBITAK under grant number 104E067.

REFERENCES

[1] Utkan Demirci, Arif S. Ergun, Omer Oralkan, Mustafa Karaman, Butrus T. Khuri-Yakub *Forward-Viewing CMUT Arrays for Medical Imaging*, IEEE Transactions on Ultrasonics, Ferroelectrics, and Frequency Control, vol. 51, no. 7, July 2004.

[2] F. Levent Degertekin, Rasim O. Guldiken, Mustafa Karaman *Annular-Ring CMUT Arrays for Forward-Looking IVUS: Transducer Characterization and Imaging*, IEEE Transactions on Ultrasonics, Ferroelectrics, and Frequency Control, vol. 53, no. 2, Feb. 2006.

[3] R. Palas-Areny, J. G. Webster, *Sensors and Signal Conditioning*, 2e, Wiley 2001.

[4] Guler U., Bozkurt A. *A Low-Noise Front-End Circuit for 2D cMUT Arrays*, IEEE Ultrasonics Symposium, vol. 1, pp. 689-692, Sept. 2006.

[5] Wygant, I. O, .et al. *An Endoscopic Imaging System Based on a Two-Dimensional CMUT array: Real Time Imaging Results*, IEEE Ultrasonics Symposium, pp. 792-795, 2005.

[6] H. Ballan and M. Declercq, *High Voltage Devices and Circuits in Standard CMOS Technologies*, Kluwer Academic Publishers, 1999.

[7] Sheng-Yu Peng et al. *High SNR Capacitive Sensing Transducer*, IEEE ISCAS, pp. 1175-1178, 2006.

[8] X. Jin, O. Oralkan, F. L. Degertekin, B. T. Khuri-Yakub *Characterization of One-Dimensional Capacitive Micromachined Ultrasonic Immersion Transducer Arrays*, IEEE Transactions on Ultrasonics, Ferroelectrics, and Frequency Control, vol. 48, no. 3, May 2001.

[9] Oralkan O., X. Jin, Degertekin F.L., Khuri-Yakub B.T. *Simulation and experimental characterization of a 2-D capacitive micromachined ultrasonic transducer array element*, IEEE Transactions on Ultrasonics, Ferroelectrics, and Frequency Control, vol. 46, no. 6, Jun. 1999.

[10] A.Bozkurt, M. Karaman *A Lumped Circuit Model for the Radiation Impedance of a 2D CMUT Array Element*, IEEE Ultrasonics Symposium, vol. 4, Sept. 2005.

[11] W. P. Mason *Electromechanical Transducers and Wave Filters*, Van Nostrand, 1942.

978-1-4244-1983-8/08/$25.00 ©2008 IEEE

Tunable All-pass Filter with a Single Inverting Voltage Buffer

Bilgin Metin

Department of Management Information Systems
Bogazici University
Istanbul, Turkey
bilgin.metin@boun.edu.tr

Oguzhan Cicekoglu

Department of Electrical and Electronic Engineering
Bogazici University
Istanbul, Turkey
cicekogl@boun.edu.tr

Abstract—**This paper presents a voltage mode all-pass filter, with a MOSFET transistor, two resistors, a capacitor and an inverting voltage buffer. It is expected to be simpler compared to integrator based MOSFET-C counterparts. The functionality of the proposed circuit is verified with SPICE simulations.**

I. INTRODUCTION

First-order all-pass filters are widely used in analog signal processing in order to shift the phase of an electrical signal while keeping the amplitude constant. In the literature several voltage-mode (VM) first-order all-pass filters have been reported with a single active element [1-7], but only circuit in [7] is an electronically tunable all-pass filter. The circuit in [7] uses small signal output resistance of the active element as a current controlled resistor for electronic tunability.

Electronically tunable circuits have been an important research area in the design of analog integrated circuits, because the tolerances of the electronic components in IC realization can be very high and thus tuning after manufacturing is always desired. MOS resistive circuits (MRC) are widely used in the design of tunable circuits [8]. The resistors are replaced with MOSFETs in some special topologies such that non-linearities due to MOSFETs are canceled. In the classical MRC approach, the filters are built out of integrator blocks each of which is individually linearized as far as its input-output characteristics are concerned [9-10].

In this study we propose a tunable VM all-pass filter using a different approach from [9-10] by using a MRC block for tunability. Also, different from classical MOSFET-C method that use individually linearized integrator blocks; the proposed circuit employs a MOS transistor and an inverting voltage buffer (IVB) to cancel even order non-linearities of the MOS transistor current [8-9].

II. ELECTRONIC TUNABILITY AND THE PROPOSED FIRST-ORDER ALL-PASS FILTER

A first order all-pass transfer function is given below where K is the gain constant and its sign determines whether phase shifting is from 0 to π or from π to 0, and τ is the time constant.

$$\frac{V_0(s)}{V_i(s)} = K\frac{1-s\tau}{1+s\tau} \qquad (1)$$

The proposed circuit is shown in Fig. 1. An IVB which is connected between the source and drain terminals of the MOS transistor is used in the circuit. This IVB/MOS structure, which is given in [8], achieves complete cancellation of even order non-linearities of the transistor current, so a voltage controlled MOS resistor is obtained. The routine analysis of the proposed circuit gives the voltage transfer function as

$$\frac{V_o}{V_i} = \frac{2G_1G_x - sC(G_2 - G_1)}{(G_1 + G_2)(2G_x + sC)} \qquad (2)$$

If the conductance of the MOS transistor is represented by G_X, and for $2G_1=G_2$, we obtain

$$\frac{V_o}{V_i} = \frac{1}{3}\frac{2G_x - sC}{2G_x + sC} \qquad (3)$$

Equation (3) is independent from G_1 and G_2. It is well known that component matching can be achieved with much higher precision than obtaining absolute values for the current IC technologies. Passive elements can easily be matched with much better precision than 0.1% [11], so

---

This work is/was supported by Bogazici University Research Fund, with the project code 05A201D

978-1-4244-1983-8/08/$25.00 ©2008 IEEE

effects of G_1 and G_2 can be ignored in (3). Then, passive sensitivities of the C and G_x are equal to unity.

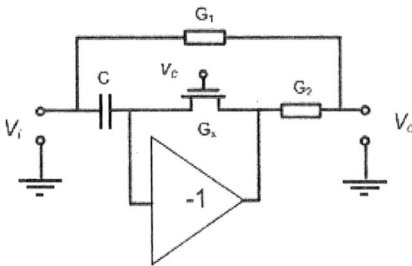

Figure 1   The proposed electronically tunable all-pass filter

III.   SIMULATION RESULTS

To verify the theoretical analyses, we simulated the circuit proposed in Fig. 1 using the SPICE circuit simulation program. The circuit in Fig. 2 is used to implement an inverting voltage buffer derived from inverting current conveyor [12] with supply voltages of V_{DD}=1.5V and V_{SS}= −1.5V and with biasing voltages of V_A= −0.1V and V_B=0.1V. For the simulations, 0.35 μm CMOS real process from TSMC are used. The dimensions of the transistors used in the circuit are given in Table 1.

In order to verify the above given theoretical analysis, the proposed all-pass filter with a pole frequency of $f_p \cong 31.8$kHz is designed with C=100pF, R_1=2kΩ, R_2=1kΩ. The MOS transistor (W=2μm, L=28μm) biased to produce 90kΩ resistor with V_C=1.65V. The frequency response of the presented circuit is shown in Fig. 3. Electronic tunability of the presented circuit by changing V_C is given in Fig. 4. The pole frequency of the presented circuit is changed between 10kHz and 56kHz successfully.

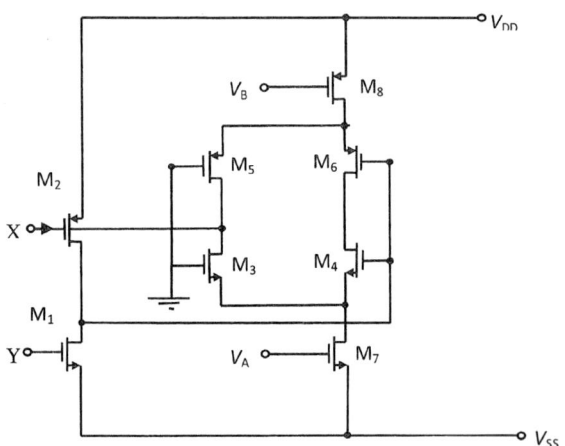

Figure 2   An inverting voltage buffer derived from the inverting current conveyor (ICCII)

To illustrate voltage swing capability a transient analysis is performed and sine waveforms are applied to the filter. The time domain output signal waveforms are shown in Fig. 5 for V_C=1.65V. The amplitude of the input signal is peak-to-peak 1.25V and its frequency is f_0=10kHz. Furthermore in Fig. 6, total harmonic distortion (THD) values are calculated for input signal amplitudes between 0.25V and 1.25V at 3kHz and 10kHz for MOSFET based and passive G_x. Simulation results agree quite well with the theoretical analysis.

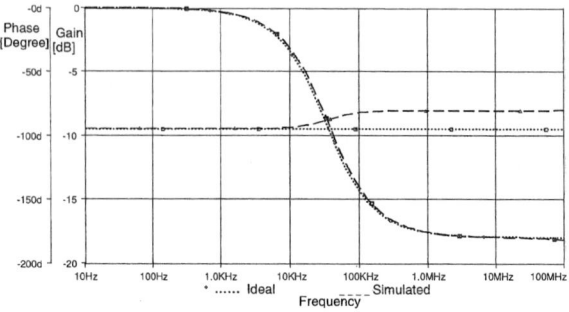

Figure 3   Theoretical and simulated gain and phase responses of the proposed circuit. Since gain of the proposed filter is 0.33, ideal gain of the presented circuit is -9dB

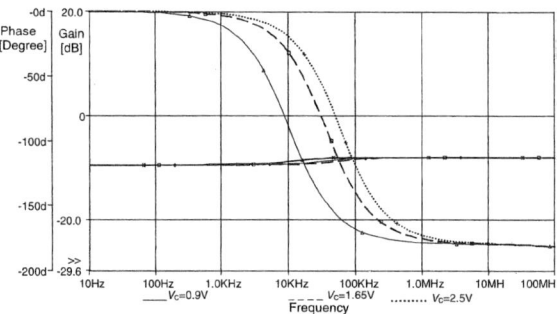

Figure 4   Electronic tunability of pole frequency with V_C control voltage

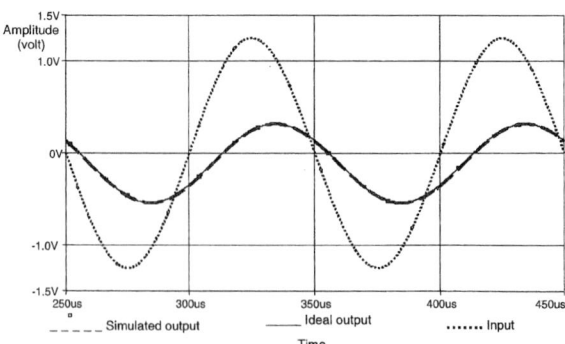

Figure 5   Theoretical and simulated transient analysis of the proposed circuit for V_C=1.65V. Since gain of the proposed filter is 0.33, output signal amplitude smaller than input signal.

978-1-4244-1983-8/08/$25.00 ©2008 IEEE

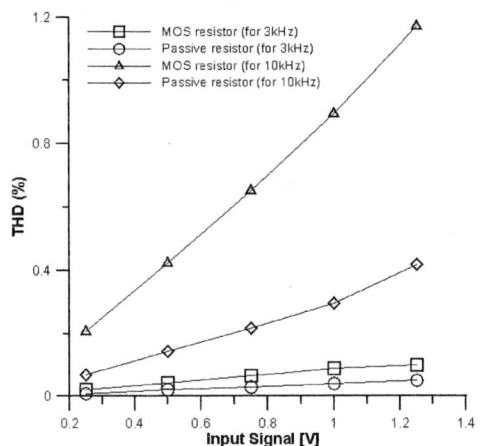

Figure 6   THD values at 3kHz and 10kHz input signal and for various input signal amplitudes at 3kHz and 10kHz.

TABLE I.   THE DIMENSIONS OF THE TRANSISTORS USED IN THE CIRCUIT REALIZATION SHOWN IN FIG. 2.

MOS Transistors	$\dfrac{W(\mu m)}{L(\mu m)}$	MOS Transistors	$\dfrac{W(\mu m)}{L(\mu m)}$
M_1	8.75/1.4	M_7	27.65/1.05
M_3	17.5/0.35	M_2	17.5/1.4
M_6	17.5/0.35	M_5	35/0.35
M_4	8.75/0.35	M_8	57.75/0.35

IV.   CONCLUSION

A novel tunable voltage mode all-pass filter is proposed employing a single inverting voltage buffer. Instead of controlling the parasitic resistance of an active element, a special MOS resistive block is used in the synthesis of the circuit. Also, the presented approach may produce simpler tunable circuit structures compared to integrator based MOSFET-C filters.

REFERENCES

[1]   Khan, I. A.; Maheshwari, S. Simple first order all-pass section using a single CCII. International Journal of Electronics 2000: 87 (3): 303-306.

[2]   Maheshwari, S.; Khan, I. A. Novel first order all-pass sections using a single CCIII. International Journal of Electronics 2001: 88 (7): 773-778.

[3]   Metin, B.; Toker, A.; Terzioglu, H.; Cicekoglu, O. A new all-pass section for high-performance signal processing with a single CCII-. Frequenz 2003: 57 (11-12): 241-243.

[4]   Higashimura, M.; Fukui, Y. Realization of all-pass networks using a current Conveyor. International Journal of Electronics 1988: 65 (2): 249-250.

[5]   Toker, A.; Ozcan, S.; Kuntman, H.; Cicekoglu, O. Supplementary all-pass sections with reduced number of passive elements using a single current conveyor. International Journal of Electronics 2001: 88 (9): 969-976.

[6]   Pandey, N.; Paul, S. K. All-pass filters based on CCII- and CCCII-. International Journal of Electronics 2004: 91 (8): 485-489.

[7]   Toker, A.; Özoğuz, S. Tunable allpass filter for low voltage operation. Electronics Letters 2003: 39 (2): 175-176.

[8]   Tsividis, Y.; Banu, M.; Khoury, J. Continuous time MOSFET-C Filters in VLSI. IEEE J. Solide-state Circuits 1986: SC-21 (1): 15-30.

[9]   Mahmoud, S. A; Soliman, A. M. New MOS-C Biquad Filter Using the Current Feedback Operational Amplifier. IEEE Circuit ans Systems-I 1999: 46 (12): 1510-1512.

[10]   Jiang, J.; Wang, Y. Design of a tunable frequency CMOS fully differential fourth-order Chebyshev filter. Microelectronics Journal 2006: 37 (1): 84-90.

[11]   Gray, P.R.; Meyer, R.G. Analysis and design of analog integrated circuits. John Wiley & Sons 1993: p. 451-452.

[12]   Awad, A.; Soliman, A. M. Inverting second-generation current conveyors: the missing building blocks, CMOS realizations and applications, International Journal of Electronics 1999: 86 (4): 413-432.

978-1-4244-1983-8/08/$25.00 ©2008 IEEE

Author Index

Abadal	p.89	Declercq	p.189, 229
Abid	p.109	Degertekin	p.101
Aboushady	p.81	Dehollain	p.189, 229
Acar vural	p.13	Dei	p.33, 205
Agnes	p.245	Delizia	p.45, 173, 193, 213
Agoyan	p.113	Demirel	p.185
Akin	p.41	Deniz	p.5
Alaca	p.125	Di guglielmo	p.149
Amoroso	p.209	Di stefano	p.53
Arslan	p.217	Dundar	p.5, 145, 233, 249
Atasu	p.153	Erario	p.245
Avci	p.69	Erdogan	p.225
Avitabile	p.165	Ferhanoglu	p.101
Azeredo-leme	p.45, 173, 213	Ferrari	p.201
Balkan	p.133	Fiscelli	p.53
Barbaro	p.121	Friedman	p.85
Barniol	p.89	Giaconia	p.53
Basaga	p.25	Giner	p.89
Baschirotto	p.45, 97, 173, 193, 197, 213	Giorgio	p.253
Batur	p.233	Giusto	p.253
Baudoin	p.157	Gok	p.117
Bayar	p.137	Gokdel	p.93
Beilleau	p.81	Gottardi	p.97
Bonizzoni	p.245	Gozzini	p.201
Bourouina	p.129	Grassi	p.197
Bozkurt	p.257	Gul	p.25
Bruno	p.197	Gurbuz	p.25, 169
Bruschi	p.33, 205	Guvenc	p.49
Burdese	p.237	Hermanowicz	p.61
Caboni	p.121	Hofer	p.181
Cakir	p.37	Holweg	p.181
Cannone	p.165	Homsy	p.121
Cappuccino	p.209	Jawed	p.97
Cascella	p.165	Jemai	p.109
Cattin	p.97	Kahraman	p.1
Cenkeramaddi	p.257	Kale	p.105
Cerid	p.249	Kallempudi	p.25
Chiheb ammri	p.109	Kayaalti	p.249
Cicekoglu	p.37, 261	Kaynak	p.169
Ciftlik	p.29	Kazan	p.57
Cito	p.45, 213	Kennedy	p.177, 241
Cocorullo	p.209	Kepenek	p.41
Collins	p.225	Kerhervé	p.185
Convertino	p.253	Klapf	p.181
Cucchiara	p.53	Koca	p.233
D'Amico	p.45, 173, 193, 213	Koivisto	p.161, 221
Daneshgar	p.177	Krichene zrida	p.109
De matteis	p.45, 173, 213	Kulah	p.29, 41, 133
De Rooij	p.121	Laabidi	p.113

Leblebici	p.77, 125		Sarioglu	p.93
Lenci	p.73		Sevim	p.17
Linder	p.121		Sezarman	p.25
Lombardi	p.197		Silay	p.229
Lopez	p.89		Summanwar	p.129
Luk	p.153		Tavares	p.45, 173, 213
Maehne	p.77		Tekin	p.169
Malcovati	p.197		Teva	p.89
Maloberti	p.245		Thoppay	p.189
Marani	p.253		Tiiliharju	p.161, 221
Marie-helene	p.21		Todman	p.153
Massari	p.97		Torres	p.89
Maunu	p.221		Torun	p.101
Mehrez	p.141		Toy	p.101
Mencer	p.153		Trang	p.21
Metin	p.261		Uranga	p.89
Missoni	p.181		Urey	p.101
Morgul	p.217		Vachoux	p.77
Mrabet	p.141		Valenta	p.157
Mulassano	p.237		Van der Wal	p.121
Murillo	p.89		Vasilevski	p.81
Mutlu	p.17		Verd	p.89
Nacaroglu	p.69		Villegas	p.157
Nannini	p.73		Voltti	p.161
Neuilly	p.129		Vural	p.57
Ocak	p.41		Wilamowski	p.145
Ozbilen	p.117		Willingham	p.105
Ozev	p.225		Yahya	p.9
Pache	p.185		Yalcin	p.145
Paruzel	p.61		Yalcinkaya	p.93, 125
Parvez	p.141		Yamaner	p.257
Patrice	p.21		Ye	p.241
Pecheux	p.81		Yildirim	p.1, 13
Perri	p.253		Yilmaz	p.125
Philippe	p.21		Yudanto	p.237
Pieri	p.73		Yurdakul	p.137
Piotto	p.33		Zeevi	p.65
Piotto	p.205		Zeki	p.49
Plana	p.185		Zervas	p.125
Pribyl	p.181		Zhao	p.45, 213
Pugliese	p.209			
Ratner	p.65			
Reis	p.45, 213			
Renaudin	p.9			
Reyneri	p.237			
Robinson	p.113			
Salman	p.85			
Sampietro	p.201			
Sari	p.133			